THE HUMAN GENOME

ELSEVIER
science &
technology books

Companion Web Site:

http://www.elsevierdirect.com/companions/9780123334459

The Human Genome
Julia E. Richards and R. Scott Hawley, Editors

Resources:

- Electronic images

- End-of-chapter questions

- Recommended reading and additional teaching tools

ELSEVIER

**ACADEMIC
PRESS**

THE HUMAN GENOME

A User's Guide

THIRD EDITION

JULIA E. RICHARDS
University of Michigan
Ann Arbor, Michigan

R. SCOTT HAWLEY
Stowers Institute for Medical Research
Kansas City, Missouri

AMSTERDAM • BOSTON • HEIDELBERG • LONDON • NEW YORK • OXFORD
PARIS • SAN DIEGO • SAN FRANCISCO • SINGAPORE • SYDNEY • TOKYO
Academic Press is an imprint of Elsevier

Academic Press is an imprint of Elsevier
32 Jamestown Road, London NW1 7BY, UK
30 Corporate Drive, Suite 400, Burlington, MA 01803, USA
525 B Street, Suite 1800, San Diego, CA 92101-4495, USA

First edition 1999
Second edition 2005
Third edition 2011

Notice
No responsibility is assumed by the publisher for any injury and/or damage to persons or property as a
matter of products liability, negligence or otherwise, or from any use or operation of any methods, products,
instructions or ideas contained in the material herein. Because of rapid advances in the medical sciences, in
particular, independent verification of diagnoses and drug dosages should be made.

British Library Cataloguing-in-Publication Data
A catalogue record for this book is available from the British Library

Library of Congress Cataloging-in-Publication Data
A catalog record for this book is available from the Library of Congress

ISBN : 978-0-12-333445-9

For information on all Academic Press publications
visit our website at www.elsevierdirect.com

Typeset by MPS Limited, a Macmillan Company, Chennai, India
www.macmillansolutions.com

Printed and bound in Canada

11 12 13 14 15 16   10 9 8 7 6 5 4 3 2 1

Working together to grow
libraries in developing countries

www.elsevier.com | www.bookaid.org | www.sabre.org

ELSEVIER   BOOK AID International   Sabre Foundation

To the writers and characters of *CSI* and *CSI Miami*, who each week do in 20 seconds what it takes working scientists days, weeks or even months to accomplish. And to the real scientists who are moving us towards a time when such wondrous speed and accuracy might actually become possible.

Contents

III
HOW CHROMOSOMES MOVE

6 Mitosis and Meiosis: How Cells Move Your Genes Around   199

7 The Odd Couple: How the X and Y Chromosomes Break the Rules   247

IV
HOW GENES CONTRIBUTE TO COMPLEX TRAITS

8 Sex Determination: How Genes Determine a Developmental Choice   273

9 Complexity: How Traits Can Result from Combinations of Factors   299

Acknowledgments

Many people have helped us to bring this book into existence, and we are profoundly grateful to each of them. First we want to thank our families, who provide us with a foundation of love and strength that is essential to everything we do. We are grateful for their patience, support and advice throughout the writing process. We could not have done this without them.

We want to thank the many people at Elsevier who made this book possible, starting with Jeremy Hayhurst, the editor of the previous edition who is the one who first worked with us to put together the idea of this edition. We want to thank Christine Minihane, Rogue Shindler, Elaine Leek, and Deena Burgess for their patience and all of their efforts that made this possible. We also want to thank Catherine Mori who co-authored the original edition of this book and was instrumental in creating its format.

We are grateful to everyone whose artwork appears in this volume. While we ourselves created many of the figures, we have also included images produced by others, as acknowledged in the figure legends. Whether it came from a publication or was prepared directly for this book, each illustration brings to the mind's eye more than we were able to paint with our words. A special thank you goes to Ed Trager, who created some of the most impressive molecular images present in the book. We want to thank Fred Rose at Elsevier for the wonderful cover design, and John Martin, a very talented artist whose exquisite image "DNA Helix with Eyeball" (available on Corbis.com) inspired the cover design.

We thank the members of our research groups who have helped in so many ways over the years, and we want to offer special thanks for all the ways in which our administrative assistants, Nina Kolich, Linda Hosman, and Emily Bird, have made our lives easier and made this book possible. Dr Carl Marrs has earned our profound gratitude by reviewing the entire book for us. Special thanks for helpful discussions during the writing go to Dr Sue Jasperson, Dr Debra Thompson, Dr Mike Erdos, Katy Downs, Dr Frank Rozsa, Dr Carl Marrs, Dr Sayoko Moroi, Dr Leslie Gordon, Colin Marrs, Dr Eric Vilain, Dr Melissa Hines, and Dr David Reed. A special thanks goes to our colleague Dr April Pyle, who contributed some of the questions and readings at the ends of the chapters. And thanks to the many seminar speakers over the years from whom we have learned so many important things that went into this book; we cannot name them all but offer as especially inspiring examples Dr Leslie Gordon, Dr Stephen Gruber, and Dr James Lupski. Scott Hawley also thanks both the Stowers Institute and its president, Dr Bill Neaves, for support and encouragement during the writing of this book. Julia Richards wants to thank her many colleagues at the Kellogg Eye Center who have answered questions, made suggestions, and supported her patiently during the writing process, and she especially wants to thank Dr Paul Lichter for all the things he has taught her over the years about the pursuit of excellence in science, in teaching, in scholarship, and in writing.

Scott and Julia owe special thanks to two superb geneticists who helped to shape their early careers. Without Larry Sandler and Herschel Roman, our research careers would likely have gone in very different directions. Since we would not have met without them, they deserve some of the credit for the existence of this book.

We often use the first person in this book, but when speaking of scientific findings ("When we do this type of experiment we find that …"), we do not mean to lay claim to the vast body of work we present here. Many researchers have expended great amounts of time and energy for more than a century to arrive at the frankly amazing body of knowledge covered here. Although we are both active researchers in the field of genetics, in this book we speak as users of our own genomes, general representatives of our field, teachers of human genetics, researchers who have done some of the kinds of experiments presented here, and continual students of this fascinating topic.

We owe thanks to the individuals and families whose stories appear in this book. Each of these stories was included not only because it makes some scientific or educational point but also because these are stories that have touched our hearts. We offer special thanks to Jim Knowles for letting us share Brenda's tale.

Others who shared their stories anonymously are just as deserving of our thanks even if we must leave them unnamed here. For some of these stories, we have simplified the tale to keep it focused on the lesson to be learned from the tale, or combined information from several similar tales. In some cases we have changed minor details to help preserve confidentiality, such as avoiding use of real names. In general, where we use no names or only first names, these are still true stories unless we have indicated otherwise. In rare cases in which we present a hypothetical situation derived from many similar stories, we try to indicate this by saying it is hypothetical or by other obvious devices such as saying, "What if we looked at a family with these characteristics?" With many of the families we encountered the hope that the sharing of their tales would keep someone else from going through the same thing that had happened to their families. If this book accomplishes that goal for even one family, the writing will have been well worth all of the effort.

Finally we thank all of our readers for their interest in this topic. We are delighted every time another student reads our guide to the genome that directs the operations of all of our cells and keeps all of our bodies running 24/7 throughout our lives.

Prologue: The Answer in a Nutshell

Our bodies contain billions of cells, intricate little factories that carry out their own internal functions as well as carrying on complex interactions with surrounding cells and the rest of the body. Each cell contains all of the molecular machinery it needs to carry out its own specialized operations; of special interest to each of us is the fact that each cell also contains all of the information needed to synthesize all of the molecular components needed to carry out those specialized operations. We refer to this body of information as our genome. This genome doesn't function as a single entity but rather consists of tens of thousands of subunits of information called genes.

Virtually all of the cells in our bodies contain the same information in the form of the same set of genes. Genes themselves are little more than a repository of information that tells the cell how to make a gene product that can carry out a function. The gene products fall into two major categories of molecules, some of which are RNA molecules, transient copies of the information contained in the genes, and some of which are proteins that are produced as a result of the cell "reading" the information contained in a specific class of RNA molecules known as mRNA. All of the other molecules produced by a cell are made as a result of the activities of those RNA and protein gene products.

Many, if not most, of the differences that exist between us reflect the fact that the information in a gene can be permanently altered by a process called mutation, and changing that information changes the gene product that is produced. Although many think of mutation as a term for something negative or harmful that can cause birth defects and genetic disease, mutations can also bring about changes that are neutral (having no effect on an individual's characteristics) or even beneficial. They can cause changes in many of the characteristics by which we recognize each other: height and build; hair color and texture; and shapes of face, ears, nose, eyes and eyebrows. Mutations can affect things that are harder to define, such as behavior. Mutations are responsible for differences that are very important even if they are invisible to us on a daily basis, such as blood type. Although mutation occurs rarely, there are an awful lot of us, and we have been breeding for a very long time. Thus there has been an ample opportunity for mutations in each of our genes to occur and in many cases to be spread widely through the population. And the altered information in the genes leads to altered gene products that are responsible for our many differences. Without mutations we would all have exactly the same set of genetic information and billions of us would all resemble each other in the same way that identical twins resemble each other. To us, the vision of billions of identical human beings is a chilling thought that leaves us quite pleased with the amount of diversity we see around us.

There are also outside influences on the things that happen inside the cell. How the cell functions can vary depending on things like the nutritional status of the cell, the temperature in the surrounding environment, or availability of oxygen. In many cases these environmental influences act by either altering the availability of the materials the cell needs to make RNAs and proteins and carry out functions, or by affecting which pieces of the genetic information the cell uses. But as we consider these environmental influences on the behavior of the cell, everything still comes back to what the cell is doing with the genes and their gene products.

So here is our answer in a nutshell: Usually no one dies because of a defect in his or her genes; they die because that defect leads the gene to produce an RNA or protein molecule that no longer performs its function correctly. This is the foundation of everything else we will talk about in this book – that information in the form of genes directs the production of gene products that actually carry out the cell's functions. And many differences we find between human beings trace back to how some function was carried out (or not carried out) by a damaged (or missing) gene product. There are some exceptions to this generalization, as there are exceptions to almost everything we will tell you in this book, but holding onto this core concept will give you a framework for many of the discussions that will follow.

In this book we hope to share the fascination with genetics that has led so many to spend their lives investigating our genomes and the ways in which alterations to those genomes impact human characteristics. We will start out historically, telling you about how the original concept of the gene was discovered, and then we will talk about how genes are inherited in human families. We will tell you about the underlying chemical nature of the genetic information and how its information is used

to create cellular structures and functions. We will discuss the many different types of mutations and how they can affect function. We will talk about how we find genes, how we test for changes in genes, and how we approach gene therapy to fix unwanted genetic defects. We will talk about simple and complex traits. And we will tell you about how the elucidation of the complete set of information in the human genome has allowed us to dramatically change how we approach looking for the causes of human traits.

But this book will go beyond the technical issues of the science to explore broader topics. We will raise questions about what constitutes "normal" as we examine the wide range of human characteristics that get evaluated in the context of such a term. We will visit the various ethical, legal and social issues that complicate modern genetics, and talk about some of the ethical cautions we can take from lessons out of history. We will explore how genetic testing, gene therapy, and other advances affect us and those around us. The emerging technologies we will discuss have tremendous power to accomplish good, relieve pain and improve peoples' lives but only if used with an eye to the physician's oath to first do no harm.

But before we can tackle such issues we need a solid understanding of the human genome. It is our human genome. It is your human genome. This user's guide is meant for each of us who live our lives every day as the final end users of the vast amounts of information located in the human genome.

Perhaps surprisingly, our story starts not in a modern lab but in a nineteenth-century monastery garden where a monk cultivating pea plants started a quiet scientific revolution …

HOW GENES SPECIFY
A TRAIT

The Basics of Heredity: How Traits Are Passed Along in Families

OUTLINE

THE READER'S COMPANION: AS YOU READ, YOU SHOULD CONSIDER

- Different historical models that explain how heredity works.

- The importance of other organisms as models for the study of human inheritance.

- What gets passed from one generation to the next that results in inheritance.

- The concept of allelic differences.

- The relationship between genotype and phenotype.

- The implications of having two copies of a gene but passing along only one.

(Continued)

The Human Genome. **DOI:** 10.1016/B978-0-12-333445-9.00001-0

- The concept of dominance and the ability of one allele to mask another.
- The concept of recessive inheritance of a trait from parents who lack that trait.
- The effects on inheritance if a trait can be caused by more than one gene.
- Epistasis and alleles that over-ride detection of what another gene is doing.
- Pleiotropy and the ability of a gene to have multiple effects.
- The difference between syndromes and traits affecting a single tissue type.
- The sociopolitical implications of putting labels on phenotypes.

1.1 MENDEL'S LAWS

In the beginning … —Genesis 1:1

We suspect that people have been curious about how heredity works ever since they figured out where babies came from. It is important to note that our current sophistication in these matters is of fairly recent origin. There is an old saying that "like begets like," but this seemingly obvious knowledge that children will be like their parents might have been surprising to some ancient Greeks who wrote about the progeny that resulted from mating members of different species, such as swans and sheep.

Farmers have long known that animal offspring often appeared to be a mixture of both of their parents. Thousands of years ago it was proposed that children resulted from a blending process, the mixing of maternal blood and paternal semen derived from blood, with all aspects of the organism being represented in the semen and the menstrual blood. Many others saw blending as involving some essential essence or particles coming from every part of each parent, but each offered a slightly different mechanism.

For long periods in our history, many people imagined that children were the offspring of only one parent (either the mother or the father). Some thought that babies were preformed in the father and sailed in sperm down the vaginal canal into awaiting uterine incubators. Drawings dating back to the seventeenth century show the tiny preformed individuals (now known as homunculi) that the early microscopists imagined they saw inside sperm. There were other schools of thought in which children were preformed only in their mothers; the father was thought to provide only a "vital spark" (much like jump-starting a dead battery). By the mid-nineteenth century, most people were willing to accept the concept that the traits observed in children were some mix of those observed in both parents and in both sets of grandparents.

In many cases, this was thought of in terms of a blending model of inheritance. Although there are many situations in which blending is not the best model of what is happening, blending was a concept that was easy to understand. If you mix red paint and white paint, you get pink paint, so why would such a mechanism not explain intermediate skin tones in someone who has one dark-skinned parent and one light-skinned parent? People imagined that there was some kind of substance, such as blood, that blended in the offspring to produce a mixture of traits in the child. (Note the term "blood relative," which implies a shared ancestry, not relationship by marriage.)

Still, there were some surprises that blending did not explain: blue-eyed kids born to brown-eyed parents, blond children of raven-haired moms and dads, kids who are taller than either parent, and so on. Blending, although it made some sense for some traits such as height and weight, did not explain many traits.

It was into this rather curious intellectual environment that Gregor Mendel was born in 1822 (Box 1.1). Like Galileo, Newton, Darwin, and Einstein, Mendel's vision would change the course of human understanding. That vision

BOX 1.1

MENDEL'S ORIGINAL WORKS

Because Mendel lived so recently (1822–1884), much is known about his life, his education, the world he lived in, and just how his discoveries were brought to light again many years after his death. Although we will describe enough about Mendel's experiments to help you understand his conclusions, and the way they formed the basis of the science of heredity, we cannot begin to fully describe either Mendel himself or his work. If you want to know more about him, or if you want to read his original scientific writings (in English or in German), check out Mendelweb at www.mendelweb.org. This site does an excellent job of making both Mendel and his science more accessible by means of annotating his works and providing links to helpful items such as glossaries, reference materials, and related sites. Even if you don't want to read Mendel's writings in detail, it is worth checking out this excellent site.

results from one simple set of experiments. As we tell you about Mendel, we will describe one of those magical moments in human cognition when a new set of concepts became beautifully obvious and clear.

What Mendel Did

Mendel was a monk with a garden plot who carried out a set of simple experiments that revolutionized our understanding of genetics. He worked with the pea plant (Figure 1.1). His experiments took years and involved more than 10,000 plants. He chose to study the inheritance (the passing of a characteristic from one generation to the next) of seven simple and obvious traits that could clearly be distinguished between different pea strains:

- Seed shape – round vs. wrinkled
- Seed color – yellow vs. green
- Flower color – white vs. colored
- Seedpod shape – inflated vs. constricted
- Color of the unripe seedpod – green vs. yellow
- Flower position – along the stem vs. at the ends
- Stem length – short vs. tall

FIGURE 1.1 Pea plants show variation in many different characteristics including height, flower color, pod color, and pattern in which the flowers are clustered on the stems. (Courtesy Edward H. Trager.)

These were simple "yes-or-no" traits and not quantitative traits such as weight that can vary over a wide range of different values. Some traits, such as stem length, can vary under

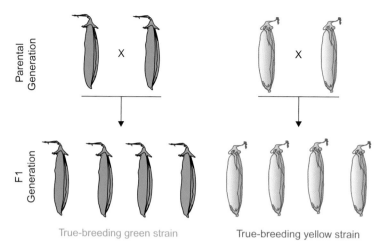

FIGURE 1.2   True-breeding pea strains represented by the peapods that the plants produce. Unlike the plants in these drawings, pea plants in most gardens often show a lot of variation in characteristics such as colors of flowers, peas, peapods, and size or shape of peas, peapods, and stems. By protecting plants from random pollination by other plants and then carrying out artificial pollination, Mendel was able to control which plants were being crossed to each other. He identified some strains that bred true for a particular characteristic, with each succeeding generation of the true-breeding strain producing progeny that were exactly like all of the preceding generations for the particular characteristic being looked at. The color of the immature peapod is one example of a characteristic for which he found true-breeding strains.

different conditions such as rich vs. poor soil, so Mendel focused on simple binary traits – traits that were "yes" (wrinkled) or "no" (smooth) – traits that did not change classification between "yes" and "no" if the environmental conditions changed. For instance, when comparing tall and short plants, the tall plants were not all the same height, but they were so much taller than the short plants that the two categories were never mistaken for each other.

But perhaps the MOST significant aspect of Mendel's experiments was that, unlike any of his predecessors, he began with pure-breeding populations of plants. For example, he had a strain of plants with yellow seedpods that produced only plants with yellow seedpods when bred to each other. Similarly, he had a strain of plants with green seedpods that produced only plants with green seedpods when bred to each other (Figure 1.2).

Mendel crossed two plants by using pollen from one plant to fertilize another plant, and then studying the characteristics of the progeny and comparing those characteristics to the characteristics of the parental plants. Please note that in the first generation, when Mendel crossed plants with green seedpods to plants with yellow seedpods, all he saw in the progeny were plants with seedpods identical in color to those of the green seedpod parent. *None of the plants had seedpods of an intermediate color* (Figure 1.3).

This experiment helped rule out several of the old ideas about inheritance. A real adherent to blending would have expected the progeny of the first generation to have yellowish-green, not true green, seedpods. Mendel's observations were incompatible with a blending hypothesis. An adherent of the vital spark or homunculus theories might have expected the offspring to always resemble just the maternal or just the paternal parent. However, it turns out that it didn't matter which way the cross was made (i.e., green males crossed to yellow females or vice versa); all of the offspring

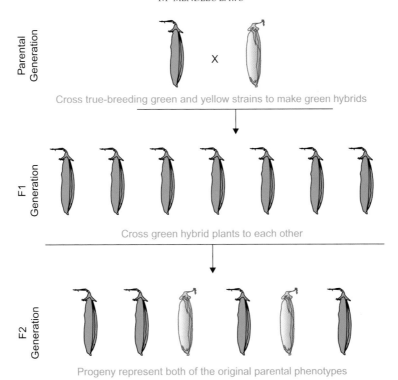

FIGURE 1.3 Making hybrid plants. If you cross plants that consistently produce green seedpods with plants that consist-ently produce yellow seedpods, the result will be a whole generation of hybrid green-seedpod plants in offspring (called the F1 generation). Breeding together plants from strains with different characteristics produces a hybrid plant. Hybrid green-pod plants from the F1 generation, when crossed to each other, produce both green-pod plants and yellow-pod plants in the F2 generation. On average, three-quarters of plants in F2 will make green pods and one-quarter will make yellow pods.

had green seedpods. So much for the theories that traits come from only the male or only the female parent!

When green seedpod plants from generation two were crossed to themselves or each other, they produced both yellow- and green-seedpod plants (see Figure 1.3). The blending hypothesis doesn't work to explain two green-pod parents making a yellow-pod offspring.

Notice that the yellow pod characteristic from generation one disappeared in generation two and reappeared in generation three. One of the things this tells us is that the yellow pod trait from generation one was passed along to generation three without being evident in gen-eration two.

How do we explain all of this? Mendel's explanation made use of several concepts, and no one of those concepts alone was enough to explain what he was seeing.

Concept: What Passes from One Generation to the Next Is Information

In the first of Mendel's three conceptual breakthroughs, he separated the *information* that produced a given trait (which we will call the genotype) from the *physical manifestation* of the trait itself (which we will call the phenotype). In the case of the pea plant, the yellow-pod recipe (genotype) produces a seedpod that appears to our eyes to be yellow (phenotype). If we were

Recipe = Genotype   Flavor = Phenotype

FIGURE 1.4 Genotype vs. phenotype. The distinction between information and what can be produced using that information is one of the most important concepts in genetics. So the recipe (genotype) is distinct from the cake (phenotype). Another key concept is that changes in the information (change the word "chocolate" to "lemon") can give a change in the phenotype (flavor of the cake). Of course, a different flavor is only one phenotypic variant that can occur.

cooking, the words of the cake recipe on the page of the cookbook would be the genotype, but the lemon flavor of the cake would be its phenotype (Figure 1.4). We can carry this analogy further to consider that some phenotypes can be complex, with more than one feature. A cake recipe specifying chocolate flavor can also specify other variable characteristics (dark or milk chocolate, one or two layers, with or without nuts) that are part of its phenotype.

Mendel argued that there were discrete units of heredity (now called genes) that were immutable pieces of information and were passed down unchanged from generation to generation. He argued that these genes specified the appearance of specific traits but were not the traits themselves. This insight gave rise to Mendel's concept of the purity and constancy of the gene as it passes from one generation to the next. In simple terms, Mendel said that genes received from the parents are passed along to the offspring in a precise and faithful fashion. So what gets passed from one generation to the next is the recipe, not the cake.

In order to explain differences in traits, Mendel supposed that genes could take different forms (now called alleles) that specified different expressions of the trait. For example, Mendel claimed that there was a gene that gave

seedpod color, and that there were two different forms or *alleles* of that gene: one specifying green color and one specifying yellow color. We will refer to those alleles that specify green color as G alleles and those that specify yellow color as g alleles. All individual plants that breed true for production of green seedpods must have only the G allele that causes green seedpods. Further, they must have gotten these G alleles from their parents and will pass them along to their offspring. Similarly, plants with yellow seedpods must have only the g allele that causes yellow pod color. They must have gotten the g alleles from their parents and will pass them along to the offspring. So let's reexamine what we saw in Figure 1.2, where we just looked at the phenotype of true-breeding green and yellow plants that produced only plants like themselves. This time, let's add in information about the genotype of those plants, which will be expressed as a listing of the alleles of the pod-color gene that are present in the plants (Figure 1.5).

Concept: Dominant Alleles Mask the Detection of Recessive Alleles

The idea of a genotype (information) that matches the phenotype (the trait produced by using that information) is easy to see and understand when the same strain of plant is bred to itself over and over, always producing plants just like the parents. Plants that only have G alleles produce offspring that only have G alleles, and they are all green. Plants that have only g alleles produce offspring that have only g alleles, and they are all yellow. However, this idea was not enough to explain what happened in Figure 1.3 when Mendel crossed the hybrid plants in generation two to each other and got back both yellow and green offspring. *Mendel explained this by saying that an individual must be able to carry genetic information for a trait it does not express.*

This was Mendel's second big insight: that this pattern of inheritance can be explained if

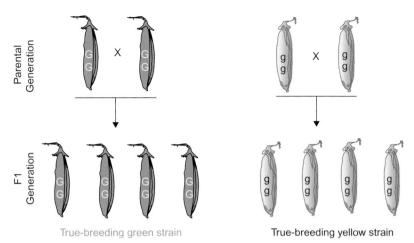

FIGURE 1.5 Genotypes of true-breeding plants. Note that every true-breeding green plant has two copies of the G allele, and every true-breeding yellow plant has two copies of the g allele.

some alleles can mask our ability to detect other alleles. Such masking alleles are known as *dominant* alleles. Thus he hypothesized that green (G) alleles could mask the expression of the yellow (g) alleles, such that individuals getting a green allele from one parent and a yellow allele from the other would be just as green as those that received only green alleles from both parents.

Mendel introduced the term *dominant* to identify alleles that predominate, manifesting their corresponding phenotype without regard to what other alleles are present. He introduced the term *recessive* to identify alleles that recede into an undetectable state, failing to manifest their corresponding phenotype if there is a dominant allele for that trait present. In this case, the G gene allele is said to be dominant (its green phenotype predominates) and the g allele recessive (its phenotype recedes into an undetectable state) because Gg plants had green seedpods that looked just like the seed pods in true-breeding plants that only have the G allele.

Individuals who carry two different forms of a gene, or two different alleles, are called *heterozygotes*. They carry two different alleles (Gg), whereas GG and gg individuals, who carry pairs of identical alleles, are called *homozygotes*.

To see an example of this kind of masking of information, let's take a look again at the creation of the green hybrid plants in the second generation. As we discussed previously, when the true-breeding green and yellow parents were bred, only green plants resulted. The color was the true green of the green parent and not an intermediate color midway between the colors of the two parents. Let's look at the genotypes that went with the phenotypes in this cross as diagrammed in Figure 1.6. A cross of the true-breeding green strain to the true-breeding yellow strain produces heterozygous green hybrids that all have genotype Gg.

Is it obvious yet how the hybrids ended up being heterozygous, with one copy of each allele? Let's look at Mendel's next insight, which explains how this happens.

Concept: One Allele Comes from Each Parent; One Allele Passes to Each Child

Mendel assumed that there is but one gene for each trait. The key to his third insight was to assume that each pea plant (or person for that matter) carries two alleles of that one gene (one from each of their parents). The offspring of

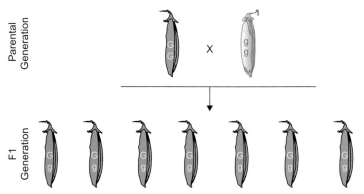

FIGURE 1.6   Masking of traits. Notice that when true-breeding green plants with only G alleles are bred to true-breeding yellow plants with only g alleles, the offspring all have one G allele and one g allele. As can be seen from the heterozygous Gg genotype of the F1 green hybrids, the green G allele masks our ability to tell that the yellow g allele is present.

this individual will receive ONLY one of these two alleles from one parent plus another copy of that gene from the other parent. Moreover, when that offspring reproduces, it transmits one and only one copy of each gene to each of its offspring, and so on. So the basic rule is: Even though each of us has two copies of each gene, our sperm or eggs each carry but one of those alleles. Half of our gametes carry the allele we inherited from our father while the other half carry the allele we inherited from our mother. And every time we make a new gamete, we have an equal chance that we will pass along either the copy we received from mom or the copy we received from dad.

So far, so good. But how do we explain the result in Figure 1.3, in which the offspring of the heterozygous hybrids produced so many more green offspring than yellow? Although the numbers in Figure 1.3 are small, we can tell you that if you do this experiment a lot of times, once the numbers are quite large, you will continue to see an excess of green offspring.

It turns out that each individual has two copies of a gene, but they put only one copy into each gamete that gets used in the formation of offspring. So an individual of genotype Gg does not make gametes with both a G allele and a g allele. Rather, that Gg individual makes some

gametes that contain only the G allele and some that contain only the g allele. Furthermore, the G gametes and the g gametes get made with about equal frequency. It turns out that when two Gg individuals mate, the odds of producing a gg offspring are only 1 in 4, or 25%. There are four different combinations of genotypes that can be produced, and each new offspring has an equal chance (1 in 4) of getting one of the four genotypes:

Mom's G with dad's G gives a GG genotype and a green phenotype.
Mom's G with dad's g gives a Gg genotype and a green phenotype.
Mom's g with dad's G gives a gG genotype and a green phenotype.
Mom's g with dad's g gives a gg genotype and a yellow phenotype.

We now have the concepts needed to explain what happened when we crossed a true-breeding green strain to a true-breeding yellow strain to create a green hybrid that was capable of producing progeny of both colors. A G allele from one parent plus a g allele from the other parent created the heterozygous Gg hybrid that was green because the G allele is dominant and masks the recessive g allele. When a Gg hybrid was crossed to a Gg hybrid, each gamete had an equally

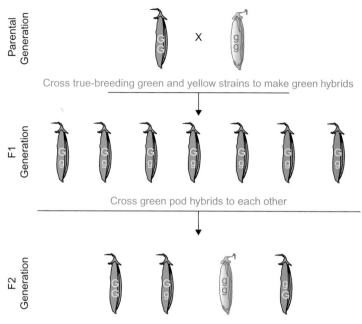

Parental Generation

Cross true-breeding green and yellow strains to make green hybrids

F1 Generation

Cross green pod hybrids to each other

F2 Generation

Progeny represent all four combinations of parental alleles:
GG, Gg, gG, and gg

FIGURE 1.7   Masking of a recessive trait. The yellow phenotype from the parent disappears in the F1 generation where the dominant G allele masks the recessive g allele and reappears in F2 when the g allele is coupled with another g allele. We expect the cross of two Gg heterozygotes to produce (on average) one green GG homozygote, two green Gg heterozygotes, and one yellow gg homozygote.

likely chance of getting a G allele or a g allele. Thus, GG, Gg, gG, and gg genotypes could all be created and result in homozygous green plants, heterozygous green plants, and homozygous yellow plants. Sometimes this is easier to follow if you look at a diagram like the one in Figure 1.7, which shows the genotypes that go along with the phenotypes.

One of the punch lines we want you to take away from this discussion of Mendel's experiments is that the genetic rules in humans, and in many other complex organisms, operate by the same three rules laid out by Mendel to explain pea genetics:

1. Genotype and phenotype are distinct, with different alleles of individual genes corresponding to different phenotypes.

2. The phenotypes created by some alleles can mask (obscure our ability to detect) the phenotypes that result from other alleles of the same gene, leading to the concepts of dominance and recessiveness.

3. An organism has two copies of each gene, receiving one copy from each parent and passing one copy along to each offspring.

Mendel's work suggests that we should be able to predict the phenotype from the genotype in humans as well, and indeed in many cases we can. We will spend much of this text discussing human traits and disease that are easily explained by the basic principles of Mendel. However, for many other traits we discover that the real-life situation is actually more complex because the trait reflects the interaction of many genes or because of interactions between genotype and the environment. Just as with alleles that produce seedpod color, some of the variant information in our two copies of the genome is bound to have a certain outcome – blue eyes vs. brown, for instance – whether our parents raised us on burgers or health food, in rich times or poor, on a farm or in a big city. However, other characteristics – how tall each of us ended up, for example – may vary depending not only on a difference in our genetic information but also on the environment in which we live. In Chapter 9 we shall return to this concept of complex characteristics that result from a complex combination of influences, genetic and environmental.

1.2 SELECTION: ARTIFICIAL, NATURAL, AND SEXUAL

Charles Darwin, one of Mendel's contemporaries, studied a very different aspect of genetics and made contributions that had an equally important impact on our understanding of biology and genetics (Box 1.2). Where Mendel started with the existence of differences between individuals as a given, and then looked at how information is transmitted in unchanged form into subsequent generations, Darwin looked at how phenotypic characteristics of a population can change over time (evolve). He proposed that the mechanism by which evolution happens (selection) is a process that makes some members of the current generation more likely to produce progeny that will make up the next generation, with this process of selection repeated over the course of many generations. There are several different types of selection that can play a role in determining which individuals will be the parents of the next generation. To examine the simplest kind of selection, let's see what would have happened if Mendel had practiced artificial selection of the parents of subsequent generations beyond generation three of the experiment in Figure 1.7.

BOX 1.2

DARWIN'S ORIGINAL WORKS

Charles Darwin (1809–1892) left us with extensive writings on evolution and natural selection: diaries and letters from his travels and more than two dozen books, including the *Journal of Researches into the Natural History and Geology of the Countries Visited During the Voyage of H.M.S. Beagle, On the Origin of Species by the Means of Natural Selection, The Descent of Man and Selection in Relation to Sex,* and *The Expression of Emotions in Man and Animal.* Darwin and his work on evolution have been seen as controversial so there are huge numbers of publications about him and his work. While many argue that his theory is unfounded, citing lack of intermediate forms, there exists a vast body of scientific information to support his model, including the existence of large numbers of examples of intermediate forms. More recent independent support for his model comes from the human genome project and other genome projects. A strong framework for understanding Darwin's model of evolution is derived from a modern understanding of molecular and cellular biology. The modern fields of biology and genetics regard the existing data as supporting evolution, with natural, artificial and sexual forms of selection as the primary mechanisms driving evolution.

Artificial Selection

In Mendel's classical experiment, the frequency of plants with green pods or yellow pods in generation three was predictable and reproducible, with the same approximate numbers coming out of experiment after experiment. But what would have happened if he had continued the experiment for more generations, selecting the parents of subsequent generations based on whether or not they had particular characteristics? If at every generation he had picked only green-pod plants as parents of the next generation he would achieve a very different result than if he had picked only yellow-pod parents in generation three and thereafter. To take it one step further, if he tested each combination of green-pod parents from generation three for whether or not they could produce yellow-pod offspring and then selected those with no yellow-pod offspring as parents of the next generation, he would shift the population of plants towards a pure-breeding line of green-pod plants. (Remember that the green-pod plants in generation three include both homozygous GG and heterozygous Gg plants so he would not arrive at a pure-breeding line in the first generation because of the presence of green heterozygotes that will not breed true.) If he selected only yellow-pod parents from generation three he would almost immediately have a pure-breeding yellow-pod line. Thus he could shift the frequency of yellow-pod plants in the population through his selection of which plants to breed in each generation.

This kind of artificial selection of organisms with a specific phenotype has been taking place for as long as humans have been breeding domesticated plants and animals, and played a role in the original domestication of our agricultural plants and animals as well as pets. While we cannot look back into the mists of time to see exactly how dogs were domesticated, or how long it took, we can look at experiments in domestication that have taken place recently enough that we can look at the details of what was done. While we might have thought that it would take long periods of time to get from the wolf to the dog, and that a lot of training of the animal would have played a role in the process, we find that domestication of the silver fox took only a few tens of generations and involved no training of the animals (Box 1.3).

Similarly, selection of which plants to breed has resulted in a modern corn plant that is amazingly different from the ancestral plant teocinte from which it is often said to have been derived. Over many centuries, different strains of corn have been bred for different characteristics, in each case by carefully selecting plants that have some resemblance to the desired trait as parents of the next generation, and watching as the population of plants gradually shifts more and more towards the desired characteristics. Different strains of corn have emerged over the course of many centuries as the result of breeding programs aimed at developing corn with increased sweetness, higher protein levels, increased lysine levels to make the protein more "complete", or resistance to pests, herbicides, infections, or drought. Interestingly, researchers have now found that the genome (the complete array of genetic information that determines the properties of an organism) of a common, agriculturally important strain of corn used in the corn genome project is 22% larger than the genome of a strain used to produce popcorn. Clearly, artificial selection of which parents will produce the next generation works to bring about major changes in the characteristics of plants and animals, sometimes over the course of centuries and sometimes over much, much shorter periods of time.

Natural Selection

While selection of the organisms to be mated is considered artificial selection, most of evolution has resulted from natural selection, a process that does not involve anyone intervening

---

BOX 1.3

ARTIFICIAL SELECTION AND DOMESTICATION OF THE SILVER FOX

A fascinating example of a recent domestication event started in 1959 when Dmitri Belyaev used selective breeding to create a line of domesticated silver foxes. He started with 30 male foxes and 100 vixen, and at each generation only a small fraction of the foxes were used as parents of the next generation. To qualify as a parent, a fox had to show receptiveness to the presence of humans through things like tail wagging and not retreating. Although about 10% of the animals showed this kind of acceptance of the handlers, they were not domesticated and special precautions were needed to prevent handlers from being bitten. There was no behavioral training of the animals, only selection for animals who naturally exhibit the desired behaviors without any training. As the breeding experiments continued, and

the fraction of the animals that were comfortable with humans increased, the breeders started selecting parents from among the animals that would actually seek out humans. By 40 years after the start of the experiment, more than 70% of the population consisted of tame foxes whose behavior resembles the behavior of dogs, including wagging tails, licking hands, and actively seeking attention from humans. Some physical changes accompanied the behavioral changes, such as developmental changes in the timing of when the body expresses some hormones associated with fear responses. Interestingly, the animals displaying the full array of tame features also look different from the ancestors Belyaev started with, with floppy ears and curled tails typical of some breeds of dogs.

---

to make decisions about who will be the parents of the next generation. Natural selection can result in a shift in which plants or animals go on to produce progeny even if there are no human beings involved. If seeds from a type of wild flower blow into an arid region, with the parents from which the seeds originate differing in their ability to survive drought, the drought-resistant plants will be more likely to survive to produce offspring. The drought-sensitive plants will be more likely to die off without leaving offspring. As survival determines which plants will produce offspring, the population will gradually shift towards having a larger representation of plants that contain genetic information that makes them more likely to survive drought conditions. In this case, what happened is exactly the same thing that happened during the domestication of the silver fox: organisms with one characteristic had more

children while the organisms with the other characteristic had less, and the outcome is the same whether the process of selecting the parents was carried out by a farmer or by environmental conditions.

Sexual Selection

We often hear talk of evolution in terms of survival of the fittest, which creates images of bigger stronger animals winning fights against smaller, weaker animals. However, the basic principle underlying evolution has nothing to do with who is stronger or whether one organism kills another. The principle of evolution is not a matter of whether a creature lives or dies. It is simply a matter of whether or not a creature with a particular trait has children, or whether that creature has more children than other members of its species who do not have the trait. While natural selection can sometimes select parents for the next generation based on who lives or dies, anything that affects how many progeny are produced is acting to affect the composition of future generations. In the case of the peacock, we see a process of sexual selection, by which males with large colorful tails are more likely to produce children in the next generation. In fact, if we look at a peacock tail it seems that there ought to be disadvantages to a huge ungainly tail that draws attention from predators, affects mobility, and uses energy and resources to produce. But female peahens preferentially mate with the males that have the elaborate tail. As we will see later in the book, the impact of evolution is not limited to lower organisms; it also applies to the human race where we can see specific examples of changes in human genotypes and phenotypes that have taken place in response to natural selection.

Although Mendel's life overlapped that of Darwin, they did not work together or share ideas. It is interesting that these two men from the same period in history each arrived at such major breakthroughs and had such a huge impact on our understanding of biology and genetics. The thing that was missing, that limited how far they were each able to take their ideas, was information on the nature of the genetic information and the processes by which the information changes to create the diversity of characteristics that they each studied.

Genetics is the study of characteristics that differ from one individual to the next and the transmission of those variable characteristics from one generation to the next. There are limits to how tall or short each of us could have become under the greatest environmental extremes of feast or famine. Those limits are set by genotype. Any feature that differs between individuals can be a valid point of genetic study, whether the variable feature is something visible, behavioral, or assayed by a biochemical test. Peas do not much resemble humans,

so what do Mendel's studies of inheritance in pea plants have to do with human genetics? In the next section we look at inheritance of human characteristics from the perspective of Mendel's laws.

1.3 HUMAN GENETIC DIVERSITY

When Scott's daughter Tara was born, Scott and his wife were immediately surrounded by the expected group of close relatives, as well as many other relatives Scott had never met. Some of them brought to his daughter's crib the most extraordinary bits of genetic folklore. He can remember one of them staring at his daughter's eyes and saying, "She has her grandfather's eyes, but then girls always get their grandfather's eyes." Wait a minute; Scott's a geneticist, and this was big news to him! Such wisdom kept coming for the next few days, a continuous stream of different bits of genetic folk wisdom. Some of it was just folklore, but some of it had good basis in fact. The point is that people have long known that traits move through families in patterns, patterns that we now call modes of inheritance. One of the tasks we face in modern genetics is figuring out which pieces of genetic folk wisdom are actually true and then understanding why they happen.

Diversity is one of humanity's greatest assets. Some of that diversity is obvious when we look around us and find ourselves surrounded by unique individuals rather than carbon copies. Much of the diversity that gets noticed the most takes the form of physical characteristics – skin color, height, weight, hair color, strength, speed, or facial features. The first thing everyone asks when a new baby is born is, "Is it a boy or a girl?" However, much important diversity has nothing to do with the kinds of characteristics that determine whether or not we get offered a modeling contract. A lot of important diversity takes place at the molecular and cellular level. Some of it seems so complex that we have to wonder if human diversity could result from the same kinds of processes that Mendel studied.

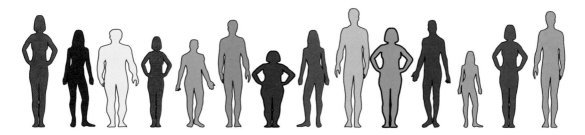

The kinds of characteristics we want to understand in humans may be as diverse as:

- Color-blindness
- Differences in the rate of aging
- Being male or female
- Risk-taking behavior
- Life-saving resistance to malaria
- Hairy ears
- Control of blood sugar levels
- Sneezing in response to sunlight
- Perfect pitch in a talented musician
- Susceptibility to a particular kind of cancer
- Schizophrenia
- Rejection of a transplanted kidney.

Which of those characteristics do you think are genetic, and which ones do you think result from things that happen to you that originate from outside sources? Evidence exists to show that genetics plays a part in everything on this list, even the behavioral traits.

While a lot of the obvious human variation is cosmetic – color of hair, shape of nose, length of legs – there is also a great deal of important variation that is not obvious to the eye. Some of the important variables are things we know about ourselves, such as blood type or cholesterol levels, and many others are things that we might not know but that our doctors could assay if the need arose. Some of these variables, such as markers of inflammation, may be biomarkers for disease, or important for being a recipient of an organ transplant, such as tissue transplantation antigens. Many of the variations

between human beings that we understand in the greatest detail are those that are associated with disease because they have received the most intense study. But there remain many other variables that differ from one person to the next that we can measure without knowing the implications for our health and well-being. Some of these variable traits seem to run in families, and one of the first steps towards understanding the underlying genetic basis of the trait is understanding the mode of inheritance.

Mendel proposed that information is different from what gets made using that information. He proposed the existence of the particles that we now call genes, and he proposed that some pieces of information are dominant over other pieces of information. All of these ideas from Mendel help us understand the processes of evolution and natural selection proposed by Darwin. It also helps to explain many things we see in human patterns of inheritance. Next we will introduce human inheritance and show how many (but not all) human traits show Mendelian inheritance, that is, transmission from one generation to the next in patterns that resemble what Mendel saw when he studied peas.

1.4 HUMAN DOMINANT INHERITANCE

For some traits in humans it is easy to see that the trait is inherited. What happens if someone has a trait that results from a dominant allele that

---

BOX 1.4

HOW TO RECOGNIZE DOMINANT INHERITANCE

Underlying concepts:

- One defective copy of the gene is enough to cause the phenotype.
- The presence of a second "good" copy cannot compensate for the defective copy.
- Quantity of gene product is critical and half as much won't do.

Rules for what we will see in a dominant pedigree:

- All affected children have at least one affected parent.
- Approximately half of children of each affected individual are themselves affected if they are old enough to have developed the trait.
- Except for rare situations discussed later in the book, unaffected parents do not produce affected offspring.

- Approximately equal number of males and females are affected.
- Affected males and females can pass the trait to their children.
- A hallmark of autosomal dominant inheritance for an early-onset trait is seeing the trait transmitted in an uninterrupted line through three or more generations, along with the other features described here.
- Children of an affected individual are at 50% risk of inheriting the trait.
- Risk to one child is not affected by how many prior births in the sibship have the trait.
- Risk can be modified by environmental factors, mosaicism, errors in meiosis, de novo mutation, non-penetrance or age-related penetrance, which we will discuss later in the book.

---

is present in most of the population? For example, if most of the population had green pods, we could introduce one plant with yellow pods and watch that yellow pod disappear for generations before it would start to emerge again when chance brought together rare heterozygotes.

But what if most pea plants had yellow pods (the recessive form) and a dominant green-pod pea plant were introduced into the population? Most individuals would have yellow pods but the rare individuals with green pods would have offspring that would each have 50% risk of having green pods. Thus a trait resulting from a recessive allele can effectively disappear for many generations in a population of individuals with the dominant characteristic. A dominant trait will be evident in generation after generation, wherever one copy of the dominant allele

is present. The dominant trait will be very easy to detect and trace through later generations if that dominant allele exists in a population in which most individuals have the recessive allele and the accompanying recessive trait (Box 1.4).

Consider the family of Jacob, who is the proband (see Box 1.5) in the family shown in Figure 1.8 where deafness has turned up in all four generations for which information is available. Individuals who inherit the family's hearing loss trait are deaf from birth. If you look at descendants of Jacob, who is deaf, and his wife Adelaide, who is not, you find that about half of their children are deaf and half are not. Like the heterozygous pea plant with both green and yellow alleles, Jacob's children have one "deaf" allele and one "hearing" allele. They have an equal chance of passing either along to their

BOX 1.5

PROBAND: FIRST POINT OF CONTACT WITH A FAMILY

It turns out that it is actually important to keep track of who the first person is by which a family is identified in a genetic study. The term *proband* is one of several terms used to indicate the first member of a family who has contact with the doctor or researchers in a study. Sometimes the proband is someone who has the inherited characteristic that is being studied, as is the case for Nick in Figure 1.8. However, sometimes the proband may be a relative who first brought a family to the attention of researchers when they went to the doctor to say, "Other members of my family have this inherited disease, and I am worried about whether I might end up getting it, too." Sometimes, someone may show up in the doctor's office because they are worried about whether they can give their children a genetic disease present in their relatives but not themselves. Look for the small diagonal arrow to find the proband in families shown in this book.

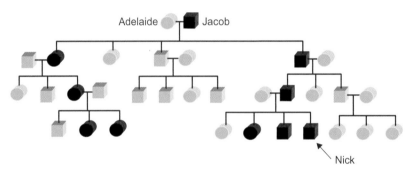

FIGURE 1.8 Dominant transmission of deafness in a hypothetical four-generation family. Each person with the trait passes the trait along to about half of their children, both sexes can have the trait, and both sexes can pass the trait along. Individuals who do not have the trait do not pass deafness along to their descendants. Every time someone deaf in this family has a child, there is a 50% chance that their child will be deaf. The chance of having a deaf child is like the chance of flipping heads or tails with a coin, an event that will happen on average about 50% of the time. Before Nick was born, his parents might have thought that the next child would probably not be deaf since they already had two deaf children. Was this valid? No. *This is a key concept: with each new genetic flip of the coin, the chance is once again 50%, and that chance is not affected by whether or not other affected children have been born previously.*

child. The difference is that individuals with one "deaf" copy and one "hearing" copy of the gene have the affected phenotype, so their children are at risk of inheriting the trait whether or not the other parent contributes a hearing genotype. In fact, in this family, about half of the children of each deaf individual will also be deaf. The transmission of information between generations is consistent with Mendel's model that half the progeny get one allele and half the progeny get the other allele, and that some traits are dominant over others.

In fact, there are dozens of different dominant mutations in various genes that can cause hearing loss which are transmitted in this manner. However, there are dozens of other mutations that display recessive hearing loss. In theory, it is easy to make the kinds of predications we make here, that the mode of inheritance is dominant and that the risk to children of affected individuals is 50%. However, in real-life situations, things are often more complicated. Small families, adoption, divorce, early death of some family members, geographic distances between family branches, and other complications can sometimes limit the information that is available to help sort out the level of risk to a new child.

1.5 HUMAN RECESSIVE INHERITANCE

As an example of recessive inheritance, let's discuss the human trait albinism, specifically a form called oculocutaneous albinism, which is manifested in people who make little or no melanin pigment (Figure 1.9).

The common perception of albinism, people with stark white skin and hair and red eyes, is over-simplified and incorrect. In fact, there is some variation in how pale people with albinism are, and some may even have yellowish hair or other signs of coloration, such as freckles. Most commonly, they have blue or gray eyes. The stereotype of red eyes is wrong, although sometimes their eyes may take on a purplish or reddish tint if the light is just right, resulting from the red tints from the retina showing through the pale coloring of the iris. This does not normally happen in most individuals with blue or gray eyes who do not have albinism, because the pigment in a pigment epithelium layer behind the iris in most blue or gray eyes normally blocks the red tints in the retina from being seen.

One form of albinism results from a defect in the gene that makes a protein called tyrosinase. The tyrosinase protein carries out a key step in making the melanin that provides color in our hair, skin, and eyes. Individuals with the classical albinism phenotype (white skin, white hair) turn out to be individuals who do not make tyrosinase, while those with some coloration are

FIGURE 1.9 Positive Exposure. Rick Guidotti's elegant photo spread in *Life Magazine* featured individuals with albinism as unique and beautiful, emphasizing the need to avoid stuffing them into some prejudicial box with a label on it just because they happen to share a genetic trait. In contrast to the attitudes during the nineteenth century when people with albinism were featured in circus side-shows, this twentieth-century article succeeded in communicating a great sense of the unique and positive value in each of those photographed, including the beautiful woman featured in this photo. At www.rickguidotti.com there is more information about Rick Guidotti's Positive Exposure campaign, which is aimed at challenging stigmatization of genetically unusual individuals and celebrating the differences that are the result of genetic diversity. (Image courtesy of Rick Guidotti for Positive Exposure. www.positiveexposure.org. All rights reserved.)

those who make too little tyrosinase or tyrosinase that does not work very well. In Chapter 4 we will talk about the mechanisms by which a gene can make a partially functional product or no product at all.

In most ways, people who lack melanin are just like the rest of us. But even though they are highly diverse in terms of a variety of traits, they do have some features that they hold in common. They also share several different vision problems. The lack of melanin during development of the eyes can cause abnormal routing of the optic nerves into the brain and

---

BOX 1.6

HOW TO RECOGNIZE RECESSIVE INHERITANCE

Underlying concepts:

- Both copies of the gene must be defective to cause the phenotype.
- The presence of a second "good" copy can compensate for one defective copy.
- The cell may have a lot of tolerance for reduced levels of gene product.

Rules for what we will see in a recessive pedigree:

- Equal numbers of males and females will be affected.
- Affected individuals will often have no or little family history.
- Parents will often both be unaffected heterozygotes called carriers with about one-quarter of the children affected.

- Carriers are unaffected, but may sometimes show minor aspects of the phenotype.
- If one parent is affected and the other parent is heterozygous carrier for the same gene, then about half of the children will be affected.
- If both parents are affected and defective for the same gene, then all of the children will be affected.
- If both parents are affected with the same trait as a result of defects in two different genes, then none of the children will be affected.
- Risk to one child is not affected by how many prior births in the sibship have the trait.
- Risk can be modified by environmental factors, mosaicism, errors in meiosis, de novo mutation, non-penetrance or age-related penetrance, which we will discuss later in the book.

---

result in inadequate development of the retina. They often use glasses, but their vision often cannot be corrected to 20/20 acuity with either glasses or surgery. They are unusually sensitive to bright light. Some are legally blind, but others see well enough to drive a car when using special lenses. In some cases, skin cancer can be a problem, especially in equatorial regions, if they don't take steps to protect themselves well enough from sunlight. And even though they are all pale in coloring, we see variation such as white hair in some and yellow hair in others.

Albinism Is Recessive

Albinism is known to be inherited in a recessive manner (Box 1.6). The fact that it is hereditary may not seem obvious if you look at someone with albinism who is born into a small family. But we do find evidence that albinism

is hereditary if we look at information from a lot of families in which someone has albinism. For an individual with albinism, about one in four of their siblings will also have albinism, so if we look at couples who give birth to a large number of children, we will sometimes find an individual with albinism who has one or more brothers and sisters who also have the trait. An individual with albinism is more likely to have a family history of albinism than someone with the pigmented allele, so examination of albinism families that are very large and have maintained good records on family history can offer clues. And if one identical twin has albinism, we normally find that the other identical twin also has it, but if a fraternal twin has the albinism trait there is only a 25% chance that their twin will also have it (Box 1.7).

When we look at inheritance of the albinism trait in Figure 1.10, we see that Mendel's rules

BOX 1.7

IDENTICAL TWINS AND FRATERNAL TWINS

Why do researchers compare identical twins (formed when one sperm fertilizes one egg and then the embryo splits to form two embryos) to fraternal twins (formed when two different sperm fertilize two different eggs) instead of comparing identical twins to brothers and sisters born at separate times? Because whenever the siblings are twins, they have something in common besides their genetic makeup – they have also shared the same environment in the uterus for nine months.

If siblings were conceived and born at different times, the differences between them might include not only their genetic differences, but also differences in their pregnant mother's nutrition, exposure to smoke, encounter with physical trauma, or consumption of medications. Twin studies that compare identical to fraternal twins are often used to try to determine how much of a characteristic can be attributed to genetics and how much can be attributed to environmental causes.

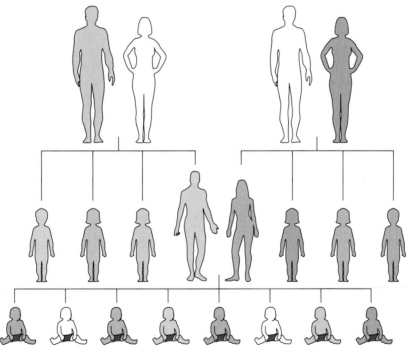

FIGURE 1.10 Albinism family tree. In the first generation, we see two different couples in which an individual with albinism marries someone who does not have the albinism trait. Notice that all of their children have pigmented hair and are carriers, individuals who lack the trait but carry the information. Marriage of two carriers leads to about one-quarter of their children having the albinism trait. In many families into which a child with albinism is born, there may be no known ancestors with albinism. In rare cases, the trait might be reported for a distant ancestor, but most often there will be no evident family history of albinism. Thus recessive information can pass through many generations before a carrier chances to marry another carrier to a produce a child with the trait. In this case, for the information to result in the trait, the child must receive an albinism allele from each parent so that they end up with two albinism alleles.

apply here. If someone has a pigmented version of the albinism gene (the pigmented allele) along with the albinism allele, the albinism trait is not manifested and they have color in their skin and hair. If the individual is homozygous for albinism alleles, skin and hair color will be white. As with the pea plants, the heterozygous individual's coloring is determined by whether or not they have a dominant pigmented allele. Their coloring is not the intermediate skin color that would result from a blended average of the two different alleles. Thus there is a human trait, absence of pigmentation, that is recessive to the dominant trait, presence of pigmentation.

Albinism and Carriers

The term *carrier* is used for people who carry a single copy of a recessive allele without having the trait. Carriers are individuals who carry the information (genotype) without manifesting the trait (phenotype), just like the heterozygous plants with green peapods. In the case of albinism, we find carriers with normal levels of skin and hair color who are heterozygous, with one allele for albinism and one allele for pigmentation. They do not have the intermediate coloration that we would expect if the albinism allele diluted out the normal allele. Albinism carriers usually will not know that they are a carrier unless they pick someone else with an albinism allele as their mate and have a child with the albinism trait.

According to NOAH, the National Organization on Albinism and Hypopigmentation, one person in every 17,000 has some form of albinism. Based on those numbers, we expect more than one in every 100 individuals to be a carrier, someone with one normal copy of an albinism gene and one defective copy of that same gene. The numbers for albinism show us something that is true and can be generalized to other traits: the frequency of carriers in a population is a lot higher than the frequency of individuals with the trait.

There are situations in which some individuals can know that they are carriers for a recessive trait even if they do not have any phenotypic features to suggest it. Since an individual with albinism carries the albinism allele on both copies of the gene, the only tyrosinase allele they can pass along to their children is an albinism allele. Thus, we know that every child of an individual with albinism has to have an albinism allele and we refer to those children as *obligate carriers*. While albinism carriers show no sign that they are carriers, there are other traits that manifest a carrier state in which a carrier who lacks the trait does have some unusual feature(s), sometimes a minor aspect of the trait, that shows that the person is a carrier.

Transmission of Albinism in Families

As with Mendel's peas, a pair of carrier parents can produce offspring with the recessive phenotype. So what happens in people? The pigmented father with genotype Aa and the pigmented mother with genotype Aa, can produce four different genotype combinations, with a child having an equal chance of getting any of the four combinations:

Mom's A with dad's A gives the AA genotype and a pigmented phenotype.
Mom's A with dad's a gives the Aa genotype and a pigmented phenotype.
Mom's a with dad's A gives the aA genotype and a pigmented phenotype.
Mom's a with dad's a gives the aa genotype and an albinism phenotype.

If we use a device called a Punnett square, we can separately identify what genotypes will be present in the gametes produced by the parents. We can then look at the different combinations of genotypes that can be produced by different combinations of sperm and egg. When we do this for two individuals who are carriers for mutations in the same albinism gene, we similarly find a prediction that about one-fourth

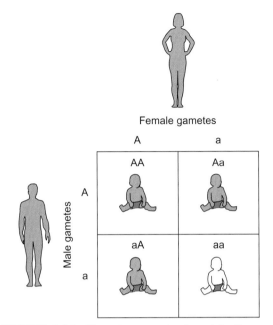

Female gametes

Male gametes

	A	a
A	AA	Aa
a	aA	aa

FIGURE 1.11   Punnett square to show inheritance of albinism in one family. Along the side, list the genotype of each type of gamete that one of the parents can make. Along the top, list the genotype of each type of gamete that the other parent can make. In each square, combine the alleles from the side with the alleles from the top to find out the genotype and phenotype of the offspring that would result from combining those gametes. When two carriers have children, the Punnett square can be used to predict the chance that they will have a child with the albinism trait. Notice that we are not specifying what color of skin and hair the parents or the normal children have, since their coloration could range from very pale to very dark and have no effect on the outcome here. The Punnett square might seem like a lot of trouble for tracking the possible combinations of two alleles from each parent, but this system can be used to track more complex combinations of gametes representing different alleles at more than one gene, in which case this exercise can take a problem that is very complex and turn it into something simple and easy to visualize.

of their children would have the albinism trait (Figure 1.11). If we consider the genotypes in the Punnett square we see that each of the brothers or sisters in this family has a 25% chance of also having the trait, a 50% chance of being a carrier, and a 25% chance that they will have inherited no albinism alleles.

Thus, for rare recessive traits, it is quite common to have a child with the trait born into a family in which no one else has the trait. If the child has a lot of brothers and sisters, another child might also have the trait, in which case the involvement of genetics might be easier to figure out. When children with albinism grow up and have children, they usually have no children with the trait except in the very rare cases in which they marry a carrier. Thus rare recessive traits often pop up unexpectedly in families with no history of the trait, or even families who have never heard of the trait in question, because the gene defect has been passed from one generation to the next without anyone being aware of it. In Sam's family, the evidence was there but did not turn up until someone searched back through family history (Figure 1.12). Often there is no recent family history of the trait, especially for something rare. Such traits may look like they are not hereditary when in fact the genetic information is getting passed along invisibly, since there often is no way to tell if someone is a carrier for a genetic defect by just looking at them.

There Are Many Recessive Traits

All of us have many recessive characteristics that never come to light because the presence of the dominant normal alleles masks the recessive alleles. For some traits such as albinism, we might find out that we are a carrier when we produce a child with the trait. Some recessive traits we might never realize we carry because they cause miscarriages at the earliest stages of life so that we never end up seeing individuals in the population who are homozygous for the allele that causes the trait. In some cases a high rate of miscarriages in a family might be a clue that we should consider a recessive trait to be lethal, but in some cases the miscarriage occurs so early in development that the mother might not even realize that she had been pregnant or suffered a loss. In other cases the allele causes a trait that only occurs very late in life, so that we

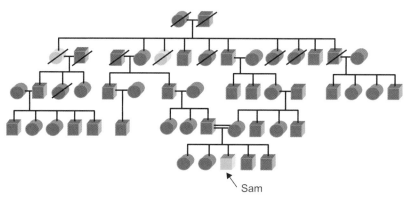

Sam

FIGURE 1.12   A hypothetical albinism pedigree shows what the family history might look like for a family of five generations in which the youngest sibship includes a child with albinism. Compare with Figure 1.11 in which pictures took up a lot of space to appreciate this more efficient notation system that makes it possible to put a lot of information into a small space. Standard sets of pedigree symbols include squares for males, circles for females, and diagonal lines for deceased individuals. Pedigrees most often use filled symbols for people with the unusual trait or disease and open symbols for people who are healthy, normal, or have the common form of the trait, but in this case white was used for individuals with the trait because of its similarity to the characteristics of those individuals. The arrow points at the person called the proband (Box 1.5), the first person in the family who was identified. If both of the proband's grandfathers migrated from Europe to the United States, would you expect that the proband would have information about those earlier generations thousands of miles away?

may well have passed on before the trait is ever manifested in our child.

If your newborn child has a trait that you have never seen in your family or maybe even in your community, how would you be able to tell that it is an inherited trait that is not the result of an infection, diet, or something else unknown that happened during the pregnancy? If the trait in question were albinism, and you did not know of anyone with the kind of white hair and skin that your child has, how would you figure out whether those features are inherited? If 1 in 17,000 individuals have this trait, you might look all around your community and not see a single case of someone else like your child. One clue can come from examining your own family history to see whether you have other relatives with the same trait, but there will often be no sign of the trait in earlier generations if the carrier rate in the populations is low.

For many inherited traits, your doctor will be able to tell you whether it runs in families

even if you can't find any relatives with the trait. Researchers working in medical genetics have learned a lot about many thousands of hereditary traits, often through studies of large numbers of families, or in some cases through studying populations of twins (Box 1.6). So when someone arrives in a doctor's office with no family history of the trait causing the problem, the doctor is not limited to the information about the patient or the patient's family but can draw on information gained from the study of many other rare individuals and families.

The study of rare families is not the only source of information that has helped doctors understand the hereditary factors contributing to many human traits. Because we don't breed people like pea plants to answer questions about the mode of inheritance (fortunately!), it sometimes takes conceptual breakthroughs from experimental work with peas and flies and other nonhuman organisms to arrive at an understanding of recessive inheritance in

---

BOX 1.8
_____

HOW TO RECOGNIZE PSEUDO-DOMINANT INHERITANCE

Underlying concepts:

- Both copies of the gene must be defective to cause the phenotype.
- The presence of a second "good" copy can compensate for one defective copy.
- The cell may have a lot of tolerance for reduced levels of gene product.
- The trait is very common and the carrier rate is high, so there is a substantial chance of carriers marrying other carriers or affected individuals.

Rules for what we will see in a pseudo-dominant pedigree:

- Equal numbers of males and females will be affected.
- If both parents are affected, all children will be affected.

- There will often be branches of the family that show three or more generations of continuous transmission of the trait, but there will also be skipping of generations typical of recessive families.
- Other branches of the family will hit a dead end, with no affected progeny from an affected individual even if there were many children.
- Affected individuals will often have a strong family history that does not exactly fit autosomal dominant inheritance.
- Risk to one child is not affected by how many prior births in the sibship have the trait.
- Risk can be modified by environmental factors, mosaicism, errors in meiosis, de novo mutation, non-penetrance or age-related penetrance, which we will discuss later in the book.

---

humans. Mendel's work may make it easier to understand how something can be considered hereditary when it has not been seen over the course of many generations of a family. Similarly, mice with cancer, yeast with enzymatic defects, and infertile flies can all tell us things that translate quite directly into a better understanding of problems faced by human beings. We hope that the things we talk about in the rest of the book will clarify just what happens at the molecular and cellular levels to bring about the patterns of inheritance that we see in the characteristics of the people in a family. And we hope that an understanding of those underlying mechanisms will leave you thinking, "Of course it happens that way. It all makes sense." Arriving at a point at which it makes sense can be especially important for recessive traits, which often leave individuals

and their families perplexed at something that seems to have just popped up out of nowhere.

Pseudo-dominant Inheritance of a Recessive Trait

An especially perplexing mode of inheritance is pseudo-dominant inheritance, which happens when a trait that is actually recessive shows a pattern of inheritance with features of dominant inheritance because the recessive allele is so prevalent (Box 1.8). If we try to classify the mode of inheritance by looking at only one large family, we are missing a critical piece of information – the frequency of the trait in the population.

For rare recessive traits, carriers heterozygous for the trait allele rarely pair up with another carrier and we expect to see a pedigree

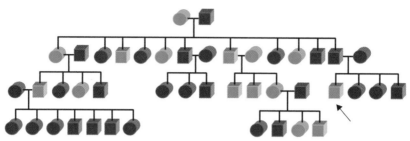

FIGURE 1.13   A Scandinavian family showing inheritance of blue eye color. Inheritance of blue eye color is seen in four generations in a row. Although this family shows some features of dominant inheritance, the proband having two brown-eyed parents and three brown-eyed siblings is more like what we would expect for recessive inheritance. We are assisted in interpreting this pedigree by knowing that blue eyes are quite common in Scandinavian populations and that Scandinavian populations have a high carrier rate for the BEY2 allele for blue eyes. This family shows so many blue-eyed individuals because the carrier rate in the population is high, so many people who marry into the family are carriers. The actual inheritance of eye color is more complex than this. There are additional genes involved in eye color, and the "brown" phenotype varies, including different shades of brown as well as some mixed colors such as hazel or blue with brown spots.

much like that shown in Figure 1.12. But when a trait is very common, we will often see cases of two carriers having children together and cases of individuals with the trait having children with a carrier. The result is a mode of inheritance like that shown in Figure 1.13.

Pseudo-dominant inheritance shows many features of dominant inheritance. The trait turns up about equally in males or females, and can be passed along by either males or females. We often see sibships in which about half the at-risk kids are affected.

So what clues do we have that this is not actually dominant? The first clue comes from the population information. Blue eyes are very common in Scandinavia, so we expect that if the trait is recessive, the carrier rate would be high. Thus there would be an elevated probability that carriers in this family would marry other carriers, or that individuals with blue eyes would marry carriers and have half their children have blue eyes. The second clue comes from the group of seven siblings with brown eyes in Figure 1.13 who are at 50% risk. The third clue comes from the blue-eyed proband whose parents both have brown eyes; if blue eyes were dominant we would not expect the brown-eyed parents to produce a blue-eyed child.

The gene shown in Figure 1.13 is the BEY2 gene. If an individual is homozygous for the blue allele of BEY2, they will have blue eyes. If the individual is heterozygous for the blue allele and the brown allele, they might have brown eyes, hazel eyes, or blue eyes with brown spots. Of special interest here, the gene found by those studying blue eyes turns out to be one of the genes found by those studying albinism! How can the BEY2 gene that causes blue eyes also be the OCA2 gene that causes albinism? Some types of mutations in OCA2 result in classical albinisms. Some types of mutations in the same gene result in atypical albinism with yellow or even red hair. And other changes result in blue eyes without albinism. In fact, while families like this one can be seen in regions where blue eyes are quite frequent, the overall genetics of pigmentation is much more complex than this, with at least 18 known genes for albinism and more than a dozen loci for skin, eye, and hair coloration listed in Online Mendelian Inheritance in Man (Box 1.9).

It can be hard to classify the mode of inheritance in a family. In some cases, what is happening in the family is complex and involves multiple factors, something we will talk about

ONLINE RESOURCES FOR INFORMATION ON HUMAN GENETIC TRAITS

Important online resources provide information about inherited human traits. PubMed at the National Center for Biotechnology Information at the National Institutes of Health offers a mechanism to search the biomedical literature for published articles, allowing access to abstracts and in some cases also providing free access to the actual article. Online Mendelian Inheritance in Man (OMIM) is a catalogue of human genetic traits that provides review articles summarizing findings on traits and genes. OMIM is a good place to start for an overview of any of thousands of different human traits. It also offers connections between phenotype and genotype information where the underlying basis is known. GeneClinics also provides review articles

that summarize what is known about human genetic traits. While they offer articles on fewer topics than OMIM, the articles may include additional clinical details and information on available sources for genetic testing relative to that trait. In some cases, searches of the literature may lead to confusing nomenclature, with more than one name used to refer to the same trait or gene, and this can be resolved through a search of the Human Genome Nomenclature database. These sites contain links to additional human genome resources. Each of these sites can be found by filling in the search box on your browser with the title: PubMed, OMIM, GeneClinics or Human Genome Nomenclature Committee.

---

later in this book. When studying human families, we routinely lack control over the availability of the information we need to sort out genetic puzzles. We face problems with missing medical information for previous generations, or lack of data on someone who died early or lost contact with the family. And we often lack information on genotype for at least some family members. Mendel was able to ensure that his green-pod plants also had green-pod genotypes by selecting true-breeding plants after many generations of breeding. We often have to figure out genotypes indirectly, by inferring genotypes of parents from phenotypes of children, for instance. Sometimes the lack of genotype information can leave us uncertain of what the real mode of inheritance is for a family.

Realistically, we also face problems with the small sizes of families. Much of genetics relies on statistical analysis and we need large enough

numbers for a study to achieve statistical significance. But we have no control over whether families we study will turn out to be large enough for statistically significant findings. If we study genetics in the mouse, we can do power calculations to figure out that we need to look at 100 individuals if we are going to achieve statistical significance, and then we can breed the 100 mice we need. But in human genetics we have no control over the size of the families that we study and we almost never end up with a family that can contribute 100 individuals to the study.

1.6 COMPLEMENTATION

Because we have one simple name for the trait called albinism, it is easy to think of it as being one single condition, but in fact there are different forms of albinism, some of which do not look at all like our classic concept of some-one with albinism. For instance, some individuals have what is called "yellow albinism," which may involve some coloration in both skin and hair. There is also a form of ocular albinism that only affects pigmentation in the eyes. Although some complex forms of albinism can involve features other than coloration, simple albinism does not cause uniformity of anything outside of coloration. Thus individuals with simple albinism are as diverse as the rest of the human race in terms of intelligence, talents, temperament, agility, strength, and health status.

We talk about albinism as a recessive trait rather than talking about pigmentation as a dominant trait because, when we are talking about an unusual or rare characteristic, we are usually trying to identify what is going on with the mode of inheritance of the unusual phenotype within a family or population. We could actually say that pigmentation is dominant over albinism. The inheritance of skin pigmentation may look simple: if you have two copies of the albinism allele, you have the coloration of the albinism trait. However, if you have at least one copy of the normal dominant copy of the albinism gene, there are actually a lot of other genes that contribute to coloration and make inheritance of skin and hair color complex.

Homozygosity for the albinism allele usually results in absence of pigment (Box 1.10). If the pigment is present, it can cause pale beige skin with blond hair or it can cause dark brown skin and black hair. The hair color and skin color genes that determine pigmentation are different from the albinism gene that determines whether there will be pigmentation. Thus mutations in exactly the same albinism gene can cause a child with white skin and white hair be born into a family of pink-skinned, Scandinavian blonds, into a family of freckled, Irish redheads, or into a family of dark-skinned, black-haired Nigerians. If you have color in your skin and hair (pink, tan, dark brown skin; blond, red,

brown, or black hair) you have the pigmented allele of the albinism gene, and your skin and hair color indicate things about which alleles you have at several other genes that determine coloration. If you have white skin and hair, you have two non-pigmented alleles of the albinism gene, and you cannot tell which alleles you carry at the genes that determine skin and hair color. However, if you were to look at your parents, siblings, and children, you might predict which alleles for those other color genes you are most likely carrying. For instance, an individual without pigmentation born into a family of African ancestors can look around him to tell that he would be more likely to pass his children genes for dark skin and black hair than genes for red hair and creamy skin, but he can only tell that by looking at his relatives and not by looking at himself.

Just to complicate this picture: there is more than one form of albinism, and the different forms of albinism are caused by mutations in different genes. There are multiple steps that lead to the formation of melanin in skin, hair or eyes, followed by additional steps that determine what color the melanin will be. This implies that Step 1 in Box 1.10 actually consists of more than one step that could be interrupted before you can get to Step 2, the determination of which color will be present. In fact, there are actually at least three different genes that we know of so far that can cause oculocutaneous albinism.

Think about what would happen if a man with two defective copies of the first gene OCA1 married a woman with two defective copies of the second gene OCA2 (Figure 1.14, A). He would provide the children with a good copy of OCA2, and she would provide the children with a good copy of OCA1. The result is that the children would be carriers for both genes but would not have the albinism trait! If the man and the woman both have defects in the same gene, OCA1, they will only be able to pass along defective copies of OCA1 and all

---

BOX 1.10

EPISTASIS

Sometimes something can have an overriding effect that keeps you from being able to detect or distinguish other characteristics. For instance, some flashing lights at intersections are red, and some are yellow. If the power for that section of the city goes out, the light does not turn on and you cannot tell what color it would be. Thus something that can block the light from obtaining electricity has an epistatic effect that blocks the manifestation of the different colors that the light could be. However, factors that affect the color of the light are separate from the factors that determine whether there is light at all. In the case of OCA1 albinism, an albinism allele blocks manifestation of a completely separate set of coloration traits, so the albinism mutation is considered epistatic. Each organism has some genes that can have epistatic effects that mask our ability to tell what other genes would be doing.

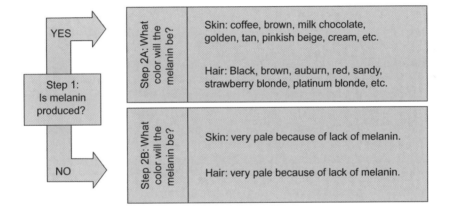

---

of their children will have the albinism trait (Figure 1.14, B). If two individuals with albinism have children who do not have the albinism trait, the parents have defects in different genes and the genes are said to be *complementary* (able to cover for each other) but not *allelic* (present in the same gene) (Box 1.11).

Most individuals with albinism have two different mutant alleles of the gene that is causing their albinism. In many cases the two different alleles of the gene may not be identical. So the great amount of variability in pigmentation and visual acuity problems in albinism result from a combination of the following:

- There are multiple different albinism genes.
- There are many different mutations in those genes, some of which eliminate the protein and some of which just make the protein not work very well.
- Different combinations of those mutations result in differences in levels of pigmentation in different individuals.

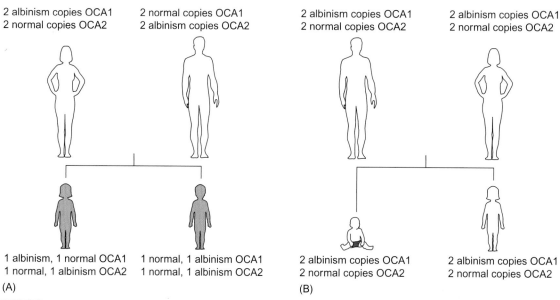

FIGURE 1.14   Complementation. (A) Affected parents with unaffected children. When mutations in different genes are responsible for the trait in each parent, the normal copy from one parent can complement the defective copy contributed by the other parent. Thus, even though both parents have the same trait, because the responsible genes are not allelic, the children end up heterozygous at each of the two genes. The alleles in this family are said to complement each other. (B) Affected parents with affected children. When both parents have the same trait due to mutations in the same recessive gene, their children will all inherit only defective copies of that gene, so they will have the trait. In this case, we would say that the gene causing albinism in the father is allelic to the gene causing albinism in the mother. Alleles in this family are said to fail to complement each other.

BOX 1.11

COMPLEMENTATION TESTING

The two families shown in Figure 1.14 represent something geneticists call a complementation test, a test used in cases where the mode of inheritance is recessive and we want to know if the two parents have the same defect. When two individuals with the same trait have children, we can look at the traits of the children to evaluate whether the parents harbor mutations in the same gene or in different genes. There are two different explanations for the family in Figure 1.14A and one explanation for what is going on in Figure 1.14B, both of whom have a recessive form of oculocutaneous albinism.

1. The mutations in the two parents are said to complement each other if the children do not have the mutant trait present in the parents, as shown in Figure 1.14A.
   - If the trait is caused by mutations in two different genes, then the husband with genotype aaBB and the wife with genotype AAbb will produce children with genotype AaBb. Because the child is heterozygous at both loci, the child is normal with regard to the trait.

2. The mutations in the two parents are said to fail to complement each other if the children

(Continued)

BOX 1.11 *(cont'd)*

have the same trait as the parents, as shown in Figure 1.14B.

- Failure to complement can happen if the mutation present in one parent is the same as the mutation (in the same gene) in the other parent (a1a1 mother, a1a1 father, and a1a1 child).
- Failure to complement can happen if the parents have different mutations in the same gene (a1a1 mother, a2a2 father, and a1a2 child).

The complementation test is one of the most powerful arrows in a geneticist's quiver. It can be far more complex than we have noted here, especially for genes whose protein product has multiple functional domains that carry out different

activities. But for most genes and most mutations, it usually works and is a very powerful tool for answering the question: just how many genes are defined by some number of mutant individuals all of whom have a similar trait.

You can imagine how important knowing whether they carry mutations in the same or different genes might be to the young couples in Figure 1.14 if they are making decisions about having more children. Later in the book we will discuss ways in which we could provide this information to them before they have children, and we will talk about the many different ways in which the information can be altered that could affect gene product presence or function. But a key point to understanding complementation is this: the test works even if we don't know the underlying cause!

1.7 EPISTASIS AND PLEIOTROPY

A gene is said to have epistatic effects on a second gene if the first gene obscures our ability to detect the phenotype associated with the alleles of the second gene (Box 1.10). If Mendel had produced a plant that did not make peapods he would not have been able to tell what pod-color alleles the plant carried. If the presence of two albinism alleles prevents the formation of melanin, then we cannot ask what color the melanin would have been if there had been any present. We consider OCA1 albinism to be epistatic because OCA1 defects block production of melanin and prevent manifestation of the effects of other eye color genes.

A gene that can have multiple different effects is considered pleiotropic (Box 1.12). We consider albinism pleiotropic because it affects skin color, eye color, hair color, and development. How can

one gene have so many different effects? One of the simplest ways this can happen is for the gene in question to play an important role in more than one cell type or at more than one point in development. We might also expect to see pleiotropic effects in the case of a gene whose product has more than one function. In Chapter 5 we will talk about factors that affect which cell types use a piece of genetic information, and in Chapter 4 we will talk about how changes to that information can affect gene product function.

For OCA1 albinism, the underlying basis of epistasis (absence of pigment) is different from the underlying basis of pleiotropy (activity of tyrosinase in multiple different cell types and at different stages of development). In other cases, epistasis and pleiotropy could both result from the same underlying mechanism.

But consider this: epistasis can sometimes involve masking of the effects of only one other

---

BOX 1.12

PLEIOTROPY

Sometimes a gene defect can show effects on multiple different organ systems in the body because it is expressed in many places in the body that each require the activities of the gene product. When a gene defect manifests itself in multiple different phenotypic features it is considered pleiotropic. Because melanin plays important roles in hair, skin, and eyes, we use the word pleiotropic to talk about OCA1 mutations that block production of melanin. The term pleiotropy does not apply when a gene function is critical in only one cell type. For example, individuals with mutations in the FAM83H gene have a simple phenotype called hypocalcified amelogenesis imperfecta which affects only the teeth. Enamel on the teeth is of normal thickness, but soft so that it erodes to leave a tooth crown composed of dentin only. No other tissues or systems in the body are affected.

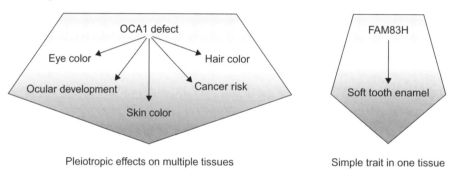

Pleiotropic effects on multiple tissues   Simple trait in one tissue

---

gene, or involve only one cell type, resulting in epistasis without pleiotropy. And in other cases, something can be pleiotropic, affecting many organs or functions, without there being an epistatic effect masking our ability to detect the phenotypic manifestation of a different gene. Some genes, such as some of those involved in formation of tooth enamel, are quite specific to a single tissue so the effects are not pleiotropic and do not involve masking of our ability to detect alleles of other genes, so they are not epistatic.

Mendel's rules did not address either epistasis or pleiotropy. He carefully selected very simple phenotypes for his experiments, and we know little about what other phenotypes he might have considered before passing over them to keep his focus on simple, true-breeding traits.

1.8 COMPLEX SYNDROMES

Many things that run in families are complex enough that family members may not have even noticed that several separate traits keep turning up together in different family members. In especially complex situations involving a variety of seemingly unrelated traits affecting multiple different organ systems or cell types, we may end up using the term *syndrome*.

Nail–patella syndrome, which affects multiple different parts of the body including the thumbnails and the kneecaps, is caused by a

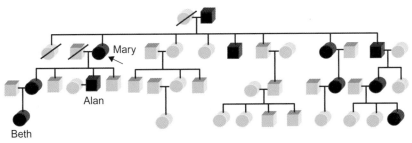

FIGURE 1.15 Transmission of the "no-nail" phenotype through four generations in a nail–patella syndrome family. As you can see, inheritance is dominant with a 50% risk of passing along an affected allele to the next generation. Both sexes are affected and both sexes can pass along the trait. Unaffected individuals do not appear to pass nail–patella syndrome along to their children.

defect in one copy of one gene. How could a defect in just one gene affect apparently unrelated things? In the case of the nail–patella syndrome gene, we know that the protein it makes carries out actions that affect a lot of other genes, so it is actually not surprising if the phenotype is complex. The nail–patella syndrome gene has *pleiotropic* effects, not only because the gene product acts on many other genes but also because it acts in many different cell types, with different consequences in the different cell types.

Consider the family in Figure 1.15, in which some members of the family are missing all or part of their thumbnails. The proband, the first person in the family who came into the study, is named Mary and she was the first source of information about the family. As you can see, Mary knows of four successive generations in which the "no-nail" phenotype has appeared, including her father's generation in addition to her children and grandchildren.

For many generations, members of this family did not realize that what they referred to as the "family thumbs" (thumbs on which all or part of the thumbnail was missing) were just one sign of a phenomenon complex enough to be considered a syndrome. The formal name for the syndrome is nail–patella syndrome, and not everyone with this syndrome has the same symptoms. However, the thumbnails are a consistent and obvious feature that every affected

person has, and family members knew that any time someone with the "family thumbs" became a parent, there was a moment of anticipation when they waited to see whether the new member of the family had thumbnails or not.

Nail–patella syndrome receives its name from the two most consistent and obvious characteristics found in affected individuals: unusual or missing thumbnails and unusual or missing kneecaps (patellae). Another consistent feature, the presence of hornlike extensions on the pelvic bones, is often hard to identify without a doctor's exam or an x-ray.

There is more to nail–patella syndrome than the nails and kneecaps. Problems with joints such as the elbows may cause restricted movement, in some cases severely restricted movement, and may sometimes call for surgery. In some cases, more severe orthopedic problems such as clubfoot may be found. Some individuals go on to develop kidney problems, and some develop an eye disease called glaucoma. Some individuals apparently don't develop problems with their kidneys or their eyes! The two consistent features, the thumbnails and the kneecaps, make it easy to study the mode of inheritance because they are features that can be identified without a doctor's visit and they are features that are obvious even in very young family members.

In fact, Mary watched for whether her children had the "family thumbs" as a point of curiosity but no concern. Her relaxed attitude about the situation changed when one of her sons, Alan, turned out to have some of the more severe medical problems that can happen to someone with nail–patella syndrome. She has watched him struggle through multiple surgeries while growing up. She has watched him decide that he will never have children because he would not be willing to pass his problems along to the next generation. Nail–patella syndrome has quit being a mere curiosity in Mary's life.

What if Mary's granddaughter Beth decides to have children? What are her chances of having a child with normal thumbnails and kneecaps? Beth has partially missing thumbnails and no kneecaps. Beth's fiancé Geoff has normal thumbnails, normal kneecaps, and no family history of nail–patella syndrome. So we know what Beth's and Geoff's phenotypes are, but what are their genotypes? Just as with the pea plants, we can use simple letter symbols to track the genotypes of different individuals. In this case we will use N to designate the "no-nail" phenotype and n to designate the common (normal) phenotype. Thus Beth has genotype Nn (n from her normal father and N from her affected mother). Geoff, with a normal phenotype and no family history of the disease, would have genotype nn. What do we expect to see among the children of Beth and Geoff? The possible combinations of alleles are:

- Beth's allele n plus Geoff's allele n gives a child with nn genotype and normal phenotype.
- Beth's allele N plus Geoff's allele n gives a child with Nn genotype and a "no-nail" phenotype.

Because Beth has genotype Nn, the n allele will be passed along to a child about half the time and result in a child with normal thumbnails, and the N allele will be passed along about half the time and result in a child with partially or completely missing thumbnails.

Geoff, with two copies of allele n, will only pass along allele n. The child has an approximately equal chance of getting either of the two genotype combinations, Nn or nn, so the child has about a 50% chance of ending with her mom's "no-nail" phenotype. We have not yet heard of a child who has genotype NN, but that doesn't mean there aren't any.

1.9 ONE MAN'S DISEASE IS ANOTHER MAN'S TRAIT

At a height of six feet tall, Julia is the short kid in her family. By the time she was in sixth grade, there were two people in the school who were taller than she – the vice principal and a fifth-grade teacher, both men. A lot of people who study genetics will tell you that height is complex, and that many genes and environmental factors affect height. They will say that extremes of height tend to move back towards average height as mixtures of multiple genes combine and get passed from one generation to the next. Yet in Julia's rather small family, height above six feet tall can be seen in nine people among four straight generations, among both men and women. Apparently, there is some difference in her genetic blueprint that makes her relatives and herself unusually tall. If she has a difference in her blueprint, does that mean she has a genetic disease? No. Her height is actually a minor thing and clearly not a disease. It is helpful when she wants to reach the top shelf and a pain when she can't buy pants that are long enough. Even the simple asking of that question raises a troubling issue. Is something caused by a change in the genetic blueprint necessarily a disease? Who gets to decide whether something is a disease and not just a trait on the continuum of human diversity? Although we would like to think that you would immediately declare, "The person who has the trait decides, of course," it is surprising how complicated the issues can become. Too often the world ends up judging us from the outside, rendering judgments about us that they would never render about themselves, often while claiming they have our best interests at heart.

One person with the apparent best of intentions asked whether Julia's height had made her feel like a freak when she was growing up. Until that moment, such a thought had never entered her head. If Julia's height can elicit a reaction like that from someone who is actually a very civil and kind person, imagine the "helpful" remarks that must assail people struggling with more complex problems than whether they can buy pants that fit.

So what have we been talking about so far, genetic diseases or genetic traits? Clearly, there appears to be a change in the genetic blueprint involved in all three traits we have discussed so far – albinism, dominant deafness, and nail–patella syndrome. This comes back to the question we raised about Julia's height: does it have to be a genetic disease just because it is caused by a change in the genetic blueprint? Frankly, the answer is no.

That answer throws us onto shaky ground as we face the question of what does constitute a genetic disease. If finding out that something is caused by a change in the genetic blueprint doesn't tell us whether it is a genetic disease, what can?

Here we come to an incredibly important concept – the distinction between a trait and a disease. The term *genetic disease* may be applied if the trait results in medical problems. If the effects are simply cosmetic, we may end up referring to it as a genetic trait instead. However, we find that this actually leaves us with many traits that occupy some middle ground, perceived differently by different people.

Whether nail–patella syndrome is a trait or a disease varies from one person to the next. For some people, the only effects are cosmetic; for others, the effects can be crippling or even lethal. We tend to talk about nail–patella syndrome as if it were a disease because the potential for the medical complications is there for all of the affected individuals, and in some cases we won't know until late in life whether someone with an apparently cosmetic case of nail–patella syndrome has missed the serious

medical consequences that could have arisen. Is the use of the term "disease" fair to the many people with unusual thumbnails and kneecaps who have no medical problems?

In the case of albinism, coloration seems like something cosmetic that should be considered a trait. However, vision problems are normally a part of albinism, and a true lack of pigment actually has medical implications in terms of susceptibility to damage from sunlight that is serious enough to make it a genetic disease. If you have a form of albinism that isn't one of the rare forms of syndromic albinism that causes other major medical problems, if you manage your vision problems and if you don't develop skin cancer, do you have a disease? If you have a trait that merely has the potential for medical problems, but those medical problems have not occurred, how might you be affected if people around you told you (or each other) that you have a disease? What if you started out from early childhood surrounded by people referring to you as someone with a disease when you had few if any health problems?

One parent of a child with albinism recently offered the vehement argument that albinism is not a disease. In many ways, this argument is valid. Even though there are some inconveniences associated with the vision problems that can be part of albinism, they can be managed so that these are people who live normal, healthy lives, and many of them might be surprised that anyone might think that they are ill. If people with the albinism trait do not consider themselves ill, the rest of us should accept this self-insight and notice how much like the rest of us they are.

In the case of deafness, which you might think would be classified as a disease because of the functional repercussions; there is an alternative perspective (Box 1.13). Some people consider deafness a genetic disease, but others consider it simply a trait. In fact, the news that a newborn child is deaf may be greeted with anguish by some families and calmness by

BOX 1.13

DEAFNESS – AN ILLNESS, A TRAIT, OR AN ETHNIC GROUP?

According to the American Association of Pediatrics, 1 in every 300 infants is born with hearing loss. There are more than 50 distinct genetic causes of deafness known so far, and hearing loss can result from a variety of non-genetic causes, including as one of the consequences of some types of infectious disease. Even as doctors work to restore hearing and researchers work to develop new technologies for those doctors to use, there are those who don't think the need for those efforts is so obvious. The deaf community has a large and thriving culture that includes its own separate language and is quite distinct from the culture of the hearing society in which this culture is embedded. Mannerisms, patterns of communication and interaction, even art forms have all emerged in unique ways that make them not just copies of the cultural patterns in the hearing world. When a deaf child is born, there can be very different reactions depending on whether the child is born into a deaf family or a hearing family. Some within the deaf community want their children to have the choice of whether to hear or not, and hearing parents usually wish there was a way to give their children the gift of sound. For all of them, the continued development of new technologies and the availability

of medical assistance are incredibly important. There are some within the deaf culture who regard medical efforts to eliminate deafness as a threat of cultural genocide – an effort to eliminate an entire separate culture and people by forcing their assimilation into a different mainstream culture for which they hold great distaste. Many others in the deaf community hold much more mild views in this era in which moderate help is available to those who seek it and not required for those who don't. With technological aids available, such as cochlear implants, it is interesting to find that some who have received these implants and gained the ability to carry on aural communication with those of us who don't speak sign language have then decided to return to the world of silence they were born into. Advances in the quality of the technological results have others who resisted cochlear implants in the past taking another look at them. If you ever find yourself saying, "Of course they should all just have their hearing restored," first spend some time reading about deaf culture and exploring the idea that in silence they may have found some things that the hearing world lacks. You may or may not end up agreeing with them, but many who hear what they have to say come away changed by it.

others. The thing that determines whether the news is distressing is often, but not always, the "hearing" status of the parents.

Some families believe that their deaf child has a disease and they want someone to come forward with a cure. Other families think that their child has a trait, and they have no interest in altering that trait. Individuals who wish they could hear view advances such as the cochlear

implant as a gift that can restore a missing sense. However, there are those within the deaf community who see the cochlear implant as a tool for carrying out cultural genocide. They see it as a technology that causes a deaf child to grow up as a marginalized individual on the fringes of a cruel "hearing" society instead of growing up safe and esteemed as a full peer within the deaf community. What a complex ethical problem to weigh and measure the gain or loss of hearing against the gain or loss of esteem and acceptance. We find ourselves wondering if there is such a thing as a right answer.

This complex set of ethical issues mirrors so many situations that we encounter in human genetics, where the answers are often terribly complex but often become at least a bit simpler if we fall back on the principle of self-determination. Julia can't tell someone deaf whether their deafness is a trait or a disease, and that deaf person can't tell Julia whether her height is a trait or a disease. Each of us knows which call constitutes the truth for our own situations. There are others who would make a different judgment call.

In this chapter, we have deliberately selected traits that are usually not seen as diseases by the people who have the trait but that sometimes are labeled as diseases by others or by rare individuals whose cases are especially severe. We have done this expressly for the purpose of making the point that one man's disease may be another man's trait. Is the trait actually causing a problem? Sometimes yes; often not. But sometimes something that is not a problem can become a problem based on how outside people treat the situation. There can be many consequences to an individual who is perfectly healthy if people around them start telling them that they have something wrong with them. This is especially true if someone is treated as if they are ill starting in infancy so that they grow up internalizing a self image that they would not have developed on their own.

As we will discuss in the final chapter of this book, some of the gravest ethical errors in genetics in the past have been made when society or medicine removed the rights of individuals to judge this for themselves. Is it broken, and should we fix it? If we look beyond the tricky issue of whether or not we *can* fix it, we find that the real answer to whether we should even try to fix it lies in the heart of the individual with the trait. One man's trait is another man's disease.

Only the individual with the trait can judge for himself or herself whether the trait is severe enough to be considered a problem, and different individuals with the same trait may arrive at very different perspectives on the question.

Study Questions

1. Which of Mendel's findings helped to rule out the vital spark theory of inheritance?
2. Which of Mendel's findings helped to rule out the homunculus theory of inheritance?
3. What are the standard symbols used to represent a male and a female in a pedigree drawing?
4. In a standard pedigree drawing, how is a deceased individual marked?
5. What is a heterozygote?
6. What is a homozygote?
7. What is the difference between genotype and phenotype, and how are they related?
8. How many alleles of a gene come from each parent, and how many are passed along to the offspring?
9. Define the term allele.
10. What is a dominant allele?
11. What is a recessive allele?
12. What are the modes of inheritance of deafness?
13. Could albinism be considered a syndrome? Why or why not?
14. What is the difference between monozygotic twins and fraternal twins?
15. What is pleiotropy and why can a defect in a single gene have pleiotropic effects?
16. Kate and Dan, two individuals who do not have cystic fibrosis, are both carriers of a defect in the cystic fibrosis gene and decide to have children together. Draw a Punnett square that shows the genotypes of the sperm and eggs they can produce and the genotypes that we would predict for their children. What fraction of their children

will be carriers and how many will not have cystic fibrosis?

17. Is the proband of a family always someone with the trait? Explain.

18. What do identical twins and fraternal twins have in common that is important to researchers doing twin studies?

19. Draw a three generation pedigree showing inheritance of a dominant form of epilepsy where the grandfather is affected, the grandmother is deceased, and a granddaughter is the proband. Use standard pedigree symbols.

20. What determines whether a genetic trait is a genetic disease?

Short Essays

1. How might our experimental outcomes be influenced by the way we think about the question and the extent to which we stop looking for further answers when we see an answer that fits our expectations? Why do we believe in what Mendel did even if his data were "too pretty?" As you consider your answer, read the article "Mud sticks: On the alleged falsification of Mendel's data" by Daniel L. Hartl and Daniel J. Fairbanks in *Genetics*, 2007;175:975–9.

2. Like Albert Einstein, Gregor Mendel lacked the stellar academic credentials that seem to be essential for anyone striving to make a mark in scientific history these days. What alternative environment allowed Mendel to pursue his questions and develop his theories outside of a university academic setting? What alternative environments currently exist for someone with talent and a shortage of official credentials to advance the examination of a question that interests them? As you consider your answer, read pages 3 through 49 of *The Impact of the Gene: From Mendel's Peas to Designer Babies* by Colin Tudge (2001, Hill and Wang).

3. How could society be structured to avoid the marginalization of deaf individuals that leads some members of the deaf community to seek a separate society outside of the hearing world? Given the increasing fraction of elderly people in our population, how might such altered social structure benefit most of us? As you consider your answer read *Everyone Here Spoke Sign Language* by Nora Ellen Groce (1985, Harvard University Press).

4. How can art be used to combat genetic ignorance and prejudice and maybe even make people safer? As you consider your answer, visit Rick Guidotti's Positive Exposure website http://www.positiveexposure.org/home.html to learn about his publications, gallery showings, and educational programs aimed at reducing prejudice and protecting people with albinism from violence.

5. Belyaev showed that it was possible to domesticate a kind of animal for which domestication had long been thought to be impossible. Why was Belyaev able to bring about this domestication in a few decades when he started with an animal that had existed for many thousands of years without showing any sign of domestication? As you consider your answer read the article "Destabilizing selection as a factor in domestication" by Dmitri Belyaev in *Journal of Heredity*, 1979;70:301–8.

Resource Project

Mendelweb.org presents original and translated versions of Mendel's paper plus supplementary information to assist students of Mendel's work at http://www.mendelweb.org/. Go to Mendelweb and look at the translated version of Mendel's paper. Write a one paragraph report on one of the pea plant traits that he studied other than pod color, and discuss the characteristic and the mode of inheritance.

Suggested Reading

Articles and Chapters

"What causes albinism? Albinos around the world face day-to-day health issues, but in Africa they have a bigger problem: being hacked to death for body parts" by Coco Ballantyne in *Scientific American*, February 2009.

"Rediscovery of Mendelism" in *Mendel's Legacy: The Origins of Classical Genetics* by Elof Axel Carlson (2004, Cold Spring Harbor Laboratory Press).

"What is classical genetics?" in *Mendel's Legacy: The Origins of Classical Genetics* by Elof Axel Carlson (2004, Cold Spring Harbor Press).

"Routes to classical genetics: Evolution" in *Mendel's Legacy: The Origins of Classical Genetics* by Elof Axel Carlson (2004, Cold Spring Harbor Press).

"Evolution in black and white" by Sean B. Carroll in *Smithsonian Magazine*, February 10, 2009.

"Darwin and genetics" by Brian Charlesworth and Deborah Charlesworth in *Genetics*, 2009;183:757–66.

"Alfred Russel Wallace arrived at the theory of natural selection independently of Charles Darwin and nearly outscooped Darwin's *The Origin of Species*" by Lyn Garrity at Smithsonian.com, January 22, 2009.

"What Darwin didn't know: Today's scientists marvel that the 19th-century naturalist's grand vision of evolution is still the key to life" by Thomas Hayden in *Smithsonian Magazine*, February 2009.

"William Bateson, the rediscoverer of Mendel" by M. Keynes and T. M. Cox in *Journal of the Royal Society of Medicine*, 2008;101:104.

"Substantial genetic influence on cognitive abilities in twins 80 or more years old" by Gerald E. McClearn, Boo Johansson, Stig Berg, Nancy L. Pedersen, Frank Ahern, Stephen A. Petrill, and Robert Plomin in *Science*, 1997;276:1560–3.

"Heredity before Mendel" in *Gregor Mendel, The First Geneticist* by Vítezslav Orel (1996, Oxford University Press).

"The Peacock's Tale" in *The Red Queen: Sex and the Evolution of Human Nature* by Matt Ridley (2003, Perennial/HarperCollins).

"Twin science: Researchers make an annual pilgrimage to Twinsburg, Ohio, to study inherited traits" by Mark Wheeler in *Smithsonian Magazine*, November 2004.

Books

Why Evolution Is True by Jerry Coyne (2009, Viking).

The Greatest Show on Earth: The Evidence for Evolution by Richards Dawkins (2009, Free Press).

Experiments in Plant Hybridization by Gregor Mendel (2008, Cosimo Classics).

From So Simple a Beginning: The Four Great Books of Charles Darwin edited by Edward O. Wilson (2005, W. W. Norton and Co.).

The Double Helix: How Cells Preserve Genetic Information

THE READER'S COMPANION:
AS YOU READ, YOU SHOULD
CONSIDER

- How an alphabet of four genetic
  letters can be used to spell out 20 amino
  acids.

- How the sequence of one strand of DNA
  tells you the sequence of the other.

- What the term complementary means.

- How DNA is replicated.

- What types of molecules make up
  chromosomes.

- What features distinguish different
  chromosomes under the microscope.

- The genetic differences between male and
  female cells.

- The genetic similarities between male
  and female cells.

- Which molecules retain long-term storage
  of genetic information.

- Which molecules carry short-term copies
  of genetic information.

- How information is transferred from the
  nucleus out to the translation machinery.

- How differences in amino acids can affect
  functions of proteins.

- How the cell tells which RNA molecules
  need to be translated into proteins.

- The structural differences between DNA
  and RNA copies of a gene.

- The differences between unspliced,
  spliced, and alternatively spliced RNA.

Light is the left hand of darkness
And darkness is the right hand of light.
—*Ursula K. LeGuin*

2.1 INSIDE THE CELL

Using whole organisms, Mendel showed that genetic information seemed to come in discrete units (you have it, or you don't have it), that each of us has two copies of each piece of information, and this information is what passes from one generation to the next. But Mendel's revelation that information is what passes from one generation to the next offered little insight into the nature of that information. He lacked the molecular information he needed to illuminate his model of how inheritance works. Since his time, advances in our ability to look inside the cell and study the biochemical processes taking place there have helped to illustrate the underlying basis of the information that Mendel has discovered.

The View Through the Microscope

Beginning more than a century before Mendel was born, the first microscopes began providing revelations previously undreamed of as primitive lenses revealed a world of microscopic structures. Peering through a lens that was little more than a melted bead of glass (Box 2.1), the early microscopists were able to bring into view the cells in cork, and later they were even able to see sperm and microorganisms. Their studies eventually led to the cell theory, the concept that the cell is the basic unit of which organisms are constructed. Before Mendel published his works, researchers realized that the human body is made up of trillions of cells, microscopic structures bounded by membranes and filled with a gel-like cytoplasm in which inclusions such as lipid droplets and complex structures called organelles are suspended (Figure 2.1), but none of these studies indicated where the genetic information was housed within the cell.

One of the most prominent structures the microscopists found inside the cell is the nucleus, a region that is usually about 5 to 10 microns (μm)

BOX 2.1

VAN LEEUWENHOEK'S MICROSCOPE

Antonie van Leeuwenhoek (1632–1723) was a Dutch tradesman who invented a microscope capable of viewing individual cells and microscopic life forms. He was not the first to invent a microscope, but he was the first to discover bacteria, microscopic parasites, and sperm. His microscopes were tiny and quite different from modern instruments found in genetics labs today. Alan Shinn has created a set of directions for construction of a replica of van Leeuwenhoek's microscope, which, as you can see in the picture shown here, is small enough to hold in one hand. Some types of cells are large enough to be distinguished with this early technology, but it will not let you see individual chromosomes. His microscopes were not powerful enough to allow visualization of DNA, but his breakthroughs provided foundations for more advanced microscopy used today to view a wide range of structures within the cell.

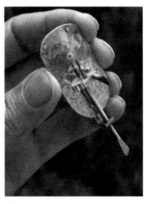

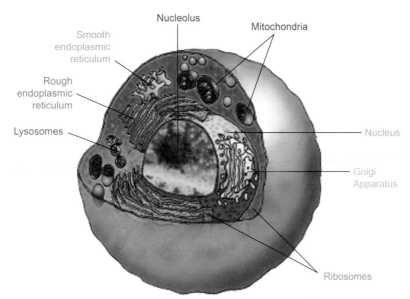

FIGURE 2.1 The major structures of the eukaryotic cell. These include an outer membrane, a membrane-bound nucleus and cytoplasm (gel-like cytosol with organelles and inclusions suspended in it). (Courtesy of Edward H. Trager.)

I. HOW GENES SPECIFY A TRAIT

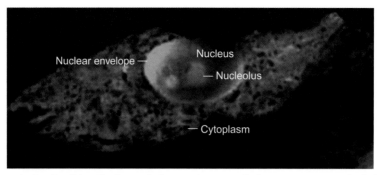

FIGURE 2.2   The nucleolus. Traditionally the nucleolus was seen as a dark structure against the lighter gray of the indistinct material within the nucleus, but modern methods allow the uses of dyes to highlight the nucleolus as shown here, as a distinct yellow region within the green nucleus. (Image from J. A. Hiscox (Hiscox Lab), *Nature Reviews Microbiology*, 2007;5:119–27. Reproduced with permission.)

across. A membrane sets the nucleus apart from the cytoplasm, the gel-like cytosol filling the cell outside of the nucleus, in which a vast collection of organelles and inclusions are suspended (Figure 2.1). The nucleolus, within the nucleus (Figure 2.2), is the site of active synthesis of molecules needed to construct the millions of ribosomes, located in the cytoplasm, that carry out protein synthesis. Mitochondria, larger structures than the ribosomes, often number in the thousands and can be found scattered through the cytoplasm. Membrane-bound containers such as the lysosomes hold molecules such as proteins that digest other kinds of molecules whose specialized functions need to be kept separated from the cytoplasm. The numbers of organelles vary from one cell type to another and under different growth conditions.

2.2 DNA: THE REPOSITORY OF GENETIC INFORMATION

The cell can be broken down into some basic classes of macro-molecules: carbohydrates, lipids, proteins, nucleic acids, and a variety of small, simple molecules such as minerals. One of the mysteries of the early twentieth century was the question of which cellular structure or class of molecule could house the genetic information.

It was understood that there must be a great deal of complexity to the genetic information to be able to encode the wide array of structures, functions, and developmental steps that make up a human being. Lipids, nucleic acids, and carbohydrates were all considered to be too simple in their structure, lacking the complexity needed to direct the running of a complex organism. Proteins were considered a good candidate for housing genetic information since proteins are made up of 20 different subunits, called amino acids, which can be arranged in a large number of different combinations.

DNA and Transformation – Griffiths

In 1928 a scientist named Frederick Griffiths made one of the first breakthroughs in our understanding of the nature of genetic information. He started with two strains of pneumococcus (a type of bacteria that causes pneumonia) that differ from each other by alleles that encode lethal or benign properties. The "smooth" strain was covered in a polysaccharide capsule which formed smooth, glistening colonies, and caused an infection that kills mice. The "rough" strain lacked the capsule, formed rough-surfaced colonies, and did not kill the mice.

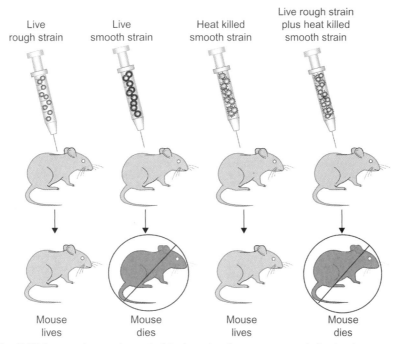

FIGURE 2.3 The Griffiths experiment. A rough (blue) strain of pneumococcus is benign but a smooth (red) strain is virulent and kills the mice. The virulent strain, when heat-killed, loses the ability to kill the mice. If the live rough strain and the dead smooth strain are combined, the live rough strain gains the ability to synthesize a capsule, grow in smooth colonies, and kill mice.

Griffiths found that if he heat-killed the smooth strain, it could no longer kill the mice. But if he injected a mouse with a mix of the live rough strain and the heat-killed smooth strain, neither of which could kill the mice when delivered alone, the mice died (Figure 2.3); moreover, when he retrieved bacteria from dead mice they produced smooth, glistening colonies even though the live bacteria he had injected had only been able to make rough colonies. He concluded that something in the dead smooth strain contained information that could transform the live rough-colony cell into smooth colonies that kill mice.

If we think in terms of Mendel's ideas about different phenotypes being the result of different alleles at a given gene, this means that the strain bearing the rough allele was transformed such

that it now carried the smooth allele! The verb "transformed" here is as accurate as it is evocative. It is exactly as if Mendel had taken one of his pure-breeding yellow-pod plants, soaked in an extract derived from ground up green-pod plants, and got offspring that now not only produced green pods but produced only green progeny: this is transformation indeed. Pea plants do not have the ability to take up the transforming principle in this way – instead they introduce a new genotype by mating – but bacteria can take up the transforming principle and actually change one allele to the other. The implication was that the information for smooth colonies and a virulent infection did not die along with the bacteria, and that something in the remains of the dead bacteria retained the ability to transmit that information to other live bacteria.

DNA and Transformation – Avery, McCarty, and McLeod

At that time, the available technology did not let Griffiths go on to pin down the chemical nature of the transforming principle present in the dead bacteria. In the years that followed, a lot of progress was made in the ability of researchers to isolate the different biochemical components of the cell. In 1944, Oswald Avery, Maclyn McCarty, and Colin MacCleod did the experiment that showed that the Griffith's transforming principle was DNA (deoxyribonucleic acid), one of several forms of nucleic acid present in the cell. They isolated a fraction of the dead cell debris that could make rough cells acquire a polysaccharide capsule and grow in smooth, glistening colonies. The chemical composition of this transforming agent most closely resembled the known composition of DNA. If they exposed the transforming agent to enzymes known to digest proteins or RNA (ribonucleic acid) the transforming activity remained, but if they treated it with an enzyme known to chew up DNA the activity reduced or disappeared. They concluded that DNA is the "fundamental unit of the transforming principle."

DNA and Transformation – Hershey and Chase

In 1952, Alfred Hershey and Martha Chase did a third key experiment. This time they used a type of virus, called a bateriophage or phage, that infects bacteria. Phage are protein capsules containing DNA, and phage can infect bacteria to make many copies of themselves, after which they lyse (break open) the cell to release the progeny phage into the surrounding environment. It was known that the T2 phage would bring about an infection that would result in many copies of the phage being made inside of the bacterial cell. When the phage infects the bacterial cell, it sits down on the outside of the cell and injects something into the cell that

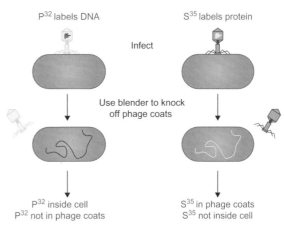

FIGURE 2.4 The Hershey–Chase experiment. Batches of T2 phage, which are made up of DNA and protein, were labeled with P^{32} or with S^{35} and then used to infect bacteria (see text). After the infection had started, the bacteria were put through a blender to knock the phage coats off the outside of the bacteria. The bacterial cells and the phage coats were collected separately and assayed for the presence of radioactivity. The presence of P^{32} inside the bacterial cell but not in the phage coats, and the presence of S^{35} outside the cell in the phage coats but not inside the cell, showed that DNA was the material that entered the cell and led to production of more copies of the phage.

contains the information needed to make new copies of the phage. The question was: what does the phage put into the cell that contains the information needed to make copies of itself? Because the phage particles are made up of proteins and DNA, they wanted to test for whether DNA or protein was getting inside of the cell.

First they made batches of T2 phage that had incorporated radioactive labels, either S^{35}, which labels proteins, or P^{32}, which labels DNA. They used each batch to infect bacteria, then used a blender to knock the phage coats off the outside of the bacteria. When they asked where the radioactive labels had gone, they found that the P^{32} label turned up inside the bacterial cell, but the S^{35} was only found in the bodies of the phage that had been knocked off the outside of the bacterial cell. If we were to do this experiment using phage labeled with both P^{32} and

S^{35} to start the infection, we would find that at the end of the infection the new phages that were produced would be labeled with P^{32} but not with S^{35}. This meant that during a phage infection, DNA entered the bacterial cell and many copies of the virus were made, but protein did not enter the cell. The implication was that DNA entered the cell and was used to make more copies of the phage. In 1969, Alfred Hershey received a Nobel Prize for this work.

While many other experiments from this era of science contributed to our understanding that DNA is the repository of genetic information, the three experiments we just presented are considered the classic examples of work on this topic. It is interesting to note that even though DNA is the genetic material for almost every organism, including human, there turn out to be some exceptions. There are some viruses that use RNA (another form of nucleic acid related to DNA) in place of DNA. Interestingly, even for these RNA viruses DNA is often important, since many RNA viruses use an enzyme called reverse transcriptase to copy their RNA genome into DNA as an intermediate template from which new copies of the RNA genome can be made.

2.3 DNA AND THE DOUBLE HELIX

So if the transformation experiments are right, and DNA houses the information needed to transform the phenotype of an organism, what is the form of that information? We want you to start out with this picture in your mind: the information in the human genome is written out in a linear chain of chemical building blocks or genetic "letters" in much the same way this book is written as a linear chain of printed symbols. These chemical building blocks are commonly referred to as A, C, G, or T, the first letters of their chemical names (Figure 2.5).

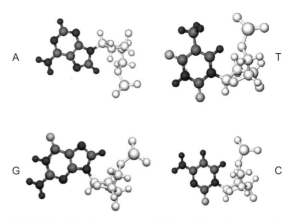

FIGURE 2.5   Structures of DNA building blocks A, C, G, and T. The chemical building blocks that act as genetic letters are called *bases*. The bases are adenine (A), thymine (T), cytosine (C), and guanine (G). The red spheres are the hydrogens, the blue spheres are the nitrogens, and the green spheres are the carbons that make up the genetic letters A, C, G, and T. The grey spheres are the atoms of the backbone that hook the DNA bases to each other. The backbone portion is the same for all four genetic letters but may look different here because they are being seen from different angles. (Courtesy of David M. Reed.)

DNA Bases – The Letters of the Genetic Alphabet

Each building block has a different chemical structure that can be "read" and recognized by the machinery of the cell with an amazingly high level of accuracy. These building blocks, which are known as *bases*, and the chains of building blocks that can be built from them, are called nucleic acids, and DNA is only one of several kinds of nucleic acids present in the cell.

The cellular machinery reads these letters at the level of the hydrogen and nitrogen and carbon and phosphorus atoms of which the letters are formed (as we display them in Figure 2.5), but we actually need concern ourselves with very little about the real molecular structures of these letters. Simply considering each base as a building block, as in Figure 2.7, lets us grasp many of the most important concepts about DNA that we will present here.

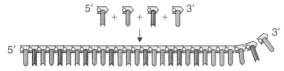

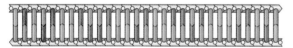

FIGURE 2.6   A backbone hooks the A, C, G, and T bases together into a long, continuous chain. Notice that the segments that make up the backbone are asymmetrical, a property that causes the DNA chain to grow in only one direction as each new base gets added to the growing chain. That is, there is effectively a beginning and an end to the chain, and bases can only get added to the end but not to the beginning of the chain as new DNA gets made. This asymmetry or directionality to the DNA chain also gets referred to as the *polarity*, sometimes called the 5′ to 3′ polarity, of the chain. The backbones (shown here in blue–gray) are made up of the gray parts of the molecules in Figure 2.5, and the information content is carried in the red, orange, green, and blue building blocks. (Courtesy of Edward H. Trager.)

FIGURE 2.7   In an even more simplified format, we show that the double-stranded structure of the DNA has backbones that connect bases to each other. Bases meet in the middle to connect the two strands to each other. In these double-stranded structures A pairs with T, and C pairs with G. (Courtesy of Edward H. Trager.)

If the genetic letters are arranged in a linear order, what keeps them in that order? The bases A, C, G, and T are not hooked directly to each other. Instead, a long backbone structure runs the length of the DNA molecule, and each base is connected to the backbone. Thus the bases form a DNA chain in the same way that the cross-ties of a railroad track make a long row of ties, not by connecting directly to each other but rather all by connecting to the same rails that run along the edges. So each base is connected to a segment of the backbone, and the backbone pieces connect to each other, which brings the bases next to each other in a line without the actual bases touching each other. As you can see in Figure 2.6, this is an easy idea to diagram without using the chemical structures.

DNA is often referred to as double-stranded. The DNA inside of the nucleus of the cell is not normally a single strand of DNA, such as that shown in Figure 2.6. Rather, it consists of a pair of DNA strands, with two backbones paralleling each other like railroad tracks (Figure 2.7). The bases are each connected to the backbones and point inwards to contact the bases on the opposite strand. As we will discuss in the next section, these are not two random strings of DNA that are stuck to each other.

If you look at Figure 2.7 you will notice that the red base always pairs with the orange base, not the blue or the green base. And this is also true for blue and green bases which pair with each other but not with red or orange bases. We now know that there are rules for how the strands pair with each other: A pairs with T, and G pairs with C. As we proceed we hope it will become evident to you why the existence of these pairing rules is so important!

The DNA Helix

Thus the backbones hold the bases together in a line, and the bases connect the backbones rather like ties connect parallel railroad tracks. Because the backbones are an equal distance apart, you might think that they would be considered parallel to each other; however, because the backbones are directional but run in opposite directions, the backbones are considered anti-parallel instead of parallel. (Because chemical nomenclature calls one end of a building block the 5′ end and calls the other end of the building block the 3′ end, and because the 5′ end of one block connects to the 3′ end of the next block, we use 5′ and 3′ to refer to and distinguish the two different ends of a string of DNA.)

The completed structure of the DNA is referred to as the *double helix*. It is called double because of the property we already described: it consists of two strands arranged so that paired

FIGURE 2.8 The double-stranded "train-track" DNA is twisted to form a helical structure like a spiral staircase. Notice that the ends on the backbone are not identical, to highlight the directionality or asymmetry of the backbone chemical structures. (Courtesy of Edward H. Trager.)

(A)   (B)

FIGURE 2.9 Molecular models of the double helix. (A) Carbon, oxygen, nitrogen, hydrogen, and phosphate atoms are each shown in a different color to show the complex structure of the helix. (B) All atoms in one strand are shown in blue and in the other strand are shown in yellow to illustrate how the two separate strands are joined together and wind around each other. (Courtesy of David M. Reed.)

bases meet in the middle and backbones run up the outside edges. The double-stranded DNA does not lay out flat like a railroad track. To see what the real structure is like, imagine picking up the railroad track structure from Figure 2.7 and twisting it so that it resembles a spiral staircase in which the anti-parallel handrails point in opposite directions (Figure 2.8).

If we look at molecular models of the double helix we can see the individual positions of each carbon, nitrogen, hydrogen, oxygen, and phosphate atom making up this complex structure (Figure 2.9). From the angle shown, the rungs of the twisted ladder look like simple linear connections but there are additional atoms behind those we can see in this picture. In fact the bases that meet in the middle of the helix form flat structures so that the helix is rather like a spiral staircase with the bases serving as steps. If you were to trace the line of the backbone as it wraps around the helix you would find that it spirals around the center of the molecule. You can similarly trace the groove between the backbones and find that the groove also spirals around the center. Enzymes that interact with the DNA reach into this groove where the protein can interact with the bases to detect the order of the sequence that is nestled between the backbones.

If we look down through the center of the helix we can see far more of the details of the complex structures that connect the backbones (Figure 2.10). We can see a kind of regularity, a repeating pattern reflected in the shapes that spiral around the center. And we can see the open space that runs all the way down the

FIGURE 2.10 Model of DNA as seen looking down through the center of the double helix. In this model, the space-filling spheres of the space-filling model from Figure 2.9 have been replaced with a model that emphasizes the bonds and angles at which the components are connected to each other. This form of ball-and-stick model opens up our view of the structure to make it easier to see down into the center of the helix. (Courtesy of David M. Reed.)

center, a space created by the fact that the helix spirals around the center without occupying it.

We tend to think of the helix as always existing in the form shown in Figure 2.7, but in fact DNA is a fairly dynamic structure that can sometimes exist in single-stranded form. As it replicates and as enzymes in the cell interact with it, double-stranded DNA can shift its form according to the molecular environment surrounding it. The most common form of DNA, shown in Figure 2.7, is called B-DNA, a structure that is tightly wound into a helix. In some conditions another form of DNA, called Z-DNA, occurs (Figure 2.11). Imagine taking hold of the helix and twisting it backwards, slightly, just enough to open up the tightly wound structure locally and make the bases at the center of the helix more accessible. Some changes in the surrounding biochemical environment can make DNA more likely to form Z-DNA, and hot spots for formation of Z-DNA have been identified at dozens of locations in the human genome where local sequence affects the tendency towards Z- rather than B- forms of DNA. An example of a type of sequence that more readily forms Z-DNA is a sequence called a CA dinucleotide repeat, in which the bases C and A are repeated over and over.

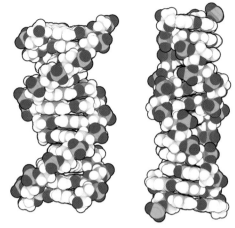

FIGURE 2.11   Models of B-DNA on the left and Z-DNA on the right. Some bases that are buried in the major groove of B-DNA are closer to the surface in Z-DNA. (Courtesy of David M. Reed.)

intrigue. It was also a fundamental leap forward that deserved and won a Nobel prize for three of the scientists whose work led to the understanding of DNA structure and replication (Box 2.2).

2.4 DNA REPLICATION

Complementary Information

When children play word games involving opposites, knowing the first word is all the clue they need to find the second word – dark/light, up/down, cold/hot, far/near. When you walk along a railroad track looking at the left-hand rail, you don't have to see or touch the other rail to automatically know its shape – the tightness of the curve, or the pitch of the slope. Think about how one side of a yin/yang symbol automatically defines the other side. What an interesting concept, that if you know part of something, you automatically know the rest of it! The unknown part of the information that you can automatically fill in is not identical to

Discovery of the Structure of the Double Helix

As you can see from the pictures we have shown, there are many different ways to diagram the DNA structures. To appreciate just how hard it was to figure out this structure that is too small to see with a traditional light microscope, consider this: in this spiral structure with a bit more than ten stair-steps to every turn, the distance across the helix is 20 angstroms, with an angstrom covering a distance that is about one hundred millionth of a centimeter!

The double-helical structure of DNA was first reported in 1953. It was an accomplishment filled with gossip, brilliance, and even a bit of

BOX 2.2

DISCOVERY OF THE DOUBLE HELIX

Although James Watson and Francis Crick are famed as the authors of the paper that announced the double-helical structure of DNA, two other papers on the subject were published along with the Watson and Crick paper in that 1953 edition of the journal *Nature*. One was from the research group of Maurice Wilkins, who later joined them in receiving the Nobel Prize in 1962; co-authors on his paper were Alex Rawson Stokes and Herbert Wilson. The third paper was by Rosalind Franklin and her student Raymond Gosling; their data helped Watson and Crick develop and confirm their model of the double helix. We will never know whether history would have given Franklin recognition along with the others since the Nobel Prize is not awarded posthumously and she was no longer alive by the time this particular prize was given. For a fuller version of the controversial story of this great discovery, we recommend the following books: *The Double Helix* by James Watson, *The Eighth Day of Creation* by Horace Freeman Judson, *The Third Man of the Double Helix* by Maurice Wilkins, and *Rosalind*

Franklin: The Dark Lady of DNA by Brenda Maddox. The three *Nature* papers publishing the initial work on the double helix, plus a follow-up paper by Watson and Crick, can all be found at www.nature.com/nature/dna50/archive/html.

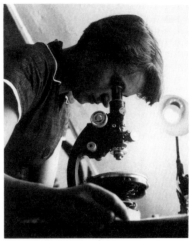

Rosalind Franklin (Photograph: MRC Laboratory of Molecular Biology, 1955. In the personal collection of Jenifer Glynn. © Henry Grant Archive/Museum of London. Reproduced with permission.)

James Watson and Francis Crick (Photograph: Barrington Brown. 1953 © Camera Press. Reproduced with permission.)

Maurice Wilkins (Photograph: George Silk, Time & Life Pictures. 1962. © Getty Images. Reproduced with permission.)

the first piece of information, but it is complementary to it.

Clearly this does not apply to many complex situations in life, but this is the key to how genetic information is disbursed into each new cell that grows in the human body as it develops from a fertilized egg. As the genetic information that each cell carries is copied, it is done using this principle – that knowing half of the information automatically tells you the other half. This concept of complementary information is one of the keys to understanding how DNA is copied, read, and used, and a critical element in the transmission of information from one generation to the next.

At the end of the paper announcing the structure of the DNA helix, James Watson and Francis Crick observed, "It has not escaped our notice that the specific pairing we have postulated immediately suggests a possible copying mechanism for the genetic material." They were right, that the uncovering of the structure of DNA (Figures 2.9–2.11), with its paired strands, pointed directly to the mechanism of DNA replication through base-pairing. The Watson and Crick model of DNA allows us to visualize the faithful replication of DNA as nothing more than simply separating the two complementary strands, the "Watson" strand and the "Crick" strand, and synthesizing two new complementary strands by using the rules of base-pairing: A pairs with T, and C pairs with G (Figure 2.12).

Because of the base-pairing rules, if you know the sequence or order of As, Cs, Gs, and Ts along one of the strands, you automatically know what the sequence of bases on the other strand must be. The term *complementary* is used to describe this relationship of the two strands that pair with each other, and you need recall little about the actual structure of an AT pair or a CG pair as long as you remember the basic base-pairing rule in Figure 2.12.

A set of two bonds hold A and T together and a set of three bonds hold C and G together (Figure 2.13), so A and T are held together more

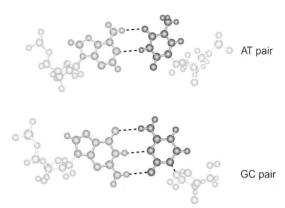

AT pair

GC pair

FIGURE 2.12   On the left, we see molecular images showing the loose type of chemical bonds called hydrogen bonds that hold together the paired DNA bases. Two hydrogen bonds hold together A and T, and three bonds hold together C and G. For most purposes, as we show on the right, we do not need to know where all of the different hydrogens and carbons and nitrogens and oxygens are located to be able to work with the basic concepts of A–T and C–G base-pairing: A pairs with T and G pairs with C. Because there are only two chemical bonds holding A to T, but three holding G to C, the GC base pairs stick together more tightly than AT base pairs. (Courtesy of David M. Reed.)

weakly than G and C. This means that it is easier to get DNA to come apart into two separate strands if it consists of paired As and Ts than if it consists of paired Gs and Cs.

DNA normally exists as paired, complementary double strands held together by the hydrogen bonds shown in Figures 2.12 and 2.13. The cell has the ability to pull these strands apart into two separate single strands that are not held together by hydrogen bonds, although the cell usually opens up very limited local regions of a large DNA molecule rather than pulling the two molecules completely apart (Figure 2.14).

DNA Polymerase Replicates DNA

Copying of the cell's 3 billion base pairs of DNA (known as DNA replication) is carried out by proteins called DNA polymerases.

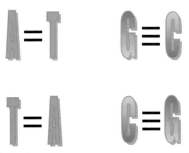

FIGURE 2.13 Pairing strength. How strongly a base pair sticks together depends on the strength of the bonds between them. There are three bonds between Gs and Cs, but there are only two bonds between As and Ts, so GC pairs hold together more tightly than AT pairs. When heat is used to separate strands of DNA, regions with a lot of AT base pairs will come apart at a lower temperature than regions with a lot of GC base pairs.

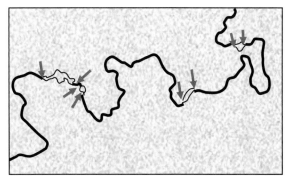

FIGURE 2.14 Replication of DNA goes hand in hand with cell division. DNA is too thin to be seen by even the most powerful light microscopes, but we can obtain an image of the DNA by shooting an electron beam at DNA coated with a heavy metal. The electron beam passes through everywhere except where the heavy metal has coated the DNA, so we are able to get an image of the DNA as a shadow of the places where the electron beam does not get through. This drawing shows what replicating bacterial DNA looks like when viewed with an electron microscope. Bubbles surrounded by thin lines are the regions in which the DNA is opened up to be replicated, and the thicker lines represent regions of double-stranded DNA where the bases are paired and replication is not taking place. Red arrows mark the replication forks, the points at which the double-stranded DNA is opening, with the fork moving forward as DNA replication advances to fill in the single-stranded material and make it double-stranded.

The double-stranded structure of the DNA comes apart into two separate strands in the region that is being copied, and the polymerase "reads" the old strand of DNA and uses the base-pairing rules in Figure 2.13 to insert the bases of the new strand in the proper order to complement the sequence on the old strand. Meanwhile, the same thing happens to the other "old" strand of DNA.

The cell can then use each single strand as a template for the creation of a new DNA strand that is paired to the old DNA strand using the base-pairing rules and hydrogen bonds, but it cannot simply start adding bases at some random point along a single-stranded length of DNA. The replication machinery requires the existence of a double-stranded region, so that it can add a new base not only by pairing it with a base on the opposite strand but also by hooking the backbone section of the base onto the 3' end to the strand that is being extended (Figure 2.15). When the cellular machinery opens up a bubble of single-stranded material the cell creates a double-stranded region from which to initiate synthesis by putting a short primer into place that is made of RNA, a different nucleic acid that is similar to DNA but

less permanent and more easily degraded and gotten rid of after it has served its purpose.

Adding one base at a time, the replication machinery builds a new strand paired to the original template strand. Each new base is added according to the base-pairing rules. As the new strand forms, each new base is added to the 3' end of the growing strand, so that DNA replication along that strand can proceed in only one direction. DNA synthesis is continuous on a strand called the leading strand, but on the opposite strand, called the lagging strand, the cell has to do many separate rounds of synthesis of an Okazaki fragment, a region of DNA a few hundred base pairs in length that begins with an RNA primer. By the time synthesis is completed,

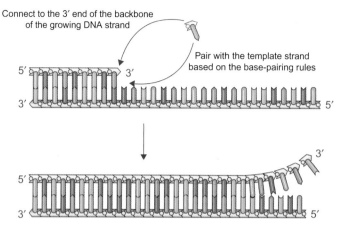

FIGURE 2.15   First step in replication. Replication requires a double-stranded region adjoining a single-stranded region. A base is added to the 3' end of the strand that is to be extended, and it is put into place paired with the first available base on the single-stranded part of the molecule according to the base-pairing rules. (Courtesy of Edward H. Trager.)

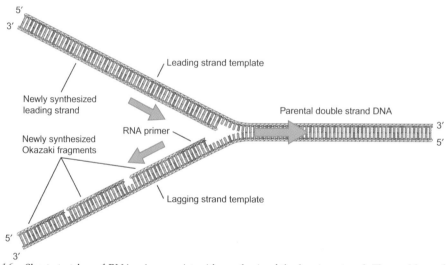

FIGURE 2.16   Short stretches of RNA primer assist with synthesis of the lagging strand. The problem of synthesizing both strands in the direction of the replication fork is solved by doing many rounds of lagging strand synthesis in short stretches. Each new RNA primer is laid down close to the replication fork with synthesis moving away from the fork creating an Okazaki fragment of several hundred bases in length. As movement of the replication fork opens up more single-stranded DNA, a new RNA primer is put in place to begin another round of synthesis moving towards the previously synthesized Okazaki fragment and away from the fork. When the synthesis of an Okazaki fragment overtakes a previously synthesized Okazaki fragment, the primer is chewed away and replaced with DNA bases. (Courtesy of Edward H. Trager.)

DNA synthesis from one Okazaki fragment has run into the primer from another Okazaki fragment, and the RNA is chewed away and replaced with DNA to join the new Okazaki fragment to the old Okazaki fragment. Thus the new DNA on the lagging strand ends up as one long continuous piece of DNA even though it was first synthesized in many separate pieces.

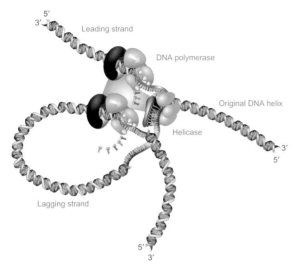

FIGURE 2.17 DNA replication. The point at which DNA helicase (orange) opens the double-stranded DNA is the replication fork, which has a double-stranded helix to the right of it and single-stranded DNA immediately to its left. Two copies of DNA polymerase (in green), hooked together by a protein called tau (blue), move towards the replication fork inserting bases to create two new double-stranded helices from the single strands adjoining the replication fork. Since polarity requires that the leading and lagging strands be synthesized in opposite directions, but both copies of DNA polymerase move towards the replication fork, the cell solves this problem by looping the lagging strand around so that the portion of the lagging strand that is being copied faces the same direction as the leading strand. (Courtesy of Edward H. Trager.)

As the DNA is copied, the bubble of single-stranded region keeps opening, with a replication fork opening ahead of the moving region of active DNA copying. At this point, the leading strand is copying towards the fork, and the lagging strand is pointing away from the fork. But the cell manages to keep everything moving towards the fork anyway by looping around the DNA of the lagging strand which is oriented to face away from the fork, so that the portion of the strand being copied is also facing towards the fork (Figure 2.17). This simple bit of geometry, twisting the lagging

strand around to face the same direction as the leading strand, lets the cell maintain a steady pace of replication that moves towards the fork without violating the rule that replication goes 5' to 3'.

Copying of DNA starts at many places along a chromosome, in areas called origins of replication. We understand a lot about such origins of replication in microbial systems, but much less is known about origins of replication in humans.

Semi-Conservative Replication of DNA

The result of replication is that where there was one helix there are now two helices that are identical in sequence to the original sequence and to each other. Because each strand of the double helix is copied by making a new strand that is complementary to it (able to pair with it based on sequence), the result is that each of the new helices consists of one old strand and one new strand (Figure 2.18). Thus the cell does not end up with an old helix and a new helix, but instead produces two new helices, a process referred to as semi-conservative replication.

DNA replication takes place during cell division, the process of making more of everything inside of the cell and then partitioning the increased amounts of cellular materials into two new cells. Before each cell division, these proteins must replicate the entire human genome *exactly* once.

DNA replication is so precise that mistakes in copying are made less than once in every 10,000,000,000 replicated bases. Copying of the DNA has to take place simultaneously at many places because the pieces of DNA inside the nucleus are very long. Otherwise, the cell would take a great deal of time to finish copying its genome and cell division would take far too long. Consider the implications for wound healing or development of an embryo if cell division were a significantly slower process than it currently is!

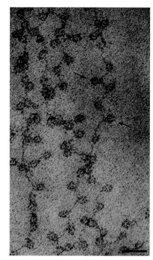

FIGURE 2.18 Semi-conservative replication. When a DNA helix is replicated, each new helix gets one new strand and one old strand, so replication of the two strands of a double helix results in two new helices, each of which contains one old strand and one new strand. We do not see replication give both new strands to one helix and both old strands to the other helix. (Courtesy of Edward H. Trager.)

FIGURE 2.19 Beads on a string. The initial level of organization in chromatin is the nucleosome, DNA wound around histone cores. Arrows point out several of the "beads" that are connected by short stretches of DNA. (From D. E. Olins *et al.*, "Chromatin history: our view from the bridge", *Nature Reviews Molecular Cell Biology*, 2003;4:809–14. © Nature Publishing Group; reproduced with permission.)

2.5 CHROMATIN

Think of DNA as if it were a long thin thread or piece of yarn. Pull on it, it breaks. Wad it up, it gets tangled. But if you wind the yarn up into a ball, it is kept in an orderly structure and is hard to break if you do not pull out individual strands to tug on. The cell solves the problem of how to keep the long, fragile DNA from breaking or getting tangled in a similar way: it uses a complex scaffolding of proteins to wind the DNA into a tightly ordered structure called chromatin that the cell can move around and act upon without the DNA ending up as a broken, tangled mess. As an interesting technical byproduct of chromatin structure, there is some variation in how tightly the "yarn" is packed and in which proteins are present. There are

differences in how much dye may be taken up when we stain DNA within the nucleus of a cell.

The structure that the cell uses to protect the DNA is called chromatin, an intricate scaffold of proteins and DNA wound in a tightly structured form. The scaffolding that the cell uses starts with DNA wound around a core of specialized proteins called histones. The result is an effect that looks a lot like beads on a string, with about 20 to 50 base pairs of DNA forming the thread between the beads (Figure 2.19). The beads, called nucleosomes, consist of 147 base pairs of DNA wound around the histone protein core.

The beaded necklace of 10 nm nucleosomes is wound to make a 30 nm thick rope. The rope of nucleosomes is then wound back and forth to form coils of up to 700 nm. Those coils are packed together to form a structure called a chromosome that contains tens of millions or,

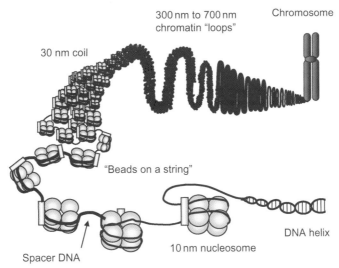

FIGURE 2.20 Chromatin. The structure of a chromosome as seen at different levels of magnification. The helical DNA is wound around a core of histones to create a structure like beads on a string, The "necklace" of beads is then coiled around to create a thick rope of "beads" and that rope then is wound back and forth in loops. The result is that a very long stretch of DNA is packed down into a very small space.

in the cases of the longest chromosomes, even hundreds of millions of base pairs (Figure 2.20).

Each chromosome is made up of many millions of genetic letters spelling out thousands of genes along the length of each chromosome. All along the length of the millions of genetic letters in a chromosome are regions that spell out discrete pieces of information that we call genes. Thus, when the cell does something to a whole chromosome, it is actually acting on a large number of genes all at once. The cell can move whole chromosomes in this way because of the tightly ordered packaging of the DNA in chromatin.

2.6 WHAT ARE CHROMOSOMES?

The First Views of the Human Genetic Blueprint

Once the beads on a string are all wound up in ropes, and the ropes are coiled, the final assembled structure is called a chromosome. Some of the earliest insights into the biological basis of genetic information actually came from types of microscopic studies of chromosomes that were going on in the late 1800s, but at the time the microscopists did not know that what they were watching under the lens of the microscope was a visual manifestation of the genetic blueprint – the chromosomes. By the late 1800s microscopists were able to see structures within the nucleus – long, thin, threadlike structures (Figure 2.21). They called these threads in the nucleus chromatin or chromosomes. They spent a lot of time looking at the changing shapes and locations of these threadlike structures within the cell during different stages in cell division. But they did not know that they were looking at the movement of the structures that contain the genetic blueprint in a kind of 23-volume genetic encyclopedia that is written into the structure of the chromosomes.

Improvements in technology let us start seeing more details in the chromosomes.

FIGURE 2.21   A nineteenth-century microscopist's view of the thread-like structures found inside of the nucleus and called chromatin. This particular drawing portrays a view of the nucleus of a salamander cell. This drawing comes from p. 24 of Edmund B. Wilson's book *The Cell in Development and Inheritance* (1896). The book is available through Google Books, and provides a fascinating portrayal of the state of cell biology before the beginning of the twentieth century.

FIGURE 2.22   Intact metaphase cell. (Courtesy of the Clinical Cytogenetics Laboratory, University of Michigan, Ann Arbor, MI; Diane Roulston, PhD, Director.)

If we stain the chromosomes we get a pattern of light and dark staining that looks almost like a microscopic bar code (Figure 2.22). At a higher magnification when we look at the structure of a replicated chromosome, we see the two sister chromatids, the two identical copies of the chromosome that are still attached to each other after the chromosome has been replicated and before the two copies separate from each other (Figure 2.23).

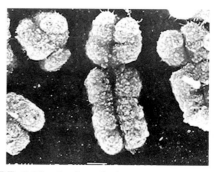

FIGURE 2.23   Replicated chromosomes. Chromosome images showing the pair of sister chromatids (the two replicated copies of the chromosome that are still attached to each other) and the constriction at the centromeres. The two chromosomes in the middle of the picture are the small Y chromosome and the larger X chromosome. (Image from indigo.com. © Indigo Instruments. Used with permission.)

Visible Landmarks on the Chromosomes

One of the most obvious landmarks on each chromosome is the centromere, a constricted region that can be found near the middle of some chromosomes and near the end of others, dividing the chromosome into longer and shorter arms (Box 2.3). The centromere constriction is especially evident in Figure 2.23, where it is not just a narrowing of the chromosome but is also the region where the two sister chromatids are attached to each other. The centromere is a large region of highly repetitive sequence, called satellite DNA, consisting of many copies of a sequence repeated over regions of tens or hundreds of thousands of bases. For example, a type of beta-satellite DNA found on chromosomes 9, 13, 14, 15, 21, and 22 has short sequence subunits repeated over regions ranging from 50,000 to 300,000 base pairs. An example of alpha-satellite sequence present on chromosome 17 consists of 500 to 1000 copies of a several thousand base sequence that is itself made up of 16 copies of a shorter sequence. Regions surrounding the centromeres have few genes and play an important role by interacting

BOX 2.3

CHROMOSOME ARMS: THE LONG AND THE SHORT OF IT

Length, centromere position, and banding pattern all help distinguish different chromosomes from each other. Chromosomes are numbered according to their size. Chromosome 1 is the longest and chromosome 22 is the shortest. Two chromosomes are left out of this numbering scheme – the X and Y chromosomes. On some chromosomes the constriction called the centromere is near the middle, and on some it is near the end. Noting the short arm (called the p arm) and the long arm (called the q arm) of a chromosome can help in telling one chromosome from another, especially in cases in which sizes are so similar that they are hard to tell apart. Thus, if someone says that a gene is located on chromosome 10p, they mean that the gene is on the short arm of chromosome 10. There is a tale that says that the names for the chromosome arms were originally supposed to be p (as in petite) and g (as in grande) but that a printer misread the g as a q, leaving us with a naming system that is not so obvious.

There is a banding pattern to the chromosomes akin to a cytogenetic bar code labeled with numbering that moves outwards from the centromere. As shown in this diagram of chromosome 9, a gene located on band 9p21 is on the short arm, and a gene located on band 9q31 is on the long arm. A gene located on band 9p23 is on the short arm of chromosome 9 and is farther from the centromere than a gene located at band 9p22. If a gene is located on band 9q21.3 its position on the long arm of chromosome 9 is farther from the centromere than a gene at band 9q21.1.

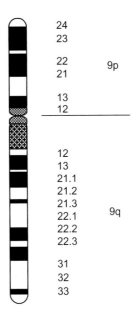

with the cellular machinery that moves chromosomes around in the cell.

At the ends of the chromosomes are telomeres comprised of long regions of short, simple sequence repeats, often thousands of bases in length. The telomeres serve as a kind of buffer zone to protect the genes of the chromosome against the natural process of chromosome shortening that is an automatic product of the way replication works. Remember that on the leading strand, DNA replication can move forward continuously all the way to the end of the chromosome. But on the lagging strand, each of the many short stretches of copying taking place initiates with a short RNA primer. How would the cell get the 5′-most bases on the chromosome copied if the only way to initiate DNA synthesis on that strand is to put an RNA primer into place that covers those bases at the 5′ end of the chromosome?

More than a dozen bases at the very end of the chromosome would be covered by the RNA primer, and there is no mechanism for converting those RNA bases into DNA. The result would be that after every cell division, one strand of the chromosome would be shorter by at least the length of the RNA primer for the very last Okazaki fragment on the lagging strand.

The cell solves this problem by having an enzyme called telomerase that can synthesize a long region of repeated sequence at the very end of the chromosome. Those telomere sequences are then available to serve as the location on which the very farthest RNA primer can be placed when synthesizing the very end of the lagging strand. The bases covered by that RNA primer do not get replicated into DNA, but those are spare bases that the cell can do without, so the loss causes no problem.

Telomerase activity does not keep going in all cell types, so gradually, over time, telomeres in some actively dividing cell types shorten. This is less of a problem for cells that are not dividing, and is not a problem at all for some specialized cell types like stem cells that have active telomerase renewing the length of the telomeres. What happens if the telomeres in a cell get to be too short? The cell enters a state called replicative senescence in which cell division is blocked so that the shortening of the chromosome cannot eat away at the sequences of the genes on the chromosome. In Chapter 10 we will talk about some kinds of cancers characterized by unstable genomes that can occur if cell division manages to resume and bypass the replicative senescence block on cell division in a cell in which the telomeres have gotten to be too short.

How to Tell One Chromosome from Another

A first look at the gross structure of chromosomes under the microscope can seem fairly confusing because there are so many of them and they look a lot alike. However, there are several physical features that let us identify the different chromosomes and tell them apart. First, the chromosomes come in different sizes, with the longest chromosome being more than five times the length of the smallest. Second, the centromere constriction that divides the chromosome into longer and shorter arms can be found near the middle of some chromosomes and near the end of others (Box 2.3).

Some chromosomes are very similar in shape, and this overall shape – length plus centromere position – does not let us tell all of the chromosomes apart uniquely. An additional trick, the use of dyes that stain some parts of a chromosome more darkly than others, produces a pattern of light and dark bands rather like a bar code specific to each of the individual chromosomes. Once the *banding pattern* is combined with other information on size and shape of the chromosome, we can tell all of the chromosomes apart from each other, even chromosomes such as 11 and 12 that are very similar in size and centromere position.

In Figure 2.22 it may not be obvious that the banding pattern is all that helpful, but if we cut out the pictures of the chromosomes and arrange them systematically, the banding pattern becomes a great aid in telling if all of the chromosomes are present (Figure 2.24). We use the term karyotype to refer to this type of display, where the cutout images of the individual chromosomes have been arrayed with copies of the same chromosome lined up next to each other.

The usefulness of banding patterns becomes a bit more apparent when the pictures of the chromosomes are cut out and arranged by size, centromere position, and banding pattern to produce this karyotype picture. For instance, looking at Figure 2.22 did not make it immediately obvious whether the individual was male or female, but once the chromosomes are lined up in pairs and by size as in Figure 2.24, it is much easier to tell that this cell has two X chromosomes that are characteristic of females (see Chapter 8). Figure 2.25 shows an idealized

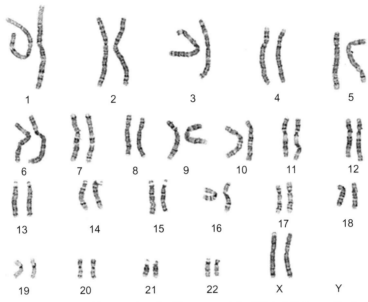

FIGURE 2.24   Karyotype of a normal human female. (Courtesy of the Clinical Cytogenetics Laboratory, University of Michigan, Ann Arbor, MI; Diane Roulston, PhD, Director.)

image of the human karyotype – the set of chromosomes arranged by size, centromere position, and banding pattern. Because the real pictures are less clear than this, karyotyping normally involves photographing and studying chromosomes from multiple cells to be sure that clear enough images have been obtained for each chromosome. To help people communicate about features they observe concerning the chromosomes, there is a system of naming for the pattern of bands that lets someone publish a geographic designation such as Xp21 and have others know exactly which band on the short arm of the X chromosome is being discussed.

Chromosomes Come in Pairs – Mostly

One of the very striking things about the carefully arrayed chromosomes of a woman shown in Figure 2.24. is the way in which the 46 chromosomes can be grouped (by size, centromere position, and bar code banding pattern)

into 23 pairs. The chromosomes that have been paired because they look alike are effectively two copies of the same chromosome, also known as *homologues*, with the same order of genes arranged along the length of the chromosome and almost exactly the same DNA sequence. Normally, the individual's mother donated one member of a pair of homologues and the father donated the other member of the pair. There are usually at least some differences between the two homologues at the level of the DNA sequence, but the gross structure of the two homologues is the same.

When researchers looked at cells from a man, they saw that only 22 pairs of visually similar chromosomes could be formed (Figure 2.26). This set of "matchable" chromosomes, which looks the same in both sexes, is referred to as the *autosomes* (chromosome pairs 1–22).

After matching those first 22 pairs, the autosomes, there were two chromosomes left over in the male cells that did not look alike. One

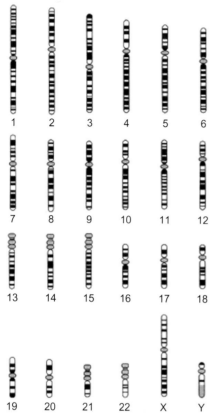

1   2   3   4   5   6

7   8   9   10   11   12

13   14   15   16   17   18

19   20   21   22   X   Y

FIGURE 2.25 Idealized diagram of a chromosome banding pattern that helps with identification of the images that are harder to distinguish in a real microscopic chromosome spread.

more recent use of brilliantly colored fluorescent dyes makes many things about the microscopic chromosomes easier to identify and distinguish. A process called *fluorescence in situ hybridization (FISH)* allows a brightly dyed piece of DNA to find and attach to a spot on a chromosome that has the same sequence as the fluorescent DNA probe. In some cases, this has been used to let investigators find out where a particular gene is located (Figure 2.27). In other cases, it provides a sensitive test of whether a particular gene is missing from a chromosome that lets researchers detect deletions that are much smaller than anything detectable by just looking at the structure of the chromosome in a standard karyotype. Normally, spots on two chromosomes would light up for most genes, but if one copy of the gene were deleted, spots would appear on only one chromosome instead.

Very recently, it has been possible to stain each chromosome a different color using probes labelled with different colors of fluorescent dyes in an impressive process called *chromosome painting* (Figure 2.28). This provides a very powerful way to distinguish different chromosomes that in some ways seems like a major improvement over the tiny black-and-white banding patterns, although some of these new color-based technologies can be problematic for researchers who are color-blind. Later on, when we talk about damaged chromosomes and cancer, we will tell you more about ways this new imaging approach can be used to study medically important processes.

This ability to see chromosomes through a microscope lens and distinguish the chromosomes from each other offers powerful opportunities to answer important questions. As we saw here, it demonstrates a fundamental genetic difference between males and females (XY vs. XX). This kind of microscopy tells us a great deal about how cells move chromosomes around and pass them from one generation to the next, something we will talk about in Chapter 5. As you will see, understanding

of the medium-sized chromosomes, called the X *chromosome*, matches one of the chromosomes present in the twenty-third female chromosome pair. However, the unmatched chromosome was a very small novel chromosome (the Y *chromosome*) not present in female cells. As we will see in Chapter 8, the structure, function, and behavior of these chromosomes help us to understand sex determination, the process that decides whether the baby will be male or female.

FISH and Chromosome Painting

Only very large changes in a chromosome can be detected using normal karyotypes. The

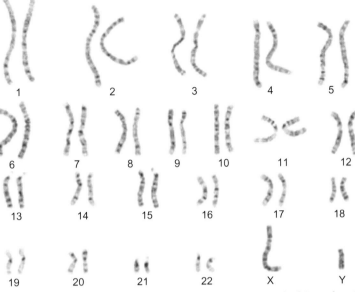

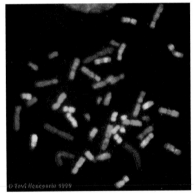

FIGURE 2.26 Male karyotype shows the two chromosomes that are not matched in males, the X and the Y, which are not included in the size-based chromosome numbering scheme. (Courtesy of the Clinical Cytogenetics Laboratory, University of Michigan, Ann Arbor, MI; Diane Roulston, PhD, Director.)

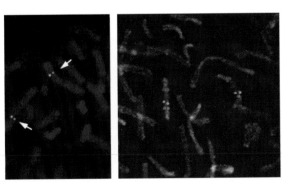

FIGURE 2.27 FISH lets us attach a brightly dyed piece of DNA to the point on a chromosome that complements the same sequences present on the dyed probe. In this picture, we can tell that the gene corresponding to this particular probe is not deleted because it lights up on both copies of the chromosome on which this gene is normally found. (Single-color FISH image courtesy of Thomas Glover, PhD, Department of Human Genetics, The University of Michigan Medical School. Multi-color FISH image comes from Octavian Henegariu at Yale University.)

FIGURE 2.28 Chromosome painting shows different chromosomes in distinct colors. (Image comes from Octavian Henegariu at Yale University.)

what happens to the chromosomes when they are passed to new cells or new generations in a human family shows us some important things about the relationship between chromosomes and Mendel's units of heredity.

2.7 EUCHROMATIN AND HETEROCHROMATIN

Some regions of a chromosome are transcriptionally active euchromatic regions. Other regions of the genome are relatively inactive heterochromatic regions (Figure 2.29). Chromosomes can be stained to distinguish the euchromatic and heterochromatic regions from each other (Figure 2.29). If we look at different cell types or cells in different metabolic states we will see some changes to this pattern, since some areas of constitutive heterochromatin contain genes but will stain as if gene-poor if those genes are not being used. However, some areas will consistently show as gene-poor (heterochromatic) areas.

Euchromatic DNA tends to possess a lot of complex sequences, sequences that use all four letters of the genetic alphabet in a complex arrangement that only occasionally includes any extensive repetition. Euchromatic DNA is densely populated with genes, but the fact that the overall region is considered transcriptionally active does not mean that all of the genes in that region are being used all the time.

Heterochromatin is largely comprised of very highly repeated (as in hundreds of millions of copies) short simple sequence repeats (for example the sequence AATAAT repeated millions of times) called satellite DNA. These runs of simple sequence elements are often disrupted by non-functional copies of types of DNA sequences known as transposable elements that (when functional) have the capacity to move about within the genome – excising themselves from one location and "jumping" to another spot in the sequence. Heterochromatin is largely devoid of functional genes. Although a small number of intrepid genes have evolved the capacity to survive and function there, for the most part, heterochromatin is incredibly repetitive and devoid of informational content. These satellite-rich regions are bound by specific heterochromatin-binding proteins which apparently serve to keep them highly condensed.

The type of heterochromatin we have just described is often referred to as constitutive heterochromatin, because it always has the basic characteristics of heterochromatin: early condensing, late replicating, and (for the most part) genetically inert. It is usually found around centromeres and may play roles in both facilitating centromere function and in mediating chromosomal associations. There is another kind of heterochromatin referred to as facultative heterochromatin, in which a normally euchromatic region or an entire chromosome is inactivated and assumes a heterochromatic state. The best example of facultative heterochromatin is the inaction of one of the two X chromosomes

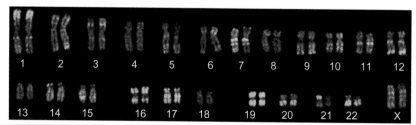

FIGURE 2.29 Staining in this picture shows where the transcriptionally active regions of chromosomes are in the cell that these chromosomes came from. Gene-rich regions are stained in green and gene-poor regions are stained in red. Note that many areas show alternating colors or an intermediate color, while other regions seem to be very solidly one color or the other. Notice that some of the most intensely red areas are located at the centromeres. (Photo by Irina Solovei, doi:10.1371/journal.pbio.0030188.g001. © 2005 Public Library of Science.)

in the somatic cells of human females. We will return to this issue in Chapter 6.

2.8 THE MITOCHONDRIAL CHROMOSOME: THE "OTHER GENOME" IN THE HUMAN GENOME

Most of the human genome is located inside of the nucleus, safely protected from the metabolic activity and the many degradative enzymes present in the cytoplasm; another part of the human genetic blueprint, the mitochondrial chromosome (Figure 2.30), is also locked

away from the cytoplasm inside of the mitochondria. The 37 genes contained on the 16,569 base pair mitochondrial chromosomes include 13 protein-encoding genes, 22 tRNA genes, and 2 ribosomal RNA genes. In fact, there are genes in the nucleus that are larger than the entire mitochondrial chromosome!

Given the role of mitochondria in the cell's energy metabolism, it is not surprising that defects in mitochondrial genes are often found to affect types of cells that are especially sensitive to energy needs, such as cells of the nervous system. Traits that show mitochondrial inheritance include a variety of neuropathies, encephalopathies, epilepsy, and a variety of complex phenotypes, including one that causes a combination of deafness and diabetes.

Mitochondrial Mode of Inheritance

It is also not surprising that many (but not all!) of the defects in mitochondrial function can be traced to genes located on the mitochondrial chromosome. The result is that transmission of such defects from one generation to the next follows a pattern of inheritance that arises quite directly from the way in which the mitochondria themselves are passed from one generation to the next (Box 2.4).

For mitochondrial inheritance, the trait is passed along by the mother but not the father (Figure 2.31). This happens because the mother passes along mitochondria to the next generation but usually the father does not. During formation of sperm the number of mitochondria is substantially reduced, so that few if any of the father's mitochondria end up represented in the offspring. In contrast, a developing egg often has more than a hundred thousand copies of the mitochondrial genome; the egg needs to carry so many mitochondria that there are enough to pass along to daughter cells during multiple rounds of cell division following conception.

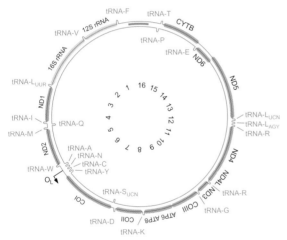

FIGURE 2.30 Diagram of the human mitochondrial chromosome based on the July 6, 2008, version of the sequence. Numbers in the center indicate position in the sequence in thousands of base pairs. Words and symbols inside the circle mark positions of protein coding and rRNA genes. Single letters at blue peaks both inside and outside of the circle represent the single letter code identification of tRNAs. (Image compliments of Edward H. Trager, after the image at Gábor Zsurka's Mitowheel site http://mitowheel. org/mitowheel.html. MitoWheel 1.2 is based on the July 6, 2008 version of the human mitochondrial genome sequence. Developed by Gábor Zsurka, gaga@gabstrakt.de (concept, programming, graphics) and Attila Csordás (ideas, testing).)

BOX 2.4

HOW TO RECOGNIZE MITOCHONDRIAL INHERITANCE

Underlying concepts:

- Mitochondria are passed along through eggs but not sperm.
- With many mitochondria in a cell a new sequence change in one mitochondrial genome may have little or no effect.

Rules for what we will see in a pedigree showing mitochondrial inheritance:

- The trait will be inherited from the mother but not from the father.
- All of the children of an affected mother will be affected.
- None of the children of an affected father will be affected.

- Equal numbers of males and females will be affected.
- The pattern of inheritance can look more complicated when the parent is heteroplasmic – having a mixed population of mutant and normal mitochondria.
- Few traits will show this mode of inheritance since there are such small numbers of mitochondrial genes.
- Risk can be modified by environmental factors, mosaicism, errors in meiosis, new sequence changes, non-penetrance or age-related penetrance, which we will discuss later in the book.

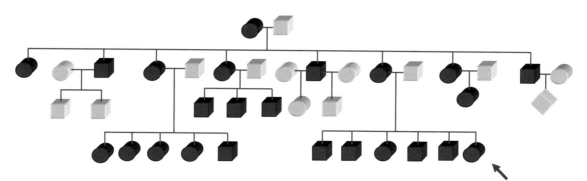

FIGURE 2.31   Large family showing mitochondrial inheritance of sensorineural hearing loss. Only four generations are shown from a larger family. In the larger overall family 34 of 18 children of affected mothers were affected, but there were no affected individuals among the 22 family members who had affected fathers. (After R. A. Friedman *et al.*, "Maternally inherited nonsyndromic hearing loss", *American Journal of Medical Genetics*, 1999;84:369–72.)

Some of the proteins needed for mitochondrial function are encoded in the nucleus and synthesized using cytoplasmic ribosomes before the gene product travels to the mitochondria. The human mitochondrial electron transport complex I is made up of proteins encoded by dozens of nuclear genes in addition to seven mitochondrial genes. Defects in activity of this complex can result from sequence chages in either nuclear or mitochondrial genes. A variety of different phenotypes can result depending on which genes have been "hit" by the sequence change. Some complex I defects show mitochondrial inheritance but others show nuclear inheritance. For instance, one of the defects that causes mitochondrial complex I deficiency leads to an autosomal recessive form of Parkinson disease.

If a mitochondrial sequence change was present in the father's genome, it is unlikely to be passed along because most or all of the mitochondria are eliminated from sperm during meiosis. But if the change happened in the female germline, it will be represented among more than a hundred thousand copies of the mitochondrial chromosome in the cell. Although a change in the female germline would leave the egg with one sequence change amidst a hundred thousand mitochondria that lack the change, it is possible for the change to be either eliminated from or enriched in the daughter's germline. The egg contains so many more mitochondria than most cells because the zygote must contain enough mitochondria to pass along to the daughter cells during the first rounds of cell division in the embryo when new mitochondrial synthesis has not yet started. Some reports indicate that as few as one in ten thousand of the mitochondria from the egg end up represented in the child. So a new sequence change will often be prevented from being passed into the germline of the next generation because it is diluted out by the huge number of normal chromosomes; however, it would be possible for a new sequence change to become highly represented within only a few generations if the change happens to be among the subset of mitochondria that make it through the bottle neck to get used in the germline cells that contribute genetic information to successive generations.

2.9 DNA *IN VITRO*

"You take a clear solution, you add a clear solution, you get a clear solution, and you call that a result!"
—*Spouse of a genetics graduate student*

Amazing things have become possible since we gained the ability to do *in vitro* some of the things that cells accomplish naturally. Our ability to replicate DNA *in vitro* formed the basis for many important technologies used in the human genome project and led to the sequencing of the human genome. It gives us tools for studying how DNA sequences vary across populations. It also gives us the tools we need to carry out gene therapy, the process by which we alter our genomes to change our phenotype.

Tools for the Study of DNA

One of the important tools available to us is the DNA synthesizer. In 1972, the first synthetic gene was created, a tRNA gene. DNA synthesis machines can produce chemically synthesized oligonucleotides of up to about 200 bases in length. Separate synthesized pieces can then be ligated together to assemble larger pieces of DNA. This process of ligating chemically synthesized oligonucleotides allows for artificial synthesis of whole genes. Lots of labs have a DNA synthesis machine on hand, and there are companies that will sell synthesized genes made to order. In addition to providing whole genes, these machines can provide short stretches of DNA that can be used as tools in some of the experiments that we will describe here.

The next critical tools we needed were enzymes. Earlier we talked about some of the important breakthroughs that opened up our understanding of genetic information – the microscope, demonstration that DNA is the genetic material, and figuring out the structure of the double helix. Another series of breakthroughs came about as different research groups identified the various proteins that act on DNA. Many enzymes in these categories are now commercially available and can be used in labs to carry out experiments involving DNA. Our array of tools for molecular biology experiments include:

- More than a dozen known eukaryotic DNA polymerases that replicate DNA.

- A variety of enzymes involved in proof-reading and repair of DNA.
- The helicase that opens up the replication fork.
- The topoisomerases that wind and unwind DNA.
- Primases the install the RNA primer.
- Ligases that can seal together the backbone pieces of two different pieces of DNA.
- Nucleases that break down DNA and RNA.
- DNA methyltransferases that modify (imprint) DNA, which we will discuss in Chapter 3.
- Reverse transcriptase, an enzyme that copies RNA sequences back to DNA.
- Restriction enzymes (made in bacteria) that cut DNA at specific sequences.

Cloning – Making Copies of DNA

Cloning first turned up in the 1970s, and in its aftermath during the 1980s and 1990s genetics swept the headlines one gene at a time, one disease at a time, one organism at a time. But what is a clone? In the context of the genes we are talking about here, *a clone is a genetically identical copy of a DNA segment produced through the use of biological technologies in vitro.*

One of the big breakthroughs that began the era of molecular biology in the 1970s was gene cloning (Box 2.5). This allowed researchers to trick bacterial cells into making them lots and lots and lots of very pure copies of a piece of DNA of interest, no matter what organism the piece of DNA originated from. Over the last 30 years, molecular cloning has led to the identification of an increasing number of important genes and proteins that were not known before cloning, culminating recently with the completion of the human genome sequence and the identification of most of the human genes. This kind of cloning involves making copies of DNA. The basic idea is that a desired section of DNA,

---

BOX 2.5

THE UPROAR OVER CLONING

In the early 1970s, when early observations on some obscure aspects of bacterial genetics led to the concept of cloning, there were grave concerns about things that could possibly go wrong if human DNA were placed into bacterial cells and allowed to replicate. These concerns ranged from simple technical issues to major fears that new pathogens would be created that could sweep through human or animal populations. Scientists involved in the early molecular cloning studies put a voluntary halt to the work and convened to discuss the ramifications of the new technology, both technical and ethical. This led to the development of a set of guidelines that put very strict physical and technical limits on the kinds of work done in the early cloning experiments. Although this slowed down the earliest progress on studies of the human genome, it allowed researchers to gain enough information about cloned genes to better determine how to safely proceed in the future. The result is that enough was learned to allow researchers to know which types of experiments continue to need a great deal of physical and biological isolation to be safely performed, and to know that some kinds of experiments can now be done safely with much lower levels of containment. Four decades of safe and successful science have proven the value of the steps taken when the scientific world first realized the potential power and potential danger of cloning and paused to develop a set of guidelines that let the work proceed safely.

often containing one or several genes, is separated away from the rest of the human genome in a way that lets researchers make many copies of that piece of DNA in a pure form.

Before we consider how we clone a human gene or a piece of human DNA, let's consider why we would want to separate a piece of DNA away from the rest of the genome and make copies of it. It might seem like a simple thing. If you want DNA, just break open a human cell, and the DNA will flow out into the surrounding solution. In fact, it is not at all simple, since the DNA that comes out of a human cell contains the entire genome, and any kind of experiment we try to do on one of those genes will be giving us back signals and answers from all of the others at the same time. The result is background noise that overwhelms any specific signal we might be trying to find. So it leaves us trying to detect a small amount of information that is buried in a large amount of very similar information. It is like looking for the proverbial needle in the haystack, when the signal you are looking for looks too much like the vast amounts of extraneous information that surround it. However, by using gene cloning we can take the piece of DNA we want out of the cell and away from the billions of base pairs of DNA that make up the rest of the genes that we are *not* trying to look at. We can make many, many copies of that one purified gene, which makes it easy to obtain clear, clean results. Instead of trying to visualize a single needle buried in straw, we find ourselves looking at a pile of identical needles on an uncluttered surface.

One of the keys to cloning a piece of DNA is the use of something called a *vector*, a piece of DNA that can replicate itself if it is put into a cellular "replicating factory," such as a bacterial or yeast cell. The DNA bases used in the human genome, the bacterial genome, and the vector genome are the same four DNA bases, A, C, G, and T. *Cloners* – people who isolate and make copies of DNA by combining them with a

vector – use a kind of biological glue, an enzyme called a ligase, to splice together the human DNA and the vector DNA in a way that leaves no seam to indicate where the vector DNA stops and the human DNA ends. The product that results is one new intact piece of DNA that combines DNA from two different sources that were originally separate. The systems that copy the vector cannot tell that part of this new large piece of DNA is human, so they happily copy the human DNA right along with the vector itself. This process of recombining DNA from two different sources into one new structure is the basis for the term *recombinant DNA* (Figure 2.32).

Once human DNA had been cut up and combined with a vector, the resultant recombinant DNA is put into a host cell (such as a bacterial cell), where large numbers of copies of the recombinant DNA construct (or clone) could be made. Each insert-bearing vector is put into a bacterial cell, and the bacterial cell is placed on an agar plate, where the cell makes many copies of itself (and the insert-bearing vector it carries) all sitting at the same position on a plate in a little mound called a *colony*. If many such clones are grown on one plate, it is then possible to make a copy of the pattern of colonies by pressing a piece of filter paper onto the surface of the plate. As shown in Figure 2.33, washing the filter paper in a solution that contains a tag that recognizes the clone with the desired sequence lets the researcher find the colony housing the clone of interest. Because the colonies all sit apart from each other on a solid surface, once the right colony is identified, it can be lifted off of the agar plate and grown in large amounts away from the rest of the clones.

Such cloning techniques have been used to create clones that can then be used for a large number of different purposes. For instance, cloning of this kind can be used to isolate cDNA copies of individual genes or to obtain pieces of DNA from human chromosomes and mitochondria. Techniques like this have been used to make copies of human genes, such as versions

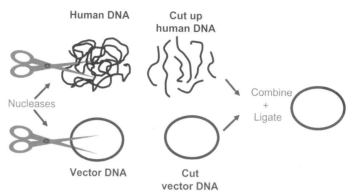

FIGURE 2.32   Recombinant DNA. Vector DNA is cut open and combined with cut-up human DNA. The cutting of the DNA is carried out by proteins called *restriction enzymes*, or *nucleases*, whose specific job in nature is cutting DNA. These nucleases act like scissors that can cut DNA at very precise positions within a piece of DNA. The two pieces of DNA are spliced together into one continuous piece of DNA that can be copied many times so that researchers can obtain large amounts of the clone in a very pure form. One of the key features of the vector is that it has an *origin of replication* used to initiate copying of the DNA. The hard part in cloning is not combining human DNA with a vector but getting the vector–human hybrid that contains the gene of interest separated away from the other hybrids containing the rest of the genes, a problem that was worked out in the 1970s.

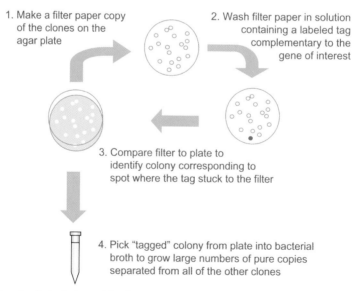

FIGURE 2.33   Finding the desired clone. After human DNA has been combined with a vector and introduced into the bacterial cell, a DNA tag is used to recognize the desired clone by base-pairing with DNA in the clone that carries the gene we want to study. There are other cloning systems, such as some that use viruses as vectors, and others that replicate the cloned DNA in yeast cells instead of bacteria. The basic principle remains the same: combine DNA that contains the gene you are after with a vector system, introduce it into a host cell system, separate the clones from each other, use a tag that lets you identify the clone you want, and remove that clone away from the rest of the clones before beginning your studies of it. Often in an experiment with a purified clone, it is common to work with many billions of copies of the clone at a time.

I. HOW GENES SPECIFY A TRAIT

of genes cloned from individuals with a sequence change on which we need to be able to carry out functional studies. Cloning has been used to make "expression" constructs in which the clone not only contains the DNA sequence of the desired gene but can even be used to produce the protein gene product of that gene. Cloned genes can be put back into human cells to determine whether addition of a particular gene can "fix" a particular metabolic defect present in a cell line. Cloning technology can be used to put a gene of interest into an experimental animal called a transgenic animal. Molecular clones were the main source from which sequence information was derived in the course of determining the human genome sequence – the order of As, Cs, Gs, and Ts of the whole length of DNA in a human cell. To accomplish this it was necessary to isolate huge numbers of clones, each containing a piece of human genome sequence, and arrange them in order, so that each clone containing a piece of human DNA could be shown to overlap with another clone containing a neighboring region of human DNA.

More than 30 years ago, when the perplexed nonscientist wife of a genetics graduate student made her declaration about our ability to see results in a test tube containing what looks like a drop of water, reading the sequence of even a small bit of DNA was technically difficult and required weeks or months of laborious effort. Some of the earliest experiments to read the sequence of DNA used enzyme-based technologies similar to what we use today, but Julia spent her graduate school days in the late 1970s and early 1980s using a rather terrible and fascinating mix of rocket fuel and carcinogens to pry the genetic spelling out of the pieces of DNA she was studying. It remains true to this day that when we want to read the sequence of a piece of DNA, we in fact start with a clear solution, add a clear solution, and end up with a clear solution. Fortunately, we now have instruments that will let us "read" the order of

genetic letters present on that piece of DNA (in place of radioactivity and x-ray films used in the first versions of DNA sequencing, and still used in some places to this day) after we have completed the enzymatic reactions that go on in that clear drop of liquid. We will tell you more about the process of sequencing when we get to Chapter 5 and talk about mutations.

For many years the hardest part of getting at the information contained in any given gene has been the problem of getting our hands on the gene we want to know about. With the information content of the genome spanning billions of base pairs and our target being only one out of the tens of thousands of genes, the process of gene discovery was traditionally lengthy, complex, and expensive. However, once a gene was discovered and its sequence known, it became relatively simple to look at that gene in many different people to see whether any of them have a copy of that gene in which the sequence is different. Now there are clones available that represent virtually any point on the human genome, and we now get to do what we call "clone by phone," by simply placing an order and having the clone delivered a day or two later.

But we now have even easier processes than "clone by phone" for getting our hands on pieces of DNA that we want to study. This process, called polymerase chain reaction (PCR), was the result of another breakthrough that was later recognized with a Nobel prize to its discoverer, Kary Mullis.

PCR – Making Copies of DNA

The biggest problem we face when we want to study a particular gene is the same problem we face when listening to someone talk in a crowded, noisy room – the signal we want to detect is there but is surrounded by too much other information that is very similar, and we can't distinguish the real signal from the background noise. With more than 3 billion base

---

BOX 2.6

IN VITRO AND IN VIVO

According to the Merriam–Webster Dictionary, the term *in vitro*, which comes from Latin and means literally "in glass," dates to the year 1894. *In vitro* actually refers to processes that can be carried on outside of the living cell, with the concept "in glass" referring to the glass flasks and glass test tubes in which so much early biological work was carried out. As science has evolved, many processes are now carried out in tubes and flasks made of plastic yet are still referred to as being done *in vitro*. The counterpart term, *in vivo*, which they trace back to 1901, also comes from Latin and means literally "in the living." Thus an enzymatic reaction taking place in the cells of your body are going on *in vivo*, but if we duplicate that enzymatic reaction in a test tube by isolating the enzyme away from any living cells, then it is an *in vitro* process.

---

pairs of sequence in the human genome, if we try to read one piece of sequence while all of the rest of the genome is also present, we cannot detect the sequence we want to read even though the signal is present. So one of the most important steps in reading the sequence of a particular gene is to get it separated away from the rest of the genes in the genome, to bring it out of its "noisy" background into a quiet, separate place where it is the only thing present that our sequencing technology can detect.

We can get our hands on a gene that is already known by making copies of the DNA we want through the use of an *in vitro* (Box 2.6) process known as polymerase chain reaction (PCR). The use of PCR lets us make billions of copies of a single gene or piece of DNA in a matter of hours at a cost of less than a dollar per sample. This effectively amplifies the signal we want (the copied piece of DNA) billions of times compared to the background noise (the rest of the genome that did not get copied).

How PCR Works

PCR copies DNA using the same systems a living cell uses to copy DNA. As we described earlier in this chapter, the secret to the replication of DNA is that the DNA polymerase enzyme copies a single strand of DNA by putting in place bases on the second strand that are complementary to the sequence of the first strand. One of the biggest problems with trying to use this approach is that we don't want to copy all of the DNA in the genome (or in the test tube in front of us on the lab bench!); we want to copy only one small stretch of DNA that is perhaps only a few hundred bases in length. So if we were to simply add DNA polymerase and copy everything in the test tube, we would be amplifying our background noise at the same rate we amplify the signal we are after.

One of the secrets to making PCR specifically copy only the piece of DNA we want, rather than copying all of the DNA in the test tube, is based on the same limitation that affects DNA replication *in vivo*. *DNA polymerase will only copy a single-stranded piece of DNA if there is a double-stranded section next to it from which to begin adding bases* (Figure 2.34). So if we can create a region of double-stranded DNA right next to the single-stranded piece of DNA we want to copy while avoiding having double-stranded DNA in other regions of the genome, we can make DNA polymerase do its job exactly where we want it.

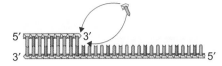

Single-stranded piece of
DNA cannot be copied
by DNA polymerase

Single-stranded piece of DNA
adjacent to a double-stranded piece can
be copied by DNA polymerase

FIGURE 2.34 Copying DNA. We cannot simply make huge numbers of copies of someone's genome by putting in DNA polymerase because it requires very special circumstances to carry out its copying function. It needs a region in which two strands of DNA are paired based on complementary base-pairing as a starting point for its copying operation. Living cells have mechanisms for creating this condition when copying chromosomes, but it does not just happen naturally in the test tube. (Courtesy of Edward H. Trager.)

BOX 2.7

DNA FOR GENETIC TESTING

Easy sources of human DNA include white blood cells from a blood sample or perhaps buccal cells that can be obtained by rubbing the inside of the cheek with a sterile swab. In some cases a sperm sample may be used. If we want to look at the DNA of a deceased person, it is often possible to get DNA from pathology samples left over from a biopsy sample taken years ago. PCR works even if there is very little DNA in the sample to be tested, and it works well even if the cells have not been stored under ideal conditions. In forensic cases, DNA can be obtained from samples that have been in storage for decades, and in anthropology studies, DNA can sometimes be obtained from very old tissue samples containing something that we sometimes refer to as ancient DNA.

We can create the conditions that DNA polymerase needs to carry out its copying function. First we need a source of single-stranded DNA that we will copy, which means we need to isolate DNA from cells of the person whose DNA we are going to examine (Box 2.7). Then we need to pull the strands of the DNA apart so that they are single-stranded, not double-stranded, DNA (Figure 2.35). This can be accomplished simply by boiling the DNA to break the chemical bonds holding the strands together.

There are more than 3 billion base pairs of human DNA sequence floating around in the test tube on the lab bench. What should we do if we only want to copy gene M, which is only 200 bases long and is the region of DNA in

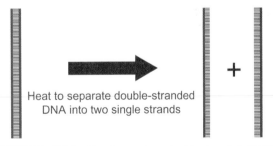

Heat to separate double-stranded
DNA into two single strands

FIGURE 2.35 Heating makes double-stranded DNA come apart into single strands.

which mutations have previously been seen to cause the disease we are studying?

We use DNA polymerase's need for double-stranded DNA to force it to copy only the DNA containing that 200 base chunk of gene M. How? If we already know the sequence of gene M, we can use the DNA synthesizer sitting on the end of the lab bench (if we have one) or we can call up a commercial DNA synthesis lab. For about 10 dollars they will make us a tube filled with huge numbers of copies of a short piece of DNA (about 18 base pairs in length) that matches a piece of sequence next to gene M.

If we add this short synthetic piece of DNA, called a *primer*, to a test tube containing single-stranded human DNA that we want to copy, the primer will find the piece of sequence to which it is homologous and will hybridize to it (stick to it based on the base-pairing rules). This creates the essential structure: a single-stranded region that we want to copy connected to a double-stranded region that gives polymerase its starting point. DNA polymerase will then start at the edge of the double-stranded piece and begin adding new bases complementary to the adjoining region, including bases complementary to the gene M exon that we want to test for mutations (Figure 2.36).

Copying DNA Between Two Primers

Now we know how to make new DNA complementary to exactly one place in the genome that we are interested in, by sticking a primer "tag" onto the human DNA at a location right next to what we want to copy. This lets us bypass copying the rest of the genome.

However, in Figure 2.36 we made only one copy. Now, instead of having one copy of gene M among tens of thousands of other genes, we have two copies. This gains us almost nothing. We need a way to make billions of copies so that we have large enough amounts to let us carry out biochemical reactions and monitor the outcome of our experiments.

PCR does not normally occur if we use one primer, instead of two. The real secret to PCR is not one primer that tags the spot you want to copy, but rather two primers that flank the

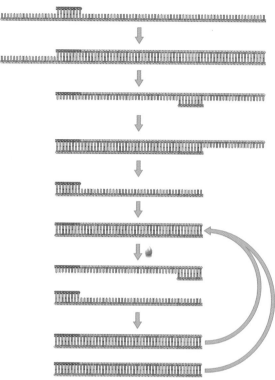

FIGURE 2.36 Making many copies of a piece of DNA. If we pick the sequence of our primer to match a spot in the chromosomal DNA right next to the DNA we want copied, we can force DNA polymerase to copy the DNA right next to it without copying the rest of the genome. DNA polymerase will begin inserting new bases right next to the primer, and the primer will become incorporated into the product so that part of the product is new DNA and part is the synthesized DNA of the primer. (Courtesy of Edward H. Trager.)

place you want to copy that let you repeatedly copy exactly the same spot in the genome. So at the end of Figure 2.36 , we had one new strand.

If we have a second primer also present that can bind to the new strand, it can now copy back across the region containing M, but now it is copying on the other strand. After several rounds of replication (Figure 2.37), we end up with a piece of DNA that is now bounded at one end by the first primer and at the other

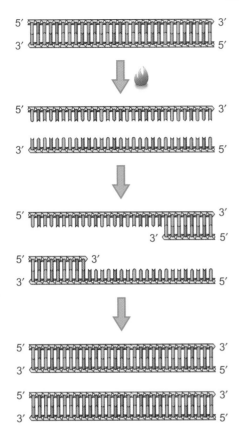

FIGURE 2.37 PCR from genomic DNA. After three rounds of synthesis we start to see the emergence of the 200 base-pair length fragments that are bounded by the primers. We will keep getting back some of the longer-length material, but the 200 base-pair fragment will quickly come to predominate. (Courtesy of Edward H. Trager.)

fragments become templates for subsequent rounds of replication. Although each round continues generating a small number of fragments that include the longer material, with each round, the fraction of pieces that are 200 base pairs long increases.

We do not always start out with long genomic DNA as the template from which we will copy when we do PCR. Sometimes we want to do something simple like make many more copies of a short fragment that we already have, in some cases even a sample of the PCR product from a previous reaction. In that case, if we start with a 200 bases fragment we immediately start getting back fragments of that length without having to wait until the third round to see them appear (Figure 2.38).

Of Hot Springs and Vent Worms

If you spend the time to follow Figures 2.37 and 2.38 through the steps, you can see how the use of two primers flanking a gene can rapidly isolate that gene (or other sequence of interest) onto a double-stranded fragment that has the primer sequences at its ends. Once this double-stranded sequence exists, it can go through the same loop – separate strands, bind to primers, synthesize new DNA, separate strands, bind to primers, synthesize new DNA – over and over. There is a big technical problem here, which is that at the end of the first round, the heating that separated the DNA into single strands killed the DNA polymerase that is needed to carry out the reaction. So we would then have to add back more enzyme. However, early on, scientists working with PCR realized that they could get around this problem if they just found a version of the enzyme that did not mind functioning at high temperatures.

A very clever solution to this problem was to go in search of organisms that live at very, very high temperatures. The logic is that, if an organism can live in a hot springs or along the edges of the hot thermal currents of the deep

end by the second primer. Because the ends of these primers were placed at positions that are 200 bases apart on the sequence of this region, the resulting piece of DNA that is produced is 200 bases long. We also end up with some DNA fragments of longer lengths that include the original template DNA and some intermediate-length pieces generated during the first round of synthesis. But by round three, we have two fragments that are 200 base pairs in length, bounded by the pair of primers. Those two

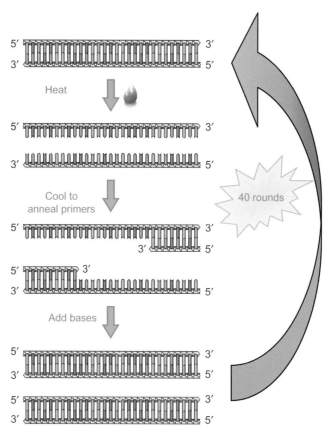

FIGURE 2.38   PCR amplification of a short fragment. If we go for 40 rounds of PCR we can end up with large enough amounts of DNA to be able to see it as a tiny pellet in the bottom of a tube if we separate it from the watery environment in which it is usually suspended. (Courtesy of Edward H. Trager.)

ocean vents, their enzymes must all be capable of surviving at temperatures close to boiling. By isolating DNA polymerase from bacteria living in hot springs and from vent worms living in superheated environments, scientists made PCR something practical and useful instead of a theoretical curiosity.

From Reaction to Chain Reaction

You don't need to spend a lot of time contemplating the mechanisms in Figures 2.37 and 2.38 to be able to get the main point here: if you can make primers that flank the piece of DNA you want, you can make vast numbers of copies of the piece of DNA that lies in between the sequences where the primers bind based on their complementary base-pairing. One double-stranded structure of this kind is copied to become two copies of the double-stranded structure after one round of PCR. After two rounds it has become four copies, after three rounds it has become eight copies, then 16 copies, and so on. After 20 rounds of PCR, we have more than a million copies.

To truly appreciate the power of PCR, consider this: After 30 rounds of PCR we have converted one copy of a sequence into more than a

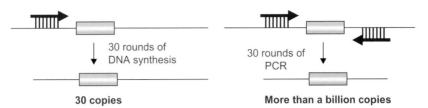

FIGURE 2.39   With one primer, 30 rounds of DNA synthesis will result in 30 new fragments of DNA. With two primers, after 30 rounds of PCR, the result will be more than a billion copies of the template DNA framed by the primers.

billion copies of that one little double-stranded piece of DNA (Figure 2.39). (In contrast, if we had used only one primer, we would have 30 new copies at the end of 30 rounds of DNA synthesis!) Usually, when we do a PCR reaction, we start out with more than one copy of the genome sitting in the test tube so our real product in the end is many times the billion copies that resulted when we started with one piece of template DNA.

Depending on how much DNA you need to end up with, a PCR reaction may often cost well less than a dollar to carry out, including the cost of isolating the target DNA. Also, PCR is fast, with some reactions taking less than a minute per round (although sometimes it takes longer). It's no wonder that PCR is felt to have revolutionized modern biology and genetics! In going from cloning to PCR, the field moved from experiments whose time line could be tallied in days, weeks or even months, to experiments whose timeline is counted in minutes. This does not mean that cloning has been abandoned, since cloning and PCR accomplish different things and let us answer different kinds of questions, but a lot of the high throughput rapid technologies of the twenty-first century rely on PCR.

The use of the simple essentials of DNA replication to carry out cloning and PCR provided very powerful tools with which to isolate genes or other DNA away from the rest of the genome and produce huge numbers of copies so that we can carry out a variety of biochemical reactions and assays to study the DNA. Even as new strategies for genetic studies have come along, the field has continued the process of improving these technologies to increase what we can do with them. In 1980 we were thrilled if we managed to see 60 or 80 bases of sequence in an experiment; in 2010, we see research articles that draw conclusions from sequencing whole genomes to get the answer. But the questions we must ask go far beyond DNA, so join us in Chapter 3 as we take a look at how the cell makes use of the information stored in the DNA.

Study Questions

1. What is an autosome?
2. What is a nucleolus?
3. What purpose do mitochondria serve?
4. What pairs with A?
5. What pairs with G?
6. What do we mean when we say that two DNA sequences are complementary to each other?
7. How rare are DNA replication mistakes?
8. What role does RNA play in DNA replication?
9. Given the DNA strand listed in the bottom boxes, use the upper boxes to specify the expected sequence of the new strand of DNA that DNA polymerase would make from this strand by using the base-pairing rules

10. What is semi-conservative replication and what does its mechanism of action tell us about which daughter cell receives the new DNA after cell division is completed?

11. What is a nucleosome?

12. What is a histone?

13. Why is DNA packed into chromatin structure instead of being left as a loose string of bases?

14. We often hear about Watson and Crick and the discovery of the double helix. Who else was involved in this discovery?

15. What does a karyotype accomplish that is not accomplished by a simple picture of the chromosomes as seen through the microscope?

16. Why can we make billions of PCR products when we use two primers but less than a hundred copies if we use only one primer?

17. As we heat DNA the two stands of DNA start to come apart. Why do the regions with the most As and Ts in the sequence separate before the regions with the most Gs and Cs in the sequence?

18. What is FISH?

19. In a family showing mitochondrial inheritance, what fraction of an affected woman's children will be affected?

20. In a family showing mitochondrial inheritance, what fraction of an affected man's children will be affected?

Short Essays

1. RNA viruses use reverse transcriptase to make a DNA copy of the viral genome as part of their life cycle. Reverse transcriptase cannot use DNA as a template, and yet we find that as much as 14% of the human genome may have had sequence altered as a result of the actions of reverse transcriptase. If reverse transcriptase only acts on RNA, how can it lead to changes in the DNA of the chromosomes? As you consider this question please read "Self interest" in *Genome: An Autobiography of a Species in 23 Chapters* by Matt Ridley (1999, Fourth Estate).

2. We find differences when we compare mitochondrial DNA sequence in one person to mitochondrial DNA sequence in another. How can these differences be used to map human migration patterns and tell where we originally come from? As you consider this question please read "Going East: New genetic and archaeological perspectives on the modern human colonization of Eurasia" by Paul Mellars in *Science*, 2006;313:796–800.

3. The position of the beads on the string in chromatin structure does not always fall at exactly the same position on the DNA. What factors affect where the DNA-histone beads occur? As you consider this question please read "What controls nucleosome positions?" by Eran Segal and Jonathan Wisdom in *Trends in Genetics*, 2009;25:335–43.

4. We have all heard about Watson and Crick and the double helix, but there were other scientists who contributed to this breakthrough discovery. What was contributed by the research groups of Wilkins and Franklin, and how essential was it to the development of the final model of DNA structure? As you consider this question please read "Maurice Hugh Frederick Wilkins CBE: 15 December 1916 – 5 October 2004" by S Arnott, T. W. Kibble, and T. Shallice in *Biographical Memoirs of Fellows of the Royal Society*, 2006;52:455–78.

Resource Project

The mitoWheel project at http://mitowheel.org/mitowheel.html presents a map of the mitochondrial chromosome and some annotation of what the different mitochondrial genes are. Go to the mitoWheel project and click on the arrow on the sequence. You will see the wheel of the map spin to bring to the top of the picture the gene that is represented at that point on the sequence, and you will get a little window that annotates each gene with its name. For some genes whose sub-units are part of multi-protein complexes, such as ND6, the annotation also indicates which complex the gene is in. Write a paragraph describing the organization of the genes on the mitochondrial chromosome, including where the tRNA genes are located relative to the other genes and where members of each complex are relative to each other.

Suggested Reading

Resources
Science Whiz DNA Kit
Molymod DNA Model Classroom Kit

Video
DNA : The Secret of Life (2005)
National Geographic: *The Human Family Tree* (2009)
NOVA: *DNA – Secret of Photo 51* (2003)
"Blame It On the DNA" from the Teaching Tools section of the *Journal of Biological Chemistry* site (www.jbc.org/site/home/teaching_tools/).
"Get Taq" on the Teaching Tools section of the *Journal of Biological Chemistry* site (www.jbc.org/site/home/teaching_tools/).

Articles and Chapters
"Routes to classical genetics: Cytology" in *Mendel's Legacy: The Origins of Classical Genetics* by Elof Axel Carlson (2004, Cold Spring Harbor Press, NY).
"Faces of Eve: Our reporter tries out a trio of genetic tests to find out what they can tell her about her identity and her ancestry" by Boonsri Dickinson in *Discover Magazine*, November 2009.

"Cataloging life: Can a single bar code of DNA record biodiversity and keep us safe from poisons?" by Bob Grant in *The Scientist*, 2007;21:36. Available at www.the-scientist.com/2007/12/1/36/1/.
"Seeing is believing: Graphical presentation strives to make sense of exploding quantities of data" by Anne Harding in *The Scientist*, 2006;20:46. Available at www.the-scientist.com/2006/4/1/46/1/.
"The nucleosome untangled" by Steven Henikoff and Bryan M. Turner in *The Scientist*, 2006;20:34. Available at www.the-scientist.com/2006/5/1/34/1/.
DNA in Life Ascending: The Ten Great Commandments of Evolution by Nick Lane (2009, W. W. Norton), pp. 34–59.
"Centromere annotation" by M. K. Rudd, M. G. Schueler, and H. F. Willard in *The Genome of Homo sapiens* (2003, Cold Spring Harbor Laboratory Press).
"mtDNA variation, climactic adaptation, degenerative diseases, and longevity" by D. E. Wallace, E. Ruiz-Pesini, and D. Michinar in *The Genome of Homo sapiens* (2003, Cold Spring Harbor Laboratory Press).

Books
The Selfish Gene (30th Anniversary Edition) by Richard Dawkins (2006, Oxford University Press).
The Cartoon Guide to Genetics by Larry Gonick and Mark Wheelis (1991, Collins Reference).
The Eighth Day of Creation: Makers of the Revolution in Biology (25th Anniversary Edition) by Horace Freeland Judson (1996, Cold Spring Harbor Press).
Rosalind Franklin: The Dark Lady of DNA by Brenda Maddox (2003, Harper Perennial).
The Double Helix: A Personal Account of the Discovery of the Structure of DNA by James D. Watson (2001, Touchstone).

HOW GENES FUNCTION

The Central Dogma of Molecular Biology: How Cells Orchestrate the Use of Genetic Information

The Human Genome. DOI: 10.1016/B978-0-12-333445-9.00003-4

THE READER'S COMPANION: AS YOU READ, YOU SHOULD CONSIDER

- The difference between DNA and RNA.
- How RNA is produced.
- How RNA is processed.
- Why alternative splicing is important.
- How the cell gets things into and out of the nucleus.
- How cells containing the same genetic information can take on different form and functions.
- What the different kinds of RNA do.
- Why some genes are used by all cells but other genes are used by only some cells.
- What types of genes would be needed by every cell type.
- What transcription factors and regulatory elements are.
- The difference between cis and trans.
- The difference between a transcription factor and a regulatory element.
- That promoter regions require the right orientation to function, but enhancers do not.
- That there is more than one way to look at the concept of normal.
- How developmental regulatory events in a fly can teach us about human development.
- How the actions of transcription factors combine to affect transcription.
- Why it might be hard to tell who has a genotype that goes with an inducible phenotype.
- The difference between genetic and epigenetic effects.

3.1 WHAT IS RNA?

The library was about to close, and Chelsea had not finished the report that was due the next day. The bound journals containing the articles she needed could not be checked out from the library, but she was not done extracting the information she wanted. Not surprisingly, she photocopied the items and went home to finish writing. After she had tossed the copies and turned in the report, the professor later asked her to rewrite part of the report. Suddenly she needed to consult one of the papers she had tossed. No problem! She simply went back to the library to consult the original bound copy that was still available for her use.

As we will discuss here, the cell similarly keeps copies of its critical information in a permanently archived form and yet can make use of transient disposable forms of that information as many times as it needs to. In this chapter we will tell you about the primary result of using the information in the nucleus (RNA) and a variety of things that we know about how the cell produces RNA, and we will talk about how all of this impacts our understanding of what does (or does not) constitute "normal."

Permanent DNA and Temporary RNA Copies

Our bodies, and the cells that make up our bodies, are composed of proteins, lipids, carbohydrates, nucleic acids, and a wide array of smaller molecules. Although all cells in the body are made of the same general classes of molecules, there are big differences between cell types in terms of many of the specifics. A red blood cell uses hemoglobin proteins that are not needed by cells in the kidney. Cells in the pancreas produce insulin that is not made by cells in the bone marrow. Fat cells can store substantial amounts of lipids that do not accumulate in the brain. In this chapter we will talk about how the cell uses RNA to make use of the genetic information in the cell and how that use of genetic information can be translated into such wide differences in cellular composition and activities.

What does the cell get from using RNA that it cannot accomplish by using DNA? First, it gains the ability to use more copies of the information than the two copies present on the two chromosomes. DNA is the permanent repository of the cell's information, stuck in the nucleus where it cannot access some of the cellular machinery, while RNA serves as a temporary copy of that information that can move out of the nucleus to take information to places that DNA never reaches. Within its DNA the cell possesses two copies of every gene, but often if a gene is being used the cell needs access to far more than two copies of the sequence. The cell can't solve the problem of needing more copies by making more copies of the DNA that comprises the chromosomes because, while a given cell type might need thousands of copies of any one gene at a given time, it might need very few or no copies of the gene sitting next to it.

Second, the cell gains flexibility. A cell does not always need the same numbers of copies. A cell that needs thousands of copies of a gene during fetal development might not need to use that information at all in the adult. By making RNA copies of genes when and where they need to be expressed, and at levels appropriate for the amount of gene product the cell requires, the cell can change how many copies of information are available depending on what it needs to do under the circumstances. Some genes, called housekeeping genes, are used by all the cells and do the same thing in all of those cells. Other genes, such as the genes that make the oxygen-carrying globin proteins in the red blood cells, show tissue-specific expression, that is to say, are used in some cell types but not others. While some cells are making RNA copies of thousands of genes, other cells are making RNA copies of a different set of thousands of pieces of genetic information contained on those same chromosomes. Some genes have the ability to respond to the environment, bringing forth information

when the body needs it and then turning off the information when the environmental situation no longer needs the information. Some pieces of information used at one stage in development may be missing at the next stage in development. But through it all, the nucleus retains two copies of each gene, whether or not they are being used.

Third, the cell gains the ability to build things that need RNA as one of the structural components. Although we often think of RNA in terms of its ability to carry information aimed at something else separate from the RNA molecule, in fact some RNA molecules are themselves the direct object of the information in the cell, serving as building blocks in cellular organelles.

Inside of our cells, we have lots of families of related molecules, things that are chemically similar but which each carry out different specialized functions. Like DNA, RNA is a nucleic acid consisting of bases held together by a backbone made of a chain of alternative sugars and phosphate groups. Three of the bases, A, C, and G, are the same bases used by DNA. One of the bases, U, is similar to the T used in DNA, and appears in an RNA sequence at the same position that would have used a T in a DNA sequence (Figure 3.1). DNA and RNA backbones are also very similar (Figure 3.2).

As with DNA, RNA consists of RNA bases hooked onto one of the backbone pieces, with the bases connected only through the backbone and are not directly connected. We most often talk about water, rather than calling it H_2O, when talking about the content of lakes and oceans. Similarly, we can usually just consider the RNA bases as simple building blocks (Figure 3.3), referred to by single letters or color coded in pictures, instead of showing the detailed molecular structures that we see in Figures 3.1 and 3.2.

In many cases, when talking about RNA we do not even show the building blocks. We simply show lines representing the long string of nucleotides, with landmarks indicating regions

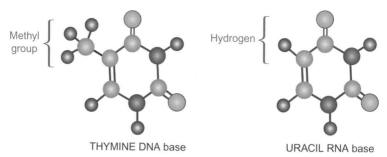

Methyl group {   Hydrogen {

THYMINE DNA base   URACIL RNA base

FIGURE 3.1   One of the important differences between DNA and RNA is that RNA has uracil (U) and DNA has thymine (T). The difference between these bases (shown in blue) is that thymine has a methyl group (a carbon (C) and three hydrogens (H)) not found in uracil.

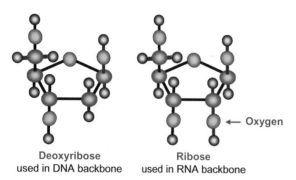

Deoxyribose
used in DNA backbone

Ribose
used in RNA backbone

← Oxygen

FIGURE 3.2   The DNA and RNA backbones. An important difference between DNA and RNA is that the backbone that fastens the bases together (like a railroad rail or the railing between posts on a staircase) is different between these two different kinds of nucleic acids. The difference seems small to the human eye – an oxygen present on the ribose sugar of the RNA backbone that is not present on the deoxy-ribose sugar of the DNA backbone. In fact, the names of the two different classes of nucleic acid stem from this difference in whether that oxygen is present; RNA stands for ribonucleic acid, a name based on the presence of the sugar ribose in the RNA backbone, and DNA stands for deoxy-ribonucleic acid, a name based on the presence of deoxy-ribose (ribose without an oxygen) in the DNA backbone.

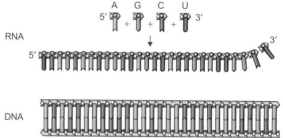

FIGURE 3.3   RNA and DNA. For most purposes, we can represent the critical DNA and RNA information with several small building blocks. Here we show the critical differences between DNA and RNA: (1) the usually double-stranded status of DNA vs. the usually single-stranded status of RNA; (2) U and T shown in different colors because they differ by the presence of a methyl group on T that is not on U; and (3) the presence of an extra oxygen on the RNA backbone pieces (green bead). (Courtesy of Edward H. Trager.)

of sequence interest. In Figure 3.4 we are looking at regions of an RNA that are conserved, that is to say they show a very similar sequence when compared between different species. The kinds of landmarks we will display on an RNA molecule will depend on the features of

the RNA we are discussing, but we can often talk about the key features of an RNA sequence without having to show the string of As, Cs, Gs, and Us of the sequence.

If it does not matter that you remember exact structures of DNA and RNA building blocks, why do we show them at all? We want you to see the extent to which the cellular machinery considers tiny changes to be drastic and profoundly important. If we compare the chemical structures of DNA and RNA we discover that

FIGURE 3.4 Simple diagram of an RNA molecule highlighting where the conserved regions of sequence are located on a transcript produced by a gene that shows conservation of sequence between different species. Yellow regions are so highly conserved that these sequences are very similar between humans, animals, plants and even yeast. Purple regions are conserved among animals. Green regions are conserved among mammals, and dark blue regions are conserved only among primates.

they are identical except for the presence of the methyl group on the T (Figure 3.1) and the oxygen on the RNA backbone (Figure 3.2). These look like small changes to the human eye, but to the cell they are critical differences. These differences determine which proteins interact with these molecules, change where these molecules are found in the cell, change what use the cell makes of them, determine whether they tend to spend time in double- or single-stranded form, and even change how permanent or transient the lifespan of these molecules might be. As we will see throughout the rest of this book, similarly small changes in chemical structures lie at the heart of human genetics and form the basis for human genetic variation and much of human health and disease.

Single-Stranded and Mixed-Stranded RNA

When we think of RNA we usually think of it occurring in single-stranded form, in contrast with DNA which we commonly think of as occurring in double-stranded helical form. However, the real story is not this simple. RNA can take on complex structures involving some single-stranded regions and some double-stranded regions. One of the classic examples of this kind of "mixed-strandedness" structure is a kind of RNA called tRNA that we will talk about later in this chapter. A tRNA has some single-stranded regions and some double-stranded regions (Figure 3.5). This comes about when one region of a tRNA molecule pairs with a different region of the same tRNA molecule, using the now-familiar base-pairing rules.

FIGURE 3.5 tRNA diagram. On the left we see a three-dimensional molecular model of a tRNA, showing both single-stranded and double-stranded regions. (Image Copyright © 2008 Gustavo Caetano-Anollés, University of Illinois Institute for Genomic Biology.)

3.2 WHAT IS RNA FOR?

Many RNAs seem like a kind of molecular "photocopy," carrying disposable information out of the nucleus to be used before tossing the photocopy into the recycle bin, while the original hard-bound book stayed safely in the library of the nucleus, available to be used again and again. The type of RNA that carries information from the nucleus to the ribosomes, called messenger RNAs or mRNAs, contains coding sequence that the cell can decode to direct the synthesis of proteins. Only part of the mRNA contains the protein sequence. There are also regions at the 5' end and at the 3' end of the transcript that do not contain coding sequence, but rather play roles in determining

TABLE 3.1   Examples of different types of RNA and the role they play in the cell

Type of RNA	Also called	Function
Messenger RNA	mRNA	Tells ribosomes the order in which to add amino acids to the growing protein chain
Ribosomal RNA	rRNA	Part of the ribosome that "reads" mRNAs
Transfer RNA	tRNA	Brings amino acids to the ribosome and helps the ribosome insert the amino acid into the growing protein chain
Heterogeneous nuclear RNA	hnRNA	Unprocessed or partially processed transcripts not yet ready for use
Small nuclear RNA	snRNA	Part of the machinery that processes RNA molecules
Small nucleolar RNA	snoRNA	60–300 bases. Help with chemical modification of rRNAs, tRNAs, and snRNAs
Micro RNA	miRNA	~21–22 bases. Single-stranded RNA binds to the 3' untranslated region of a specific mRNA to block translation
Small interfering RNA	siRNA	20–25 bases. Double-stranded RNA becomes part of the silencing complex that degrades targeted RNAs that are homologous to the siRNA

processes such as the localization of the mRNA in the cell and its stability.

For a long time we thought of other types of RNA as mostly playing support roles for this messenger function. But the field of molecular genetics has become increasingly aware that there are a wide array of different kinds of RNAs, not just mRNAs and helper RNAs playing supporting roles for the use of mRNAs. It turns out that many different kinds of RNA qualify as non-coding RNA, that is to say, as RNA that does not contain a sequence to be decoded by the ribosomes or direct the synthesis of proteins (Table 3.1). What are some of these non-coding RNA functions?

In some cases RNA is a structural molecule that carries out functions that the cell needs. Some of the most important examples of structural RNAs are the ribosomal RNAs. These RNAs combine with ribosomal proteins to make up the ribosomes. When the cell is metabolically active it needs a lot of ribosomes to be able to make the many kinds of proteins being used by the cell. Thus an actively dividing cell transcribes ribosomal RNA very actively

(Figure 3.6), at such a high rate that it causes the region called the nucleolus to be visibly distinct within the nucleus.

In other cases, RNA serves as an enzyme, a type of molecule that catalyzes a chemical reaction. An example of an important class of enzymatic RNA would be those RNAs that play a role in splicing of other RNA molecules.

Some RNAs are part of a composite molecule that combines some RNA with some other types of molecules such as proteins or a given amino acid. A good example would be the transfer RNA, or tRNA (Figure 3.5), that forms a composite molecule consisting of an RNA molecule with an amino acid fastened to it. As we will discuss in Chapter 4, the tRNA molecules play a critical role in reading and using the information contained in the mRNAs.

Of increasing importance to our understanding of cell biology are the many different kinds of non-coding RNAs (the ones that don't get used to make a protein). Several kinds of small RNAs can interfere with target RNAs to regulate the final level of RNA or protein product. For some of the recently discovered non-coding RNAs

FIGURE 3.6 Transcription of RNA is carried out very actively in a cell that is getting ready to divide. The genes are organized in a cluster, and a new transcript from a gene begins before the previous transcript has been completed. The result is a cluster of RNA molecules that appear to stream from the chromosome like branches on an evergreen tree. The chromosome is the vertical line from which the branches radiate. (Image by Professor Oscar Miller/Science Photo Library. © All rights reserved.)

functions have been identified but for many others the function has not yet been worked out. The result is that even as we feel like most human genes have been identified, we are finding ourselves faced with new questions about how many kinds of genes there are.

The information in the DNA of the chromosomes does not itself directly cause the phenotypes we observe. Rather, the phenotypes result from functions that do or do not take place and from structures that are or are not correctly made. Structures and functions of the cell involve many different types of molecules. Although some of these molecules come from sources outside of the cell, such as the

diet, most of the molecules are produced as a result of the use of the genetic information on the chromosome. As we have discussed in the previous sections, the molecules that result most directly from use of genetic information are the RNA molecules that are direct copies of the information contained on the chromosomes. But one of the most important types of molecules are proteins, the molecules that directly synthesize many of the other classes of molecules in the cell such as lipids and carbohydrates. Proteins are the molecules that result directly from the cell "reading" the information encoded in the messenger RNA molecules.

Now that we have introduced you to RNA, we come to the next question: how does the cell make RNA?

3.3 TRANSCRIPTION OF RNA

The Process of Transcription

Transcription (making an RNA copy of DNA) is like *replication* (making a DNA copy of DNA) in several ways. Transcription uses base-pairing rules to put a new RNA "letter" into position along a growing RNA chain by picking out the RNA "letter" that can pair with the DNA "letter" present on the DNA chain. In DNA, A pairs with T, T pairs with A, G pairs with C, and C pairs with G. The pairs are the same between DNA and RNA, except that the DNA A pairs with the RNA U instead of T (Figure 3.7). So the DNA sequence ATGCTTCGA will end up as AUGCUUCGA in the RNA molecular that is made by "reading" off of the template strand TACGAAGCT shown in Figure 3.7. The cellular machinery uses these rules to let it "read" the DNA into RNA. RNA molecules are synthesized from DNA molecules by a process called *transcription*.

During transcription, only one strand of the two DNA strands that make up a given gene (the one known as the template strand)

DNA strand ATGCTTCGA   AUGCUUCGA   RNA strand
DNA strand TACGAAGCT   TACGAAGCT   DNA strand

FIGURE 3.7 Pairing of RNA bases with DNA bases works like pairing of DNA with DNA, except that RNA uses the base U in place of the base T, and the backbones that connect the bases to each other are different. In each case, if you know the order of bases along the strand of DNA that is being copied or read from, you know what the sequence of the new strand will be.

is used to direct the synthesis of a strand of RNA (Figure 3.8). If RNA had been made from the opposite strand we would call it anti-sense RNA, because it would be information identical to the sense strand (of course, with the use of U instead of G), but not usable by the cell for making the protein product of that gene or carrying out whatever other purpose the RNA from that gene normally performs. As we will learn below, there are times when the cell (or researchers) make use of the anti-sense information from a gene for regulatory purposes, but the primary role of a gene is carried out through its sense strand.

Copying occurs as the double-stranded DNA is opened up by proteins that move along unwinding and opening the double helix, much like what we saw happen for DNA replication. An enzyme called RNA polymerase moves along the single-stranded DNA inserting new RNA bases according to the RNA base-pairing rules. As we saw for DNA replication, the polymerase is only part of the story, since other proteins are needed to handle the DNA. Transcription is taking place amidst the complex, wound, coiled structure of the chromatin, and the DNA would normally not be accessible to the polymerase. The transcription initiation complex of proteins (Figure 3.9) contains a complex mix of proteins, including gyrase, which opens the DNA ahead of the polymerase.

For each gene that is transcribed, RNA polymerase will produce a single-stranded RNA molecule that is complementary to the bases that compose the template strand of that

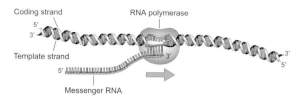

FIGURE 3.8 Transcription. When making RNA, the cell uses the same kind of base-pairing system used in DNA replication. Transcription uses RNA polymerase to insert RNA bases into the growing RNA strand according to the RNA base-pairing rules. The RNA comes apart from its DNA template, the DNA returns to double-stranded form and the single-stranded RNA moves away from the chromosome to be processed and used by the cell. Some RNA will travel to the cytoplasm where it will direct the synthesis of protein, but other RNAs will stay in the cell where they will. (Courtesy Edward H. Trager.)

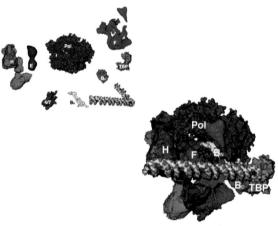

FIGURE 3.9 Proteins involved in transcription. Proteins shown on the upper left, including RNA polymerase II (Pol), come together with the DNA helix (red, white, and blue) to form the protein-DNA pre-initiation complex (shown on the lower right) that is waiting to begin transcription. (From Roger D. Kornberg, "The molecular basis of eukaryotic transcription", *Proceedings of the National Academy of Sciences of the USA*, 2007;104(32);12955–61. © 2007 National Academy of Sciences, U.S.A.)

gene, so RNA molecules for different genes have different sequences, that is they each "look" like the genes they are produced from but "look" different from each other.

Like DNA replication, the process of transcription is directional. The new RNA chain only grows in one direction from the start site. Once the first RNA base has been put in place by pairing with the DNA base, only one end is available to have another base added to it. Unlike what happens with DNA replication, only one of the strands gets copied into RNA during transcription. The RNA that results is single-stranded, containing the sequence present in only one of the two DNA strands.

Genes and the RNA molecules produced by them are relatively short compared to the great length of the DNA molecules that encoded them. Transcription is very rapid. On average it proceeds at almost four thousand bases per minute, which is to say at a rate that vastly exceeds the ability of any of us to write or type the letters corresponding to that sequence! As we will discuss later in this chapter, the amount of transcription at any given gene can be carefully controlled by the cell.

3.4 ORCHESTRATING EXPRESSION

Each cell in the body carries the same genetic blueprint, the same nucleus full of information, yet liver cells and blood cells and brain cells are amazingly different in both form and function. If all of the cells have the same genes, how can they be so different? The answer is that not every gene is always expressed, that is to say, transcripts do not get made from every gene in every cell. Some genes that are expressed in all cells at all times are called *housekeeping* genes. Each type of cell also expresses a distinct *cell-type specific pattern* of genes that gives that cell its special properties and functions. Expression of some *developmental regulatory* genes is specific to certain stages during the growth and development of the person. *Inducible* genes are genes that are not normally expressed but that can be expressed in response to something to which the cell is exposed.

Let's look at some examples. All cells need to produce the enzymes that generate energy to run the cell. However, only red blood cells need to produce hemoglobins, the proteins that carry oxygen from the lungs to the tissues and return carbon dioxide from the tissues to the lungs. Only the cells of the retina in the eye need to produce the light-sensitive proteins, such as rhodopsin, that permit us to see. This is not to say that liver cells lack the genes for hemoglobin or rhodopsin; rather, liver cells simply do not express these genes (i.e., they do not make RNA transcripts from those genes). Instead, the liver cell expresses its own specific repertoire of genes needed to make the proteins that carry out the specific functions of the liver.

Indeed, the development of a human being from conception to death is the result of a complex, preset program of expression of genes that get used in some cell types and not in others or get used only at specific times during development of the fertilized egg into a living, breathing baby. The cell selectively accesses the information it needs while ignoring the information that is meant for some other cell or situation.

The effect is similar to that achieved when an orchestra uses the same set of notes to generate vastly different pieces of music (Figure 3.10). The same set of instruments achieves effects as different as Beethoven's Fifth Symphony or "Classical Gas" depending on which of the approximately eighty available notes get used and in what order. Similarly, orchestration of the use of the complete set of genes (the *genome*) can achieve the profound differences between a muscle cell and a nerve cell (patterns of cell-type-specific gene expression) depending on which genes get used, the order in which the genes are transcribed, how often they are transcribed, how long the mRNA endures once it is produced, simultaneous expression of multiple genes, and coordinated regulation of expression of those genes. When the wrong notes play in an orchestra, the melody is changed or dissonance occurs; when the wrong notes play in a genetic symphony, the results can be as profound and perplexing as having legs appear in place of antennae on the head of a fly. Just as

FIGURE 3.10   An orchestra offers an interesting model for the combination of spatial and temporal differences in gene expression that take place at different stages in life in different types of cells in the body. This picture shows the University of Michigan Life Sciences Orchestra, which is made up of members of the life sciences community from throughout the campus. (Courtesy of the University of Michigan Health System Gifts of Art program. Photo by Lan Chang.)

the orchestra dropping a few bars of Mozart into a Beethoven piece could greatly change the pattern, so would expressing leg genes where antenna genes should be expressed.

The ability of a cell to control *gene expression* depends on sequences within the DNA itself known as promoters and regulatory elements and on a set of regulatory proteins that bind to those DNA sequences to control the expression of each gene. The *promoter* lies at the beginning of the transcribed region for each gene and defines the site at which transcription is begun. It is the binding site for RNA polymerase. *Regulatory elements* may lie upstream, within, or downstream of a gene, and they determine how accessible the promoter is to RNA polymerase (and thus the extent of transcription). Genes are switched on or off by the binding of proteins known as *transcription factors* to these regulatory elements. The combined action of these proteins and DNA sequences allows each cell to express a specific subset of genes at various times in the life of the organism. This exquisite

control of gene expression allows the development of complex life forms such as ourselves by facilitating the development of thousands of different types of cells. To understand this mechanism we need to now consider the biology of each of these players in transcriptional regulation more carefully.

The Concept of a Promoter

As we saw in Figure 3.8, transcription begins by the binding of a transcription initiation complex, which includes a very large enzyme called *RNA polymerase*, to a site on the DNA next to where transcription of RNA will begin. The region from which the RNA polymerase initiates transcription is called the *promoter*. The promoter is usually located close to the beginning of the RNA transcript. Once the polymerase is bound to the promoter, it moves along the DNA. As it moves, it makes a single RNA copy from only one strand of the DNA double helix. It adds new bases to the growing RNA strand by using the rules of base-pairing to insert the bases that complement the bases present on the DNA. When the RNA polymerase reaches the end of the DNA that comprises the gene, it detaches from the DNA and releases the newly made RNA molecule to have a polyA tail added and be spliced if it is a spliced gene.

One of the primary levels of regulation of gene expression happens at the level of transcription by controlling whether or not an RNA copy of the DNA sequence gets made. Regulation can also be imposed by modulating how often an active gene is transcribed (i.e., how many RNA copies of that gene are made in a given interval of time), or how stable the RNA is (how long the RNA copies stick around to be reused). So when we think about regulating transcription, we can think about whether or not the polymerase has access to the DNA that is to be transcribed, and we can think about what affects how frequently or rapidly transcripts get made.

Regulators

We can think in terms of two different things that affect transcription: DNA sequences that can control transcription and protein regulators that can bind to those DNA sequences. Control of gene expression may act like a rheostat so that under some conditions few copies are made and sometimes many copies are made. The points in the DNA sequence that can act as rheostats that can turn expression up or down are called *regulatory elements* (Figure 3.11). A gene normally has a control panel of multiple regulatory elements adjoining the transcription start site and sometimes has additional control elements located elsewhere in the vicinity of the gene. Thus control of whether the gene is transcribed and how much RNA gets made is normally not controlled by a single, simple switch; rather, control of transcription of any given gene is the product of the combined effects of multiple regulatory elements clustered in a regulatory region.

If the amount of transcription that takes place is the product of action of multiple regulatory elements in the DNA sequence, what controls whether they are on, off, up, or down? The answer is that regulatory proteins called *transcription factors* control the action of the regulatory elements by binding to them (Figure 3.12). When a regulatory protein binds to a regulatory element, it effectively changes the setting of the regulatory element's "switch." For some regulatory elements, binding of regulatory proteins can turn the setting up while binding of other regulatory elements turns the setting down. Similarly, if a regulatory protein is normally bound to the regulatory element but then stops binding the regulatory element and comes off the DNA, that can change what is happening to transcription, too.

As with locks and keys, the regulatory proteins are very specific to the switches. Thus there is not one type of protein in the cell that regulates all of the "on" switches for all of the genes. In fact, a particular DNA sequence that acts as a regulatory element may actually be

Transcription
start site

FIGURE 3.11 Level of expression of a gene may be affected by regulatory proteins called transcription factors that bind to regulatory elements in the DNA sequence to affect binding of the polymerase to the DNA and the subsequent transcription event. These act like rheostats whose combined activities affect how much transcription can take off from the nearby transcription start site.

found in the vicinity of the promoter region of many different genes. When the cell makes copies of the regulatory protein that binds to that regulatory element, that regulatory protein will bind to and change the state of not one but many genes. Thus some regulatory proteins that are present in the eye can turn on (or off, up, or down) multiple different genes expressed in the eye if they all have the same regulatory element to which that regulatory protein binds. If that regulatory protein is not also found in red blood cells, those same eye genes might not be expressed in the red blood cells. However, some other regulatory proteins found in the eye are also found in other cell types.

So patterns of tissue-specific gene expression are the result of complex combinations of regulatory proteins acting to turn the switches next to the genes into their correct positions of up or down. If a gene is being expressed in a cell that only makes regulatory proteins that bind to the positive switches for that gene, the gene will be transcribed. That same gene will not be transcribed in a cell that is making only the regulatory proteins that bind to the negative switches for that gene but none of the regulators of the positive switches.

Cis and Trans

An important pair of concepts – cis and trans – were already known from studies of genetics,

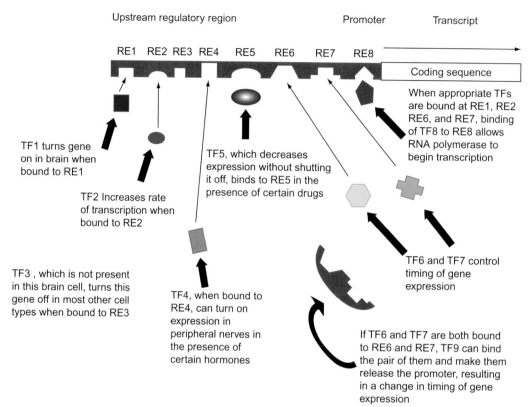

FIGURE 3.12   A simplistic theoretical model of how regulation of transcription of a brain-specific transcript would work. Regulatory elements (specific bits of sequence in the regulatory region before the beginning of the transcribed sequence) and transcription factors (proteins that direct the level and timing of transcription of the gene) can act in concert by binding to the DNA and/or each other to turn gene expression on, off, up, or down in a way that may differ in different cell types or at different points in the growth and development of the individual. A schematic diagram of the upstream regulatory region for a brain-specific transcript is provided. TF, Transcription factor; RE, regulatory element to which TFs bind.

but they have been substantially clarified by our understanding of promoters and regulatory elements.

The word cis is derived from a Latin word that means "on the same side." When thinking about gene expression, a model for something that acts in cis is something that can act only on something that is on the same chromosome it is on, where the things it is going to act on has to be physically bound to the thing that acts. An example of something that acts in cis is a promoter, which directs transcription of the copy of the gene that is on the same copy of the chromosome with the promoter but cannot direct transcription of the copy of the gene that is on the other copy of the chromosome. Most commonly things acting in cis will be DNA sequences that are present on the same copy of the chromosome as the gene or sequence being interacted with.

The word trans is derived from a Latin word that means "on the opposite side." When thinking about gene expression, a model for something that acts in trans is something that can act on both copies of a chromosome, without the requirement that it act on something that is contiguous with it. An example of something that acts in trans is a transcription factor, which can act on the promoters for both copies of the gene. In fact, many of the best examples of things acting in trans will turn out to be the proteins that interact with the DNA.

There are other uses for the terms cis and trans in biology and chemistry. The various usages keep turning out to be consistent with these concepts of being on the same side or being on the opposite side.

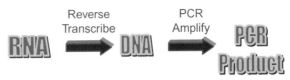

FIGURE 3.13 RT-PCR. The reverse transcriptase enzyme makes a DNA copy of the RNA. Use of a pair of primers homologous to different points on the RNA (and DNA) sequence lets us carry out PCR to make huge numbers of copies of a piece of the RNA. Notice that reverse transcription seems to violate the general rule that DNA goes to RNA goes to protein.

3.5 MONITORING GENE EXPRESSION

RT-PCR: One Gene at a Time

One of the next major breakthroughs in RNA detection was a form of PCR that lets us look at RNA (Figure 3.13). One of the keys to this technology is an enzyme reverse transcriptase (RT) that can copy RNA back to DNA; we then use that DNA in a PCR reaction called RT-PCR. This technology is quick and cheap. It can be used to study even very small amounts of RNA, but it usually only looks at one gene at a time.

Northern Blots: One Gene at a Time

One of the tools that helps us study RNA is size fractionation of the RNA, separation of different RNA molecules from each other based on how long each RNA is. When we have a lot of RNA available we can use a gelatin-like material (gel) through which smaller RNAs move faster than large RNAs. The many different sizes of RNA in the cell spread out into a smear on the gel (Figure 3.14). All of the RNA of a particular size will run at the same position on the gel, so if there are a lot of RNAs that are the same size, they will run at the same position on the gel and form a distinct line against the background smear of RNA. The cell makes many copies of the two ribosomal RNAs that are

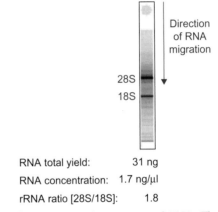

RNA total yield: 31 ng
RNA concentration: 1.7 ng/µl
rRNA ratio [28S/18S]:   1.8

FIGURE 3.14 Size fractionation of RNA. The smear of gray represents the whole RNA content of the cell, and the two darker bands represent the large (called 28S) and small (called 18S) ribosomal RNAs. This technology lets us measure RNA isolated from a tiny sample and still have some left over to do an experiment. This picture shows size-fractionation carried out using a technology that lets us look at amounts of RNA as small as a few nanograms.

major components of the two ribosomal subunits. As we can see in Figure 3.15, the cell makes such an excess of the two main ribosomal RNAs (the 28S and 18S RNAs) that they stand out as individual bands against the background smear of all the RNAs.

Another tool that helps us study RNA is hybridization, a process by which a probe (dye-labeled DNA) can bind to RNA by the base-pairing rules. When we want to study a specific

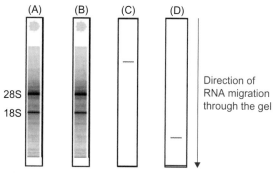

FIGURE 3.15   Northern blot. (A) RNA that has been size-fractionated. After transferring RNA onto filter paper, the filter paper looks blank but we can stain it with a dye that non-specifically stains RNA (B) so that we can see an image that looks just like what was on the gel. If we hybridize using a probe labeled with fluorescent dyes the probe will only stick where homologous RNA is present on the filter, so we will see a band (C or D) that corresponds with the gene the probe detects. The red lines in C and D mark the positions where dye-labeled probe is bound indirectly to the filter by being paired with RNA that is bound to the filter. The probe in C finds a band corresponding with RNA that did not run as far as the 28S band so probe C detects an RNA bigger than the 28S ribosomal RNA. The probe in D finds a band corresponding with RNA that ran farther than the 18S RNA so probe D detects an RNA smaller than the 18S ribosomal RNA. This process of size fractionating RNA, blotting it onto filter paper, and hybridizing to a labeled probe is called a Northern blot.

gene that is lost in the smear of RNA on the gel, we do something called a Northern blot. We transfer the size-fractionated RNA out of the gel onto a piece of filter paper, effectively making a copy of what is in the gel (Figure 3.15B). We hybridize using a dye-labeled probe homologous to the gene of interest, and the probe binds to the filter wherever there is RNA homologous to the probe (Figure 3.15C and D). The result is a band at one spot on the filter that corresponds to the location of the RNA of interest.

Sequencing the Complete Transcriptome

The techniques we just talked about let us look at one gene at a time, or sometimes at a very small number of genes in one experiment. However, as we are approaching knowing all of the genes in the human genome, we are reaching a point where we want to do whole genome experiments that let us look at large numbers of genes (preferably all of the genes) in one experiment. The process of surveying expression levels of large numbers of genes (or a whole genome worth of genes) is called expression profiling. The term transcriptome is used to refer to the complete array of RNAs expressed by the complete complement of genes from throughout the whole genome.

Currently the most direct way to carry out expression profiling of the whole transcriptome is to simply sequence each of the separate RNAs present in a sample – to read the order of As, Cs, Gs, and Ts along part of the RNA molecule so that we identify the RNA from its sequence. This gives the most precise tally available of which RNAs are the most abundant. It is an expensive technique but as the price on next generation sequencing is rapidly falling, sequencing is starting to push out some of the other recent technologies.

Microarrays

One of the most common approaches to expression profiling during the first decade of the twenty-first century has been microarray technology. It lets us look at the level of expression of tens of thousands of genes in one experiment (Figure 3.16). Each microscopic spot on a slide or chip contains DNA representing one of the genes. In this experiment, RNA labeled with a fluorescent dye sticks to DNA on the chip through the base-pairing rules. Bright spots indicate genes that are expressed at a high level. Pale spots indicate genes that are weakly expressed. Blank spots indicate genes that are not expressed. In a sample screen by microarray technology, an RNA that is present in ten copies in our sample will give a much weaker signal than an RNA that is present in 1000 copies

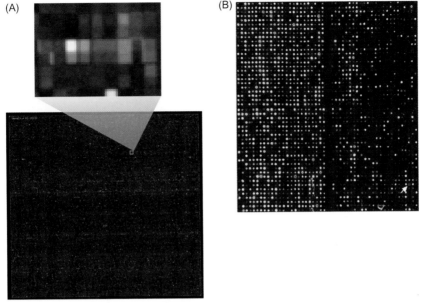

FIGURE 3.16   Chips containing tens of thousands of genes. Each spot contains DNA corresponding to one gene. (A) When testing a single RNA sample, bright spots correspond with genes that are expressed in the tested RNA sample, and dark spots are genes that are not expressed in this sample. A small region inside the blue square has been expanded to show the different signal levels we see for genes that range from very highly expressed to unexpressed. (B) When comparing two different RNA samples, we can label one RNA sample with green and the other with red. If a gene is expressed in only one of the two samples, the spot will be either green or red, but if it is expressed in both RNA samples the spot will be yellow. As for the experiment in (A), if the spot is blank then the RNA for that gene is not expressed in either sample.

(Figure 3.16A). In a similar experiment, we can compare two different RNA samples by labeling them with different colors and hybridizing both RNA samples to the same chips, so that we can use the colors of the spots to help us identify genes that are expressed in one sample but not the other (Figure 3.16B).

When we do whole-genome sequencing or microarrays, we see how easy it is to evaluate a whole genome at once. So why would we not just do microarrays for all of our RNA experiments? First, it is much more expensive to do a whole-transcriptome experiment. And there are other questions we have about RNAs besides whether or not the RNA is expressed, so sometimes we still need to just do a gel or a Northern blot or other experiments that look at the RNA from an individual gene.

What do we have at the end of an expression profiling experiment? We have a list of genes that are expressed in a particular tissue or under a particular set of conditions. Often it is hard to draw many conclusions from just looking at the list of hundreds of genes that show differences in our experiment.

Use of bioinformatics lets us arrive at generalizations about the groups of genes found in an experiment. We can use a process called gene ontology to arrange the proteins into functional groups. Gene ontology can tell us which genes are member of the same families of genes (genes with related sequences and functions) and which genes are involved in related cellular processes. For instance, in the experiment in Figure 3.16 gene ontology tells us that under the tested conditions we are seeing a decrease in expression of genes that encode proteases and an increase in expression of genes that encode protease inhibitors. This suggests that under the tested conditions we might expect to find an overall decrease in the rate of proteolysis (the process of breaking down proteins). This generalization lets us design experiments to test whether the actual process of proteolysis has been changed in these cells.

There are lots of other tools available to us that let us study RNA and gene regulation. For instance, we can do tests for where the DNA has been methylated. We can determine which transcription factors are bound to a piece of DNA. We can look at the rate at which the RNA from a particular gene is being transcribed or degraded. We can look at where the RNA is located within the cell.

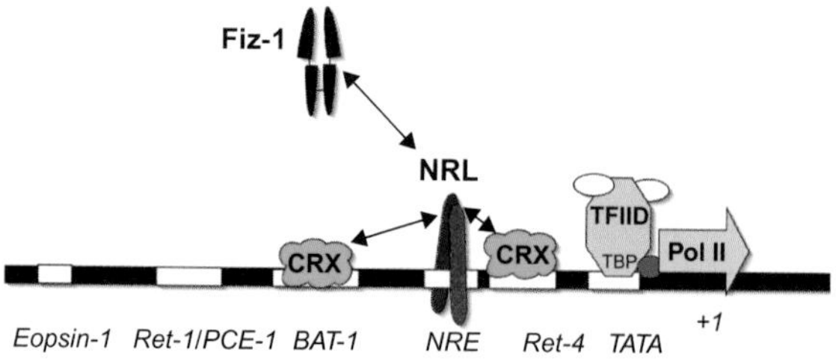

FIGURE 3.17　Known transcription factors binding to known regulatory elements in the well-studied rhodopsin promoter. Binding of transcription factor NRL to this region can activate transcription even if transcription factor CRX is not there. Similarly, binding of CRX can activate transcription even if NRL is not present. However, binding of both NRL and CRX at the same time results in synergistic interaction that substantially increases expression over and above what either of them brings about separately. (Courtesy of Ken Mitton, Oakland University Eye Research Institute, Rochester, MI.)

3.6 INTERACTION OF TRANSCRIPTION FACTORS

An Example of Two Regulatory Factors That Play a Critical Role in the Eye

We each see the world around us through the complex processes taking place in the eye. Light passes through the pupil to carry the image of the outside world to the retina in the back of the eye, a specialized tissue consisting of multiple layers of cells of different types of cells. Some of these cells are photoreceptor cells that have the ability to detect light. Other cells in the retina are nerves that are connected to the photoreceptor cells and have the ability to send the message along to the brain that says, "The photoreceptor cell saw light." Some of these photoreceptor cells are rod cells that detect faint light at night, like starlight. Other photoreceptor cells are cone cells that detect different colors of light that we see during the daytime. Thousands of genes are expressed in photoreceptor cells, and some of these genes are transcription factors that affect the expression of many other genes. Two of these transcription factors, CRX and NRL, are not only *important for expression of the photoreceptor genes, but they are also important for the developmental processes that lead to the production of normal photoreceptor cells. When we find rare human beings who do not make either NRL or CRX we find that they do not have a normal retina and they suffer from retinal degeneration. By studying gene expression in photoreceptor cells that do not make one of these critical transcription factors we can begin to understand the complex processes that keep the retina healthy and functional.*

Regulation of any one gene actually involves multiple transcription factors acting in concert. In many cases, some combination of negative and positive regulation may be going on. In the case of CRX and NRL in the eye, interactions of two proteins enhance expression over what either of those proteins would bring about on their own (Figure 3.17). The action of the products of these genes is important for normal development of the retina in the eye and is also important for regulation of the expression of the gene that makes rhodopsin, the protein in the eye that detects faint light at night.

BOX 3.1

TRANSGENIC ANIMALS

We can carry out genetic engineering to create transgenic animals that have had a gene added or removed. If we insert a gene we call the resulting mouse a "knock-in" animal. If we remove a gene, then the animals are called "knock-out" animals. We can also do "conditional knock-outs" – mice that are born with the gene in place but constructed in such a way that we can delete the gene at some later time. And we can target specific mutations into genes that are already there, which lets us identify a muta-

tion in a human gene and then put that mutation into the comparable gene in the mouse to see what effects the mutation would have. In cases where the alteration is present from conception, we can study the effects of the alteration on development. If we do a conditional deletion of the gene at a later time we can study the effects in the adult animal without having messed up embryonic development; this is especially important if we want to study a gene whose loss is lethal in the embryo.

What are these transcription factors doing? While some ocular transcription factors are directly regulating expression of some obvious gene like rhodopsin or the color opsins in the eye, the story is in fact far more complex than that. We have technologies available that let us reduce the level of RNA from specific genes. We can do this in cells that we grow in flasks in the lab (cultured cells) or we can do it in genetically engineered animals (Box 3.1). This lets us look at how cells, or even whole animals, respond when the expression levels for one gene are altered.

When researchers at Washington University in St Louis wanted to ask what happens when mouse Crx levels are altered, they started out with transgenic mice that make too little or too much Crx (Box 3.2). They then used microarrays to ask how the whole transcriptome had responded to the change in Crx levels. They found that almost three hundred genes showed a different level of expression in mice with altered Crx levels as compared to the wild type (normal) mice.

If they constructed mice with altered levels of Nrl, they found that more than four hundred genes responded with altered levels of expression. If we compare the lists of genes that change expression in response to altered Crx

and the list of genes that changed expression in response to altered Nrl, we find just over one hundred genes that appear on both lists. This experiment tells us that changing the levels of one transcription factor can affect expression of many other genes. Another interesting thing we learn from this experiment is that there are a lot of genes whose expression levels respond to changes in either of these genes, but there are many more genes that respond to only one or the other of these two genes (Figure 3.18). In this experiment, researchers also found that changing the level of expression of one transcription factor can affect the level of expression of other transcription factors.

What these experiments tell us is that Crx and Nrl each regulate the expression of other transcription factors, each of which has its own network of downstream effects on expression of many other genes. If we look at 16 transcription factors that change expression in response to Nrl changes, we find that a half-dozen of them change expression in response to Crx changes.

If we were to go on to do additional experiments of this kind for each of these genes, we could gradually build up a detailed picture of a complex network of the interactions that take

---

BOX 3.2

GENETIC CHANGES AND EPIGENETIC CHANGES

Genetic changes take place through a process called *mutation* that permanently alters the DNA sequence so that subsequent generations receive faithful copies of the altered sequence without changing it back. If an A has been changed to a T, it stays changed in subsequent generations and the cell now treats the genetic letter at that position as a T, with no "memory" that it used to be an A. Epigenetic changes leave the order of As, Cs, Gs, and Ts unaltered while making other temporary and reversible modifications to the DNA that have local effects on the ability of

that sequence to be used by the cell to produce a phenotype. In the imprinting situations we are talking about, the effect is like that of flipping an "on" switch to the "off" position, an "off" switch to an "on" position, or turning a rheostat up or down or changing where in the house we position the light before we turn it on. The lamp controlled by that switch is exactly the same lamp, but it will act differently depending on the position of the switch controlling it, and the position of that switch can be changed.

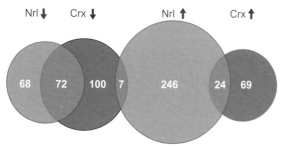

FIGURE 3.18 Genes that respond to changes in levels of expression of Crx or Nrl in mouse. Inside the red circles are the numbers of genes that change only in response to Nrl changes. Inside the blue circles are the numbers of genes that change only in response to changes in Crx. Numbers in the regions of overlap tally those genes that respond to changes in expression of either Nrl or Crx. Numbers in the areas of overlap between circles tally the genes that change in response to changes in both transcription factors. (After T. H-C. Hsiau *et al.*, "The cis-regulatory logic of the mammalian photoreceptor transcriptional", *Network PLoS ONE*, 2007;7:e643.)

If we use a process called RNA interference to reduce expression of these transcription factors in human retinas donated for use in research after death, we see how many similarities there are between these networks in mouse and human. One clue that tells us about how similar the roles of these transcription factors are in mouse and human is the finding that a mutation in the human NRL gene can cause retinal degeneration, which suggests that it plays a similar role in mouse and human.

Enhancers – Another Level of Regulation

Transcription not only needs the regulatory activity taking place in the promoter region, but it also requires the participation of another regulatory region called an enhancer. The proteins of the transcription initiation complex can sit at the starting line, ready to race along the DNA, churning out a growing line of RNA, but they must interact with the enhancer for transcription to take place. The enhancer is a region of sequence that is essential to transcription – without its activity the transcription initiation complex will just sit on the starting line and not go anywhere.

place between these regulatory networks in the adult retina. In fact, one of the things that we see just from studying these two genes is that Nrl is among the genes regulated by Crx!

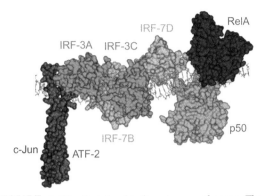

FIGURE 3.19 Proteins binding to an enhancer. These pictures show how a group of proteins come together to form a complex structure that interacts with the DNA of the IFN-β enhancer. The resulting DNA-protein structure is called an enhanceosome. (From Panne, D., Maniatis, T., and Harrison, S. C.,"An atomic model of the interferon-β enhanceosome," *Cell*, 2007;129:1111–23. © 2007, with perimission from Elsevier.)

If a group of proteins bind to the enhancer (Figure 3.19), then the DNA can bend around to bring the enhancer close to the promoter.

Once the enhancer and the promoter are close together do they interact? No. But the proteins bound to the enhancer interact with the proteins bound to the promoter. If we watch a video of this happening, it looks rather like a distant spot on the DNA loops around to momentarily kiss the promoter region before swinging away again. As soon as the proteins bound to the enhancer region kiss the protein complex at the promoter, the RNA polymerase takes off, racing along the DNA at a pace of thousands of bases per minute, while the growing RNA strand trails behind it.

What are the differences between a promoter and an enhancer? Enhancers may be located near a gene or within an intron of the gene it helps regulate or even thousands of bases away from the promoter for that gene, while promoters need to be located right next to the transcription start point. A promoter is directional. If you cut it out, flip it over, and put it back in it won't work. An enhancer is not directional, so if you flip it over it still does its job. While promoter regions may be spread across thousands or even tens of thousands of bases of sequence, enhancers can be quite small.

RNA Turnover

So, transcription sounds like it is controlled at some fairly straightforward levels. The proteins bound to the enhancer interact with the proteins bound to the regulatory elements in the promoter region, and the RNA polymerase is off to the races. But the process of regulating where a gene is used, when it is used, and how much gene product is made is not a simple matter of just the on and off switches in the promoter leading to increases and decreases in the amount of RNA being made.

The issue of how long each RNA lasts is another important issue. If the RNA molecule is a temporary copy of the information, how long does it last? Does each RNA get used once? Or do they all last for a fixed amount of time? In fact there is a lot of variation in how long an RNA stays around in the cell, with some RNA molecules being recycled quite promptly and others hanging around to be used many times. How long an RNA molecule lasts is very tightly regulated by the cell. Some kinds of sequences, such as AU-rich regions towards the 3' end of the transcript, help to determine how long it will last. An example of a kind of RNA that may be rapidly recycled by the cell can be found in some cytokine mRNAs. Cytokines are produced by cells of the immune system and the body needs to be able to produce them rapidly in response to infections and to get rid of them again rapidly when the need for an immune response is past. The rate at which the RNA is degraded is determined by a combination of sequences present on the RNA and proteins that bind to the RNA to protect it or degrade it. This makes the mRNA inherently fragile and susceptible to degradation but lets the cell provide temporary protection for the mRNA while it needs to keep it around.

There are quite a number of other levels at which things can be regulated, including not only the rate at which RNA is produced but also the rate at which it is gotten rid of. Since RNA molecules are the temporary messengers rather than the permanent information repository, the cell has a regular process of RNA breakdown going on to get rid of messages that are no longer needed. Some RNA molecules are used briefly before being discarded, but others are more stable, that is, they stay around in the cell for a longer period of time before the cell breaks them down and gets rid of them. If the developmental stage advances or something changes in the cell's environment so that the cell finds that it has a lot of RNA present from a gene it no longer wants to use, the cell might just wait for the natural decay rate for that RNA to take its course.

Sometimes, however, the cell may need to be able to get rid of that RNA rapidly without waiting for the gradual loss of the RNA. One of the ways that cells cope with the need to reduce the amount of an RNA that has already been made is through RNA interference, using small double-stranded interfering RNAs (siRNAs) that contain a sequence complementary to the sequence in the RNA to be eliminated (see Table 3.1). Pairing of the siRNA with the RNA signals to the cell that this is an RNA to be eliminated without waiting out the normal life span of the RNA.

Similarly, sometimes the cell needs to be able to get rid of a protein that has already been made by using the gene's information by processes we will discuss in Chapter 3B when we talk about proteins. In some cases, a protein to be eliminated may be digested by a *protease* that the cell makes for just that purpose. In other cases, the cell may conserve its resources and keep the protein around in an inactive form so it can use it again later. One of the ways it does this is by sticking a chemical tag onto the protein that is supposed to be active and then removing that chemical tag when it wants the protein to shut down for a while.

Thus the cell is actually orchestrating a very complex array of events that control the production of the mRNAs, the persistence and reuse of those mRNAs, the amount of the protein present in the cell, and the activation status of proteins that are kept around even though they are temporarily not needed.

Of course, there are many other levels of regulation of events taking place in the cell. There are factors in the cell that affect how long a protein gene product will stay around to carry out its activities, and other factors that affect whether or not the protein is currently in an activated state or whether it is present at the location where it would carry out its activity. Whether we are talking about hormone signaling, trafficking of a protein to the right location in a cell, or phosphorylation to activate a protein, all of the complex steps in running activities inside of the cell are tightly regulated and carefully controlled.

Regulation of gene expression is carried out by the coordinated efforts of the regulatory proteins bound to the regulatory elements of the promoter, and the proteins bound to the enhancer. Changes in regulation of gene expression not only determine specific differences between cell types in the body but also between different stages of life. Thus the human genome is carefully orchestrated, with a symphony that begins with a frenzy of fluctuating gene expression throughout fetal development and childhood and then settles into a slightly more stately adagio in which gene expression continues to evolve as we age. This cascade of gene expression changes, which seemingly starts with fertilization of the egg, really begins a step before that, at the point when it is determined what genetic material goes into the egg and into the sperm.

3.7 INDUCIBLE GENES

Many other external factors can affect expression of human genes, including infections, inflammation, allergies, injury, nutrition, medications, temperature, and emotional reactions. In many cases, several of these effects may be going on at once.

Many of the changes in gene expression that take place in response to external factors are designed to help us heal or adapt to our environments. During wound healing, expression of collagen genes produces collagen proteins that become part of the scar that eventually seals off the wound site. However, some changes in gene expression can turn out to be maladaptive, such as those that take place during inflammatory processes that can cause further damage after some kinds of injury.

Hormones

Like an electrical light that is not on because the light switch has not been flipped, some regulatory proteins may be present in a cell but not in an active state. What turns on, or activates, these regulatory proteins is not electricity but rather binding to another molecule, such as a hormone.

Hormones are small proteins that are made by some cell types to be used as signals to send messages to other cell types. The way the recipient cell detects the hormone is by making a protein called a receptor that binds to the hormone.

Let's start with the example of testosterone and the androgen receptor that detects its presence. As we will discuss in Chapter 7, production of testosterone by cells in the testes sends signals to other tissues in the body to develop secondary sexual characteristics appropriate to a male, but this works only in individuals who have a functional receptor for testosterone present in the right tissues.

How does the receptor respond to encountering the hormone, and how does this encounter with the hormone cause changes in the cell? The hormone is small and enters the cell where it binds to the receptors (Figure 3.20). The receptor responds by changing shape, which activates it. The cell then transports the receptor into the nucleus where two copies of the receptor come together to form a transcription factor.

There are many promoters in the cell that have sequences to which this receptor-transcription factor can bind, so we expect that lots of genes will show expression changes. However, as with other transcription factors, usually it takes the action of multiple transcription factors working in concert to bring about an expression change, so we will see expression changes only in those genes that bind the androgen receptor and also have the other needed transcription factors bound. This means that even though many different cells in the body will experience the flood of testosterone, they will not all show the same pattern of gene expression changes in response to the hormone.

There are other kinds of hormone receptors that do not act as transcription factors. They

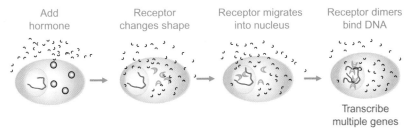

FIGURE 3.20 How one type of hormone works. Hormone binds to the receptor, the receptor changes shape and migrates into the nucleus where it pairs up to form dimers that bind DNA in the promoters of multiple genes and bring about gene expression.

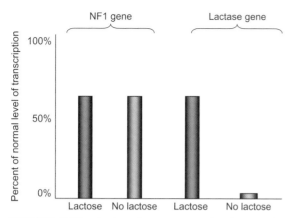

FIGURE 3.21   As for other instances of induction, the induction of the lactase gene is a specific process that does not affect most genes. If we provide lactose to the cells, it will induce expression of the lactase gene but there are many other irrelevant genes that will show no changes in expression.

sit on the surface of the cell. When the receptor binds the hormone, the receptor changes shape (as happened for the androgen receptor) but the receptor does not migrate to the nucleus. Instead, the change in shape lets it carry out a process called signal transduction, whereby the receptor gains the ability to interact with proteins inside of the cell that tell the cell to change what it is doing.

Hormones are not the only things that can induce a change in gene expression. Many of the substances that we are exposed to – diet, medications, and pollution – can cause changes in cells that result in altered gene expression. Gene expression is important to our ability to process and metabolize foods that we consume. Some genes involved in such processes can change expression depending on whether or not that food is part of the diet. Lactose is a sugar that is not always present in the diet, and if someone goes for a long time without consuming lactose they turn down the level of expression of the gene. If they start consuming lactose again, the level of expression of the gene can increase. Some of these effects are very specific,

so we see that in response to lactose the lactase gene (responsible for digestion of lactose) turns up expression but an irrelevant gene like NF1 does not (Figure 3.21).

3.8 EPIGENETIC CONTROL OF GENE EXPRESSION

In 1944 and 1945 the Dutch population struggled through a severe famine. The babies of pregnant mothers who experienced significant dietary deprivation during the early to middle part of their pregnancy were likely to be small and to suffer from a complex set of health complications as they grew and aged, including obesity and insulin resistance. Enough time has gone by that the study of this situation has been extended to the grandchildren of these women, who also turn out to be likely to show a smaller birth weight. Study of the DNA showed that as long as sixty years after birth, the children have an alteration to the DNA of their IGF2, a gene considered relevant to their insulin resistance phenotype. Does this mean that the children ended up with a mutation as the result of their mother's starvation during pregnancy? No, in fact the order of As, Cs, Gs, and Ts in the IGF2 gene remained unchanged. So what was different about IGF2 in these children of famine, and why would this difference cause the reported phenotype in the children and grandchildren?

Sometimes individuals with the same DNA sequence across an entire region (including promoter and coding sequence) can show differences in levels of expression of a gene contained in that region. So far we have talked about effects on gene expression in terms of transcription factors and changes in which transcription factors are available to bind to the DNA of the promoter. But there are other factors that can also affect gene expression, and sometimes these effects can persist through more than one generation. Do we consider them hereditary? Not exactly. After all, they are not permanent and they can change under a variety of conditions. But the story of the Dutch famine children

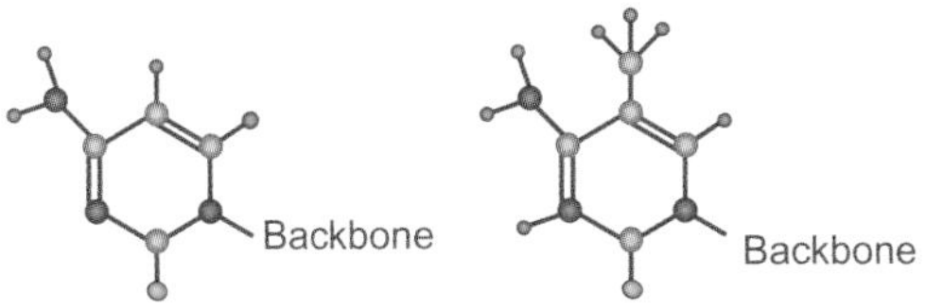

FIGURE 3.22   Two versions of one letter in the genetic alphabet. On the left we show C, and on the right we show 5 methyl C. Notice that the only difference is a group of three atoms (carbon and three hydrogens) at the top of the 5 methyl C in place of a single atom (one hydrogen) at the top of C. For purposes of replicating DNA and transcribing RNA, the cell treats these two as if they were the same base. But in regulatory regions, the cellular machinery reads them as if they were two different letters. Thus some transcription factors will bind if one of these is present at a key position in a promoter, but will not bind if the other one is present at that position instead.

tells us that the effects can sometimes persist for more than one generation.

What is the nature of this change if it does not alter the DNA sequence? It turns out that the cell can modify the DNA base C by adding a methyl group to it, yielding a modified base that we will call 5mC (Figure 3.22). For purposes of DNA replication, the cell reads 5mC as if it were a plain old ordinary C. But for purposes of gene expression, some of the proteins in the cell distinguish between C and 5mC. The result is that presence or absence of this methyl group can affect expression of the methylated gene. Notably, methylation happens to Cs that occur before Gs, and not Cs found before any of the other bases. And even Cs before Gs are not all methylated. In addition, the cell can modify the histone proteins associated with a gene in ways that can affect gene expression. Changes that result from things like methylation of C or modification of the histone proteins are considered epigenetic, a term for changes that affect the phenotype without altering the DNA sequence (Box 3.2).

There is a region within the IGF2 gene called the differentially methylated region (DMR), where we find dinucleotide pairs that have a

C before a G under some situations and have a 5 methyl C before G in other situations. When researchers look at the methylation status of these Cs in the children of the Dutch famine mothers, they find specific Cs that are methylated significantly less often in the "famine children" than in individuals whose mothers were not starved. This results in changes to the expression of the gene, so one model for what might be going on is that changes in methylation of DNA are responsible for the level of expression of the gene. At this point, many studies of many genes support the idea that epigenetic modifications such as methylation of key Cs in a gene's promoter or modifications to histone proteins (remember chromatin?) can affect the level of expression of that gene.

There is another level of epigenetic effects that involve the histones. The phrase "histone code" is sometimes used to describe the pattern of modifications to the histones that form the primary scaffold for the chromatin structure. There are at least five different kinds of chemical modifications that can be made to histones. For instance, the presence of acetyl chemical groups on the tails of histone proteins seems to correlate with the gene being "on," which is to say transcribed. The state of modification of the histones may affect things like the ability of the cellular machinery to loosen and open the chromatin structure, a step that is essential to transcription.

It had long been thought that the pattern of methylation present in adult DNA gets wiped clean in the sperm and the egg, so that a new methylation pattern is laid down in each new embryo that is starting at the beginning of a developmental program that is controlled partly by transcription factors and partly by epigenetic changes such as methylation. But the Dutch family offers one of several lines of evidence that is forcing us to rethink how epigenetic effects work and consider that perhaps at least some of the epigenetic "marks" on chromatin (methylated Cs or modified histones) are retained and passed along to the next generation. The fact that the grandchildren show the "famine child" phenotype of the children suggests that at least some epigenetic marks on DNA can carry over for at least a couple of generations. So for many genes, this process of wiping out the memory of the methylation pattern and laying down a new pattern in the embryo may be happening for many genes, but there are clearly some genes that succeed in carrying forward a "memory" of what happened to the DNA in previous generations. It is not considered a mutation because it is reversible and thus not a permanent change.

If we look at the Avy yellow-coat color allele in the agouti mouse, we see that these mice suffer from obesity, diabetes, and cancer, and have a copy of the Avy gene that has very little methylation. When these yellow mice are fed a substance that causes increases in levels of DNA methylation, the Avy gene becomes methylated, expression of the Avy gene switches off, and the resulting progeny have dark coat colors and none of the typical health problems found in the yellow parents.

Epigenetics gives a lot of additional flexibility to the cell's orchestration of gene expression. It may seem a bit messier than we would like, but as we sometimes say, *evolution is not Michelangelo and the Sistine chapel. It is a teenage kid with a broken car and no money. It just does whatever works!*

3.9 WHAT CONSTITUTES NORMAL?

As usual, the school morning started out with crabby protests of, "I don't feel good, Mom." The cheerful camaraderie that Ari shared with her mother in the early evenings had disappeared the night before, around bedtime, with Ari's first complaints of a stomach ache. Now the downhill spiral of their morning interactions progressed, as always, as Ari objected to the clothes her mother had laid out and glared at her mother's efforts to bring some kind of order to her mass of curls. And then, as usual, it was time to leave and the crisis erupted. "Mom, I don't feel good; can't I stay home?" Her mother frowned and said, "Not again. Do you have a test today? Is there some problem at school?" Ari hunched over the thin arms that crossed her stomach protectively and said in a small, quiet voice, "No, I don't have a test, nothing is wrong at school. I like school. I just have a stomach ache, OK?" Her mother shook her head in aggravation, breaking out the antacids and the analgesics and wondering what she was supposed to do now. Psychologists, doctors, nurses, and teachers over the course of several years had all been mystified at what to do about this seemingly healthy, self-confident, smart, academically successful child for whom bedtime and leaving for school regularly turned into a stomach ache. Weekends and early evenings she seemed happy and well adjusted. Bedtimes and school mornings, she felt ill and asked to stay home. "Separation anxiety," they said, "and distress about going to school." It all seemed like a mystery until her mom had a guest to dinner who declined the ice cream, saying, "I can't eat dairy products. Milk gives me a stomach ache because I have lactose intolerance." Milk. Stomach aches. Lactose intolerance. She thought about Ari's weekday routine, including a great big glass of milk at dinner, hours before the stomach aches began. After several doctor's visits and some tests, a diagnosis of lactose intolerance revolutionized Ari's life by letting her avoid milk or use enzymes in pills to break down the milk sugar called lactose. Ari's stomach aches were banished, and so were the questions about whether she harbored secret anxieties about separating from her mother or going to school. Interestingly, the story of lactose intolerance offers us two different fascinating lessons, one genetic and one societal. So let's find out what is causing Ari's lactose intolerance, and then let's consider what an understanding of lactose intolerance can tell us about the concept of being "normal."

Ari's lactose intolerance results from her inability to make enough of an enzyme called lactase. According to the American

Gastroenterological Association, as many as 50 million Americans may suffer from symptoms of lactose intolerance, including more than three-quarters of those with African, Middle Eastern, or Native American ancestry and more than 90% of those with Asian ancestors. In populations around the world, babies routinely use an enzyme called *lactase* to break down the milk sugar *lactose* in their mother's milk into glucose plus galactose (Figure 3.23). They maintain the ability to make that enzyme and break down lactose until they are weaned. What happens next can be quite different depending on what population this child was born into. If he was born in Thailand, he probably lost the ability to break down lactose by the time he was two years old. If he was born into a Caucasian family in England, he might continue being able to use lactose for the rest of his life. Symptoms of lactose intolerance can include abdominal pain, gas, and diarrhea. Consuming milk can make a lactose-intolerant adult uncomfortable, but resulting dehydration can be a complication with consequences beyond discomfort in very young children. It is interesting that in each of these populations, some people retain the ability to use lactose throughout their lives because they keep making the full amount of the enzyme that digests the lactose. In Scandinavia, such individuals are quite common; in Southeast Asia, they are quite rare.

So why do people lose the ability to make the lactase enzyme that digests lactose? Babies need this enzyme because their milk-based diet contains high amounts of lactose. Consider this: if your diet beyond that point no longer included lactose, why would your body want to continue wasting energy making large amounts of lactase? In fact, even in lactose-intolerant individuals, some residual lactase is still present, just not the levels needed to cope with large amounts of milk. So this ability to regulate lactase production, and to stop making so much of it after weaning, would appear to result in savings of energy and resources for the cells that normally make the

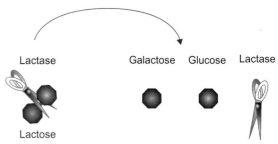

FIGURE 3.23 Metabolizing lactose. The complex sugar lactose is made up of two other simpler sugars that are joined together, galactose and glucose. Notice that, after the enzymatic reaction, the substrate of the reaction (the sugar lactose) has been changed into two separate molecules, one of galactose and one of glucose, but that the enzyme itself is unchanged and ready to carry on the same reaction again if more lactose turns up. Lactose intolerance results when an individual does not make enough of the lactase enzyme, and thus ends up dealing with lactose instead of galactose and glucose.

enzyme. Unless, of course, you live on a dairy farm in Wisconsin or England, in which case the savings in energy and resources would hardly be a reasonable tradeoff for the losses in dietary advantage you would have if you couldn't drink milk.

The Problem with Diagnosing an Inducible Phenotype

In the case of Ari's family, what we see looks a lot like recessive inheritance of a disease gene (Figure 3.24), but it can sometimes be hard to tell if someone has lactose intolerance if they are not exposed to a lot of lactose. Her father's family, with western European ancestry and a standard American diet full of dairy products, appears to be full of lactose-persistent people. However, it is actually harder to tell about people on her mother's side of the family, who consume a more standard Middle Eastern diet. Although they also use dairy products, they eat recipes based on yogurt more often than milk, and drink tea or wine or water more often than

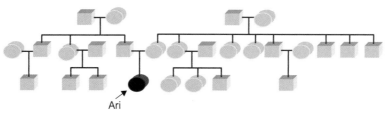

FIGURE 3.24   Recessive inheritance of lactose intolerance in Ari's family. The dark symbol marks the one known case of lactose intolerance on both sides of the family. It is clear that her father's family has many individuals with lactase persistence, but the phenotype is unclear for most of her mother's side of the family because of their relatively low-lactose diet in which milk products tend to be yogurt-based. We do know that Ari's mother and her father can both drink milk without problems, so they both appear to be carriers.

milk. Lactose intolerance can be hard to diagnose even in someone who is exposed to a lot of milk, as happened with Ari; someone who is consuming a nearly lactose-free diet simply because of cultural context may not even know whether milk gives them problems. We know from many studies of other families and individuals that lactase persistence is dominant over lactose intolerance.

It is often difficult in many genetic studies to tell just who is or isn't affected with a particular problem, but especially so with situations such as lactose intolerance that have an environmental component. A genetic disease called favism is detected only in individuals with the defect who eat fava beans. Normally, detection of malignant hyperthermia happens because someone undergoes general anesthetic for surgery. Individuals susceptible to steroid glaucoma will only develop this potentially blinding eye disease if they are exposed to certain corticosteroid medications. So for many of these complex traits with both genetic and environmental elements, many of us have no idea what our phenotype or genotype might be because we have not encountered the conditions that would elicit the trait in someone with the predisposing genotype.

This process of bringing about the expression of a phenotype in response to exposure to something is called *induction*. Although many inherited phenotypes are congenital (present from birth) or developmental (develop at a particular stage in the course of development and aging), there are many phenotypes that are inducible. Tanning is the process of inducing more skin melanin in response to sunlight (or tanning beds). Certain allergies might be said to be induced by exposure to the allergen, such as penicillin or a bee sting. In genetic terms, we usually use the term *induction* to refer to a process that brings about increased expression of a gene.

Induction and Gene Regulation

Although we tend to think of mutations (changes to the sequence of the genetic blueprint) in terms of changes to the sequence of the RNA molecules or proteins that are produced, in the case of *hypolactasia* (a state of reduced lactase activity resulting in lactose intolerance) it looks as if we might be dealing with a different type of mutation, a mutation in the promoter region that affects regulation of expression of the gene.

So what is causing Ari's lactose intolerance? Theories have evolved over the years regarding the cause of lactose intolerance, but it was not until 2002 that a possible cause was identified. A group of researchers in Finland, where about 18% of the population is lactose intolerant,

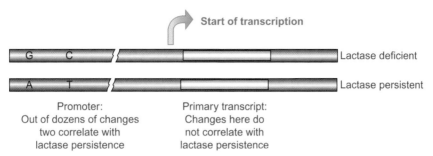

FIGURE 3.25   *Persistent* and *deficient* versions of the gene that makes the lactase enzyme. Persistence of expression of lactase in the Finnish population seems to correlate with two sequence changes in the promoter region that are many thousands of base pairs away from the transcription start site. Different changes in the promoter seem to be involved in other populations. Most of the people in the world have a G and a C at these positions rather than the T and A found in the lactase-persistent individuals.

studied the gene responsible for making the lactase enzyme. The gene that makes the lactase enzyme (called lactase-phlorizin hydrolase) are spread over more than 30,000 base pairs of sequence on the long arm of chromosome 2. If you compare DNA from a lot of different people, you find that there are many differences in the DNA of the transcribed region that contains the coding sequence of the gene, but none of the differences anywhere in the transcript can account for the differences between people who keep being able to drink milk and those who can't. However, when researchers from Finland looked in the promoter region of this gene, at two places that are located thousands of bases before where the gene begins being transcribed, they found two different changes that seem to turn up repeatedly in people in Finland who can drink milk, but not in people with lactose intolerance (Figure 3.25).

So some people in the world have a $G_{22,018}/C_{13,910}$ genotype (G at a position 22,018 base pairs before the transcription start site, and C at a position 13,910 before the transcription start site) and others have an $A_{22,018}/T_{13,910}$ genotype. Those in the Finnish population with the $G_{22,018}/C_{13,910}$ combination lose the ability to make adequate levels of lactase, usually some time between the ages of 10 and 20 years old.

Those with the $A_{22,018}/T_{13,910}$ combination keep making lactase and continue being able to consume milk. If we look in other populations, the sequence that correlates with lactose intolerance is also $G_{22,018}/C_{13,910}$, but the sequence associated with continuing to make lactose may be different in other populations. So we arrive at a model that says that the $A_{22,018}/T_{13,910}$ sequence is in fact causing the change in gene expression and sugar utilization (see Box 3.3), but it is really just a model, and more work will be needed to prove that regulation of the gene works just this way.

So What or Who Is Normal Anyway?

So what is the "normal" genotype, or for that matter, what is the "normal" phenotype? There is a tendency to think that the genotype that makes you sick must be the mutant genotype, and that the phenotype that involves illness must be the mutant phenotype. Certainly, the way lactose intolerance and lactase persistence are talked about further contributes to this idea. People who get sick when they drink milk are said to have lactose intolerance; it is something they go to the doctor about, perhaps even something they use "medication" for (the enzyme tablets to digest the lactose). You will notice that, earlier,

---

BOX 3.3

PROVING IT'S A CAUSATIVE MUTATION

In fact, the evidence offered so far strongly suggests that $A_{22,018}/T_{13,910}$ is causing some people in Finland to retain high levels of expression of the lactase gene. Does this kind of association (the genotype is there when the phenotype is there and missing when it is missing) actually prove that this is the cause? No. It is considered to be highly significant evidence, but you can have things co-occur without one causing the other. For instance, what if there is another mutation even farther away from the transcription start site that is the actual cause and they just haven't found it yet? If $A_{22,018}$ and $T_{13,910}$ are each part of the genetic fingerprint of the chromosome on which the causative mutations took place, we would expect that "affected" individuals who are still making lactase as adults would have the mutant alleles at all three positions in

the sequence if they are not too far apart, even though only one of those changes is causing the phenotype. Why? The chance of a recombination event falling between two things is proportional to how big the distance is between those two things is. In genetic terms, a few tens of thousands of bases is really a rather small chromosomal region that can sometimes be transmitted over a surprisingly large number of generations without recombining. So what can we do to tell whether $A_{22,018}$ or $T_{13,910}$ or the combination of the two is the cause of lactose persistence in Finland? The research to prove this model is actually likely to go in several different directions. Studies of additional populations, as well as studies of families, may help to extend the generalizations arising from the studies that first found these sequence changes in Finland.

---

we asked what is causing Ari's lactose intolerance, and it seemed like a perfectly normal question. We would bet that you did not find yourself saying, "No, that's the wrong question. The question is: Why are there people who don't get sick when they drink milk?"

In fact, the right question really is, "What makes some people persist in making high levels of lactase long after being weaned?" Why is that the right question? Because the "normal" state appears to be lactose intolerance. Studies of the genetic fingerprints of the region of chromosome 2 surrounding the lactase gene show that the lactose-intolerant genotype $G_{22,018}/C_{13,910}$ is a much older genotype that has been around long before the lactase-persistent $A_{22,018}/T_{13,910}$ arose.

More generally, if we look outside of the Finnish population, it has long been suspected

that humankind started out lactose intolerant and over time natural selection favored an increased representation of lactase persistence in some populations, beginning about 10,000 years ago at the time of the introduction of dairy farming. The argument is that children in a dairy farming culture would experience improved nutrition, not only in terms of calories, protein, etc., but also in terms of vitamin D in northern climes, where scarcity of vitamin D in the diet can be a potential problem. This mechanistic model for how certain populations came to have a much larger representation of lactase-persistent individuals is not yet proven, but it makes sense and fits the information available.

So what we conclude at this point is that the original genotype in most of humanity was apparently the $G_{22,018}/C_{13,910}$ genotype, and that the normal nutritional state of the human race was use of milk in babies and toddlers followed by loss of the ability to use milk after weaning. Most people descended from those lactose-intolerant $G_{22,018}/C_{13,910}$ ancestors throughout much of Africa and Asia still have that genotype. The mutants, then, are the rare Africans and Asians, the more frequent Middle Easterners, and the very common Western European Caucasians who are able to continue metabolizing the lactose in milk throughout life.

You might think our point would then be that the lactose-intolerant $G_{22,018}/C_{13,910}$ genotypes must be the normal ones and the lactase-persistent $A_{22,018}/T_{13,910}$ genotype is abnormal.

This leaves us with different ways that we can define normalcy. We might consider it the healthiest or most advantageous phenotype. We might consider it to be the most common phenotype. Or we might consider it to be the original genotype/phenotype before a more recent mutation changed it. As you can see from our discussion here, for some traits the same phenotype (or genotype) might be considered to be the normal variant by one definition and yet not to be the normal variant by another definition.

But normal vs. abnormal is not actually what either the genotypes or phenotypes are about. The real point is that there is tremendous diversity among the different populations of the earth, and whether something is normal is really just a matter of whether it is more common, and not whether it is maladaptive. Whether a particular genotype is helpful or harmful depends on a lot of factors.

If you were to ask Ari if she is ill, she might answer "yes" because she is surrounded by foods she must avoid, and even when trying to avoid things she knows about, she still periodically ends up with a stomach ache if she eats crackers that she did not realize contained whey (a milk product in which lactose has been concentrated). However, if you went to Asia and tried to identify individuals who are lactose intolerant, might you have a hard time even telling who is and who is not lactose intolerant? If you are surrounded by a culture in which meals are focused on rice and fish and vegetables instead of dairy products, would you consider yourself to have an illness just because you don't happen to still be making an enzyme you have no use for?

Under most circumstances, if you were lactase deficient in a lactose-free environment, you would never even find out which genotype or phenotype you have. Frankly, in that environment, your body might even have a very slight advantage of avoiding wasting resources making a protein that won't be used.

So the fact that we know which individuals are the mutants, that is, which individuals have the version of the sequence that is more recently arisen, does not actually mean that we know anything about what constitutes normal, about which individuals might be ill, or even whether the mutation has beneficial or negative impact on the lives of the people of any given genotype. Some mutations cause problems. Some mutations have so little effect that we can't even detect a phenotype that results from the mutation. And some mutations, like the ones we have just been talking about, can actually be beneficial. Knowing who is a mutant does not tell you who is normal or not normal, and it does not tell you who is ill or not ill.

As we learn more and more about the correlations between genotype and phenotype, we see more and more that the word *normal* may not even be a useful term. If normal is simply whatever is usual, and relative advantages or disadvantages vary with the environment, we are eventually going to have to learn to remove the judgmental tone that goes with the word. Being normal might offer you some assistance under some circumstances and not under others, but there is nothing that says that having the most common genotype is inherently either better or worse.

Study Questions

1. What is a promoter and what is it used for?
2. What is an enhancer?
3. What is an inducible gene?
4. What is a housekeeping gene?
5. What is a hormone?
6. What role does RNA polymerase play?
7. What is a "transcription factor," what is a "regulatory element", and what is controlled by their interaction?
8. What is ectopic expression?
9. What are two similarities between DNA and RNA?
10. What are two differences between DNA and RNA?
11. What are three different aspects of gene regulation that can be controlled by transcription factors?
12. In the top box place the sequence of a strand of RNA that RNA polymerase would make from this strand by using the base-pairing rules and reading the bottom strand.

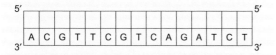

13. What is the difference between cis and trans?
14. What is the difference between RNA polymerase and a transcription initiation complex?
15. Why might we consider lactase persistence to be the normal phenotype and why might we consider lactose intolerance to be the normal phenotype?
16. What effect can result from having a methyl group added to a cytosine, and why is this not considered a mutation?
17. Why might it be difficult to diagnose an inducible trait?
18. When a hormone binds to a receptor in the cytoplasm, a change in gene expression can result. How does this happen?

19. The rate at which a gene is being transcribed is not the only factor that can affect the amount of an RNA that is present in the cell. What else affects the amount of RNA present?
20. What is the difference between a "knock-in" animal and a "knock-out" animal?

Short Essays

1. The products of human genes are normal proteins found in the human body, so why do international sports organizations worry about the use of gene products as drugs? And how might they be able to tell if such drugs are being used? As you consider this question please read "Gene doping and sport" by Theodore Friedmann, Olivier Rabin, and Mark S. Frankel in *Science*, 2010;327:647–8.
2. Francis Crick's Central Dogma says that the information flow goes from DNA to RNA to protein. Is there a sense in which we can think of the information flow going from proteins backwards? As you consider this question please read the section on The Central Dogma Revisited in "The early years of molecular biology: Personal recollections" by Robin Holliday in *Notes and Records of the Royal Society of London*, 2003;1957:195–208.
3. In a world where genetic testing is increasingly available, can we simply read a sequence and tell what we are going to turn out like? As you consider this question please read "Why your DNA isn't your destiny" by John Cloud in *TIME Magazine*, January 6, 2010.
4. Lamarck proposed the inheritance of characteristics acquired from the environment. If epigenetic changes can pass from one generation to the next but do not persist across many generations, does this or does this not qualify as inheritance? As you consider this, read about the agouti mouse in the article "Should evolutionary theory evolve?" by Bob Grant in *The Scientist*, January 2010, pp. 25–30.

Resource Project

Go to the GeneCards site at www.genecards. org/and look up the ZEB1 gene. Write a one-paragraph report on the tissues in which it is expressed and the type of information GeneCards provides on the subject. If you use only the information on this site is it obvious why this gene would cause disease in the cornea of the eye? Why or why not?

Suggested Reading

DVD
NOVA: *Ghost in Your Genes* (WGBH Boston, 2008).

Articles and Chapters
"Proteins by design" by David Baker in *The Scientist*, 2006;20:26; www.the-scientist. com/2006/7/1/26/1/.

"A reinnervating microRNA" by Robert H. Brown in *Science*, 2009;326:1494–5.

"Now showing: RNA activation" by Elie Dolgin in *The Scientist*, 2009;23:34; www.the-scientist. com/2009/05/1/34/1/.

"The shape of heredity" by Susan M. Gasser in *The Scientist*, 2009;23:34; www.the-scientist. com/2009/07/1/34/1/.

"Beyond the Book of Life" by Stephen S. Hall, *Newsweek*, July 13, 2009; www.newsweek. com/id/204233.

"Persistent epigenetic differences associated with prenatal exposure to famine in humans" by Bastiaan T. Heijmans, Elmar W. Tobi, Aryeh D. Stein, Hein Putter, Gerard J. Blauw, Ezra S. Susser, P. Eline Slagboom, and L. H. Lumeyin in *Proceedings of the National Academy of Sciences of the U S A*, 2008;105:17046–9.

"Histone methylation is making its mark" by Brendan A. Maher in *The Scientist*, 2003;17:27.

"MicroRNA polymorphisms: The future of pharmacogenomics, molecular epidemiology and individualized medicine" by Prasun J. Mishra and Joseph R. Bertino in *Pharmacogenomics* 2009;10:399–416.

"The sea change that's challenging biology's central dogma" by Gary Taubes in *Discover Magazine*, October 2009.

"Hormones in concert: Multiple hormones act in concert to regulate blood sugar and food intake. The idea has already led to a new diabetes therapy; will it also yield new strategies for obesity?" by Christian Weyer in *The Scientist*, 2009;23(12):34; www.the-scientist. com/2009/12/1/34/1/.

Books
Epigenetic Principles of Evolution by Nelson R. Cabej (2008, Albanet Publishing).

Ecological Developmental Biology by Scott F. Gilbert and David Epel (2008, Sinauer Associates).

The Epigenome: Molecular Hide and Seek by Stephan Beck and Alexander Olek (2003, Wiley–VHC).

The Genetic Code: How the Cell Makes Proteins from Genetic Information Encoded in mRNA Molecules

THE READER'S COMPANION: AS YOU READ, YOU SHOULD CONSIDER

- What the genetic code is.
- How the cell uses the genetic code to make proteins.

- Which parts of an amino acid are the same for all amino acids.
- Which parts of an amino acid differ between amino acids.
- What key properties of amino acids make them functionally different from each other.

(Continued)

The Human Genome. DOI: 10.1016/B978-0-12-333445-9.00004-6

- What makes different proteins so different from each other.

- How an mRNA differs from other forms of RNA.

- What role ribosomes play in translation.

- What a tRNA is and how it is used.

- How RNA molecules are processed.

- How use of a stop codon differs from use of the other codons.

- How the cell knows which RNA molecules to translate.

- How molecules get into and out of the nucleus.

- What direction information flows in the cell.

- How the cell can make more different proteins than there are genes.

- How modular genes can let the cell use the same protein function in more than one way.

- The many different levels at which functions of a protein product are regulated.

- What sequences mark where a splice site will be cut.

- Why it is hard for bioinformatics to predict exactly where a gene will be spliced.

4.1 THE GENETIC CODE

What hath God wrought?—**Samuel Morse**

On May 24, 1844, using pulses of electricity traveling through a 41-mile-long telegraph line from Baltimore, Maryland, to Washington, DC, Samuel Morse transmitted his first message using a code consisting of combinations of three characters: a short

electrical pulse written out as, and called, a "dot," a long pulse written out as, and called, a "dash", and a space. When these symbols are sent over a wire as sound pulses, the dash sound is equal to about three times the length of the dot sound. The dots and dashes are separated by spaces of various lengths depending on whether they denote a gap between characters, between letters, or between words. Using two symbols plus spaces, Morse could indicate all 26 letters of the English alphabet, many letters specific to other languages, the numbers zero through nine, and the common punctuation marks. The most familiar signal to people who don't know Morse code is the distress code SOS, transmitted as ...– – –... (three dots, three dashes, then three dots). Morse's original message over the first telegraph line would have taken the form of

.– – .... .– –   .... .– – .... – –. – – –   –..

.– – .–. – – – ..– – –. .... –..– –..

which says "What hath God wrought?" when it is decoded. Morse revolutionized communication in the nineteenth century not only because of his engineering inventions that allowed for transmission of signals but also because of his realization that large amounts of complex information can be encoded and transferred from one place to another with even such a limited primary alphabet as the three Morse code characters, dot, dash, and space.

The key to many of the important processes in the cell is information flow. We have told you that information is permanently archived in the nucleus, and that temporary copies of that information are made by carrying out transcription to create RNA. We have also told you that some RNAs can act directly as enzymes, structural molecules or even transport units that carry amino acids; but the information flow in the cell goes far beyond that. The last major piece of the puzzle of what the cell is doing with its information archive has to do with proteins and the genetic code.

Remember that scientists originally thought that DNA could not be the repository of genetic information because it is too simple, containing only four genetic letters. They knew that the cell is incredibly complex, with large numbers of different kinds of molecules, and especially with vast numbers of different kinds of proteins. But they could not understand how an alphabet of four letters could manage to spell out the level of complexity that we know is present in the human cell. We have talked about the fact that the RNA "encodes" the information in the protein, but how does it actually do that?

The answer is something many of us have read about in newspaper articles or heard about on television: the *genetic code*. The genetic code uses only four letters but manages to contain far more information than can be spelled out with four letters by using the same type of encoding mechanism used by Morse code: combining three DNA bases together and reading them as if they were a single letter. Clearly the DNA code is not specifying letters of the English language and is not transmitted via electrical signals on a telegraph.

Codons and the Code

One breakthrough in understanding the genetic code came when scientists figured out that each "letter" in the genetic code is spelled out by three bases in a row. Just as the 26 letters of the English alphabet can be designated by Morse code combinations of dashes and dots, four genetic building blocks when combined in groups of three can produce 64 different letters in the alphabet of the genetic code. Each three-unit letter, called a codon, specifies one of the building blocks that make up proteins, building blocks called amino acids.

In fact, since there are 64 codons and only 20 different amino acids, there is actually some redundancy in the system. Thus some amino acids get designated by more than one codon. When the cell "reads" an RNA transcript with bases ACUAGA, it does not read A and then C

and then U and then A and then G and A and so on. Instead, it reads ACU as one letter (codon) and then it reads the next three bases, AGA, as a different letter (codon). Thus the first three bases of the coding region of an mRNA will encode the first amino acid of the protein, the second three bases the second amino acid, and so on. *Each codon specifies the incorporation of one and only one amino acid and the process of reading a string of codons on an mRNA to produce a protein is called translation.*

We end up with 64 codons if we write out all of the different ways in which a four-base alphabet can be used to write out letters that are three units long; however the cellular machinery that translates the code uses only 20 amino acids as protein building blocks. Thus, in some cases, different codons must code for the same amino acid (thus the code is said to be *degenerate*). So some amino acids such as methionine can be encoded by only one codon, AUG, but some amino acids such as leucine can be designated by as many as six different codons.

Note that no codon encodes multiple different amino acids, but some amino acids can be encoded by multiple different codons. Using Table 4.1, which contains a key for translating the genetic code, you can find out what amino acid will result from any of the 64 possible codons. You can also see that the amino acid encoded by UGU can be represented by the full-length name, cysteine, or by a three-letter symbol, Cys, or by the single-letter symbol, C. Some amino acids that are relatively rare are encoded by only one amino acid (such as tryptophan or methionine). Other amino acids such as arginine may be encoded by as many as six different codons. Notice that Table 4.2 uses the version of the code that is spelled out with the bases used in RNA: A, C, G, and U.

In Table 4.1 it is easy to look up an amino acid to see what codons go with it, but it is cumbersome to try to look up any particular codon in Table 4.1 because they are not arranged in order by codon. Several types of

TABLE 4.1   The genetic code uses a three-base codon to specify each amino acid

Amino acid	Three-letter symbol	Single-letter symbol	Codon
Alanine	Ala	A	GCA, GCC, GCG, GCU
Arginine	Arg	R	AGA, AGG, CGA, CGC, CGG, CGU
Asparagine	Asn	N	AAC, AAU
Aspartic acid	Asp	D	GAC, GAU
Cysteine	Cys	C	UGC, UGU
Glutamic acid	Glu	E	GAA, GAG
Glutamine	Gln	Q	CAA, CAG
Glycine	Gly	G	GGA, GGC, GGG, GGU
Histidine	His	H	CAC, CAU
Isoleucine	Ile	I	AUA, AUC, AUU
Leucine	Leu	L	UUA, UUG, CUA, CUC, CUG, CUU
Lysine	Lys	K	AAA, AAG
Methionine	Met	M	AUG
Phenylalanine	Phe	F	UUC, UUU
Proline	Pro	P	CCA, CCC, CCG, CCU
Serine	Ser	S	AGC, AGU, UCA, UCC, UCG, UCU
Stop codon		*	UAA, UAG, UGA
Threonine	Thr	T	ACA, ACC, ACG, ACU
Tryptophan	Trp	W	UGG
Tyrosine	Tyr	Y	UAC, UAU
Valine	Val	V	GUA, GUC, GUG, GUU

TABLE 4.2   DNA→RNA→Protein

The DNA sequence:	ATTAGGTACGTATGTGAT
	TAATCCACGCATACACTA
Results in an mRNA:	AUUAGGUACGUAUGUGAU
Which gets read as:	AUUAGGUACGUAUGUGAU
To produce the protein:	Ile-Arg-Tyr-Val-Cys-Asp

keys have been developed to try to make it easy to look up either item, a codon or an amino acid, and find its counterpart codon or amino acid. An especially nice format is shown in Figure 4.1, which uses a wheel-like structure to show the correspondence between the codons and the amino acids.

If you know the codon and want to find out the amino acid, Figure 4.1 lets you start with the letter in the center and move outwards to

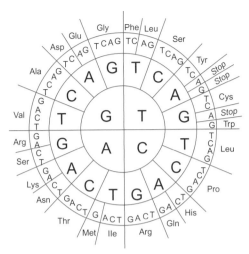

FIGURE 4.1 The genetic code. It is easier to translate codons via this table than when using Table 4.1. (Derived from H. Gobind Khorana in *Chemical Biology: Selected Papers of H. Gobind Khorana* (with introductions), World Scientific Series in 20th Century Biology, Vol. 5.)

The mRNA is read, or *translated*, from one end to the other, beginning at the start and proceeding one codon at a time toward the end of the mRNA. For our purposes, we will consider that translation begins at a *start codon* (AUG) encountered by the ribosome as it reads along the message. At the point at which the start codon is encountered, the ribosome begins the chain with the insertion of the amino acid methionine that corresponds with AUG. In reality, the AUG that starts the translation process is not always the first one, but for now we will just consider the cases in which the first AUG is used.

The AUG start codon directs the addition of the amino acid methionine (Met). As each successive codon is read, the ribosome incorporates the indicated amino acid into the growing protein. Translation stops when the ribosome encounters one of the three stop codons (UAA, UGA, or UAG) that do not specify the incorporation of an amino acid. Once the translation of the mRNA is completed, it is in some ways also a "translation" of the DNA sequence back in the nucleus even though the DNA and the ribosome never encountered each other (Table 4.2).

assemble a codon, and then look at the amino acid on the outside ring that sits next to the third base of the codon. Thus we can see that TGG encodes tryptophan (Trp). If we start with G and then T, the third position of the codon can be occupied by any of the four bases to result in a codon that encodes valine (Val). Notice that this key uses T instead of U. Even though the cellular machinery reads RNA that uses U, many technologies for determining the sequence of a gene "read" the sequence from DNA derived from the chromosome rather than "read" the sequence from the RNA copy.

Three codons do not specify the incorporation of any amino acid. Instead, they are placed at the end of the coding sequence contained on the mRNA to tell a ribosome to stop translating the message and release the assembled protein. These codons, referred to as UAA, UAG, and UGA in Table 4.1 and as TAA, TAG, and TGA in Figure 4.1, are appropriately called *stop codons* because they stop the translation process and cause the message and protein chain to be released from the ribosome.

4.2 MOVING THINGS IN AND OUT OF THE NUCLEUS

One of the problems (and benefits) of a eukaryotic cell is the presence of the nuclear membrane that separates the contents of the nucleus from the cytoplasm. Why is it a problem that the cell keeps the DNA locked up in the nucleus? The archive of genetic information is inside the nucleus, but the cellular machinery for using information from mRNAs is located in the cytoplasm.

How does the cell get mRNAs out of the nucleus to where the genetic code can be translated? The cell carries out carefully regulated transport of information into and out of the nucleus. Microscopic pores through the nuclear membrane allow for transport of molecules

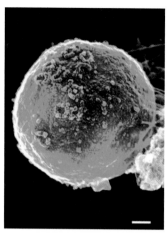

between the nucleus and the surrounding cytoplasm (Figure 4.2). The nuclear pore is not a simple hole through the membrane, but is in fact a space filled with a complex of proteins that carry out a very active process of regulating what goes into or out of the nucleus.

Why does the cell have the translation machinery for reading genetic information separated from the main repository of genetic information? The answer is: to protect that repository of genetic information, which has to last for the lifetime of the cell. The first rule for running a healthy eukaryotic cell is, "Never let your DNA wander out into the cytoplasm, where it can experience very bad things, such as being chewed up by enzymes!" Out in the cytoplasm, many enzymes protect the cell from invasion of genetic information from foreign sources such as viruses by degrading foreign nucleic acids entering the cell before they can get to the nucleus to compromise the genetic content of the cell. Clearly this defense mechanism does not always work, since some kinds of viruses do succeed in infecting our cells.

Once the cell has read the genetic information in the mRNA and used it to make a protein, some of those proteins need to get back into the nucleus. For instance, some mRNAs direct the synthesis of proteins such as transcription factors, activator proteins, gyrases, helicases, ligases, histones, and polymerases that all have to be able to get back into the nucleus to act on the DNA and RNA that live there.

4.3 THE CENTRAL DOGMA OF MOLECULAR BIOLOGY

The cell carries out a set of steps to get from the archived information in the nucleus to the final production of a protein: transcription to produce mRNA followed by translation to make a protein. The two separate steps are the critical linked elements of the Central Dogma of Molecular Biology:

This direction of information flow is of critical importance. The information originates out of the permanent DNA copy in the nucleus. Transcription produces the transient copies that the cell needs on a temporary basis. And the cell uses some of those transcripts as messenger RNAs that direct the synthesis of proteins.

4.4 TRANSLATION

The table and figure presented above tell us how to translate a piece of RNA sequence that we see written on the page, but that is not how the cell does it. What happens in the cell that lets a set of four different bases in an RNA chain tell the cellular machinery how to make a completely different kind of molecule called a protein?

Translation Requires an Adaptor Molecule Called tRNA

The codons in an mRNA molecule cannot and do not directly recognize the amino acids whose incorporation they direct. The cell uses an important organelle called a ribosome to carry out protein synthesis, and the ribosome reads the mRNA sequence through the use of an adaptor molecule called a *transfer RNA* (*tRNA*). Basically, one end of this adaptor recognizes one of the codons on the mRNA and

the other end of the adaptor has the amino acid that goes with that codon. The way the adaptor recognizes the codon is by having an anticodon, a set of three bases on the tRNA molecule that can base-pair with the codon in the mRNA (Figure 4.3). Each tRNA has an anti-codon at one end and the corresponding amino acid attached at the other end. It turns out that there is a specific tRNA molecule for all but three of the possible codons.

Three codons do not have a tRNA. These are the stop codons, and when they occur there

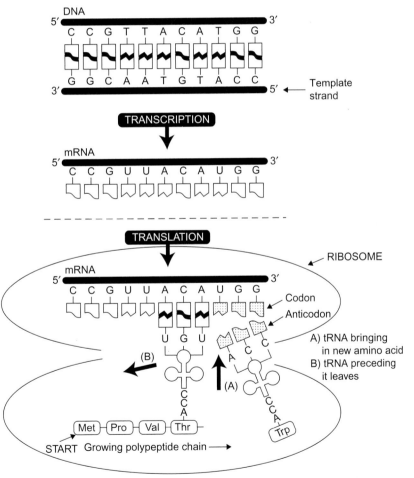

FIGURE 4.3 The process of translation. As a threonine tRNA sits in place with its threonine attached to the chain of amino acids that have already been added, an expended valine tRNA will have just left, with no amino acid attached, and a tryptophan tRNA is moving into place to add a tryptophan to the growing protein chain.

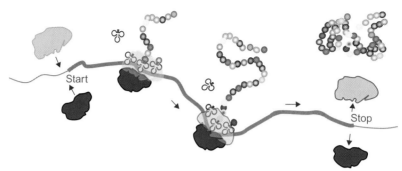

FIGURE 4.4   Ribosome assembly and transcription. The large and small subunits come together on the RNA and begin reading the coding sequence at the start codon by inserting a methionine at the first position of the protein chain. The ribosome moves along the mRNA (or the mRNA moves through the ribosome, depending on your perspective), and as each new codon on the mRNA moves into position the corresponding tRNA clicks into place, precisely positioned to allow the amino acid attached to it to become unattached to the tRNA and to become attached to the growing protein chain. Meanwhile a new tRNA is moving into place so that a new amino acid can be added. When the stop codon is reached, there is no corresponding tRNA to move into position and the ribosome subunits come apart and release both the mRNA and the completed protein chain.

is no tRNA that can fit into place and drop off an amino acid. This lack of a tRNA moving into place signals the ribosome to stop translating the content of this particular mRNA and to release the mRNA back into the cytoplasm where it can potentially be picked up by another ribosome to start the translation process again. The stop codons are UGA, UAA, and UAG.

How does the process start? The two parts of the ribosome, a large subunit and a small subunit, exist separately in the cytoplasm. When an mRNA is available to be translated the two pieces of the ribosome come together around the mRNA to form an intact ribosome in the process of carrying out translation (Figure 4.4). The ribosome moves along the mRNA, using tRNAs that match the codons on the mRNA as the mechanism for adding amino acids to the growing protein chain, so that each new amino acid added corresponds to the next codon on the mRNA. The result is that the order of amino acids in the protein is directly determined by the order of the codons on the mRNA. Once the ribosome reaches the stop codon it separates into its two subunits and releases the mRNA. Although human and bacterial ribosomes are quite similar in function, they have enough specific differences that it has been possible to develop some important antibiotics that target bacterial ribosomes while leaving the human ribosomes alone (Box 4.1).

4.5 MESSENGER RNA STRUCTURE

Now that we know how to read an mRNA to make a protein, let's tackle another question: out of the many types of RNA in the cell, how does the cell know which ones to use in making proteins and which ones to use for other purposes? Since all of the RNA molecules are strings of As, Cs, Gs, and Us, how does the cell tell which ones it should be using to direct the synthesis of proteins and which ones should serve non-coding purposes? The cell actually marks the RNA strands that it wants to use as mRNAs through two modifications to the RNA. At the 5' end of the RNA strand that will become mRNA the cell adds an unusual nucleotide referred to as a cap. At the 3' end the cell cleaves the RNA strand and adds a string of As to create what is called a polyA tail (Figure 4.5).

BOX 4.1

THE RIBOSOME

The human ribosome is built by bringing together a small ribosomal subunit and a large ribosomal subunit. Each ribosomal subunit is made of about 65% RNA and 35% protein, and the actual active parts of the organelle are the RNA components so this is sometimes classified as a ribozyme. The role of the ribosome is to bring together molecules in a very precise alignment that enables a biochemical reaction to proceed, so it has sometimes been classified as a molecular assembler. Hundreds of proteins are involved in the assembly of a ribosome.

Although human and bacterial ribosomes read the same code and have many structural similarities, there are just enough key differences that some very important antibiotics are those that target the bacterial ribosomes while leaving the human ribosomes alone. In 2009 the Nobel Prize in Chemistry was awarded to Venkatraman Ramakrishnan, Thomas A. Steitz, and Ada E. Yonath for their work showing how different antibiotics interact with the three-dimensional structure of ribosomes.

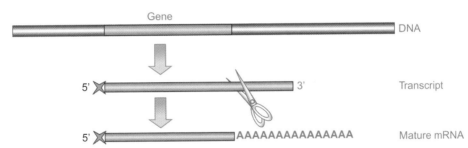

FIGURE 4.5   Processing of an mRNA. A gene encoded in the DNA of a chromosome is transcribed to create a primary transcript which is then capped at the 5′ end and cleaved at the 3′ end before the addition of a polyA tail that tells the cell that this is an mRNA intended to direct the synthesis of a protein.

One of the easy ways to identify a non-coding RNA is to look and see that it lacks a polyA tail.

The next question is: where does the ribosome start reading the RNA to start building the protein chain? Here we encounter an important concept: *the translation from RNA to protein does not start at the first base on the RNA molecule; it starts at an AUG start sequence after reading past other bases present on the RNA strand before the AUG.* After assembly of the growing protein chain begins, the ribosome keeps reading until

it encounters a stop codon, which can be UAA, UAG, or UGA. Again, this is an important concept. *The translation does not keep going until the end of the RNA molecule.* The result is that the coding sequence, the part of the RNA used to direct synthesis of protein, is sandwiched in between two regions of non-coding sequence that precede and follow the coding sequence (Figure 4.6). How much of the mRNA is coding sequence and how much of it is untranslated can vary a lot from one gene to the next.

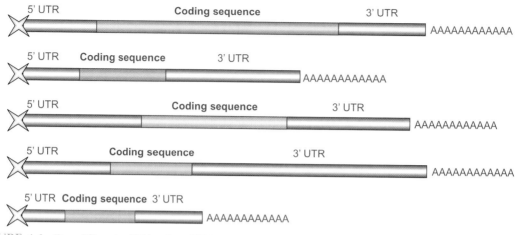

FIGURE 4.6   Five different mRNAs. An mRNA starts out with non-coding sequence at the beginning, followed by a start codon, the coding sequence, a stop codon, and more non-coding sequence. The 5′ end has a cap and the 3′ end has a polyA tail. This is the universal structure of a fully processed mRNA whether the gene is large or small. However, there is no rule regarding just how long the 5′ untranslated region (5′UTR) or the 3′ untranslated region (3′ UTR) will be relative to each other or relative to the coding sequence.

RNA Turnover

If RNA is a temporary copy of the information, does it get used once and get discarded? Or does it get used many times? Or does the RNA last for a fixed amount of time? In fact there is a lot of variation in how long an RNA stays around in the cell, with some RNA molecules being recycled quite promptly and others hanging around to be used many times. How long an RNA molecule lasts is very tightly regulated by the cell. If an mRNA loses its polyA tail or its 5′ cap it will be degraded by RNAses, enzymes whose specific function is recycling RNA that has been used. Some kinds of sequences, such as AU-rich regions, present in the untranslated regions of an mRNA, determine how long it will last. An example of a kind of mRNA that may be rapidly recycled by the cell can be found in some cytokine mRNAs. Cytokines are produced by cells of the immune system and the body needs to be able to produce them rapidly in response to infections and to get rid of them again rapidly when the need for an immune response is past. The rate at which the RNA is degraded is determined by a combination of sequences present in the 3′ untranslated region and proteins that bind to the RNA to protect it or degrade it. This makes the mRNA inherently fragile and susceptible to degradation but lets the cell provide temporary protection for the mRNA while it needs to keep it around.

4.6 SPLICING

Splicing

One of the most interesting things that can happen to an RNA molecule is a process called splicing. This process removes a piece of the transcript while joining together the sections of transcript that flank the removed piece. Some genes are spliced at multiple locations, some genes are spliced only once, and some genes are not spliced at all.

We have saved a discussion of splicing for this section instead of including it in the RNA

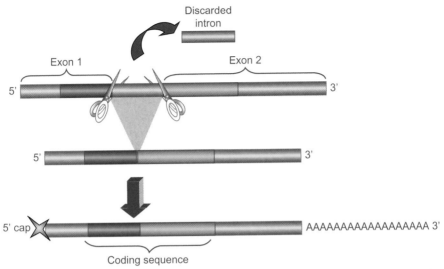

FIGURE 4.7 Processing a transcript. To process a primary transcript into a mature mRNA, remove the intron, add a cap at the 5′ end, splice the exons together, and add a polyA tail. The polyA tail and the cap help to protect the mRNA from being degraded by cellular enzymes, and the polyA tail signals to the cell that this is a messenger RNA due to be transported out of the nucleus to be used in making a protein. Notice that exon 1 contains both non-coding information at the 5′ end and coding information used in the making of the protein. Similarly, exon 2 has both coding information and the non-coding information at the 3′ end.

chapter because it is actually highly relevant to our discussion of proteins. Many of the transcripts that get spliced are mRNAs and there are some things about mRNAs that we have to take into account when looking at how an mRNA is spliced. Splicing alters the RNA sequence that is available to be read by the ribosome, and when we consider that some genes can be spliced in more than one way (alternative splicing) we will see that it can actually expand the number of different kinds of proteins the cell can make from a single gene. Thus, the cell can make more proteins than the total number of genes present in the cell.

In the simplest case of splicing, a piece is cut out of the transcript. The two pieces flanking the deletion are brought together and rejoined so that what is left is one contiguous piece of RNA. The piece that is cut out is called an *intron*. The pieces flanking the intron are called *exons*. As we can see in Figure 4.7, an exon in an mRNA can include both coding and non-coding sequence.

Splicing actually happens in a stepwise manner and makes use of several important sequences along the RNA molecule (Figure 4.8). The first step is cleavage of the RNA at the donor splice site, followed by the free end of the intron curling around to attach to a sequence called a branch sequence (or lariat sequence from the lariat-like shape that is created by binding to it) out in the middle of the intron. Then the RNA is cut at the acceptor splice site, the lariat-shaped intron is removed, and the two exons are spliced together.

Genes with Multiple Introns

In more complex cases, there may be more than one intron within a transcript that needs to be removed (Figure 4.9). Introns are not all removed simultaneously, so for a gene with many introns the cell ends up with multiple different RNAs that are all derived from the primary transcript, some missing one of the

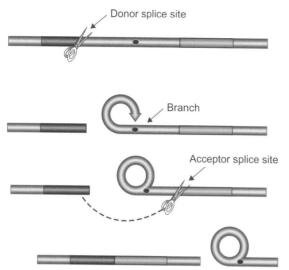

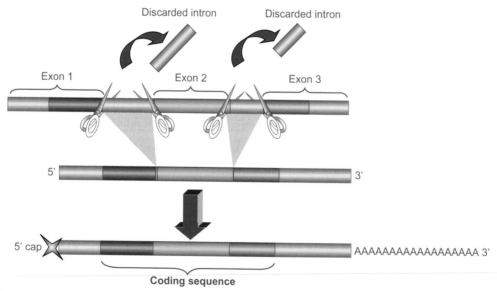

FIGURE 4.8   The three steps of splicing. First cut the RNA at the donor splice site. Second, bend the cut end of the intron around to bind it to the lariat sequence out in the middle of the intron. Third, cut the acceptor splice site and connect the two flanking exons to each other.

introns, some missing several introns, and some missing all of them. Note that in Figure 4.7 the exons at the beginning and the end both have a combination of coding and non-coding sequence, but in some genes that is not the case. Sometimes there will be an intron contained entirely within non-coding sequence so that the exon that mixes coding and non-coding might be second in from the end. We have diagrammed splicing of an mRNA, because many mRNAs are spliced, but there are some non-coding mRNAs that are also spliced.

There is a simple key to keeping track of what the bits of the spliced transcript are called. The parts of the transcript that get left in the final mRNA and *ex*ported from the nucleus are called *exons* and the parts of the mRNA that get left *in*side of the nucleus and eliminated are called *introns*.

Splicing happens within about 10 minutes after synthesis. It proceeds this quickly even in

FIGURE 4.9   Splicing of a gene with multiple introns. The structure of the gene on the chromosome shows the two introns. In the middle we see the removal of the introns, and at the bottom is the form of the final mature mRNA with the introns removed and a 5' cap and polyA tail added. The lighter purple sections are non-coding sequence and we can see the standard final structure in which non-coding sequence is found at both the beginning and the end of the transcript.

FIGURE 4.10   Donor and acceptor splice sites. Each donor splice site and each acceptor splice site is defined by a region of fewer than 20 bases, most of them inside of the intron. The sequences of the donor splice sites (in blue) found in different introns and different genes are similar to each other but not identical; for each position in the donor recognition sequence there are some bases that are more likely to be found at that position, but the only bases that are always the same in all of the donor splice sites are the first two bases at the beginning of the intron which are always a G followed by a U. The sequences around the donor splice site are rich in As and Gs, but there can sometimes be a C or U at some positions. A sequence that would be predicted to be likely to be used as a donor splice sequence would be CAGGUAAGU. The sequences of the acceptor splice sites (in pink) found in different introns and different genes are similar to each other but not identical; for each position in the acceptor recognition sequence there are some bases that are more likely to be found at that position, but the only bases that are always the same in all of the acceptor splice sites are the first two bases at the beginning of the intron which are always an A followed by a G. For the acceptor splice site, 12 of the 13 bases in the intron prior to the AG are all either Cs or Us. A sequence predicted to be a good acceptor splice site would be CCCUCCUCCCUUGCAGG. Because implied RNA sequence is often determined by sequencing the DNA from which that RNA came, we often hear the donor splice site sequence mentioned as GT rather than GU.

genes that have introns as large as hundreds of thousands of bases in length!

Where Does Splicing Cut?

The average human gene has more than a half dozen introns that get removed by splicing, so how does the cell know where to cut the RNA to splice it? Splicing is carried out in the nucleus by the spliceosome, a complex of proteins and small RNAs. The splice site at the 5′ end of the intron is called the donor splice site, and the splice site at the 3′ end of the intron is called the receptor splice site. As shown in Figure 4.11, the donor splice site always uses the bases G and U as the first two bases of the intron (the first two bases removed from the spliced RNA). The receptor splice site always uses the bases A and G. But the problem is we find cases of G followed by U at a large number of positions in most RNA molecules, and there are cases of A followed by G at a large number of positions in most RNA molecules. Splicing takes place at locations on the RNA that are determined by the sequence of the RNA, but there is no simple unique splice site sequence that tells the cell "always cut where you see this sequence and never cut if you don't see this sequence." Even though there is no absolute

splice site sequence, there are certain patterns of bases that are most likely to be found surrounding the splice sites (Figure 4.10).

There is no unique sequence that is always used as a lariat sequence, the site at which the cut RNA forms a branch attachment. There is always an A to which the chemical bond of the branch point is attached, and this A is embedded in a short surrounding sequence that is similar for branch points in different genes but is not always the same. Part of why it is hard to predict where a lariat site will be found is that many introns are so large that this sequence may be found more than once. Something that helps us identify probable lariat sites is that they are expected to happen about 20 to 50 bases before the acceptor splice site, so we don't have to consider every sequence in the intron that looks like a lariat site.

We have little expectation that a splice site would cut exactly at the junction between the 5′ untranslated region and the coding sequence, rather than falling to the left or the right of that point. We have no expectation that it would cut exactly after the last base of the coding sequence. Why can we be so sure? Because we know things about the sequences that help to determine whether splicing will take place and we know things about the kinds of sequences that signal the end of coding sequence, we can tell that they are different sequences. You can have a stop codon signaling the end of coding sequence, or you can have a donor splice site, but they cannot both sit at the exact same position in a gene. The result is that we expect that at least some part of the last exon that includes coding sequence will also include some amount of non-coding sequence.

4.7 MODULAR GENES

When Jared and Mei bought their latest computers, they each found themselves faced with a variety of options: different speeds, different amounts of memory, and different peripherals such as speakers and scanners. By the time they were done making their selections, their computers had many things in common, but Jared the database manager had selected an automated tape-drive backup system that Mei didn't think she needed, and Mei the graphic artist had selected high-end graphics card options that Jared knew he didn't need. Because construction of computers is so modular, it is easy to optimize the features present on each computer without computer companies having to maintain separate lines of machines with every possible combination of features.

The genome makes use of similar efficiencies by designing some genes to be modular, allowing different cells to use the same gene in different ways or in some cases letting one gene make more than one protein structure from the same gene. Do we mean that the genome uses some genes together in a modular fashion? Well, yes, the genome does that, making multiple different proteins from different genes and then assembling the proteins into a complex structure made up of several subunits. But that is not what we are talking about here. When we talk about the modular gene, we mean a gene that is made up of a set of separate pieces in such a way that not all of the pieces always get used, just as computers don't all end up with

every option on the manufacturer's list for that particular computer. Sometimes genes do the same thing, picking and choosing which parts to include and which parts to leave unused. In fact, more than half of all human genes undergo some alternative splicing.

The key to alternatively spliced, or modular, genes is that a transcript from a gene with multiple introns does not always get spliced in the same way. One intron might be removed under one circumstance and then a different intron might be removed under other circumstances. So, under some circumstances, the cell uses some parts of a primary transcript to make the mRNA that will dictate the protein sequence, and under other circumstances, the cell uses different parts of the primary transcript. In these cases, the two different *splice variants* usually share most of the same sequence and differ in only some places.

What is the functional effect of such alternative arrangements? Imagine that the cell has an enzymatic function to perform, and that two copies of the protein have to be linked together to allow them to carry out the enzymatic function. Then suppose that sometimes it needs to carry out the enzymatic reaction while stuck to the surface of the cell that made the protein, and at other times there are other situations in which the cell wants that protein to be able to leave that cell and go to another part of the body to carry out its enzymatic reactions. What does our cell need to be able to carry out the enzyme reaction under both sets of circumstances?

- A gene encoding an enzymatic module plus a linker plus a cell surface anchor.
- A gene encoding an enzymatic function plus a linker plus an export vehicle.

In some cases the cell accomplishes this by having two different genes that are very similar to each other. Both genes would have sequences to make the enzymatic part of the protein and the linker portion that holds the two copies of the

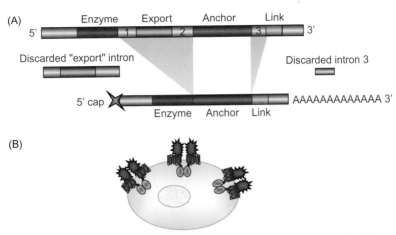

FIGURE 4.11   Selective use of some exons. (A) One splice variant of this gene removes the "export" exon along with the intronic sequences on both sides of it. Another intron is removed between the anchor module and the linking module. (B) The resulting protein has sections that can carry out the functions associated with the three exons that remained in the final mature mRNA, so this cell has proteins anchored in its membrane that can carry out the enzymatic function.

protein together. One gene would have combined those sequences with sequences that encode the export module while the other gene would combine them with sequences that encode a section of protein that can anchor in the cell membrane.

Often, instead of maintaining separate genes for each version of the protein, the cell efficiently uses one gene for several purposes (see Figures 4.12 and 4.13). Since we know about splicing, it is easy to conceive of what the structure of such a gene would be. There would be four different functional regions of the protein encoded by the gene, and alternative splicing would select different combinations of exons depending on what is needed. Enzyme plus anchor gives us a version of the enzyme that stays with the cell (Figure 4.11). Enzyme plus vehicle gives us a version of the enzyme that travels to other locations where the enzymatic activity is needed (Figure 4.12). If splicing were to remove the linker section then the two proteins could not come together to form a dimer and the enzymatic activity could not happen even if the protein made it to the right place.

Why not just make two different genes, one gene that makes an anchored version of the enzyme and a different gene that makes a traveling version of the enzyme? In fact, sometimes the body accomplishes these multiple goals by having different genes that share some functions but not others. However, sometimes the genome is especially efficient and uses the modular approach to get more uses out of one gene, so that the same gene can serve a local purpose, such as putting a particular enzyme function on the surface of the cell that makes the protein or exporting that enzyme function to the bloodstream to carry out its function elsewhere in the body (Figure 4.12).

The Implications of Modular Genes

Some genes are not spliced. What you see in the DNA tells you what you get in the mRNA and the protein. Some genes are spliced, but the gene structure is simple and the final mRNA made from that gene will always be the same. Some genes are alternatively spliced, with some

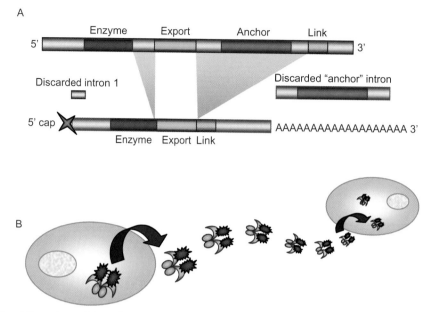

FIGURE 4.12 Alternative splicing allows use of a different combination of exons and their associated functions. This variant removes the "anchor" module of the gene to leave a transcript that will direct synthesis of a protein that gets exported from the cell.

genes using two alternatively spliced forms and some using many. These alternatively spliced genes are the modular genes.

What are the implications of modular genes? First, the total number of proteins that the human organism can make is effectively much larger than the total number of genes. Second, because sometimes the alternative splicing takes place differently in different cells, this means that cells with different needs can make different uses of the same genes. Third, it adds another level of complexity to the cellular processes for converting information in the DNA into a final set of proteins and functions.

4.8 WHAT ARE PROTEINS?

We have told you about the DNA archives of genetic information, and the RNA copies that are made from that archive. We have told

you that many RNAs either encode proteins or play a role in helping the cell turn the mRNA sequence into a protein sequence, and that alternative splicing can greatly expand the number of proteins that can be made from the available number of genes. We have also told you about the central dogma of molecular biology, the critical direction of information flow from DNA to RNA to protein. What is so important about proteins, and why is so much of the cellular machinery geared to the production of proteins? To begin understanding this we must first answer the question, "What are proteins?"

Remember that scientists originally thought that proteins must be the repository of genetic information because they are so much more complex in their structures than DNA. Two important factors that contribute to this increased complexity are (1) proteins are made up of 20 different building blocks called amino acids, resulting in a vast number of different

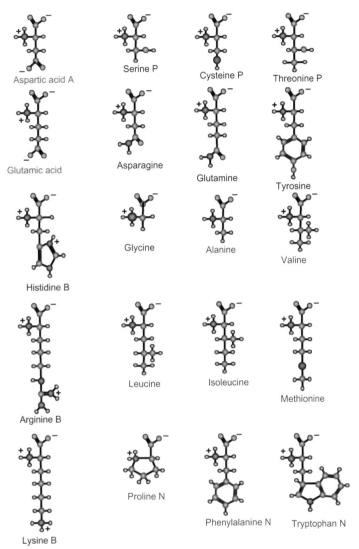

Aspartic acid A

Serine P

Cysteine P

Threonine P

Glutamic acid

Asparagine

Glutamine

Tyrosine

Histidine B

Glycine

Alanine

Valine

Arginine B

Leucine

Isoleucine

Methionine

Lysine B

Proline N

Phenylalanine N

Tryptophan N

FIGURE 4.13   The amino acids. These protein building blocks have substantially different properties. Those on the left are either basic (brown lettering) or acidic (green lettering) because they have either an extra positive charge or an extra negative charge. Six amino acids with blue labels do especially well interacting with water in areas like the cytoplasm (they are hydrophilic), and eight amino acids with pink labels do not like to interact with water (they are hydrophobic). The smallest amino acid, glycine, can fit in well in either type of environment.

combinations of the order in which those amino acids can be arranged, and (2) the amino acids have very different properties including differences in size, shape and charge (see Figure 4.13), while the four building blocks that make up DNA are very similar to each other.

These substantial differences in properties of amino acids stand in sharp contrast to the nucleic acids, which, as we saw in Chapter 2, seem so similar in their size, shape, and chemical properties. The result is that the many different arrangements of amino acids called proteins tend to fold into a huge range of vastly different shapes with widely different functional properties, while DNA forms itself into the highly regular form of the double helix.

To understand the overall structure of a protein, we do not need to keep track of every detail about the chemical structures in Figure 4.13. As shown in Figure 4.15, amino acids have two molecule sections that are the same in all amino acids, called the amino terminal and carboxy terminal sections of the amino acid, and there is one section called the R group that is different (sometimes substantially different) between the different amino acids. We can make some important generalizations about protein structure from understanding the three parts of the amino acid building blocks. The amino terminal end often has a positive charge that is balanced by a negative charge at the carboxy terminal end. A chemical side chain called an R group is attached to the core structure of the amino terminal and carboxy terminal sections of the amino acid (Figure 4.14).

When a chain of amino acids is formed to make a polypeptide, the cell joins amino acids together in a chain. To connect two amino acids, the cell brings two amino acids close together and removes several atoms at the junction point of the two molecules (Figure 4.15). A molecule of water is released and a bridge between the two amino acids is built, a chemical bond called a peptide bond. The parts of the amino acids that interact to build the peptide bond are always the same, so the formation of the peptide bond is not affected by the presence of different side chains.

The shape of the polypeptide can vary wildly depending on which amino acids are joined together. As you can see in Figure 4.16, some combinations of four amino acids take up substantially more space than others. In addition, other molecules in the cell will interact very differently with the polypeptide with the two extra charges (lysine and histidine) than with the polypeptide that has no extra charge.

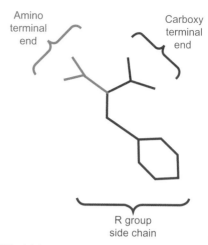

FIGURE 4.14 Simplified diagram of a phenylalanine amino acid. An amino acid is a complicated assembly of atoms. Fortunately, we often need to show only just enough information to communicate three simple aspects of an amino acid: the R group which is a side chain that is the part that differs between the different amino acids, and two regions that are shared in common by all the amino acids and that are the areas in involved in connecting amino acids to each other: a region called the carboxy terminal end (that has a carbon and two oxygens), and a region called the amino terminal end (that has nitrogen and hydrogens). A lot of what is missing from a diagram of this kind is a lot of hydrogens that are unnecessary for our purposes, and implied and obvious to those who need to understand the chemistry.

While we can diagram a chain of amino acids (a polypeptide) as shown in Figure 4.16, for some purposes we need only list single letters that symbolize the names of the amino acids – keeping the information content while not having to worry about the detailed chemical structures. Thus an alanine alanine dipeptide might simply be written out as AA and a tetrapeptide of cysteine, proline, methionine, glycine might be written as CPMG. We normally start listing the sequence at the amino terminal end of the protein and stop at the carboxy terminal end of the protein.

As we later start to consider ideas such as bioinformatics, where computers can model and handle analysis of biomolecules, we can

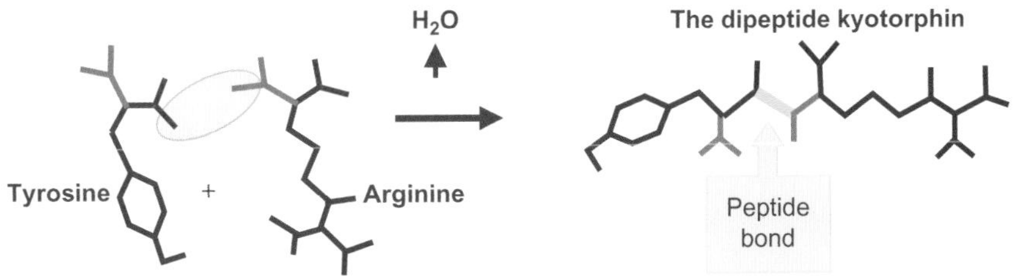

FIGURE 4.15   Peptide bond formation. Amino acids tyrosine and arginine come together to form the dipeptide kyotorphin which is synthesized in the brain and plays a role in regulation of pain signals. Note that when the atoms inside of the yellow circle are removed, the atoms that are removed form a molecule of water (H_2O) and a new bond called a peptide bond is created that holds together what is left of the amino acids after the two hydrogens and an oxygen are removed. Notice that we can diagram the amino acids using these simplified forms which keep all of the key information while implying other information that does not make a big difference to our efforts to interpret how a particular amino acid will affect protein function.

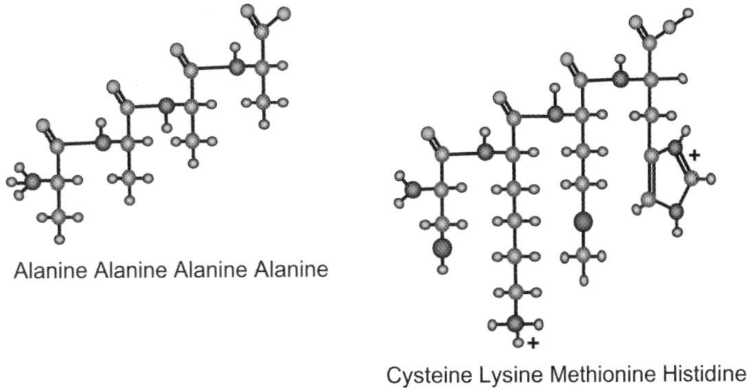

FIGURE 4.16   Two different combinations of four amino acids joined together with peptide bonds to form tetrapeptides. Notice the big difference in overall size and shape of these two tetrapeptides, one small and fairly regular in its form and the other much larger and more complex. Also notice that the sequence of these tetrapeptides (which amino acids are there and in what order) is what determines that shape.

feed information into the computers in such simplified form – a simple list of single letters that identify the order of amino acids in a protein or the order of bases in DNA or RNA – and let the computer keep track of some of the specifics about the characteristics those amino acids will give to the chain of amino acids. We do need to keep in mind some of the fundamental differences, such as whether an amino acid is charged or hydrophobic, large or small, but for our purposes we will most often just talk in terms of the sequence of amino acids as a string of letters or the shape of the folded up chain

that results, such as the image in Figure 4.17 where the OTC protein appears in the form of a ribbon.

As with the other figures in which we present molecular details, the main point of these amino acid diagrams is to show you a concept, in this case the idea that the different amino acids have great differences from each other. Of the amino acids used in human proteins, the largest amino acid, tryptophan, has about three times the mass and four times the volume of the smallest amino acid, glycine. Some of the hydrophilic amino acids do well

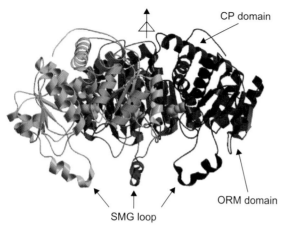

FIGURE 4.17 Information dictates form and function. The ornithine transcarbamylase protein made by the OTC gene assembles three copies of itself (with the three copies appearing here in three different colors) to make the functional protein that carries out a step in the urea cycle that keeps ammonia from accumulating in the human bloodstream. The straight and looped sections of the ribbon in this picture indicate how the protein is folded at that point in the protein. This particular image shows the conformation of the protein when it is complexed with the molecule it acts on. (Reproduced with permission, from D. Shi, H. Morizono, X. Yu, L. Tong, N. M. Allewell and M. Tuchman (2001) "Human ornithine transcarbamylase: crystallographic insights into substrate recognition and conformational changes", *Biochemical Journal*, 354:501–9. © The Biochemical Society.) The shape of this protein is determined by the sequence. Notice how much this image shows about the affect of the amino acid composition on the shape of the protein, and how much more we are seeing about the protein than we get by just looking at the primary sequence:
MLFNLRILLNNAAFRNGHNFMVRNFRCGQPLQNKV
QLKGRDLLTLKNFTGEEIKYMLWLSADLKFRIKQKG
EYLPLLQGKSLGMIFEKRSTRTRLSTETGFALLGGHP
CFLTTQDIHLGVNESLTDTARVLSSMADAVLARVYK
QSDLDTLAKEASIPIINGLSDLYHPIQILADYLTLQEHY
SSLKGLTLSWIGDGNNILHSIMMSAAKFGMHLQAATP
KGYEPDASVTKLAEQYAKENGTKLLLTNDPLEAAHGGN
VLITDTWISMGQEEEKKKRLQAFQGYQVTMKTAKVAAS
DWTFLHCLPRKPEEVDDEVFYSPRSLVFPEAENRKWTIM
AVMVSLLTDYSPQLQKPKF.

some are small, some are positively charged, some are negatively charged, and some are neutral and have no charge. A positive charge on one amino acid may interact with a negative charge on another amino acid elsewhere in the same protein or in a different protein, and two amino acids with the same charge may tend to move away from each other. Some amino acids are larger, taking up more space within the protein. One amino acid is known for putting bends or kinks into the protein chain wherever it occurs, and another amino acid has properties that let it connect different parts of a protein together by forming a specialized chemical bond.

The order of amino acids determines critical properties of the protein. It dictates whether the protein has sections that can interact with other proteins. The order of amino acids can determine which parts of the molecule are chemically bonded to other parts of the molecule. Overall, the order of the amino acids ends up determining the folded-up three-dimensional shape of the protein and the location of positive and negative charges within the three-dimensional structure. All of this determines the protein's function. Even the substitution of one amino acid for another can sometimes completely inhibit the ability of a protein to function. Thus, there are many different proteins and the differences between the functions of proteins are the direct result of the difference in the order of amino acids that make up the proteins

As we will discuss in Chapter 5, mutation can result in substitution of one amino acid for another. Some amino acids are similar enough in properties that they can usually substitute for each other, so that if a mutation results in the placement of an arginine at a particular point in a protein in place of a lysine, there is a good chance (but no guarantee) that there will be little functional impact of the change. Putting a proline (with its bend) or cysteine (with its bond-forming properties) in place of other amino acids will often cause functional changes.

in water-based environments such as the cytoplasm. Other amino acids (considered hydrophobic) prefer water-free environments such as lipid membranes. Some amino acids are large,

Modification of Proteins

As described in Box 4.2, many proteins are changed beyond their original form that emerges from the ribosomes. In some cases the cell takes a large protein and cuts it into multiple smaller proteins that each has its own function. An important example of this, the pre-pro-opiomelanocortin protein made by the POMC gene, is cleaved to produce separate proteins: ACTH, three different melanocyte stimulating hormones, and beta-endorphin. Individuals who are defective for POMC have a complex phenotype, including red hair and severe obesity. This phenotype reflects the fact that knocking out POMC alters one protein that affects coloration and another protein that alters appetite. For many genes, cleavage leads to only one functional protein product, but cleavage is needed to make the protein functional; in these cases the cell cuts a small piece off the front end of the protein.

Some proteins act alone, but in many other cases proteins become active when they join together to form complex structures. In some cases these structure are formed by multiple copies of the same protein coming together. In other cases different proteins come together to interact.

One of the most common chemical modifications to proteins is phosphorylation. Adding or removing a phosphate chemical group can activate or inactivate the protein. Another common modification to proteins is the addition of carbohydrate groups to the protein.

Finally, the cell can affect activity levels of a protein by controlling access to the protein. In some cases the cell may sequester proteins or export them from the cell.

What Are Proteins For?

Proteins perform a vast array of different functions in the cell. There is a lot of variation in how abundant any protein is depending on what it is doing. Many proteins play structural roles in the cell, in the extracellular matrix of proteins that surround the cell, or in structures of the body such as bones, tendons, ligaments, hair, nails, and teeth. Some of these proteins persist for amazing numbers of years, such as proteins in bones and teeth. Other proteins, such as some classes of hormones and neurotransmitters, carry out signaling between cells. There are classes of proteins involved in transporting other molecules within the cell or between cells, and other classes of proteins that sit on cell surfaces waiting to detect signals coming to the cell. Proteins made by white blood cells provide immune responses to infectious agents and proteins in red blood cells carry oxygen to cells in the body and carry away the carbon dioxide they need to get rid of.

Many proteins carry out fundamental processes essential to all cells, such as transcription. We use the term *housekeeping genes* to refer to genes whose products are expressed in and necessary for basic functions in all cells types. Other genes and proteins are tissue-specific, carrying out functions that are specific to a particular organ or cell type, such as the NRL transcription factor that shows expression that is specific to the retina. It is surprising how many genes are expressed in many different cells and tissues that are not housekeeping genes and are not expressed everywhere.

4.9 GENE PRODUCTS AND DEVELOPMENT

We now have all the different pieces to the puzzle – the DNA, the RNA, the proteins, what they are made of, how they are produced, and some of what determines when, where, and how much of any gene product is made. What happens when the fertilized egg pulls all of this together and uses the genetic information in the nucleus?

How We Become Human

Embryonic development starts with the fertilized egg that is nothing like a finished human

---

BOX 4.2

MODIFICATION OF PROTEINS

Many proteins are inactive when they are first made, and need to be modified before they can carry out their functions. In some cases, a short stretch of amino acids called a signal sequence is cut off the front (amino terminal) end of the protein. In some cases specific kinds of carbohydrates are attached to specific points in the protein sequence. In other cases function of the protein is affected by the addition of a phosphate group by a kinase, or the removal of that phosphate group by a phosphatase. The existence of many different types of protein modifications points to a whole new level of regulation of the levels of the final functional gene product. Even if the cell is producing a lot of RNA from a gene and translating it into protein, the phenotype of the cell remains that of a cell that has not produced any RNA from that gene up until the time when the protein becomes active and begins to carry out its functions. So the cell maintains a set of kinases and phosphatases that can turn proteins "on" and "off" in a very specific, regulated manner. Whether a protein is acted on by one of these modifying enzymes depends on whether the protein contains one of the sequences of amino acids recognized by the enzyme, and it also depends on whether that recognition sequence is located on an accessible part of the protein that is "visible" to the enzymes seeking to modify the protein. In recent years, kinases and phosphatases have become important new targets for the development of novel therapies for a variety of traits including some cancers. So in some cases if we cannot make a particular gene product go away, then we have to ask ourselves if we can just turn it "off" so that it does not matter whether or not it is present in the cell.

---

body. How does development begin? When a sperm fertilizes an egg, it causes a set of changes that initiates a complex series of developmental events. Among the first events are changes that the cell makes to existing proteins that have been sitting in the egg waiting for this moment. One of the critical changes is the addition of phosphate groups to some of the existing proteins, but there are lots of other ways in which the activity levels of a protein can be altered (Box 4.2). Other changes taking place following fertilization include changes in transcription of some genes, including up-regulating expression of some genes and down-regulating expression of others.

Among the results from calcium-mediated modifications to proteins in the cell are two different key types of events – inactivation of some of the factors that have worked to actively keep the cell cycle shut down in the unfertilized egg, and activation of some of the initial signals that tell the cell to move forward with cell division and subsequent steps in differentiation of different cells into different functional categories. Interestingly, simply adding calcium into an egg can start off some of these events, although the haploid egg will not go on to develop into a viable embryo because it cannot survive with only one copy of each chromosome.

There follows an amazing progression not only in the number of cells present but also in the shape of the embryo and the gradual addition of new organs and cell types. Rapid and dramatic changes in gene expression accompany the changes in the embryo. The actions of transcription factors such as those we have been describing gradually distinguish cells that will become the outsides of our bodies from cells that will make up our "innards."

As the combinations of transcription factors that are present in a cell change during embryogenesis, cells that initially are specified only as to general regions of the body – inside or outside, towards the top or towards the bottom – gain more and more specific instruction on what they will become. Thus changes in transcription factors bring about the gradual differentiation of cells. A cell that has initially been "told" by a combination of factors only that it will be internal to the body then gets a set of signals telling it to become part of the top of the organism. Later, as the embryo continues to divide and make more cells, some of them get signals indicating that they are now fated to be something neuronal, then that they are to be part of the brain, then that they have become the type of brain cell that receives signals that come from the eye.

If genes regulating gene expression during development are not expressed exactly when they are supposed to be expressed, the embryo may not be able to go on to later stages of development and the effects can be lethal. Some of the genes expressed early on determine profound things such as which end of the organism will be the head rather than the feet. Major errors in laying out the basic pattern of the embryo tend to be lethal quite early.

Developmental Regulatory Events

As a stream of CO_2 sent the swarming cloud of fruit flies drifting off to sleep on the side of the glass vial in which they had been raised, Stacy peered at the label to confirm that she had the right bottle. She then gently tipped the flies out onto a white plate and adjusted the microscope to bring the sleeping flies into focus. She then separated the flies into two groups, one containing wild type flies and the other containing flies homozygous for a recessive mutation. The eyes of 15 wild type flies gleamed back

at her from their ruby-red, multifaceted surfaces. She briefly scanned the 30 eyes and 30 little antennae that perched on the 15 normal little fly heads to confirm that nothing was out of the ordinary. Then her eyes turned to the dozen mutant flies that were of interest to her project. Those flies had no eyes at all! Later that afternoon she looked at flies carrying a dominant mutation. As she had expected, these flies had developed without antennae – instead of antennae there were tiny legs sticking out of the tops of their heads where their antennae should have been (Figure 4.18).

As expected? Legs on their heads? Is this a joke? Why would Stacy expect to find that her fly bottles held flies with no eyes or flies with legs growing out of their heads? How did they end up like that? And what does this have to do with humans? To understand what Stacy was looking at and what this has to do with the human genome, let's talk about gene expression, the selective use of different genes by different cells in the body.

Study of *developmental regulatory* genes tells us that specific gene regulation events can set off a cascade of changes in gene expression that correspond with changes in the developmental program for a set of cells, taking cells that were all destined to become parts of the front end of the organism and committing some of those cells to become parts of the eye while committing other cells to become parts of the brain. Some of the most dramatic lessons on this subject come from the study of fruit fly mutants like the *Antennapedia* mutant in Figure 4.18. These types of mutants, called homeotic mutants, result in changing the developmental fate of a set of cells, so that cells that were destined to become one body part become a different body part instead. It is now known that in humans as well as flies, the *homeotic* genes responsible for some of these cell fate commitments are transcription factors that bind to the DNA to regulate the expression of many other genes. Interestingly, mutations that cause expression of a homeotic gene in the wrong place with

FIGURE 4.18   On the left is an image of a normal fly head, seen straight on. On the right is the same view of a head from an *Antennapedia* mutant in which a change in gene expression causes production of legs where the antennae should have been. We talk more about the mutations in Chapter 5. (Images of male and female *Drosophila melanogaster* from *The Physical Basis of Heredity* by Thomas Hunt Morgan [Philadelphia: J. B. Lippincott Company, 1919].)

the resulting production of an organ not normally seen in that location (*ectopic expression*) can cause a very different set of characteristics than loss-of-function mutations in that same gene. The dominant *Antennapedia* mutation alters gene regulation in such a way that cells that should turn on the battery of genes, whose functions lead to antenna development, instead activate a set of leg-building genes.

Some genes are responsible for left–right asymmetry, such as specifying the internal asymmetry of the human organs – heart on the left, liver on the right, and so on. Some people who have their hearts on the right instead of the left have two damaged copies of a gene called the *situs inversus* gene (Figure 4.19). You might think that this means that the damaged copy of the *situs inversus* gene makes the heart end up in the wrong position. You might also think that hearts end up on the right unless the *situs inversus* gene is working correctly. The story is more complicated and interesting than that. What actually happens is that when the *situs inversus* gene product is missing, the developing body does not know whether to put the heart on the right or on the left and so assigns the sidedness randomly. So half the people with two damaged copies of the *situs inversus* gene have their

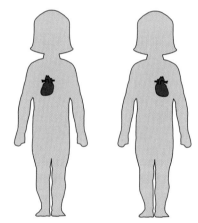

FIGURE 4.19   Each of these children has two damaged copies of the *situs inversus* gene. The lack of the *situs inversus* gene product removed a regulatory step that normally would tell the body which side to put the heart on, so in each case the laterality got assigned randomly. In one case, one time the heart happened to get assigned to the left where it belonged, and in the other case it happened to get assigned to the right. In fact, in these people, more organs than just the heart are affected. Imagine how confusing it would be to sort out the mode of inheritance for this gene for which half of the affected individuals with the gene defect have the normal phenotype. It is an example of a situation in which studying animal models of disease can greatly simplify our understanding of what is going on.

hearts on the right, and half the people with two damaged copies of the *situs inversus* gene have the hearts in the normal position on the left. Lack of *situs inversus* protein does not cause right-sided placement of the heart; it causes a random decision to be made as to whether to go to the right or the left.

As the embryo develops, changes in gene expression gradually cause more and more specialized events. The very first sets of gene expression in the egg direct cells to divide and migrate and establish specialized pools of precursor cells. Very early actions by transcription factors set off cascades of gene expression changes downstream of the initial transcription event, including production of other transcription factors that are each followed by their own complex array of gene expression changes. Genes specify where the organs are located, what form they take, and what functions they can carry out. For example, regulatory proteins that affect expression of genes involved in formation of the eyes must be expressed in just the right amount and at just the right time, or the baby might end up with damaged eyes or even no eyes at all (similar to the flies without eyes).

So we see that the human body is made up of highly specialized organs and cells, each with their own specific patterns of gene expression running chronically or sometimes in response to some stimulus from the environment. Liver cells produce certain key enzymes in response to transcription factors binding to the regulatory elements in the genes that produce the enzymes. Similarly, signaling molecules in nerves, ion channels in kidneys, hormones carrying messages from one cell type to another, and structural molecules making up muscles and bone all originate in the activities of transcription factors interacting with regulatory sequences for those genes. However, there is another layer of complexity to the makeup of a human being, and that is the cascade of gene regulation events that causes some cells to take on the structure of a nerve and gain the ability to express neurotransmitters even as a cell in a different developmental lineage is taking on the ability to make immunoglobulin molecules that help protect us against infection. The orchestration of this complex series of gene expression events is initiated when the sperm fertilizes the egg, but the end of this orchestrated piece is not birth or even completion of adolescence because, frankly, we continue changing our gene expression capabilities throughout the aging process. If we look at our makeup at the level of gene expression, we are truly in a process of "becoming" throughout the full lengths of our lives.

Eye-building Genes in Flies and Humans

So what about the missing eyes in the third set of flies she examined? Flies homozygous for loss of function mutations in the PAX6 gene arrive in this world with no eyes. The PAX6 gene encodes a transcription factor (or regulatory protein) that binds to regulatory elements in multiple different genes that are important for eye development. In the absence of PAX6, some of the switches to which this regulatory protein should bind are left empty. For genes that need PAX6 to operate their switches, regulation of transcription changes when PAX6 is missing or if there is not enough of it.

Does this really have anything to do with humans? Sometimes babies are born with no eyes or very tiny, nonfunctional eyes. Since it turns out that humans have a version of the PAX6 gene, we have to wonder whether damage to the PAX6 gene could be the cause of the lack of eyes in these babies. For more than a decade it has been known that human beings who have one damaged copy of PAX6 make about half the usual amount of PAX6 transcript. They often suffer from a trait called aniridia, in which part or all of the iris of the eye is missing but most of their ocular structures and functions are normal. Does this mean that PAX6 has nothing to do with missing eyes in human beings? No. In fact, one child has been found in whom both copies of the PAX6 gene were damaged so that the child could make no normal PAX6 protein. This child was born too ill to survive long, with head and face deformities, central nervous system problems, and no eyes. Such extensive problems were not the result of any massive trauma or infection before he was born. Rather, the problem was an error in gene expression due to the lack of one specific regulatory protein that was needed to operate some critical regulatory element switches. It seems a terrible thing that a change in the blueprint too small to see even under the most high-powered microscope should be able to have such devastating consequences in this newborn child.

Where Do All of the Differences in Proteins Come From?

Are mutations only present in those who have some distinct phenotype such as cystic fibrosis or abnormal eyes? Are the changes in phenotype always a simple matter of a protein that is missing or can't work? As we will see in the next chapter, there are many different kinds of mutations, and they can have amazingly different functional consequences. There are also a variety of different factors that can lead to mutations happening in the first place, and mutations have been happening throughout human history. So join us in Chapter 5 to learn about why we consider each and every one of us to be a mutant.

Study Questions

1. What is an mRNA and what role does it play in the cell?
2. List the sequence of amino acids that would be assembled based on the following mRNA sequence:
   5'-AUGGGCUACCCCGUGAUUGCUGA-3'
3. What happens to the exons and the introns after the cell has finished splicing the transcript?
4. What are we referring to when we talk about the redundancy of the genetic code?
5. Draw a diagram of a primary RNA transcript from a gene containing two introns and label the diagram.
6. What processing steps must be carried out to change the primary transcript from question 5 into the mature mRNA that can be exported from the nucleus?
7. What is a codon?
8. Where does the heart end up located in an individual who has defects in both copies of the *situs inversus* gene, and why?
9. What is ectopic expression?
10. Draw a diagram of a tRNA used by the ribosome to add a lysine to the growing protein chain.
11. What is a branch site or lariat site?
12. What is a modular gene?
13. How is transcription like translation? And how is transcription different from translation?
14. The number of different proteins made by the cell is larger than the total number of protein coding genes. How does the cell accomplish this?
15. How can mutation of a single nucleotide in an intron cause an alteration in the protein product of that gene?
16. The genetic code has only four letters. How can it be used to specify 20 different amino acids?
17. What does Figure 4.1 tell us about which codons are used to specify a particular amino acid that is not obvious from the way the information is shown in Table 4.1?
18. Gene A produces an enzyme whose function is critical for the health of the cell, and Gene A is expressed in every cell type in the body. Gene A has a defect in exon three that renders the enzyme incapable of

carrying out its usual enzymatic reaction, yet only cells of the nervous system show signs of deteriorating health in the affected individual. Why are the non-neuronal cells healthy while the neuronal cells are ailing in this individual?

19. Why might changing a glycine to a phenylalanine be expected to have an impact on the function of a protein?

20. Why might changing glutamic acid to histidine be expected to have an impact on the function of a protein?

Short Essays

1. Biologist who are asked how many amino acids are encoded by the human genetic code consistently answer "20". And yet a very small number of human proteins include a 21st amino acid, selenocysteine, which is only made and used when selenium is present in the environment. If there is a 21st amino acid, why do we all so blithely talk about 20 amino acids and codons that encode 20 amino acids? Since there is no codon dedicated to selenocysteine, how does it get incorporated into proteins? What happens when there is no selenium around? As you consider these questions please read "A forgotten debate: Is selenocysteine the 21st amino acid?" by Robert Longtin in *Journal of the National Cancer Institute*, 2004;96:504–5.

2. Drugs used to be developed by trial and error, as happened with the discovery of the ability of willow bark, and the aspirin contained in it, to reduce pain and fevers. Recently rational drug design makes use of information on protein structure to predict which compounds would be good to try as drugs. How is this done? As you consider this please read "Drug discovery: Pulled from a protein's embrace" by William L. Jorgensen in *Nature*, 2010;466:42–3.

3. We have talked about the genetic code used by ribosomes to translate the information contained in mRNA molecules, but scientists are starting to talk about a second kind of code. What is this newer version of the code and what is it used for? As you consider this question please read "The code within the code" by Heidi Ledford in *Nature*, 2010;465:16–17.

4. We often mention several key forms of post-translational modification of proteins such as phosphorylation, but new forms of modification are continuing to emerge from the ongoing investigations of the diverse forms of proteins carrying out cellular functions. What is one of the new forms of modification that has been found, and how far back in the evolutionary tree do we see signs of this form of modification? As you consider this question please read "Ampylation is a new post-translational modification" by Melanie L. Yarbrough and Kim Orth in *Nature Chemical Biology*, 2009;5:378–9.

Resource Project

The Jena Library of Biological Macromolecules at the Institute for Molecular Biology in Jena, Germany, houses a huge body of information on structural biology and the structures of large biomolecules. Go to the IMB Jena site www.imb-jena.de/IMAGE.html and look around at some of the images of macromolecules. Then look under the General heading of the section on Basic Information on Biological Molecules to find information on The Very First Three-dimensional Biopolymer Structures. Which protein was the first for which the first three-dimensional structure was worked out? And which were the first DNA-drug complex and the first DNA-protein complex with elucidated three-dimensional structures? When were each of these structures published?

Suggested Reading

Articles

"Genome at 10: The hunt for the 'dark matter'" by Peter Aldhous and Michael Le Page in *New Scientist*, 2010;2765 (25 June 2010).

"Protein diversity from alternative splicing: a challenge for bioinformatics and post-genome biology" by D.L. Black in *Cell*, 2000;103(3):367–70.

"Alternative splicing in the postgenomic era" by Benjamin J. Blencowe and Brenton R. Graveley (2007, Springer).

"Homeobox genes and disease" by E. Boncinelli in *Current Opinion in Genetic Development*, 1997:7(3):331–7.

"Prion protein misfolding" by L. Kupfer, W. Hinrichs, and M.H. Groschup in *Current Molecular Medicine*, 2009;9(7):826–35.

"Cancer genome sequencing: a review" by Elaine R. Mardis and Richard K. Wilson in *Human Molecular Genetics*, 2009;18(R2):R163–R168.

"Dark matter transcripts: sound and fury, signifying nothing?" by Richard Robinson in *PLoS Biology*, 2010;8(5).

"Illuminating the dark matter of the genome" by Magdalena Skipper in *Nature Reviews Genetics*, 8;2007:490–1.

"Maps, codes, and sequence elements: can we predict the protein output from an alternatively spliced locus?" by S. Sharma and D.L. Black in *Neuron*, 2006;52(4):574–6.

Books

Molecular Cell Biology (6th Edition) by Harvey F. Lodish and Arnold Berk (2008, St Martin's Press).

Protein Synthesis and Ribosome Structure: Translating the Genome edited by Knud H. Nierhaus and Daniel Wilson (2004, Wiley–VCH).

We Are All Mutants: How Mutation Alters Function

The Human Genome. DOI: 10.1016/B978-0-12-333445-9.00005-8

THE READER'S COMPANION:
AS YOU READ, YOU SHOULD
CONSIDER

- How the loss of function of a gene product relates to the mode of inheritance.

- How the gain of function of a gene product relates to the mode of inheritance.

- That the term mutation is both a process and the result of that process.

- That mutations may have silent, deleterious, or even beneficial effects on function.

- Differences in usage of the terms mutation and polymorphism.

- The different causes of mutation.

- How mutations that do not substitute an amino acid can alter protein sequence.

- How chromosomal aberrations such as translocations can bring about a gain of function.

- How mutations can happen in the absence of environmental exposure.

- The different types of mutations that can alter coding sequence.

- Why copy number matters.

- How changes in a repeat length can affect protein products.

- How gene size affects mutation target size.

- Why some kinds of deletions are not easily detected by standard mutation screening.

- How mutations outside of coding sequence can alter protein structure.

- Why men are at increased risk of passing along certain kinds of mutations as they age.

- Why apparent anticipation may not always be real.

5.1 WHAT IS A MUTATION?

Julia was born with blonde hair. Scott was born with red hair. While time has converted those colors to gray, the DNA sequences that caused those different colors remain. The great variety of hair colors in the population result from numerous mutations that have occurred in the genes that control hair color. The ongoing process of mutation over history makes us unique through creation of many diverse alleles among the approximately 24,000 human protein-encoding genes. Thus, while we work on problems that have been created by harmful mutations, we are pleased that many benign mutations combine to make each of us different. We chuckle over the idea that many of the things that we most like about ourselves are visible indicators that we are mutants in a very positive sense that offsets the typical Hollywood movie stereotype of mutants as scary creatures.

We are going to make a daring assertion: we are all mutants, every single one of us. Those of us who struggle with ailments and curse our appearance in mirrors are mutants. Those of us who beam with pride over our fine health, good looks, or talents are mutants. Every human being carries a large number of mutations that alter the function of a large number of genes. In that sense we are all mutants – at not one, but many genes.

Actually, every cat, dog, horse, chicken, armadillo, dolphin, lobster, gecko, fruit fly, slime mold, corn plant, piranha, wildflower, and redwood tree is a mutant. If we were to look at the sequences of different individuals from any one species we would find a vast sea of differences in the sequence of one individual as compared to another. In the midst of the many sequence differences that give us our individual appearances and talents we also find a surprising load of mutations scattered throughout our DNA. Some of the mutations are benign. Some combine to give each of us our unique characteristics, contributing to our virtues and our flaws, be they medical, cosmetic, or behavioral. Other mutations are carried silently, some having no potential to cause harm and others waiting unrecognized to be manifested only if they come together with another allele of that same recessive locus. It is estimated that each of us is a carrier for multiple mutations that would cause severe problems if both copies were knocked out. Fortunately, many of those harmful recessive mutations are rare, and we never find out that we have them unless we end up having children with someone else who is a carrier for that same defect (Box 5.1).

In Chapter 1, we talked about Mendel's observations that true breeding strains of a

---

BOX 5.1

BIRTH DEFECTS

The vast majority of babies are born healthy. In populations in which the parents are not related to each other, more than 97% of the children can be expected to be born without birth defects, and more than 99% will be free of major birth defects. This reflects, at least in part, the rarity of dominant alleles that can cause birth defects, but it also reflects just how often the hidden genetic defects (the recessive alleles) in one person are different from the hidden genetic defects in another. So what happens when people who are related have children together? They have an increased chance of sharing defective alleles, and the rate of birth defects may double. But it is important to note that most babies are born healthy.

According to the March of Dimes, some of the most common birth defects include cerebral palsy, spina bifida, cleft lip/palate, lack of one or both kidneys, obstruction of the small intestine or urine passage, diaphragmatic hernia or abdominal wall malformations, chromosomal anomalies, of which Down syndrome is the most common, and limb malformations. Although there are genetic components to some birth defects, many people born with birth defects go on to have normal children, and good prenatal care reduces the chance of some types of birth defects. One of the most noticeable advances in the prevention of birth defects came from the observation that addition of folic acid to the diet of pregnant women reduces the frequency of spina bifida.

TABLE 5.1   Alternative definitions for mutation and polymorphism

	Mutation	Polymorphism
Formal definition	Any DNA sequence change, without regard to whether it has functional consequences OR The process by which a DNA sequence is altered	Any sequence variant present at a frequency of 1% or higher in a population, whether or not it has functional consequences
Other common usage	A DNA change that has functional consequences, i.e., a change that causes a phenotype	Benign sequence variants (those that do not affect phenotype) even if the frequency is unknown or less than 1%

plant would continue to breed true. We can follow the transmission of a trait from one generation to the next because in most cases the information being transmitted is stable and passes in unchanged form. But very rarely the information changes, and once changed the new version of the information thereafter passes along in a stable and heritable fashion.

There are a variety of different terms used when we talk about changes in the DNA sequence and not everyone uses these terms in the same way. The term *sequence variant* is the most neutral term for changes in the DNA sequence. It implies nothing about allele frequencies or causation. In some cases we may use more specific terms that will be discussed in this chapter, terms that also imply nothing about whether or not the change is causative and results in the phenotype. Sequence variants may also be referred to using terms such as single nucleotide polymorphisms (SNPs) that change only one base of the DNA sequence. The term mutation also refers to the molecular process that alters the information. *Mutate* is the verb used to describe the action of changing the DNA sequence, but we will sometimes use other terms to describe specific actions that result in a mutation, terms like *alter*, *substitute*, *insert*, or *delete*.

Some terms for sequence variants have definitions that have been formalized by some branches of the field of genetics. In other cases the terms are in common usage for situations that are not covered by the formal definitions

(Table 5.1). Formally, a mutation is any change in the sequence, without regard to whether or not it is causing a phenotypic change, but there are many papers in the literature that use the term mutation to specify only those sequence variants that cause a phenotypic change. Formally, a polymorphism is any sequence variant present in more than 1% of a population, but many papers in the literature use the term for any sequence change without regard to frequency; some papers reserve the term to specify only those sequence variants that are known to be non-causative. Proper usage in cases where the sequence variant does not cause a phenotypic change would be "benign polymorphism."

Because these different definitions are in regular usage it is important to take note of the definition being used when reading books or papers that use the terms mutation and polymorphism. Often the authors of whatever you are reading will not state the definition they are using, so it is important to look at the way the term is being used and the information that accompanies the term to tell which definition is being employed. It is also important to be careful how you yourself use or define these terms when you are using them. Remember that the term sequence variant will always be the most neutral, implying nothing about causation or allele frequency.

A human mutation takes the form of a difference in the sequence of the DNA. Most of the mutations are on the chromosomes that are located in the nucleus, since the vast majority of the DNA is there, but some of the mutations can also be found in the mitochondrial DNA. Mutations include changes from one base pair to another (for example, AT to GC), deletion of one or more base pairs, or insertion of one or more bases.

5.2 THE PROCESS OF MUTATION

If you compare the sequence of any two human beings, you will find large numbers of differences between them. However, if you consider what fraction of the whole DNA sequence shows differences, the number is tiny. In fact, any two human beings share about 99.9% identity in their DNA sequence. This is a very high level of similarity, but much of our attention in the field of genetics is focused on the differences – on average one difference per 1000 bases when we compare any two individuals. When we consider the size of the human genome – more than 3,000,000,000 bp – we find that one in a thousand bases adds up to millions of differences between any two individuals.

Where do all of these differences come from? Some differences come from environmental exposures, but as we will discuss, some of them happen without any external factors at all.

Chemicals Can Cause Mutation

The world around us is filled with chemicals. Some are naturally occurring molecules present in the complex foods we consume. Some are synthesized molecules with long complex names that appear on the labels of products we consume. Some are present in products that bear labels that warn against consumption. Some chemicals bear warning labels because they will cause irritation or allergies or nausea, but some bear warnings because they have the potential to cause mutations.

How would we test a chemical to find out whether it has the potential to cause genetic damage? A test called the *Ames test* (Box 5.2) was developed to identify things that have the potential to cause mutations. Some chemicals called mutagens can cause mutations, and most mutagens are also carcinogens – things that can cause cancer if they mutagenize a cancer gene, something we will discuss in Chapter 10. Thus the Ames test not only tells us which chemicals can cause mutations but also tells us which chemicals have the potential to cause cancer.

An Example of a Chemical (EMS) that Causes Mutations

There are many chemicals that are known to cause mutations to occur at high frequencies – but perhaps the best known, and most commonly used, of these chemicals is ethyl methane sulfonate (EMS). EMS is a small molecule that is easily taken up by most organisms. Once inside a cell, EMS has the ability to chemically modify the bases (most notably, guanine) in the cell's DNA. This matters because the chemically modified guanine no longer pairs with cytosine, but rather pairs with thymine. Usually, modified bases will be removed (and replaced with the normal version of the base) by the very efficient repair enzymes that are constantly patrolling the cell's DNA looking for breaks or abnormal bases. Replication of normal DNA faithfully produces two identical replicates of the original sequence according to the base-pairing rules.

But if the modified base, we'll call it G*, persists until replication then a mutation can result. When G* is introduced into one position in the sequence, replication of that DNA will lead to normal replication of the strand that contains the C but the strand that contains the G* will undergo an abnormal event that pairs T with G*. As we can see in Figure 5.1, that abnormal pairing resolves after the next replication to yield two pieces of double-stranded DNA with the original sequence, one with the abnormally paired G*T, and one containing a new AT pair where the GC pair had previously been.

BOX 5.2

THE AMES TEST FOR CHEMICAL MUTAGENS

The Ames test uses bacteria as a very sensitive biological indicator of whether or not a substance can cause a change in DNA sequence. Dr Bruce Ames started out with a bacterial strain with a mutation in a gene required to make the amino acid histidine. Because of the defect in this gene, the bacteria can only grow on food that provides histidine. When he exposed the bacteria to a filter paper disc containing a *mutagen* – something that can cause changes in the DNA sequence – many of the bacteria ended up with a new mutation in the gene required for making histidine. This new mutation restores the ability of the bacteria to grow on food lacking histidine. If he exposed the bacteria to a disc containing something that is not a mutagen, very few changes

took place in the histidine production gene and the bacteria did not grow on the histidine-deficient food. We think of the Ames test as a way to check out "chemicals," as if knowing the chemical structure and name of something makes it potentially more harmful; in fact, some chemicals are mutagens and some are not. There are many naturally occurring compounds that also test positive as mutagens in the Ames test, including substances found in moldy peanuts and over-cooked hamburgers. Just asking whether something is tagged with the dread words "chemical" or "additive" or asking whether it comes from some safe-sounding source such as an herbal compound or a health-food store does not tell us whether it is in fact safe to consume.

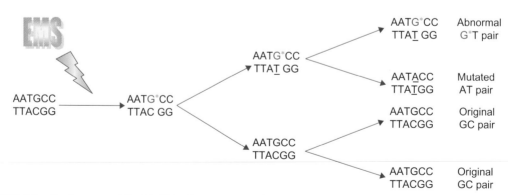

Disc with harmless non-mutagenic substance placed in petri dish of his- bacteria causes no mutations so bacteria do not grow.

Disc with mutagen placed in petri dish of his- bacteria causes mutant bacteria to grow closer to the disc where the concentration of the mutagen is the highest.

FIGURE 5.1   Replacement of G with G* leads to presence of an AT base pair in place of the GC base pair of the original sequence. This process will also lead to some copies of the sequence that retain the original GC base pair. So some daughter cells will receive the original sequence and some will receive the mutated sequence.

If the change from the GC base pair to an AT base pair occurs in a coding region it can change the codon, which in some cases will change the amino acid that is inserted in the growing protein chain. For example, EMS could change the codon GCC (alanine) to ACC (threonine). At other points in the sequence, GC→AT mutations might create premature stop codons, disrupt splicing, or damage sequences that bind critical regulatory proteins. EMS can also cause other types of DNA damage, and thus other classes of mutations, but GC→AT changes are the most common result.

In most organisms EMS is a very potent mutagen. For instance, when EMS is used to induce mutations in fruit flies in Scott's lab, the average frequency of new mutations is approximately one new mutation in a given gene in every 500–1000 progeny. Precisely because EMS is so potent a mutagen, and thus a potent carcinogen, it must be handled with extreme care, and only by people with extensive training in its use. The same can be said for many other chemicals that are potent mutagens.

Radiation Can Cause Mutation

In 1946, Hermann Muller was awarded the Nobel Prize in Physiology or Medicine for his work in the 1920s showing that X-rays can cause mutations. X-rays are only one of many different forms of radiation. Some forms of electromagnetic radiation such as radio waves and visible light do not cause genetic damage. Some substances that emit alpha particles are a hazard to genetic material if they are ingested or inhaled, but the path the radiation travels is so short that the particles will not penetrate to internal organs from outside the body. Other substances emit gamma particles that can penetrate tissues to break DNA; broken DNA can lead to harm from mutations or major chromosomal rearrangements called translocations (see Chapter 6), or can be helpful when used

for radiation therapy of cancer. In some cases breakage of DNA can lead to the kinds of rearrangements of large sections of chromosomes that we discuss in Chapter 6. Ultraviolet light can change the DNA by causing adjoining thymines to become chemically bonded into abnormal dimers; sometimes the cell's DNA repair mechanisms succeed in cutting out the thymine dimers and sometimes a mutation results.

We often hear about radiation as something that comes from such things as nuclear reactors and nuclear disasters (Box 5.3). But we are also naturally exposed to chronic low levels of radiation produced by our environment. Sunlight includes ultraviolet light. Many types of rocks, such as granite, give off low levels of radiation because they contain very tiny amounts of radioactive elements that are part of the rock's composition. Radon gas that contaminates some basements is a more significant radiation concern and is considered a major cause of lung cancer. Since all of us live our lives around low levels of radiation from things like rocks, the question is not so much one of whether radiation can cause mutations but rather one of how much radiation is a problem. Standards for what constitutes safe levels of radiation exposure have been set through studies of mutation rates resulting from different kinds of radiation exposure (Box 5.4).

Fortunately, our cells possess natural mechanisms for fixing some kinds of damage, such as certain types of breaks and unusual structures that radiation can cause in DNA. Do these exist to cope with natural background levels of radiation? Perhaps, but many of the cellular mechanisms that deal with radiation are the same cellular mechanisms that deal with the normal processes for DNA replication. The cell regularly evaluates and fixes sequence differences during proof reading steps used to keep error rates low during normal DNA replication. The cell is also designed to repair single-strand breaks in DNA that are a natural part of the process of DNA replication.

BOX 5.3

RADIATION CAN CAUSE MUTATIONS

In 1986 an accident destroyed one of the nuclear reactors at Chernobyl in Russia with a resulting release of tons of radioactive materials into the air. Thirty-one people are reported to have died from radiation exposure, and millions of people were exposed to radiation dispersed across the surrounding countryside. There are reports of increased thyroid cancer and detectable DNA alterations in children of parents who were exposed to the radiation. Government reports project a several percent increase in cancers in response to radiation exposure in the Chernobyl region. And there are reports of increased mutation rates in plants in the region. All of this supports what had been found previously: that in the case of nuclear events, we need to be concerned not just about the immediate high level exposures of a terrible nuclear disaster, we must also focus attention on chronic exposure to radiation during the aftermath, which can result in mutations at a level that should be of serious concern.

BOX 5.4

HOW DO WE MEASURE MUTATION RATES?

Some studies measure mutation rates by looking at sequences of specific genes, while others look at things like increased cancer risk. More recently we have gained the ability to compare complete genome sequences between parents and children to look for mutations. Researchers struggle to sort out just how much radiation under what conditions will cause a problem. *The bottom line is: radiation can cause permanent changes in the sequence of DNA if someone is exposed to enough of it for a long enough time, but fear of radiation may not always be warranted.* Some kinds of exposure, such as medical X-rays, are purposefully designed to use low enough levels to be safe. We are all exposed regularly to background levels of naturally occurring radiation that cannot be avoided, but our bodies are adapted to live with this. There are some situations involving radiation that we should consider very carefully before putting our genetic blueprints in harm's way through exposure. Problematic situations might include living around radon, smoking, excessive exposure to solar radiation via sunburns, or (in case it ever comes up in your neighborhood) joining in a cleanup crew after a radioactive accident if we are not trained to work with radiation and do not have appropriate protective gear. In cases where the risk of exposure is substantial, there are measures that can be taken such as radiation shielding, radiation monitors, and badges that can detect levels of personal exposure. So radiation does not necessarily have to be harmful if we are in a position to take appropriate precautions when working with it, but it can be a major problem in large-scale environmental scenarios such as Chernobyl where many people live around radiation without the appropriate tools and information to protect themselves.

BOX 5.5

MUTATIONS WITHOUT ENVIRONMENTAL EXPOSURE

We cannot blame all of our mutations on exposures to things around us. A certain amount of mutation goes on no matter what we eat or drink, apparently as a simple result of the rate of errors made by the machinery of the cell as it copies the DNA. Even the most pristine lifestyle will not protect against the fact that the natural machinery by which the genetic blueprint replicates has built into it limits on how perfectly it can carry out its copying functions. As the polymerase moves along the DNA making a new copy, it must correctly read a base and put its correct complement into place, every time, over and over again, for more than a billion bases of sequence if it is going to correctly replicate the entire genome once without making any mistakes. Frankly, without any exposure to chemicals or radiation at all, sometimes the polymerase gets it wrong. At some points in the DNA the rate at which polymerase makes errors is increased because a naturally occurring chemical modification (that only gets made to some of the bases in the DNA) makes the base "look" like a different base to the polymerase as it comes through making copies. When you take antioxidant vitamins, you are helping your cells to repair oxidative damage to your DNA that your cell machinery must work constantly to repair. So there is already a base-line rate of mutation going on before we ever add in additional insults such as smoking to further aggravate the ability of the cell to get it right, every time, over and over and over.

Mutations Can Happen without Environmental Exposure

Finally, we have to note that some amount of mutation will happen without any chemical or radiation exposure of any kind (Box 5.5). So in cases where a new mutation arises, we will not always find some smoking gun in the form of a chemical spill or a food additive to blame.

The natural process of DNA replication is not perfect. Sometimes an error will take place during replication. As the cell replicates DNA, it has to correctly read each of more than 3 billion bases of sequence. But DNA polymerase has an inborn error rate independent of outside exposures. Of the millions of polymorphisms in our genomes, many are old, having arisen in some distant ancestor. But we can compare the sequence of parents and child to see that new mutations are arising on a regular basis. The cell has proof reading processes to catch these "typos" in the blueprint and correct them, but some changes still slip through. Only rarely, such as in the case of cancer, will a new mutation lead to a phenotype so dramatic that we will realize that something new has happened.

So even in a perfect world with no mutagens at all, we would all still have some amount of mutation going on in our cells over the course of a lifetime (Box 5.5). Fortunately, many of those mutations fall between genes, or within introns, or fail to alter coding sequence, or happen in a skin cell that then dies and is sloughed off. Much of the time, even if a mutation takes place in one of our cells, we will not pass it along to our children unless the mutation happens to take place in the germline – in the lineage of cells that produces the eggs or sperm that carry the blueprint along to the next generation.

Most of these differences in the DNA sequence have been handed along to us by our parents, but new mutations can arise and be passed along to our children. Although we cannot control what we received from our parents, we can affect the chances of creating a new genetic problem that will plague our descendants. So you are going to be stuck with some level of mutation going on no matter how pristine an existence you live, but working around radiation without taking appropriate protective precautions or living on a toxic waste dump site can actually increase the chances that you will pass a new mutation along to your children. So the next time you find yourself wanting to roll your eyes at what some environmentalist is saying, stop and ask yourself how much you know about what causes mutations and what effects those mutations can have. In some cases the answer will be that the particular environmental situation is actually already safe enough, but sometimes the answer will be that there is something that needs to be cleaner, not only for our protection but also for the protection of future generations.

How Often Do Mutations Happen?

To some degree, mutation can be considered a spontaneous process. DNA polymerase, the enzyme that executes DNA replication, is an unbelievably accurate enzyme. Still, it inserts the wrong base at a frequency of about one error in every 10 billion bases replicated. That number may seem small, but remember that every time the human genome is replicated, DNA polymerase must copy approximately 6 billion base pairs. Thus, on average, slightly less than one new base pair mutation will occur every time the human DNA complement is replicated. If that number seems small to you, remember that a human being will go through more than 1 quadrillion complete replications of his or her DNA (cell division) in his or her lifetime. Thus many cells in our bodies might be expected to carry at least one base change mutation, but only some of those changes will actually make

any functional difference to the gene product and cellular functions.

Realize that the vast majority of new mutations will occur in somatic cells and that even those deleterious mutations that do occur will likely have little or no effect because they only exist in the single cell in which the mutation event occurred. In most cases, the resulting impairment in gene function will be "covered" or "masked" by the un-mutated copy on the normal homolog. Moreover, even if such mutations were to result in the death or impairment of a single cell and its somatic descendants, it is unlikely that the loss of a single cell or cluster of cells would be terribly deleterious to the organism. (We will, however, consider a rather dramatic exception to this generalization when we discuss the genetics of cancer in Chapter 10.)

Perhaps of more interest to us is the frequency of mutations in the germline, mutations that get passed along to a child who then carries the mutation in every cell in the body. What fraction of human gametes might be expected to carry a new mutation that will impair or prevent the proper function of a given gene? Based on assaying the frequency of those new mutations in known genes that have phenotypic consequences, scientists have concluded that each gene in the human genome will be mutated (that is to say functionally altered, not just changed silently) only once in every 100,000 gametes.

By this measure, Mendel does not seem to have been so far off – mutation is a very rare process indeed. Remember, however, that the actual frequency of germline DNA changes is much higher because of the types of silent mutations mentioned above. The observed mutation rate that we usually deal with in terms of clinical risk estimation measures only those changes that dramatically alter gene function. Accordingly, whenever you think about mutation rate, stop and ask yourself whether you are looking at all heritable changes happening to the DNA (many) or whether you are looking at changes that are detectable as a phenotypic change in a living person (a much smaller number). And remember that some agents in the environment can greatly increase the mutation rate above the baseline level of errors made by the polymerase during normal copying of the DNA (see Boxes 5.2 to 5.3), so the rate of mutation going on in some individuals experiencing substantial environmental exposures may be higher than in other individuals with little environmental impact on their mutation rate.

Efforts to discuss functional *mutation rates* – the rate at which a mutation can cause a change that will alter some function or structure in the cell – are complicated. The mutation rate relative to a particular disease or gene depends on what size of gene we are talking about, what type of mutation we are talking about, what region of the genome we are looking at, and whether we are looking at a mutation in somatic cells, sperm or eggs. It also depends on whether we are asking about mutations with detectable biochemical effects, mutations that alter the amino acid sequence, or mutations that simply change the DNA sequence. The mutation frequency for a gene that is 900 base pairs long is likely to be much lower than the mutation frequency found for a gene that is 9000 bp in length. The mutation frequency will be different for deletions vs. point mutations. Even point mutation rates can vary more than tenfold, with an occasional hot spot in the genome showing an unusually high mutation rate that might be a thousand times that of some other bases in the genome.

One mutation in any given gene per 100,000 people might seem like a very rare event, but consider this: if you only had one gene, that mutation rate would mean that your chances of passing a new mutation along to your child would be about the same as the chance of winning one of the big lotteries. One change per gene per 100,000 people seems less rare when we are looking at the ~24,000 protein coding genes and many more non-protein coding genes that exist in each of us.

Our ability to determine the mutation frequency is improving as whole genome sequencing is coming into practice. In the past, we started out with data on sequencing of a selected sample of genes and then projected those rates across the genome or did estimations of mutation rates per base pair or relative to different gene target sizes. Now we can measure the mutation rate directly at each position in the human genome sequence. As more whole genome sequencing takes place we will begin to see where mutation hot spots or cold spots exist in the genome. Initial experiments using whole genome sequencing to evaluate the mutation rate per base for all mutation types, not just causative mutations, shows a rate of about one change per 30 million bases. Interestingly, the mutation rate seems to be similar for humans, nematodes and fruit flies, and most new mutations are benign, causing no detectable phenotype.

An important implication of such mutation rates is this: a continuing supply of new mutations provides the genetic variation underlying phenotypic variation. The millions of years of evolution that led to the human race and human phenotypic variation could not have happened without the continuing supply of mutations that create genetic variation underlying the phenotypic variants on which natural selection acts.

5.3 HOW WE DETECT MUTATIONS

Light microscopes can view whole chromosomes, and electron microscopes can see smaller structures such as the beads on a string image of DNA wound around histone cores. But even the highest-powered microscope cannot read the order of As, Cs, Gs, and Ts in the DNA. Yet clearly in a world where headlines trumpet news about the human genome sequence, someone is managing to read the order of genetic letters contained on those chromosomes we were looking at. How?

Sequencing technologies have evolved in a long, gradual manner. Almost 30 years ago, reading even a small bit of DNA sequence was technically difficult and required weeks or months of laborious effort. Now we have simple enzymatic methods that let us "read" the order of genetic letters present. In Chapters 11 and 12 we will talk about how we find a particular gene that we want to study. Here we will consider how we look at the sequence of that gene once we have found it.

Sequencing

Sequencing is basically a set of biochemical reactions that let us determine where the bases are located along a DNA chain. We count the number of letters away from the beginning of the sequence and ask what letter is at that position.

We will start out by looking at how to tell the sequence of a piece of DNA that has been produced by PCR amplification, a process we described in Chapter 2. For instance, if our DNA fragment has the sequence GCTACCGCTTTCGACTGATGGCAT, we can perform a sequencing reaction that tests for where the Ts are in this sequence. The biochemical reaction we carry out will generate copies of the sequence that are likely to stop where there is a T in the sequence. So our sequencing reaction that reads the Ts in the sequence will make a lot of copies of our target fragment, some stopping at the first T, some stopping at the second, and so on. The T sequencing reaction will produce all of the fragments shown in Table 5.2.

In the test tube, where the biochemical reactions go on within a clear drop of liquid, many copies of each of the above fragments are floating around in the solution. So we have carried out a biochemical step that creates new DNA fragments that end wherever there is a T in the sequence, but all we see is something that looks like a drop of water. How do we get useful information out of the fact that these different DNA fragments, these different "relatives" of

TABLE 5.2   The T sequence reaction creates fragments that end wherever the sequence contains a T

Sequence of fragments generated by the T reaction shown in Figure 5.2	Fragment size
GCT	3
GCTACCGCT	9
GCTACCGCTT	10
GCTACCGCTTT	11
GCTACCGCTTTCGACT	16
GCTACCGCTTTCGACTGAT	19
GCTACCGCTTTCGACTGATGGCAT	24

the target fragment, are floating in solution? We separate the fragments based on size and then ask how long each fragment is.

To separate the DNA fragments based on size, we create a gel matrix (that looks a bit like a giant, square, unflavored Jell-O Jiggler!) made up of cross-linked molecules with spaces between them through which the DNA can run. Because DNA has a lot of negative charges on it, it will move towards the positive pole of a battery or power supply (Figure 5.2). If we cut a hole (called a well) into the gel, put in our DNA samples with all of the different sizes of fragments, and turn on the electrical current, the DNA will move through the pores of the gel. The important part is this: the smallest fragments will move the fastest, and the largest fragments will move more slowly.

The use of dyes, in some cases fluorescent dyes, lets us tell where the DNA is located in the gel. If there were only one copy of each length of fragment, we would not be able to see the DNA, but since there are many copies of each length fragment, and since things the same size all run at the same position in the gel, use of dyes that detect DNA let us see where the different-sized fragments are in the gel (Figure 5.2). At a position where a group of same-sized DNA fragments runs together, they create a signal that is strong enough that we can visualize

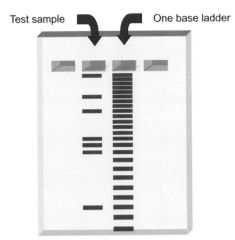

FIGURE 5.2 DNA fragments of different sizes run through gels at different rates. The smallest fragments run the fastest, and the largest fragments move more slowly. We can figure out the size of our DNA fragments by comparing our test DNA to fragments of known sizes (in this case a one base pair ladder of fragments that differ in size by one base pair). Although we use this method to size many types of DNA fragments, this is not how we do it when sequencing DNA, as we will see in Figure 5.3. A real gel would be partially transparent and the DNA would be stained with a fluorescent dye.

the DNA as a bar that looks like one of the bars in a bar code (although we actually call these bars bands).

Of course, we don't just want to know where the Ts are in the sequence; we want to know the order of all four bases. By doing four separate reactions, with each reaction designed to stop at one of the four bases, we can read the placement of each base in the sequence. If we make a gel just like the one in Figure 5.2, we can load the A reaction in the first well, the C reaction in the second well, the G reaction in the third well, and the T reaction in the fourth well, then turn on the electrical current and let the small fragments outrace the larger fragments. A mock-up of such a gel in Figure 5.3 shows us how easy it is to read the sequence. We see that the G lane contains the smallest fragment, so G is the first genetic letter in this genetic word we are trying to read. The next-smallest fragment is in the C

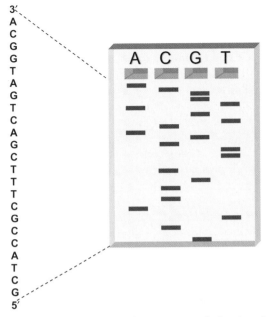

FIGURE 5.3 Diagram of a sequence gel showing the positions of DNA bands created by carrying out A, C, G, and T sequencing reactions on the piece of DNA with the sequence GCTACCGCTTTCGACTGATGGCA. The band closest to the bottom of the gel represents the smallest fragment (in this case a G) and the fragments closest to the top of the gel (the wells) are the largest fragments. Starting at the bottom of the gel and reading upwards generates the sequence listed to the left of the gel.

lane, so C must be the second letter, and the T lane holds the third-smallest fragment, making T the third letter in what we are trying to spell.

Sequencing of this kind is often done on very large gels with as many as 96 wells allowing for sequencing of many samples in one experiment. Instead of staining the DNA with dyes, the DNA is tagged with radioactivity so that the image of where the DNA is on the gel can be captured on a piece of X-ray film (Figure 5.4).

The versions of sequencing we just showed are cumbersome, but they are very good for showing how sequencing works. Do four sequencing reactions that create fragments that end with the four bases, then separate the fragments by size, and start "reading" the sequence

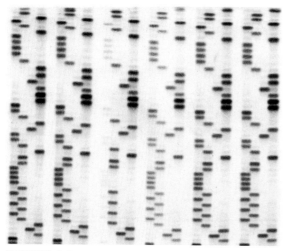

FIGURE 5.4 Picture of an X-ray film image of the same region of DNA sequenced from six different samples. Notice how the pattern from the first four lanes is repeated again in every cluster of four lanes.

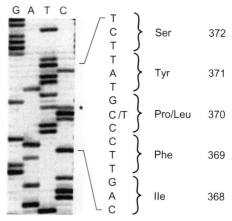

FIGURE 5.5 Sequence of exon three of the MYOC glaucoma gene, showing a mutation that changes amino acid 370 from a proline, in the normal protein, to a leucine in an individual who is affected with glaucoma. The mutation is detected as a point on the gel at which a band in the T lane runs as far as a band in the C lane so that the two bands appear side by side. (Courtesy of Frank Rozsa.)

by seeing which lane contains the smallest fragment, then seeing which lane holds the next smallest and so on.

This type of sequence reaction and display via X-ray film can be used to detect mutations in human DNA samples. If someone is heterozygous for one of the bases in the region being sequenced, two different fragments will be generated that correspond to the two different "base terminating" reactions (Figure 5.5).

Although some sequencing is still done this way, more recent technologies do not require radioactive labeling or X-ray film. Four different colors of fluorescent dyes label the reactions that end at each of the four bases. The combined batch of all four colors of fragments are size fractioned and a laser-based system reads off the color of each fragment as it moves past the monitor used to read the sequence directly into a computer. The smallest fragments arrive at the monitor first, so if we know that a green dye tags fragments ending in A, and a green fragment reaches the monitor first, we know that the first letter in the sequence is A. By

knowing the order in which different colors flow past the monitor we know the order of the DNA bases in the strand of DNA that we just sequenced. The computer graphs each fragment it detects as a peak on a chart (Figure 5.6), so for our fragment that has A as the first letter, the first peak on the charge will be green. The computer also provides a printout of the letters that correspond with the peaks so that we receive both the graphic image of the peaks and a printout of the sequence in letters.

The presence of two different peaks of different colors at the same position in Figure 5.7 signals the presence of a mutation in heterozygous form. The computer "reads" this double peak as the letter "N" instead of putting an A, C, T, or G at that position. However, the computer will sometimes list an "N" in a position that represents a technical artifact rather than a real mutation, so accuracy of sequencing is increased by human review of the sequence traces to decide whether an apparent mutation is real.

Because there are technical limitations to the sizes of DNA fragments we can measure

A T TG AC T TGGC T GT G G A TG A A GC A GGC CT C T

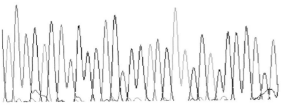

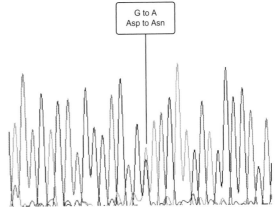

FIGURE 5.6 This DNA sequence was read by a laser-operated system that enters the data directly into a computer. It is read from left to right. Green = A, blue = C, black = G, and red = T. The letters across the top indicate which bases the computer has called for that position in the sequence. Notice that there is one dominant color of peak at each position. (Courtesy of Rosa Ayala-Lugo.)

A T TG AC T TGGC T GT G N A TG A A GC A GGC CT C T

FIGURE 5.7 DNA sequence showing a missense mutation. This shows a sequence of the same region of the genome shown in Figure 5.6. However this sequence was determined from a DNA sample from an individual with a mutation in this region of the genome. Green = A, blue = C, black = G, and red = T. At the position marked, a green peak and a black peak appear at the same position. This means that we are seeing both copies of this piece of DNA in someone who has A on one copy and G on the other copy. This change in the DNA causes some proteins made from this gene to have asparagine in the protein at a position usually occupied by aspartic acid. (Courtesy of Rosa Ayala-Lugo.)

accurately, sequence reactions often yield hundreds of bases of sequence, or perhaps even more than a thousand, but not millions. So the sequence of a whole gene is often assembled by sequencing lots of separate pieces and then assembling those partial sequences back together into the completed large sequence.

The latest sequencing technologies (Box 5.6) have actually gone in the other direction, moving towards very efficient, very rapid production of huge numbers of very short sequence reads that can be mapped back against the known genome sequence to build up an assembled sequence made out of overlapping data from huge numbers of very short sequence reads. The result is that any one spot in the sequence may well be read many dozens of times in one sequence reaction. Any one copy of the sequence generated by the sequence reaction has a small chance of making a mistake in reading any one base in the sequence, but it is unlikely that the sequencing reaction will make exactly the same mistake when that same spot in the sequence is read many times. So if we read each spot in the sequence 64 times, and see that 61 of those reads say that the base is a T, two of the reads say it is an A, and one of the

reads says it is a C, then we conclude that the correct base at that point in the sequence is T. If we had read the sequence only once, the chance that that particular base would have been read incorrectly is very small, but the chance that one of the bases somewhere in the sequence would be read incorrectly approaches a level of possibility that should concern us. However, with these new approaches, if we read each spot dozens of times we can detect the rare mistakes. Thus redundancy increases accuracy.

Another common method for detecting specific individual sequence changes makes use of enzymes called restriction enzymes which cut the DNA any place the DNA has a copy of the short sequence that the enzyme recognizes.

BOX 5.6

NEXT GENERATION SEQUENCING

Recent developments in sequencing allow for high throughput sequencing, very rapid readout of sequences from very large regions or even sequences of whole genomes. The complete known sequence of a genomic region or a whole genome can be re-sequenced through assembly of the information from a massive number of very short sequence reads that may produce data sets containing many billions of bases of DNA from one experiment. Use of robotic technologies to carry out technical steps combined with reading of sequence data directly into a computer and bioinformatics to analyze data make management of such large data sets possible. Use of high levels of redundancy (many separate reads across the same point in the sequence) helps protect against errors, and helps to distinguish between those cases that represent

an error in a single read of the sequence at that point and those cases that represent real differences in the sequence due to heterozygosity. Some of these sequencing reactions are carried out in liquid. Some are carried out using DNA templates anchored to beads or silicone chips. We are reaching the point where re-sequencing of a known genome can be done for a price in the range of tens of thousands of dollars rather than millions, and we are moving towards the day of the thousand dollar genome. When we reach that point, we can start considering how to incorporate genome sequencing into the panel of diagnostic tests available in clinics. At that point the hardest part will not be obtaining the sequence of someone's DNA. The hardest part will be figuring out what the sequence is telling us about disease risks and optimal treatments.

The sequence recognized by one enzyme can be different from the sequence recognized by another enzyme, and the enzyme cuts the DNA wherever its recognition sequence occurs. Some restriction enzymes with very short four base-pair cut site sequences occur very frequently, while others with cut sites as long as eight bases in length are found much more rarely. If sequencing reveals a mutation that happens to change a restriction enzyme cut site, then it is possible to screen many additional samples for the presence of that sequence change through testing for whether or not the restriction enzyme can cut the DNA at that point. After the DNA from each sample is cut with the enzyme, the cut fragments of DNA are run out on a gel to determine the sizes of the DNA fragments that result from the enzyme activity. If we have copies of a gene from two different sources and one of them cuts with the restriction enzyme we are using

and the other does not, then we have a restriction enzyme polymorphism (RFLP) (Box 5.7).

Sequence Tagged Sites and the Bioinformatics Revolution

The combination of PCR and sequencing brought about a massive *bioinformatics* revolution in biology. To carry out PCR, you need two primers flanking the item you want, each primer usually about 18 bases long. So if you want to study a gene, you do not even have to start out knowing its sequence – all you need to know is the sequence of two primers that flank what you want. Usually, that means you need to know less than 40 bases of sequence total to be able to get your hands on the DNA you are after.

So the big information revolution that went along with PCR was the development of the

---

BOX 5.7

HOW TO TELL IF RFLP TESTS ARE VALID

One problem with RFLPs is the need to tell the difference between a DNA fragment that fails to cut because the sequence is different and a DNA fragment that fails to cut because the enzymatic reaction didn't work. Often the piece of DNA used in the RFLP test has been generated by PCR so we can choose the piece of DNA to be tested by choosing where we place the primers. If we place the primers so that the piece of DNA between them contains not only the cut site that we want to test but also contains another cut site for that enzyme that does not vary between individuals, then we can be sure that our test fragment will have to cut at least once if the enzymatic reaction has worked. In such a case, if the invariant cut site has cut but the site we are testing did not cut, then we can be sure that the enzymatic reaction did work and we know which allele of the RFLP is present on the DNA fragment.

---

concept of the *sequence tagged site* (STS), a small piece of sequence that will let you tag the piece of DNA you want even if you don't yet know the sequence of the entire piece. When the concept of the STS was refined to its most efficient usage, researchers studying a particular gene no longer had to send a test tube containing a particular gene to another lab that also wanted to study that same gene; all they had to do was send them the sequence of the gene. In fact, we often do not even need the sequence of the gene, just the sequence of a pair of primers that can be used to PCR amplify the DNA containing the gene. Once sequence databases became available, researchers did not even have to send the sequence to the other research group. They just posted the sequence in one of the online databases such as GenBank.

The result is that experiments that used to take weeks, months, or years at the lab bench may now occupy a few hours of computer time that may or may not require some follow-up lab work at the bench. In fact, some researchers now do "virtual PCR," running their tests electronically in the databases containing the human genome sequence instead of (or before) doing experiments with tubes and chemicals.

Researchers now refer to these computer-based experiments as *in silico* experiments.

Such virtual experiments have not eliminated real lab work, but they have increased the rate at which we can accomplish the same amount of lab work. It means that more time can be spent on the meaningful parts of the experiments, such as finding out what a gene does, and less time gets spent on trying to come up with the raw materials for the experiments.

5.4 BASIC MUTATIONS

Missense Mutations Change the Protein Sequence

One type of mutation that is easiest to understand is the *missense mutation* (Figure 5.8), which is usually a change of one base in the DNA sequence but can involve more than one base. A missense mutation changes the amino acid specified to be used at that point in the protein. Since the amino acids have different properties – different size, different shape, different charge, different polarity – a change in amino acid can have a wide range of effects. Some missense

A MISSENSE MUTATION

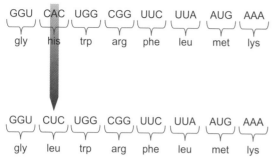

FIGURE 5.8   A missense mutation. A single base change in the mRNA due to a base change mutation in the DNA results in the incorporation of a different amino acid, leucine in place of histidine.

mutations alter the gross structure and folding of the protein. Others will change more local properties, such as charge or the ability of that point on the protein to interact with some target molecule. Many missense mutations are what we call point mutations because only one point (or genetic letter) in the sequence was changed. Some missense mutations make the protein non-functional while others may be gain-of-function mutations giving the protein a new ability, such as binding a receptor it did not previously bind.

A disease that often results from missense mutations is phenylketonuria (PKU). Babies born with PKU used to be doomed to profound mental retardation and in some cases experienced additional problems, such as epilepsy. Their inability to convert phenylalanine to tyrosine would lead to accumulation of excess phenylalanine, with especially drastic consequences to the development of the brain in infants and children. Once the cause of PKU was identified, it became treatable by dietary approaches. However, the answer is not as simple as cutting out all of the phenylalanine, since damage will also result if the phenylalanine levels get too low. According to the National Newborn Screening and Genetics Resource Center, newborn screening for PKU now takes place in all 50 states of the United States so that infants can be placed on a low-phenylalanine

diet as promptly as possible to minimize nervous system damage. Unfortunately, there are still many other places in this world where such children will have suffered irreparable damage before anyone figures out that they need to avoid phenylalanine. It may sound easy to say that putting these children on a low-phenylalanine diet fixes the problem, but it is incredibly difficult to keep the phenylalanine levels low enough, and all too often there is still some damage in the course of growing up. In addition, a woman with PKU has to go back on the severely restricted diet during pregnancy or her child will be harmed, an outcome that is far too common because of the difficulty inherent in keeping the phenylalanine levels low enough to protect the baby.

The common cause of PKU is defective phenylalanine hydroxylase, the enzyme that carries out the conversion of the amino acid phenylalanine to a different amino acid called tyrosine. Out of more than 60 different PKU alleles listed by Online Mendelian Inheritance in Man, more than three-quarters of them are missense mutations and many of the others either delete an exon or affect a splice site. Also, there are many different missense mutations, hitting different locations spread across hundreds of amino acids, rather than one or a few specific mutations found repeatedly in many different individuals. The frequency of PKU varies from one population to another, and in some populations the existence of a very small number of mutations in all of the PKU cases suggests the existence of founder effects in which many cases alive today descend from a very small number of shared ancestors. In other populations, there are many different mutations indicating more heterogeneity. All of this suggests that mutations causing PKU have arisen independently many times over the course of human history.

Nonsense Mutations Truncate the Protein

Some mutations have a very different effect from a missense mutation. Three of the 64 codons

A NONSENSE MUTATION

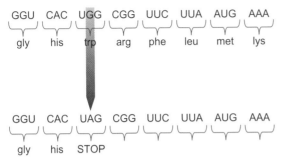

FIGURE 5.9   A nonsense mutation. A single base change in the mRNA due to a base change mutation in the DNA results in production of a stop signal that tells the ribosome to stop adding amino acids to the growing protein chain.

are stop codons that instruct the ribosome to stop adding amino acids to the growing protein chain. If the codon designating an amino acid is mutated to designate a stop codon instead, it is called a *nonsense mutation* (Figure 5.9).

Nonsense mutations have the effect of truncating the protein, leaving a shorter protein that is missing its tail end. In some cases, if the nonsense mutation happens very early in the gene, most of the protein might be missing. In some cases this means that the cell has a bunch of truncated protein to deal with, but in some cases the cellular mechanisms for dealing with problematic proteins kick in and the "bad" protein is either gotten rid of or in some cases not made at all.

For which cases would we predict that missense and nonsense mutations would have very similar phenotypes? Both types of mutations can result in a completely inactive protein product (loss-of-function), either through changing a critical amino acid or causing truncation that causes inactivity. We would predict similar phenotypes for cases in which missense and nonsense mutations are both causing the disease via the same mechanism – loss-of-function. Thus, for many loss-of-function diseases, such as cystic fibrosis, we can see missense or nonsense mutations causing disease of comparable severity, and we see many mutations in each category.

Missense mutations and nonsense mutations in the same gene will not always cause the same phenotype. This will happen in cases where the nonsense mutation causes the protein to become non-functional but the missense mutation causes the protein to gain a new function, whether a negative function such as poisoning the cell or a positive function such as binding to a different hormone receptor.

For many disease genes, both missense and nonsense mutations are found. In the case of the gene that makes the *myocilin* protein, missense mutations towards the end of the gene most often cause an early-onset, severe form of *glaucoma* (Figure 5.10). As children, teenagers, or young adults, the patients experience a great increase in pressure inside of the eye, followed by gradual death of the nerves in the retina that carry visual signals back through the optic nerve to the brain. The nerve cells apparently die in response to the pressure, and the pace of nerve death can be rapid and cause substantial visual deficits if left untreated.

In contrast to these missense mutations, there is a nonsense mutation that is actually the most common disease-causing mutation in the myocilin gene. It also causes glaucoma, but the glaucoma that results is usually a much later-onset disease, usually starting in middle age or later, involving much less pressure elevation in the eye and a much longer time period over which damage to nerves manifests.

So why would a nonsense mutation in myocilin cause a different phenotype than what we see for the missense mutations? It has been suggested that myocilin missense mutations that cause severe disease do so by actively causing a problem rather than through a lack of the gene product or function. So for myocilin, does less severe glaucoma result from a nonsense mutation because a loss-of-function mechanism is less damaging than some function added by the missense mutation? That remains to be seen.

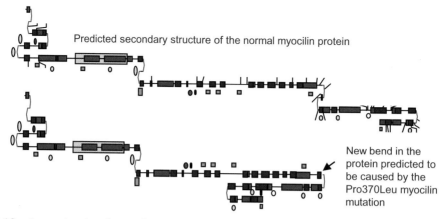

Predicted secondary structure of the normal myocilin protein

New bend in the protein predicted to be caused by the Pro370Leu myocilin mutation

FIGURE 5.10   A mutation that changes how a protein folds. The upper picture shows a computer prediction of the secondary structure of the normal myocilin protein, with boxed areas representing features such as regions of random coiling or pleated sheets, symbols showing locations of chemical modifications, and black lines showing positions of some known mutations. The lower picture shows the same protein produced by a mutant copy of the gene, with the arrow pointing to a new fold caused by a mutation that causes an early-onset form of glaucoma. The real three-dimensional structure of the protein would be much more complex. (Courtesy of Frank Rozsa.)

Mutations Don't Always Change the Protein

Notice that point mutations are not always missense mutations or nonsense mutations – that is, sometimes you can change a single base within the coding sequence and get no change in the amino acid or the protein. Why? Remember that there are 64 different codons that designate which amino acid will be used, and there are only 20 amino acids. This means that there is *redundancy* in the code. So some changes in the DNA sequence will replace one of the codons specifying leucine with a different codon specifying leucine. We joke sometimes about finding a leucine-to-leucine mutation, meaning a change in the sequence that does not affect the protein. These are called *silent mutations* (Figure 5.11).

Even when a missense mutation does change the protein sequence, sometimes this causes no change in the protein function and thus does not change the phenotype. This is especially easy to understand in cases in which an amino acid of very similar size, shape, charge, and polarity replaces the original amino acid. So missense

A SILENT MUTATION

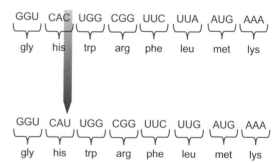

FIGURE 5.11   A silent mutation. A single base change in the mRNA due to a base change mutation in the DNA results in no change in the amino acid incorporated. Because a lot of the redundancy in the code occurs in the third position of a codon (in this case UUA goes to UUG), it is most common to see silent mutations "hit" the third position. However, arginine can be encoded by codons that begin with either C or A, and a number of other amino acids can be encoded by alternative codons differing at either position one or position two.

mutations often cause functional problems, but just knowing that there is a change in amino acid is not enough to tell you that this change is necessarily bad enough to affect the phenotype.

BOX 5.8

BENIGN POLYMORPHISMS

One of the things that we see in a lot of genes is that there are some categories of sequence change that are considered benign sequence variants – points in the sequence of that gene that differ between individuals in the population without showing any association with a trait. Sometimes these are also referred to as *polymorphisms*, but there is disagreement in the field about how the word is or should be defined; some see it as meaning any sequence change present in more than 1% of the population, and others see it as meaning only sequence changes that do not cause a trait, no matter what the frequency. Similarly, the term *mutation* does not mean the same thing to everyone. In some branches of study, a mutation is any change to the order of bases in the sequence; to other branches of study, mutation tends to be used to refer only to causative changes, that is to say, changes that result in a trait. We find the use of terms *causative mutation* and *benign sequence variant* as ways to bypass some of the conflicts in definition, but we can often be found falling back on the use of the easier and more familiar, but ambiguous, terms.

In one disease gene with more than 50 known missense mutations, only about half of them seem to cause the disease. The other half appear to be what we call *benign polymorphisms* (Box 5.8).

Supposedly Silent Mutations Can Sometimes Alter the Sequence

Most molecular biologists will ignore a newly discovered silent mutation, one that does not cause an amino acid substitution even though it is located in the coding region of the gene. After all, if mutations act by changing something about the protein, and a silent mutation does not change the amino acid sequence, then how could it possibly have any effect?

Some mutations that are presumed to be silent can affect the amino acid sequence by altering how the gene is spliced. In the case of one silent mutation in the gene that causes Marfan syndrome, a silent mutation *did* turn out to be the cause of the disease. Does this overthrow the things we have told you about mutations that cause gain or loss of protein function? No, but this silent mutation that does not cause a stop codon or alter the amino acid used by that codon does affect the splicing of the gene (Figure 5.12). Even though this sequence change is nowhere near the splice boundaries, it causes a new alternative splicing pattern that deletes an exon that is normally always kept in the transcript. So we only realize that this supposedly silent mutation is causative if we go beyond thinking in terms of missense and nonsense mutation categories and go on to ask about splicing.

In other cases, a mutation may create a new splice site. If the new splice site is present within an intron, this could have the effect of putting a shorter version of that exon into the final transcript. In some cases, the new splice site might engage in alternative splicing to a different set of exons than the old splice site connected to. Sometimes the cell is actually smart enough to ignore the new splice site and still use the old one. At this point, it is still difficult to predict what will happen when a mutation affects a sequence involved in splicing.

Insertions and Deletions

Sometimes, instead of changing one base in the sequence for another, a mutation adds

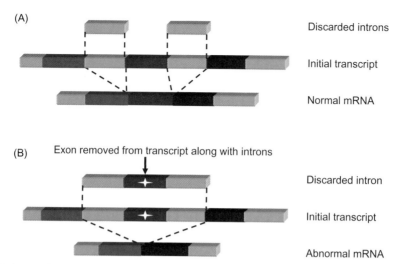

FIGURE 5.12   Splicing of fibrillin in a case of Marfan disease. (A) The normal splicing pattern removes two introns and leaves three exons in the final spliced mRNA. (B) Splicing of the transcript with the mutation removes a region containing the two introns plus the mutated exon that lies between them in one single piece. The result is a spliced mRNA that lacks the mutation but is also missing the central exon. Lavender sections are untranslated regions and introns, and the darker-colored sections are exons. The star notes the position of the mutation.

A DELETION MUTATION

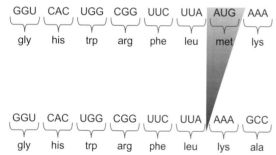

FIGURE 5.13   A deletion mutation. Removal of three bases (a codon) or a multiple of three bases results in removal of the accompanying amino acid(s), but the reading frame is not thrown off so that the sequence beyond the deletion is preserved and results in the incorporation of the same amino acids farther downstream even though the protein is now shorter. An insertion of three bases (or a multiple of three bases) will similarly affect only the sequence right at the point of the insertion but leave the sequence farther along the protein unaltered.

(inserts) or removes (deletes) bases (Figure 5.13). The DNA code is read three letters at a time, so inserting or deleting a number of bases that is a multiple of three will cause the simple addition or removal of one or more amino acids corresponding to those triplets. Whether that causes a problem will then depend on how the size, shape, charge, and polarity of that amino acid affect the protein's folding and functions.

Insertions and deletions can cause much larger problems if they affect numbers of bases that are not multiples of three. Why? Because the genetic code is read in units of three, any change involving three, six, or nine bases will only affect the one, two, or three amino acids encoded by those codons.

As shown in Figure 5.14, a change involving one, two, four, or five bases will not only affect the codon being altered but will also throw off the frame in which the code is being read beyond

A FRAMESHIFT MUTATION

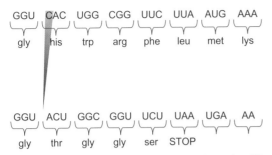

FIGURE 5.14   A frameshift mutation. Through adding or deleting a number of bases that is not a multiple of three, a frameshift mutation throws off the reading frame, rather as if spacing in a document joined the last half of a word to the first half of the next word so that the new information makes no sense. Notice in this case that the frameshift not only changed the sequence of amino acids beyond the point of the shift, but it also resulted in a stop codon farther downstream in the reading frame, with no translation shown beyond that point because the ribosomes stop adding amino acids after a stop codon.

the deletion. The resulting mutation, called a *frameshift mutation*, can not only change the local sequence plus the reading frame but also often results in the use of a newly created stop codon at a different point in the sequence, which can cause the new form of the protein to be either longer or shorter than the normal protein. Thus a frameshift mutation (from either deletion or insertion) often results in a protein that is a different length than the original protein, with a new section of seemingly random amino acids attached to the end of the protein that have nothing to do with the sequence of amino acids that was there before. Thus the effects of a frameshift mutation seem as if they ought to be even more substantial than the effects of a simple nonsense mutation.

If an entire exon is deleted the effect can be the same as if a splice site had been mutated – an exon missing from the final protein product. In some cases the consequences may be severe, but in some cases the deletion of an exon may mimic the effects of alternative splicing, resulting in a protein product that is one of the naturally occurring variants normally produced by the cell.

Splice Site Mutations

Since the supposedly silent mutation discussed above can alter splicing, it is not surprising that changes in the donor or acceptor splice sites themselves can affect splicing. A mutation in a donor or acceptor splice site sequence will sometimes block the cell from recognizing that splice site. What happens then? For a donor splice site mutation, a common outcome would be that the splicing machinery bypassed the splice site that contains the mutation, leaving the final mRNA structure containing intronic material that should have been removed (Figure 5.15).

For an acceptor splice site mutation, the cell will skip using that splice site and go looking for another splice site to use. Some times the cell will simply use the next splice that is normally used for that gene, resulting in skipping of an exon (Figure 5.16). In other cases the cell may go looking for another sequence in the region that resembles a splice site but is not normally used as one. The new splice site may be at a new location in an intron, resulting in the inclusion of some intronic sequences in the mRNA, or it may be located in coding sequence, resulting in cutting part of an exon out of the mRNA.

While some mutations in splice sites cause altered splicing, many changes to splice site sequences do not change the splicing pattern. In other cases, the splicing machinery may not simply jump over a bad splice site to the next splice site that is normally used in that gene; instead, altered splicing may sometimes use a potential splice site in the region being spliced, electing to use one of the other sequences that resembles a splice site but is normally not used. Remember from our discussions in Chapter 3 that there is no absolute, invariant sequence that determines which sequences will be used as splice sites, and often there are other potential splice sites present in the vicinity of the splice sites that are normally used.

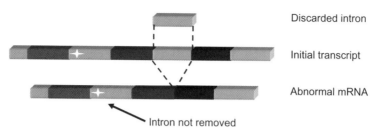

FIGURE 5.15 Splicing of a transcript with a mutation in a donor splice site. If the gene in Figure 5.12 had a splice site mutation in the first donor splice site, the slicing machinery would skip using that splice site and leave the first intron in the final version of the spliced mRNA. The result is a spliced mRNA that has retained an intron that would normally have been cut out, and that still has the mutation.

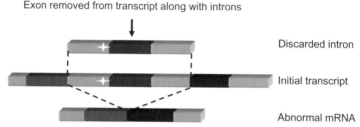

FIGURE 5.16 Splicing of a transcript with a mutation in an acceptor splice site. If the gene in Figure 5.12 had a mutation in the first acceptor splice site, the splicing machinery would skip that splice site and might instead use the next acceptor splice site. The result is a spliced mRNA that lacks the mutation but is also missing the central exon, which is the same structure that resulted from the silent mutation in Figure 5.12B.

5.5 MUTATIONS IN DNA SEQUENCES THAT REGULATE GENE EXPRESSION

As we saw during our discussion of lactose intolerance in Chapter 3, some mutations with important functional effects cause changes in gene regulation instead of changing the structure of the gene product. To identify a regulatory mutation we need to know where to look for it, we need to identify a sequence change, and we need to determine whether the sequence change is a causative mutation. Identification of causative regulatory mutations is often more difficult than identification of some of the mutations we have discussed earlier in this chapter. And one of the most difficult first steps in the process of looking for regulatory mutants for a particular gene is

defining the extent of the functional promoter that adjoins the transcription start site. As we discussed in Chapter 4, the problem is not only one of determining the length of the functional promoter, but is also one of determining whether the gene has more than one promoter region. In addition, if we really want to understand which sequence changes are playing a role in regulation of the gene in question, we also need to identify enhancer regions and screen them as well.

Finding regulatory mutations is a challenge even if we have well-defined promoter and enhancer regions to screen. The functionally important sequences within a promoter are small, sometimes as small as four or six base pairs in length, and such sequences may be scattered across thousands of bases of sequence. Recall that in the case of the lactase gene the functionally relevant sequence variants were two SNPs that were separate from each other and both located more than 10,000 base pairs away from the transcription start site. So when we find a sequence variant near a transcription start site we need to ask whether it is close enough to be within the functional promoter and we need to ask whether it has hit one of the short sequences to which transcription factors bind.

Finally, we need to know whether the sequence change is having any affect on the rate of transcription. In some cases we can evaluate this by looking at the level of mRNA for the gene in cells from someone with the mutation and in cells from someone without the mutation. Sometimes, we can test RNA levels in readily accessible cell types in the body, such as white blood cells present in a blood sample or cheek cells present in a saliva sample, to see whether the gene is expressed there.

But sometimes the gene is expressed in cells that we cannot access for testing. So if a gene is only expressed in the brain or the pancreas, we might elect to do functional testing in cultured cells to look for transcription. We can create a piece of DNA containing the sequence of the promoter attached to a reporter gene that makes an easily detectable gene product such as luciferase, a bioluminescent protein such as the protein that makes fireflies light up. We can make two copies of this construct, one containing the normal version of the sequence and one containing the mutation that we want to test. We can get the cultured cells to take up the DNA constructs, and we can use a luminometer to read the levels of light being put out by the luciferase produced. If the normal construct makes more luciferase than the mutant construct, the luminometer will detect more light coming from the cells that received the normal construct.

Assays of mRNA from individuals with and without the mutation, and functional tests of constructs with and without the mutation, help to confirm functional effects of mutations. While such tests are done in the course of research studies on promoter mutations, high throughput screening of large numbers of samples would be substantially slowed up by the need for validation steps that take this much time and customization.

As you will recall from Chapter 3, mutations can affect regulation in several different ways. A mutation may be located in the copy of the gene being regulated (in cis) in the promoter or enhancer region. A mutation located in cis will usually only affect regulation of the transcription region located on the same copy of the chromosome. Alternatively a mutation can be located elsewhere in the genome (in trans), altering the sequence of a gene whose product interacts with the promoter or enhancer of the gene being regulated. Mutations that act in trans are expected to affect regulation of both copies of the gene being regulated.

5.6 COPY NUMBER VARIATION: TOO MUCH OR TOO LITTLE OF A GOOD THING

In some cases a deletion is large enough to remove a whole gene, or even a small region containing several genes. When someone has only one copy of a given gene they are considered to be hemizygous for that gene. Hemizygosity often results in the cell producing only half the normal amount of the gene product, because it is producing transcript and gene product from only one copy of the gene. In the case of the neurofibromatosis phenotype, disease symptoms such as neurofibromatous tumors can result from a missense mutation, a nonsense mutation, or deletion of one whole copy of the NF1 neurofibromatosis gene. In one case a deletion of this kind was identified when screening of genetic markers showed that someone had inherited genetic markers around NF1 from one parent but not the other. In other cases deletions were identified through the use of fluorescence in situ hybridization (FISH).

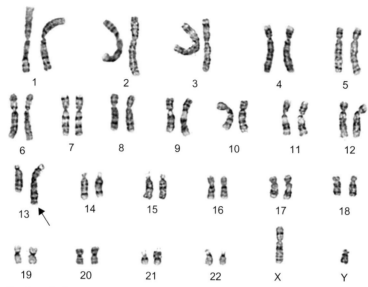

FIGURE 5.17   Duplication of a large section of the long arm of chromosome 13 becomes obvious in a karyotype image. Note the large difference between the two copies of chromosome 13 and the way in which the banding patterns are used to assist in correctly pairing up the different images. (Courtesy of the Clinical Cytogenetics Laboratory, University of Michigan, Ann Arbor, MI; Diane Roulston, PhD, Director.)

In addition to deletions, there are also cases where a gene, or a set of contiguous genes, may be duplicated. Sometimes the extra copy is located right next to the original copy. Some deletions or duplications are large enough to be seen in a karyotype (Figure 5.17). Someone who has one normal copy of the gene on one copy of the chromosome and a duplicated copy of the gene on the other copy of the chromosome ends up with three copies of the gene. This will commonly result in production of excess gene product. Gene dosage differences for a particular region of a human chromosome are often referred to as copy number variants (CNVs).

CNVs were originally studied in the context of individuals with changes specific to a particular gene and disease. The Rieger syndrome phenotype (iris, tooth, and umbilical anomalies that may be accompanied by glaucoma) can result from missense mutation or nonsense mutations, but can also result from changes in the copy number of RIEG1. Interestingly, Rieger syndrome can result from either losing a copy of RIEG1 or gaining a copy of RIEG1!

Contiguous gene deletion syndromes take out the same chromosomal region in many different individuals, usually from only one of the two copies of the chromosome. In the case of Williams syndrome more than 99% of individuals with this syndrome have a contiguous gene deletion of a region of chromosome 7 containing several dozen genes, among them the gene that makes elastin, a protein that helps some tissues be more elastic. Individuals with Williams syndrome share many features, including an elfin facial appearance and a characteristic set of cardiovascular and endocrine abnormalities. They also demonstrate mild mental retardation, an unusually sociable personality, a strong sense of empathy, a tendency towards anxiety, and an affinity for music. According to the Williams Syndrome Association this trait is found in about 1/10,000 births.

As we have gained the ability to carry out a variety of different kinds of genome-wide analysis, we have gained the ability to look at copy number variation across the whole genome. Copy number variation can sometimes be benign, and can sometimes affect phenotype through changes in gene dosage. Some events that create CNVs can disrupt genes at the boundary of the event, or cause a position effect, a difference in gene expression that results when a gene is placed into a new sequence context or fused with a new promoter region.

More than 38,000 CNVs in excess of 100 bp in size have been found so far in the human genome. Some range well above a million base pairs in size. The smallest CNVs are the most common. When data from two different individuals are compared they will often have many CNVs in common. Whole genome sequencing has identified many smaller CNVs that have not been detectable by technologies with less resolution such as high density screening of SNPs. CNVs may well represent a more frequent and more important contribution to human phenotypic variation than what results from the many SNPs in the genome. CNVs have been reported to play roles in many different traits including some simple Mendelian traits (pituitary dwarfism, hemophilia, mental retardation) as well as some common, complex phenotypes such as autism, schizophrenia, and Parkinson's disease.

5.7 EXPANDED REPEAT TRAITS

When Alex first started having major health problems in her late seventies her children didn't know about it. She had gradually become more and more rigid about her daily routines, and she had developed a tendency to restate the same thing several times in a row, but her behavior changes were *attributed to her advancing age. Although Alex had become a bit shaky as she aged and had had several falls, no one was terribly concerned because her sister had shown much more severe movement problems without anyone ever assigning it any medical significance. Besides, the movement problems seemed minor when compared to the disturbing problems that surfaced when Alex's daughter took a trip with her, including severe anorexia, obsessive compulsive behaviors, and hallucinations. Initial efforts to diagnose what was wrong led to discussions of Alzheimer disease as one of the possibilities, but the big surprise was the final answer. Alex has Huntington disease, a disease that usually starts in early middle age and is most famous for the irregular jerking movements known as Huntington's chorea. Alex did go on to develop the characteristic movement problems, but the first signs that she had a problem were mostly psychological. Alex's case is not so unusual, since some cases of Huntington disease manifest psychiatric and cognitive symptoms in addition to the movement disorder. Alex's case is quite unusual because her symptoms started so late in life. So what causes Huntington disease, how did the doctors figure out that that is what she has, and why did Alex remain free of symptoms long past the age at which many Huntington disease patients die?*

Huntington disease (also sometimes referred to as HD) offers us insights into a very different type of mutation. This trait involves a long, slow process of neurological degeneration. Onset is usually in the thirties or forties, although first signs of disease can show up in very young children or the elderly. Death occurs on average about 17 years after the disease starts. Symptoms include the progressive development of uncontrolled or jerky movements. Although cognitive and psychiatric effects are quite common, some individuals and even whole families can remain free of the typical dementia even at late ages and stages of disease. HD is also known as Woody Guthrie disease, after one of its most famous victims. About one in 20,000 human beings have Huntington disease, and it shows an autosomal dominant mode of inheritance.

By the shores of Lake Maracaibo in Venezuela, thousands of individuals mark the days of their lives by the stages of the disease that we call Huntington disease, and which they call *El Mal de San Vito*. This is the greatest concentration of Huntington disease cases found so far in the world, and they are all descended from one Venezuelan woman who had ten children in the early 1800s. Eventually a family tree showing more than 17,000 of her descendants has been built.

For the members of this family, neurological manifestations that many of us have never witnessed are a commonplace part of everyday life. Many of them struggle to wrest a marginal living from activities such as fishing in a community marked by the poverty that results when so many of the adults are too ill to work. Because the mode of inheritance of HD is autosomal dominant, on average half of the descendants of anyone who is affected will also develop the disease. The half that are spared from developing the disease themselves often find their lives dominated by it as they work to help their loved ones who fall prey to the disease.

As you might guess, we are talking to you about Huntington disease not only because of its medical importance, but also because it offers us another set of genetic lessons. An international consortium, headed by researchers at Harvard and Columbia Universities, set out to find the gene. Many members of the Lake Maracaibo HD family participated in the research study that first mapped the location of the gene on chromosome 4, and then identified the gene itself. What did researchers find at the end of one of the longest disease gene hunts of the late twentieth century? What they found in their search provides a framework with which to understand a whole family of genetic diseases. These diseases do not share the same symptoms, do not involve any one gene or biochemical pathway. Rather, they are all caused by a type of mutation that is quite different from the other mutations we have been talking about.

Expanding and Contracting Repeats

Simple *tandemly repeated sequences* occur throughout the genome, with more than 50,000 copies of some repeat categories found scattered around the genome. Most often, they are found between genes and in introns but can also be in promoter regions, in untranslated sections of transcripts, and occasionally even in coding sequence. These tandemly repeated sequences can take the form of mononucleotide repeats (e.g., AAAAAAAAAAAAAAAA), dinucleotide repeats (e.g., CACACACACACACACA CACA), trinucleotide repeats (e.g., CATCA TCATCATCATCATCATCATCATCATCA TCATCAT), or repeats with a longer subunit length. They share the property that sometimes runs of a simple sequence repeat can change length – undergo expansion and contraction (Figure 5.18).

Most sequence changes in simple sequence repeats tend to be additions or subtractions of whole repeat units, with standard mutational mechanisms causing changes such as missense and nonsense changes happening at a much lower frequency than addition or subtraction of full repeat units. So if the repeat length is TAGT, the cell will most often alter this sequence by adding another unit of TAGT within the repeat.

The presence of mononucleotide repeats in coding sequence creates a problem. If a mononucleotide repeat unit of A occurs 16 times in a row and a single A is added or removed, the result is a shift in frame. If multiple copies of the repeat unit are added, most will also cause frameshifts, unless the number of bases added is a multiple of three (Figure 5.19). A mononucleotide repeat of a single base, in this case A, will cause a frameshift if one (A), two (AA), four (AAAA) or eight (AAAAAAAA) bases are added; there will be addition of more copies of the amino acid encoded by the AAA codon but no frameshift if the number of copies added is a multiple of three (AAA, AAAAAA, etc.).

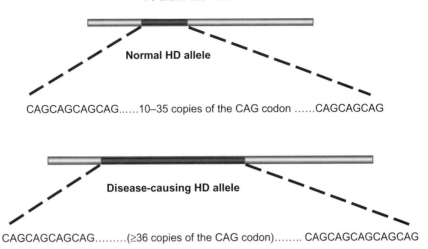

FIGURE 5.18   Triplet repeat expansion at the HD locus. Dark region represents CAG repeat region of the gene.

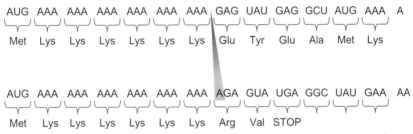

FIGURE 5.19   Mononucleotide repeat expansions. Small changes in number of repeats units will throw off the reading frame unless the number of bases added is a multiple of three. For multiples of three, additional codons are added within the repeating section of the coding sequence, leading to an increase in size of the run of repeated amino acids. But the sequence beyond the run of repeated amino acids will be conserved unless the reading frame is thrown off.

Other repeat lengths that are not multiples of three are similar problems. Frameshifts will result from adding or subtracting a repeat unit that is not a multiple of three, including additions or subtractions of two (CA), four (CATG), five (CGTGA) bases or more in length. The most common simple sequence repeat is the CA repeat (Figure 5.20). There are more than 50,000 places in the human genome that have CA repeats. Adding or subtracting two (CA), four (CACA), eight (CACACACA) or ten (CACACACACA) bases of a dinucleotide repeat will shift the reading frame of a CA repeat. Adding or subtracting six (CACACA) or 12 bases (CACACACACACA) will keep the reading frame intact. So some repeat length changes will not cause a frameshift if they happen to land on a multiple of three, but the majority of them will.

For a dinucleotide repeat, if the repeat length change was an increase (but not a multiple of three) there will be a frameshift

DINUCLEOTIDE REPEAT EXPANSION

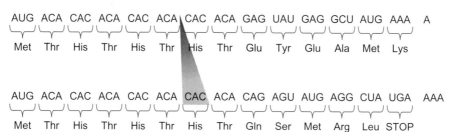

FIGURE 5.20   Dinucleotide repeat expansion. Because the repeat length is not a multiple of three, changes in repeat length throw off the reading frame. CA repeats like this exist all over the genome, and any one repeat at a particular position in the genome shows different lengths in different individuals. The longer the repeat length, the greater the variation in length between individuals. Many CA repeats demonstrate as many as 18 different alleles in a population, but repeats with a longer unit (tetramers with four bases, pentamers with five bases, etc.) show less variability.

TRINUCLEOTIDE REPEAT EXPANSION

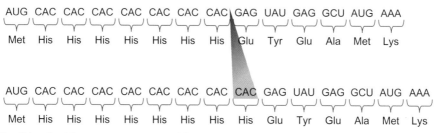

FIGURE 5.21   Trinucleotide repeat expansion. Adding another CAC to the trinucleotide repeat in this piece of coding sequence does not cause a frameshift, so the change in the length of the protein will be changed only by as much as the number of amino acids corresponding to the number of repeat units added.

(Figure 5.20) but that does not mean that the new protein will necessarily be longer. When such frameshifts happen, the altered amino acid sequence no longer uses the original stop codon. The new stop codon may be either closer to the transcription start or farther away, so the new protein sequence may be either longer or shorter than the original. We expect that the biggest effect on the protein length will most often result from the frameshift and not from the small number of added or lost nucleotides in the repeat.

The repeat units in a *trinucleotide repeat* are the same length as the length of a codon – three bases. If a trinucleotide repeat within a coding sequence gains or loses one or two subunits, as commonly happens, it gains or loses one or two copies of an amino acid in the protein sequence (Figure 5.21). It will not cause a frameshift, so the amino acid sequence beyond the end of the repeat will not be altered.

More than a half-dozen different genetic diseases, many of them neuromuscular disorders, involve expansion of trinucleotide repeats located in coding sequence (Table 5.3). Thus trinucleotide repeat expansions, if the change in number is small, have a small effect on the protein, whereas repeat unit lengths that are not a multiple of three can cause even a small change in DNA sequence to result in a major frameshift mutation with resulting major alterations to the sequence and structure of the protein.

TABLE 5.3   Examples of traits caused by expansion of simple sequence repeats

Trait	Gene	Repeat unit	Repeat location within the gene
Dentatorubral-pallidoluysian atrophy	DRPLA	CAG	Coding sequence
Huntington disease	HD	CAG	Coding sequence
Spinobulbar muscular atrophy	AR	CAG	Coding sequence
Spinocerebellar ataxia 1	SCA1	CAG	Coding sequence
Spinocerebellar ataxia 2	SCA2	CAG	Coding sequence
Spinocerebellar ataxia 3	SCA3	CAG	Coding sequence
Spinocerebellar ataxia 6	SCA6	CAG	Coding sequence
Spinocerebellar ataxia 7	SCA7	CAG	Coding sequence
Myotonic dystrophy 2	ZFN9	CCTG	Intron
Fredreich ataxia	X25	AAG	Intron
DM1-associated cataract	SIX5	CTG	Promoter
Progressive myoclonus epilepsy	CSTB	CCCCGCCCCGCG	Promoter
Myotonic dystrophy 1	DMPK	CTG	3′ untranslated region
Spinocerebellar ataxia 8	SCA8	CTG	3′ untranslated region
Fragile X	FRAXA	CCG	5′ untranslated region
Fragile XE	FRAXE	CCG	5′ untranslated region
Spinocerebellar ataxia 12	SCA12	CAG	5′ untranslated region

However, sometimes trinucleotide repeat expansions can be large, with truly drastic consequences, even though the resulting sequence has remained "in frame" and retains the normal sequence beyond the region of the repeat. As we discuss below, the expansion of CAG repeats leads to the production of an expanded stretch of glutamine amino acids at some point in the offending protein.

However, not all of the trinucleotide repeat expansions take place in coding sequence (see Table 5.3). In these noncoding cases, models for how the repeat expansion causes disease include effects on RNA metabolism and changes in gene expression; in one case, the final effect is accumulation of iron in the mitochondria. Much still needs to be learned about such expansions.

Huntington disease is caused by one of the CAG repeat expansions in coding sequence. As shown in Figure 5.18, the mutation underlying Huntington disease is due to the amplification of the triplet codon repeat CAG (which encodes glutamine). Normally, this gene includes a stretch of many copies of the glutamine codon arranged as one long, tandemly repeated array. Different individuals have different numbers of glutamine codons at this point in the gene. The number of copies of the repeat predicts the individual's risk of developing HD, with individuals who have 36 or more copies being at risk of developing the disease. As shown in Table 5.4, there are different risks associated with increasing lengths of the CAG repeat.

The age at onset of HD correlates with the number of repeats, although the length of the

TABLE 5.4   HD phenotype of different CAG repeat copy numbers

No. CAG repeat copies	Phenotype
10–26	Normal
27–35	Normal with increased risk of having a child with HD
36–39	Some individuals with HD, some individuals normal
≥40	HD

*Data from GeneReviews article "Huntington Disease" by S.M. Warby, R.K. Graham and M.R. Hayden.

repeat appears to account for only part of the variance in the phenotype. Those with the longest repeat lengths show the earliest onset of disease. Those with the shortest (causative) repeat lengths show the latest onset of disease. If we think back to the story we told you about Alex, we find that she showed very late onset of HD, and that in her family the CAG repeat length is 40. In others with longer repeat lengths, onset of disease is earlier and some of the earliest onset cases show a more rapid and severe course of disease.

While the length of the repeat is strongly predictive of age at onset of disease, it does not completely account for age at onset. After a search for modifier genes that can modulate the primary phenotype caused by the repeat length, sequence variants in several other genes were found that seem to account for some of the variation in age at onset between individuals with similar repeat lengths.

The Effect of Repeat Length Differences

Why would an increase in the number of copies of the glutamine codon lead to such a devastating disease? The Huntington protein has a stretch of glutamines, and when the repeat in the gene expands, the number of glutamines in the protein increases. Very long stretches of polyglutamine turn out to be sticky, and the ability of polyglutamine to stick to other proteins is thought to play a role in the creation of aggregated proteins in dying nerve cells. In addition to the non-specific process of creating aggregates, this stickiness property can tie up some specific proteins needed by the cell. This stickiness pulls the proteins into aggregates and prevents the cell from making use of the functions those proteins perform. The active presence of a long polyglutamine stretch all by itself may be enough to explain the disease, although we cannot rule out additional effects from loss or decrease of some aspects of function of the HD protein itself. Reinforcement for the idea that longer polyglutamine stretches may have an actively toxic effect on the cells comes from a finding that neurological disease develops in transgenic mice expressing a protein that consists of little more than a long stretch of polyglutamines (see Box 5.9). Many other neurodegenerative diseases are caused by a similar increase in CAG repeats, with accompanying increase in the length of the polyglutamine stretch, but the various proteins involved in those diseases have nothing in common other than the polyglutamine stretch.

Some disease genes have their expandable repeats located in non-coding sequence (Table 5.3). For myotonic dystrophy the CAG repeat is located in the 3′ non-coding region, so expansion of the repeat in the RNA cannot be encoding an expanded polyglutamine region in the protein. Researchers have found that some of the genes with expandable repeats outside the coding region have transcripts being produced off of both strands. So even as the cell is producing a coding strand RNA with an expanded CAG repeat, it is also producing a non-coding RNA strand with an expanded CUG repeat.

In the case of the DMPK gene, as this CUG repeat gets longer, it gains the ability to alter the splicing of transcripts from other genes. The result is that a mutation in the DMPK gene leads the cell to alter the splicing, and thus the protein product produced, by other genes in the cell that do not have the disease-causing mutation. This implies that some of the disease pathology might actually be the result of alterations to proteins (and their functions) when there is no mutation in the genes encoding those proteins!

Thus, there appear to be at least two different mechanisms of action of disease-causing repeat expansions. In cases such as HD, we see CAG repeats in coding sequence that create sticky polyglutamine regions that affect levels of other proteins in the cell. In cases such as myotonic dystrophy, we see a different mechanism, accumulation of toxic non-coding RNA with expanded CUG repeats, that influences the splicing of other genes. In either case, the complex disease pathology appears to involve effects on proteins encoded by genes that do not themselves harbor a disease-causing mutation.

Researchers continue investigating how long sections of polyglutamine in a protein can have toxic effects that lead to neurodegeneration. Interesting areas of investigation include approaches that would block the stickiness of the polyglutamine region, break up the aggregates, or restore adequate levels of specific proteins that have been pulled out of action by the sticky aggregates.

Current therapeutic approaches being tried in humans include use of antidopaminergic agents to deal with the choreiform movements and use of anti-depressants and anti-psychotics to deal with psychiatric symptoms, but none of this targets the underlying cause of the disease so none of them slows or stops progressive neurodegeneration. Transgenic models of the disease are letting researchers test out both standard pharmaceutical approaches and gene therapy before trying them in humans (Box 5.9), and some therapies are making their way through the process of clinical trial testing (Box 5.10).

Genetic Testing and HD

The lengths of CAG repeat associated with risk are known, so the genetic test for HD is one of the simplest genetic tests available – a PCR assay. The fact that the test is simple, however, does not mean that everyone at risk runs right out to get themselves tested.

Because there is not yet any cure available for Huntington disease, the discovery of a genetic marker that could be used accurately to identify people carrying HD mutations was a bit of a mixed blessing. For years before such testing was available, members of Huntington disease families asked if there wasn't some way to test and find out who would end up being affected. Once the test became available, some who had asked for the testing backpedaled and indicated that they didn't want to be tested yet or that they didn't want to be tested at all. Why might someone decide not to be tested?

Dr Nancy Wexler, who played a leadership role in the international consortium that cloned the HD gene, has applied the name *Tiresias complex* to the dilemma of making the choice regarding whether to be tested for something for which there is no cure. The name comes from the blind seer Tiresias who, in *Oedipus the King* by Sophocles, said, "It is but sorrow to be wise when wisdom profits not." In describing the Tiresias complex, Dr Wexler asked, "Do you want to know how and when you are going to die, especially if you have no power to change the outcome? Should such knowledge be made freely available? How does a person choose to learn this momentous information? How does one cope with the answer?"[1]

This is a problem faced by everyone at risk trying to decide whether or not to be tested for HD. While many relish the idea of taking the test with a vision of getting back the answer that they do not have the defect and can get on with their lives without further worry about HD, when they are faced with actually taking the test, they have to face the fact that the answer

[1]"The Tiresias Complex: Huntington disease as a paradigm of testing for late-onset disorders" by N. S. Wexler in. *FASEB J*, 1992;6:2820–5.

BOX 5.9

TRANSGENIC ANIMAL MODELS OF HUMAN DISEASE

Researchers have made *transgenic animals*, animals that have been altered by putting in or taking out genetic material, of great importance to Huntington disease research. By constructing mice that have an expanded repeat in the mouse copy of the Huntington disease gene, they have made an animal model of the disease (the HD mouse) in which it is possible to study the same kinds of cellular and neurological processes seen in the Huntington disease patients. By making a different kind of transgenic animal called a "knock-out" mouse that lacks the Huntington disease gene, researchers showed that animals cannot remain healthy without Huntington protein, so a gene therapy approach that simply removes the Huntington disease gene seems unlikely to work. Animals with altered or missing copies of a gene provide a valuable tool for the first stages in drug discovery. Although we most often think of transgenic animals as being mice, other kinds of organisms can be genetically modified through similar technologies. The most important other animal model of Huntington disease currently seems to be a transgenic fruit fly – the HD fly! The HD mouse and the HD fly have each been used in testing compounds that seem as if they have a chance of protecting against some aspect of the disease pathology. HD mice treated to make their Huntington less "gluey" lived longer and had fewer symptoms than the untreated mice. Testing drugs in animal models allows researchers to identify which drugs and strategies are safe enough and work well enough to consider for testing in humans. Looking at the disease processes in the genetically modified animals helps researchers understand the basic underlying processes of the disease. And some of the first steps towards gene therapy for Huntington disease are taking place in transgenic animal models. However, since things that work in mice and flies do not always work the same way in humans, researchers need to take very careful steps as they work to find out whether a successful treatment of a transgenic animal is safe and effective and can be applied to human patients.

BOX 5.10

CLINICAL TRIALS TESTING OF POTENTIAL NEW TREATMENTS

The process for drug approval starts with phase I trials that screen dozens of subjects to test for any major harmful effects. Phase II trials screen hundreds of subjects to test for whether the compound actually does any good. The process finishes up with phase III trials that screen thousands of individuals to find out whether the drug is both safe and helpful when tested on large numbers of people, and to help pin down the optimal dosages and better understand the side effects, if any.

might not come back the way they want it. So if you are 18 and at risk, finding out that you do not have the genetic defect would let you get on with life's plans, but if you are 18 and find out that you carry a causative CAG repeat length, you are faced with years, perhaps decades, of knowing what is to come and knowing that you can do nothing about it. This is an issue that the human genetics community is struggling with for Huntington disease and many other fatal disorders that we can currently diagnose but not cure. As research continues, we move towards a time when the finding of a CAG repeat length of 40 or more might be accompanied by a doctor saying, "And here is what we are going to do to keep the disease from developing." Until that day, each at-risk individual has to come face to face with the dilemma of the Tiresias complex. There are quite a number of people who end up deciding that they do not want to test before the failing health of their bodies forces the information on them.

A test for an HD defect is not only about the adult individual's hopes and fears for their future health. One of the other major issues at stake is that of having children. In the past, most at-risk individuals had to decide whether to have children before they were old enough to know whether they carried the HD defect, which is to say, before they knew whether they carried a defect that they could pass along to their children. Do at-risk individuals have to be tested, and risk news of impending HD, to be able to make an informed decision about having children? These days, there is an option that lets them have children without the HD defect without finding out whether they carry it themselves. When *in vitro* fertilization is used to create embryos, preimplantation genetic testing allows the identification of embryos that do not carry the defect before any embryo is selected to be implanted in the mother. The doctors can do the test and implant only the ~50% of the embryos that lack the defect. And if the parents have requested it, they can do so without telling them whether there were any embryos that tested positive for a length of the CAG repeat that could cause HD. Why wouldn't all at-risk parents take this approach to having children? *In vitro* fertilization is expensive; many people cannot afford it and do not have health insurance that will cover it. The hormone treatments involved in harvesting the eggs can be hard on the mother's body. *In vitro* fertilization has a limited success rate. And finally, many pregnancies that happen naturally were not planned.

Long Repeats Beget Longer Repeats

The mechanism by which new mutations arise in the HD gene remains unclear. There are various models for how the cellular machinery can make errors in copying a simple sequence repeat, some involving the act of replication, some involving the mechanisms that repair errors, and some involving the local chromatin structure which can be affected by things like methylation of the DNA. It has even been suggested that the mechanisms leading to error may differ from one cell type to another.

Regardless of the mechanism by which instability occurs, it is clearly present and it gets worse as the number of copies of the repeat rises. Using a technique called *polymerase chain reaction* (PCR), which was described in Chapter 2, the HD gene mutation rate in different individuals was measured in terms of the expansion or contraction of the repeat number for this gene by analyzing the genes in a single sperm (Figure 5.22). Normal- or average-sized HD alleles (15 to 18 repeats) showed three contraction events (reduction in the number of repeats) among 475 sperm. Even at that low level, the rate of expansion or contraction is an astonishing 0.6%. However, when they looked at a man bearing a normal allele with 30 triplet repeats, the mutation rate in terms of expansions and contractions went up 11%, that is, 11% of all of the sperm carrying this allele carried a variant copy of the HD gene. (Remember what was said earlier, the standard mutation rate is on the order of 1 in 100,000, so it would appear that the

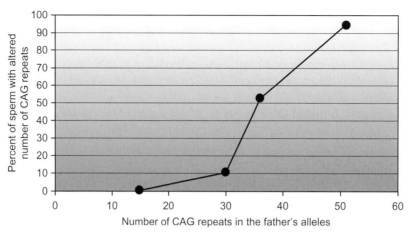

FIGURE 5.22   Rate of repeat length expansion depends on the repeat length in the man producing the sperm. A man who has more repetitions of CAG in the HD gene sequence will have more sperm in which the length of that CAG repeat has changed, that is, mutated to a new length. Individuals with a higher number of copies within the normal length will produce some sperm in which the new CAG repeat length is long enough to cause disease. (Data from Leeflang *et al., Human Molecular Genetics*, 1995;4:1519–26.)

mutation rate here is 10,000 times higher than that!) An allele with 36 repeats showed a mutation rate of 53%, and in fact 8% of the sperm bearing this allele ended up with expansions so large that they would have caused disease. Disease-causing alleles, with 38 to 51 repeats, showed expansions or contractions in more than 90% of the sperm carrying these alleles.

Simply put, as the number of repeats increases, so does the frequency of changes in repeat length, including expansions of the repeat. (Curiously, the frequency of contractions also increases up to 36 triplets but falls off as the copy number of the allele increases above 36.) So one can imagine how these mutations arise: one small increase makes a second increase more likely, and that increase further increases chances of an increase at the next generation, and so on. Realize that the mutation rate from a normal allele (15 to 18 repeats) to an intermediate and unstable allele (say an increase from 15 to 30 repeats) is quite low; however, once the repeat number gets above 30 repetitions, the mutation rate to a disease-causing allele is much higher. So someone with a repeat length of 60 repetitions is expected to transmit the "wrong" repeat

length to the next generation more frequently than someone who starts out with 16 copies.

Anticipation

A juvenile-onset form of HD manifests before age 21, and often in childhood, and shows some additional clinical features such as seizures. Juvenile-onset HD occurs in about 7% of cases. The earliest-onset cases of HD show repeat lengths in excess of 60 copies. In one of the most severe cases reported, a child with onset before four years of age was found to have more than 200 copies of the repeat!

This phenomenon, in which a much younger age at onset seems to occur among the younger generation, is called *anticipation* (Box 5.11). For many diseases it can be hard to tell whether apparent anticipation is real or whether improvements in medical care, including more aggressive pursuit of diagnosis and better diagnostic tools, are actually responsible for earlier identification of affected people.

Anticipation seems to be real in the case of Huntington disease. In many families the age at which the disease starts may appear to be fairly

BOX 5.11

ANTICIPATION AND APPARENT ANTICIPATION

Anticipation is earlier onset of a trait in subsequent generations. For some traits, such as Huntington disease, anticipation is real and represents a fundamental genetic change from one generation to the next. But often, apparent anticipation is a technical artifact that does not actually represent a biological difference between the late onset parent and the early onset child. In some cases, a trait may be inducible and the age at diagnosis will simply reflect the age at which the environmental exposure occurred. For some traits, if we were to look at many different individuals we would discover that for all of the late–early parent–child pairs there are just as many early–late parent child pairs; the trait is simply highly variable in its age at diagnosis and shows no specific trend towards earlier onset. In other cases, increased awareness of a trait in a family may lead family members to start consulting a doctor at an early age in life regarding the trait, with age at diagnosis correlating with when the patient asked, "Am I affected?" and not with some biological difference. Sometimes improvements in diagnostic technologies let us detect a trait at a much earlier stage of the disease. This points to a critical concept: the important difference between age at onset and age at diagnosis.

BOX 5.12

MYOTONIC DYSTROPHY – AMPLIFICATION IN THE EXTREME

More than a dozen different genes in the human genome can cause neurological damage as a result of the expansion of a region of trinucleotide repeats. Myotonic dystrophy, the most common form of muscular dystrophy, can be found in 1 per 8000 individuals and is inherited as an autosomal dominant trait. As for the HD gene, the mutations in the DMPK gene that cause myotonic dystrophy involve a triplet repeat expansion. However, the DMPK triplets are located in the noncoding part of the transcript beyond the stop codon, so the effect cannot be explained in terms of changing the number of repetitions of an amino acid in the protein. Although a normal individual only has five copies of the triplet, affected individuals can have 100 or even 1000 or more copies of the repeat. Although expansions in the HD gene are more likely to come from the paternal line, expansions in the DMPK repeats seem to take place in the female germline.

consistent, with middle age onset in most families and late age onset in some. Anticipation is not all that common, but it gets attention because it is so alarming to see a child developing disease symptoms of a late-onset disease. This type of anticipation has been observed in others of the trinucleotide repeat expansion disorders and is not specific to Huntington disease (Box 5.12).

Before the HD gene had ever been found, it was already known that individuals with juvenile-onset HD more often inherited the disease from their father rather than their mother. Why would very early onset cases of HD more often be found to have inherited the trait from their father rather than their mother? It turns out that mistakes in correct copying of the repeat length occur more often in males than in females. It has been proposed that this may be in part related to the state of methylation of the DNA. A different pattern of methylation takes place in the male and female germlines, and methylation of a region affects its local chromatin structure. How methylation and chromatin structure affect errors in repeat length replication is still under investigation.

5.8 THE MALE BIOLOGICAL CLOCK

We often hear talk about a female biological clock, a limitation on how old a woman can be and still expect to be able to become pregnant or have a healthy child. Rarely, if ever, do we hear about a male biological clock, a reproductive limitation that men should consider when deciding whether to wait until late in life to father children. While androgen deficiency in aging men (ADAM) may be a concern comparable to the female concern about declining fertility while progressing through middle age, there is another biological clock that each must be concerned about – the rate at which genetic defects accrue in the germline.

Note that while both sexes show an increasing rate of genetic anomalies in the children as the parents age, the type of age-associated genetic defects coming from mothers is not the same as the type of age-associated genetic defects coming from fathers. In Chapter 6 we will tackle the female biological clock and the rate at which chromosome anomalies develop with increasing age of the mother. Here we will talk about men and the fact that the frequency of mutations in sperm increase as men age.

One of the most profound examples of increasing genetic risk with age of the father is that of repeat length expansion. Age at onset of Huntington disease is inversely proportional to age of the father from whom the disease was inherited, that is to say, earlier onset cases tend to have older fathers. This factor of paternal age adds to the other factors we have discussed that can affect repeat expansion: length of the pre-existing repeat in the father, chromatin structure in the region of the repeat, and modifying alleles in other genes.

Simple base changes such as missense and nonsense mutations also turn up more frequently in sperm as the age of the father increases. The rate at which mutations are produced in sperm increases with the age of the father. A paternal age effect on mutation rate can be seen in the case of Apert syndrome, in which early fusion of plates of the skull leads to problems with normal skull and brain growth, and some fusion of fingers or toes occurs. There appears to be a mutation hot spot that is subject to this paternal age effect. The disease can result from mutations at either of two positions in the FGFR2 gene which makes the fibroblast growth factor receptor 2 protein, and the incidence of de novo mutations at these two positions in the FGFR2 gene increase with increasing paternal age. It has been hypothesized that there may actually be selection in favor of these particular FGFR2 mutations when they appear in the male germline. This paternal age effect may be specific to the germline since a comparable mutation rate is not seen in white blood cells in aging men.

Interestingly, small deletions and insertions do not show an increase with age. The rate of such changes also does not vary between males and females.

5.9 MUTATION TARGET SIZE

Duchenne muscular dystrophy is an illness that forever changes anyone who comes in contact with it. You might expect that Scott's brief encounter with two boys dying of this disease would offer only lessons in stark reality, but his memories of this experience also retain impressions of hope and dignity. Back when he was a high school student in the late 1960s, Scott did a brief stint of volunteer work for the Muscular Dystrophy Association. He did what a high school student could do: he answered phones for the telethon and visited two brothers who were both suffering from this disease at a local convalescent home. The reality of their illness defies description; this is truly a terrible disease in which alert young people slowly waste away to death in their late teens. Amidst the realities of the prisons their bodies had become, those two boys wanted to talk of just one thing or, more correctly, of one person. They both idolized Elvis Presley. For his part, Mr Presley had gone to some great lengths to return their affection. He had flown them to one of his concerts in Las Vegas and met with them before and after. There had been cards of best wishes and, as Scott recalls, a phone call or two from Mr Presley. Despite what this disease was doing to their bodies, Mr Presley's kindnesses had made these two boys feel quite special and, as one of them told Scott, even quite lucky.

What Is the Largest Gene?

Human genes come in an amazing array of different sizes. The average length of an mRNA is around 1500 bases and much shorter transcripts are seen in the form of small regulatory microRNAs a few dozen bases in length. There are two different genes that vie for the title of largest human gene. Both genes encode important muscle proteins and defective versions of both genes are implicated in disease phenotypes affecting muscles.

Some consider the largest gene to be the DMD gene, which is more than 2,200,000 base pairs in length and encodes the 3685 amino acid dystrophin protein. It covers the longest stretch of genomic DNA of any gene, covering about 1.5% of the length of the X chromosome. It is larger than the entire genome of the bacterium *Haemophilus influenzae*. Splicing out the 79 introns reduces the millions of bases of the primary transcript down to a spliced mRNA that is just over 14,000 bases in length. The DMD gene is so large that it takes more than 16 hours to transcribe. Alternative splicing produces at least seven distinct splice variants, and the use of eight different tissue-specific promoters leads to transcripts with at least eight different 5' end sequences. Defects in the DMD gene cause Duchenne muscular dystrophy in about one in every 3500 males and in much smaller numbers of females (Box 5.13).

Although the DMD gene covers one of the largest stretches of chromosomal DNA, the TTN gene is sometimes mentioned as being the largest human gene because it makes a mammoth mRNA of more than 80,000 nucleotides and has more than 360 introns (Figure 5.23). It produces a primary transcript of more than 280,000 bases and the titin protein is 34,350 amino acids long, almost three times the size of the dystrophin protein. Extensive alternative splicing also produces many other smaller transcripts and proteins that are all derived from the same very complex primary structure of the gene. Defects in the TTN gene cause muscle weakness phenotypes called titinopathies, some of which include accompanying heart defects. Observed phenotypes include tardive tibial muscular dystrophy (TMD), familial hypertrophic cardiomyopathy (CMH9), dilated cardiomyopathy (CMD1G), a form of limb-girdle muscular dystrophy (LGMD2J), or early-onset myopathy with fatal cardiomyopathy (EOMFC). Interestingly, some mutations that cause limb girdle muscular dystrophy show an autosomal dominant pattern of inheritance while other mutations that cause the same phenotype are inherited in an autosomal recessive manner.

Does Large Size Affect the Genetic Behavior of These Genes?

A gene the size of the DMD gene represents a huge target for mutation. The types of mutations observed in the biggest genes commonly

BOX 5.13

DUCHENNE MUSCULAR DYSTROPHY

Duchenne muscular dystrophy (DMD) is a well-known disorder that results in death in the late teens. The DMD gene encodes a protein called *dystrophin* that is required for muscle maintenance and to prevent muscular atrophy. Most affected individuals are boys who lack functional copies of this protein. Inheritance of DMD is X-linked recessive so most affected individuals are males whose only copy of the DMD gene is defective. Women are not normally affected because the good copy of the DMD gene on the second copy of the X chromosome protects them. Affected boys are normal at birth, develop muscle weakness at age four to five years, are confined to a wheelchair by their early teens, and succumb to either respiratory or cardiac failure. Girls with DMD are very rare. They may happen when a woman who is a carrier has children with a man who has a DMD mutation in his germline. They are also found in cases where a new mutation early in development turns the second good copy into a defective copy. They may be found in very rare cases of X inactivation that inactive most of the copies of the X carrying the good copy. And they may happen in some cases where a translocation interrupts the good copy of the DMD gene.

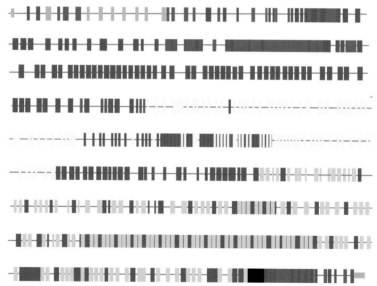

FIGURE 5.23   Structure of the titin gene. Extensive alternative splicing produces many different transcripts and many different protein variants from this one gene. The rectangles are exons and the lines are introns. Blue marks regions of unique sequences and the other colors mark regions containing repeated sequences and different types of protein motifs. (Modified by permission from M. L. Bang et al. "The complete gene sequence of titin, expression of an unusual approximately 700-kDa titin isoform, and its interaction with obscurin identify a novel Z-line to I-band linking system", *Circulation Research*, 2001;89:1065–72.)

include chromosome aberrations such as translocations, duplications, and deletions in addition to point mutations such as nonsense and missense mutations.

Newly arisen mutations cause DMD in about a third of the DMD boys born into families where no one else has DMD. The DMD mutation rate is about one new mutation in every 10,000 gametes, a rate 10 to 100 times greater than that observed for most human genes, an increase that is due to the larger size of this gene. Large deletions, from those removing just one exon to cytologically visible deletions of the entire gene, account for more than 60% of the loss-of-function DMD mutations. Most, if not all, of these deletions completely block production of dystrophin protein. The largest deletions remove the whole gene. Smaller deletions either alter the reading frame or block proper splicing. DMD nonsense mutations cause truncation and insertions shift the reading frame. The one theme that is missing from the population of DMD patients is disease-causing missense mutations. Even in those cases that do not involve large deletions, the predominant theme is failure to make dystrophin protein rather than alterations to the sequence of the protein.

Mutations that Don't Eliminate Dystrophin

With a gene as big as DMD, all the different types of mutations should hit such a large target. Although some missense mutations can cause DMD, the number of missense mutations is quite low relative to the size of the gene. Since we expect that missense mutations are happening on a per base pair basis at a comparable level to the rest of the genome, this implies that we are missing some missense mutations if we look only at individuals with DMD. Another class of mutation that is under-represented in DMD is moderate-sized deletions that leave the reading frame intact and remove only a small local section of the gene. So if the missense

mutations and small in-frame deletions are happening but not causing DMD, we have to ask ourselves who we are missing that has the missense mutations. Are they benign sequence variants present in the normal population? While some are, many are not.

Missense mutations and small in-frame deletions that cause reduced function rather than loss of function produce a muscular weakness disorder with a very different phenotype known as *Becker's muscular dystrophy (BMD)* (Box 5.14). BMD is first picked up clinically at 6 to 18 years of age, and patients often aren't confined to a wheelchair until 20 to 30 years old. Men with BMD live to 40 to 50 years of age and often produce children. Indeed, the ability of men with BMD to produce children explains the observed fact that most cases of this disorder are inherited. Unlike DMD, where approximately one third of the cases are due to new mutations, less than 10 percent of BMD cases result from new mutations.

Like DMD, BMD results from defects in the DMD gene (Box 5.15). However, most mutations that lead to BMD arise from either single-base missense mutations or from small deletions that do not disrupt the reading frame of the protein. Thus, unlike DMD patients, they have a problem with dystrophin, but they do produce at least some dystrophin protein with at least some level of activity. Even this very small amount of dystrophin activity appears to significantly deter muscle wastage and thus greatly ameliorates the phenotype, at least in comparison with DMD. You might think of the difference here as being that a DMD individual makes *no* functional gene product, a BMD individual makes a gene product that doesn't work very well.

5.10 ABSENT ESSENTIALS AND MONKEY WRENCHES

When Mendel conceived of different alleles of a gene being responsible for the differences

BOX 5.14

DMD: ONE GENE, THREE DIFFERENT DISEASES

DMD is both more common and more severe than BMD, even though they both result from changes in the same gene. Boys with DMD are usually diagnosed during the preschool years. They are usually delayed in learning to walk, become wheelchair-bound by middle-school age, and often die before 20 years of age even with the best medical care now available. Death is commonly the result of either breathing problems or dilated cardiomyopathy, which results from lack of the dystrophin protein in the heart and skeletal muscle. Boys with BMD become wheelchair-bound as young adults and tend to die in middle age. Almost half of the heterozygous carriers for Duchenne's and Becker's muscular dystro-

phies show some cardiac involvement, including dilated cardiomyopathy in about a fifth of them. In addition, cases of X-linked dilated cardiomyopathy with no skeletal muscular dystrophy have been attributed to mutations in this gene. What makes the difference? The difference is that individuals with dilated cardiomyopathy make normal skeletal muscle dystrophin; they make an abnormal version of the protein but only in heart muscle. Individuals with BMD make defective dystrophin protein. Individuals with DMD make effectively no dystrophin protein. Ongoing research is aimed with great hope towards development of gene therapy approaches to providing functional dystrophin to muscles that lack it.

BOX 5.15

A CRUCIAL LESSON

Mutations at the same locus can produce different traits, depending on the type of mutation and its position within the gene. The effect of producing an altered product may be quite different from the effect of producing no product at all, and mutations in different functional domains of a protein may change one aspect of the protein's function while leaving other aspects of the protein's function unchanged. So mutations that eliminate dystrophin cause an early onset severe phenotype, and mutations

that compromise dystrophin function without eliminating it cause a milder, later onset trait. And among those mutations that retain some dystrophin function, mutations in some regions cause earlier onset of cardiomyopathy, while mutations in other regions cause later onset of cardiomyopathy. So sometimes the mutation type or location predicts the trait, and sometimes it predicts a particular feature of the trait depending on what aspect of a complex function has been compromised.

in the observed traits, he had no model for the form the differences in information might take nor for why some forms of the information would manifest themselves as dominant traits

while others would appear as recessive traits. He could not explain why some alleles would turn out to be recessive while others would turn out to be dominant. Recessive inheritance can be seen in many diseases with serious or even potentially lethal consequences, such as cystic fibrosis, phenylketonuria, and Lesch–Nyhan syndrome. Dominant disorders include comparably severe illnesses such as Huntington's disease, Lou Gehrig disease (amyotropic lateral sclerosis), and Marfan syndrome, which is believed to have affected Abraham Lincoln. Why do changes in some genes cause a dominant problem when the information in that gene is altered, when changes in other genes cause a recessive problem? Next we will talk about the problem of trying to correlate mode of action of a mutation with mode of inheritance of that mutation.

Absent Essentials and Monkey Wrenches: The Relationship Between Mutation Mechanism and Mode of Inheritance

On some fundamental level, mutations can be divided into two classes:

- *Loss of function*: Mutations that result in the absence of something essential, whether a structure or a process, might be thought of as absent essentials or by the more commonly used term "loss of function."
- *Gain of function*: Some mutations produce an abnormal protein that actively does something wrong, and in doing so disrupts an essential cellular function. We sometimes refer to these mutations as monkey wrenches because of the English expression "Throwing a monkey wrench into the works" which implies that something has been actively added into the equation with accompanying negative consequences.

Classically, people have described mutations in terms of mode of inheritance. They often equate loss of function mutations with recessive inheritance because that combination is often seen and the combination makes sense. Similarly, they often equate gain of function

mutations with dominant inheritance. As shown below, although these correlations are often observed, there is no guarantee that a particular mode of inheritance will correlate with a particular mode of action (gain or loss of function). We gain in understanding when we maintain the distinction between the mode of action (gain or loss of function) and the mode of inheritance (dominant or recessive). We gain important insights when we find cases where a loss of function is not recessive or a gain of function is not dominant.

The Expected Case: A Loss-of-function Mutation Produces a Recessive Trait

"Now I get to grow up."—*Thank you note from a child with cystic fibrosis to researchers who helped find the CF gene*

When the gene responsible for cystic fibrosis was cloned, it was a landmark event in molecular genetics. Researchers all over the world had inched their way towards an answer to what was causing this killer disease. Finally, an international collaboration of doctors and molecular geneticists used cutting-edge technologies, traded resources, shared information, and pooled ideas to make the breakthrough and find the gene. At that time, many of today's advanced technologies and resources were not available, and some steps that are now done in a few hours with computers took years of laborious slogging through experiments at a lab bench. The tale of the cloning of the cystic fibrosis gene was really one of determination, heroism, and hope, as shown in the flood of mail that the researchers received after the announcement that the gene had been found: letters of congratulations, letters of hope, and letters from small children writing to thank them for being given the opportunity to grow up.

Mutations in the *cystic fibrosis* gene (Box 5.16) act as we expect – loss of function leads to recessive inheritance. The cystic fibrosis

---

BOX 5.16

CYSTIC FIBROSIS – A RECESSIVE LOSS-OF-FUNCTION DISEASE

According to the Cystic Fibrosis Foundation, one in every 30,000 people in the United States has cystic fibrosis, and more than 10 million people are carriers. Vigorous percussion of back and chest, accompanied by regimens of antibiotics, anti-inflammatories, and mucous thinning drugs are all ways to try to help cut down on infections that can threaten the patient's life. Enzymes, vitamins, and diet all help with digestive problems that affect some individuals with cystic fibrosis. With many medical advances in recent years, more and more children with cystic fibrosis are going on to become adults. Current research on treatments includes nutritional studies and investigations of antibiotics, drugs to change salt transport in the lungs, and even gene therapy trials.

---

gene makes a protein called CFTR that has to be present and working correctly for the lungs and other organs to stay healthy. The CFTR protein is an enzyme that transports salt (technically chloride and bicarbonate ions) across the membranes of several tissues including the lungs, pancreas, liver, intestines, and reproductive tract. The most commonly known feature of cystic fibrosis is congestion and infection in the lungs. If salt that is supposed to leave the cells of the lungs instead stays inside them, there will not be enough fluid outside the cells. Without enough fluid, the mucus in the lungs gets too thick. This in turn leads to inflammation and the possibility of chronic respiratory infections. These repetitive infections can be fatal, although improved treatments are allowing more and more children with cystic fibrosis to grow up.

Let's imagine a child, May, with moderately severe cystic fibrosis. When we examine her genetic blueprint, we find that both of her copies of the cystic fibrosis gene have a mutation that creates a stop codon, in place of amino acid 553. May is missing more than 60% of the length of the protein, including regions of the protein that carry out important functions. Thus she appears to be lacking functional CFTR protein. This child's recurrent bouts of illness do not really occur because a change in a couple of bases of the sequence in her blueprint causes a problem. Instead, she is ill because lack of CFTR function is caused by the absence of the CFTR protein itself brought about by the typo in her blueprint. On the most fundamental level, she is ill because the functions that the CFTR protein should be carrying out are not being carried out. A look at her family finds that she is the only one who has cystic fibrosis but that several of her relatives are carriers who have one defective copy of the gene and one normal copy of the gene (Figure 5.24). Looking at her family structure, with her parents sharing a great-great-grandfather who carried the mutation, also shows us how she might have come to have both copies of her CFTR gene share the same mistake.

Now consider this: what if her brother Jeff carries both a normal copy of the CFTR gene and a defective copy? Most texts and even some professors will glibly tell you that Jeff will be fine simply because the CFTR mutation is a loss-of-function mutation and that loss-of-function mutations are recessive (and thus the normal allele is dominant). We have to wonder,

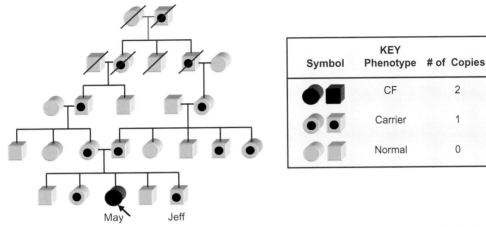

| | KEY | |
Symbol	Phenotype	# of Copies
	CF	2
	Carrier	1
	Normal	0

May   Jeff

FIGURE 5.24   Loss-of-function mutation. In this case a mutation in the cystic fibrosis gene shows the classical recessive pattern of inheritance that is often but not always seen for loss-of-function mutations. Here we show only the immediate line of descent from the shared great-great-grandfather who passed the mutation along to both sides of the family and leave out many of the spouses and other branches of the family. Also notice that since no one else in this family developed cystic fibrosis, no one realized that this mutation was present in the family until May was born.

though: Why is Jeff fine? Does one normal copy of a gene per cell produce enough good protein to carry out the needed function and prevent damage or illness when the second copy is defective? Does this imply that cells normally produce extra CFTR, so they don't mind losing some of the protein, or that CFTR is such an efficient protein that one half the normal levels of this protein can apparently manage enough salt and fluid transport to maintain health?

For different genes, there may be different mechanisms by which one good copy can compensate for a damaged copy. An important benefit of being diploid is that needed processes still take place even if there is a problem with one copy of a gene. So we expect that for some recessive diseases that only one copy of the gene makes enough protein and two copies make an excess. In other cases the body has the ability to compensate for a reduced amount of protein by putting in an order for increased production of protein from the normal copy of the gene. This is a system that works well until both copies of the gene are gone. But there is no way to turn up the rheostat on expression of a gene

if both copies are missing, and it does no good to turn up expression of a gene whose product does not work.

The Surprising Case: A Loss-of-function Mutation Produces a Dominant Trait

Although it is true that absent essentials often result in recessive traits, there are important examples of loss-of-function mutations that create severe phenotypes even in the presence of a normal allele. Some interesting examples of this unexpected situation include genes that produce structural proteins rather than enzymes, and genes that produce certain kinds of regulatory proteins that control the expression of other genes. Part of why loss-of-function mutations in these genes turn out to be dominant is because these are the genes for which the amount of gene product in the cell is critical. The cell apparently cannot tolerate a reduced amount of protein, and the cell has no ability to get the one normal copy to up-regulate its expression enough to compensate for what is missing.

BOX 5.17

OSTEOGENESIS IMPERFECTA I – DOMINANT LOSS-OF-FUNCTION ALLELES

Osteogenesis imperfecta is often inherited as an autosomal dominant disorder. Although some children may inherit it from one of their parents, about 25% of osteogenesis imperfecta cases appear to be new mutations. Types II, III, and IV osteogenesis imperfecta may involve an abnormal form of collagen being made, but in the case of type I, the problem is that not enough collagen is made. Type I osteogenesis imperfecta is considered on average the least severe form of the disease. Altogether, the Osteogenesis Imperfecta Foundation estimates that there are more than 20,000 people in the United States with some form of osteogenesis imperfecta, and type I is the most common of the four forms. Children with severe osteogenesis imperfecta may be born with multiple fractures and suffer numerous additional broken bones in the course of growing up. One of the most heartbreaking commentaries on this disease is the fact that the

Osteogenesis Imperfecta Foundation includes a whole section on their web page about the problems of good, loving parents who are mistakenly arrested for child abuse. In the cases of children who turn up with broken bones in situations in which the injury that is described does not match the severity of the damage that results, the answer might actually be child abuse, but the answer can also be that non-abusive parents are dealing with a child with bones that are too fragile for even gentle, loving care by parents who do not realize that their child has such fragile bones. Fortunately, there are doctors who can do genetic and biochemical tests that can help distinguish a child with a genetic disorder from a child who has been injured by an abusive parent, but we have to wonder how many parents in earlier times protested their innocence to deaf ears because they and their accusers both did not know that this disease existed.

Many structural proteins, such as type I collagen, are required in large amounts. Moreover, the demand for these building blocks of biological structures is usually so high that having half as much as normal is insufficient. When just one normal copy is present along with a damaged copy, the presence of one such nonfunctional allele results in a severe disease called type I ontogenesis imperfecta, a disease that causes brittle bones and early onset deafness (Box 5.17). So it is a loss-of-function mutation – the cell makes less of a structural protein than it needs. However, presence of disease in heterozygotes shows us that the disease allele predominates in the presence of the normal allele, so we consider this case of loss of function to be dominant.

The Expected Case: The Proverbial Monkey Wrench Results in a Dominant Trait

In some cases the abnormal proteins that show a gain of function are essentially poisonous, and in other cases they exert their effects by doing the normal things they are supposed to do, but doing them at a time when or in a place where those functions should not be taking place. Some monkey wrenches that actively harm the cell are called *gain-of-function* mutations. That gain of function can be something very specific, such as binding to a different hormone than the one to which the protein usually binds. It can also be as nonspecific as poisoning

---

BOX 5.18

MARFAN SYNDROME – WAS DEFECTIVE FIBRILLIN LINCOLN'S MONKEY WRENCH?

Some people believe that Marfan syndrome was the cause of Abraham Lincoln's lanky build, although finding out for sure has been held up by controversy about whether testing available samples of Lincoln's DNA would constitute violation of his privacy. On the one side is the important consideration that dying does not mean that you have lost your privacy. On the other hand, it has been argued that Lincoln would not have minded having this question asked. He was, after all, a public figure who was not particularly protective of his personal privacy. It has also been argued that knowing that he had Marfan syndrome would not hurt him, but would likely help the cause of people with genetic defects in general and Marfan syndrome specifically by demonstrating that someone with such a mutation or disease can be so successful and admired. What a strange concept, that in an era when cardiac surgery and modern heart medicines had not yet been heard of, a cardiac time bomb could have been waiting to fell him even if an assassin's bullet had not. Perhaps we will never know, if the ethical issues regarding this DNA test are never resolved. Even if the test showed that he had the defect, we still would not know if he would have faced heart failure, since severity of genetic diseases can so often vary from one person to another. More on this subject and other issues arising from modern genetics can be found in *Abraham Lincoln's DNA and Other Adventures in Genetics* by Phillip R. Reilly (2000, Cold Spring Harbor Laboratory Press).

---

the cell. Some monkey wrenches that act by interfering with the normal protein's ability to do their job are called *dominant negative* mutations. Proteins, like ballet dancers and bank robbers, often do not truly act alone! Instead, many function either as *dimers*, an associated pair of identical proteins, or as parts of large macromolecular assemblies that are composed of many proteins. Imagine then a gene whose protein product is assembled into such a large structure. Now imagine that when a missense mutation occurs, a misfolded version of the protein cripples the structure (kind of like the proverbial weak link in a chain). Such mutations can and do create a cellular disorder/defect, even in the presence of a normal gene product. The presence of the bad copy can make the structure defective, but the presence of the good copy cannot undo the damage caused by the bad copy.

An example of a monkey wrench is Marfan syndrome (Box 5.18), which is inherited as an autosomal dominant trait. As you will recall, the fact that this trait is dominant means that only one defective copy is needed to cause disease. Skeletal features of Marfan syndrome include unusual height, long limbs, and joint laxity. Some of these individuals have spinal deformities such as scoliosis, eye problems that include myopia, or even life-threatening cardiovascular defects. Individuals with Marfan syndrome have been shown to have a defect in the gene that makes *fibrillin*, a protein that combines with itself and with other proteins to form microfibrils. According to the National Marfan Foundation, there may be 200,000 people or more with Marfan syndrome or something related to it.

This broad array of problems – skeletal, cardiac, and ocular – results when defective fibrillin causes abnormal protein complexes in cartilage, tendons, blood vessels, muscles, skin, and various organs. It has been proposed that the defective fibrillin has a dominant negative effect, interfering with the formation of correct microfibrillar structures even though there is also normal fibrillin present.

A large number of different missense mutations in fibrillin can cause Marfan syndrome, as can some deletions and splicing defects, including the one shown in Figure 5.12. From a list of the known mutations we would have predicted that the mechanism would be gain of function rather than a loss of function, based on the many missense mutations and the scarcity of nonsense and frameshift mutations.

The Surprising Case: A Gain of Function Results in a Recessive Trait

So aren't mutations that create disruptive or poisonous proteins always going to be dominant? Once again, things are not that simple. The answer is simply no, they won't always be dominant. *In fact, while a mutant protein that is actively doing something new will normally act as a monkey wrench and cause problems, sometimes the new function of the protein can even be beneficial.*

This point is made most clearly by considering the case of a disorder that can be serious or even lethal in homozygotes but can actually be beneficial or even life-saving in heterozygotes. The disease in question is *sickle cell anemia*, a severe blood disorder that can cause serious illness or even death (Box 5.19). The genetic basis of this disease is the presence of two defective copies of the gene that makes a *hemoglobin* protein essential for adult red blood cells to successfully carry out their role of transporting oxygen. A *missense mutation* causes the production of hemoglobin S (also called HbS), which has the wrong amino acid, valine, at position 6, which contains a glutamic acid in the normal variant, hemoglobin A (also called HbA). Some other genetic variants are also known that can

cause sickling of red cells, but HbS is the most common cause.

Sickle cell anemia is a recessive disease that occurs when both copies of the gene are defective. The HbS protein becomes insoluble and forms aggregates. Cells that contain only the abnormal HbS tend to become rigid and deformed, taking on a sickle shape that tends to get stuck in the capillaries and break. Among the complications that can result are severe pain, infections, leg ulcers, delayed growth, and eye damage. Some of the more severe complications can include strokes, lung congestion, and pneumonia. The consequence of this incorrect amino acid is an abnormal hemoglobin molecule that causes red blood cells to become rigid and deformed (shaped like a sickle) and to block the capillaries. Over time, lung and kidney damage can accumulate. Treatments include antibiotics, vitamins, avoiding dehydration, carrying out transfusions, and, in rare cases, even bone marrow transplants.

The gene that makes fetal hemoglobin (HbF) is a different gene from the one that makes HbS, and normally the job of fetal hemoglobin gets taken over by HbA in an adult. How far "off" the HbF expression gets turned after birth apparently varies from one person to the next. And how far "off" the HbF gene gets turned also apparently affects how severe a case of sickle cell anemia will be. So one of the most interesting treatments recently developed involves manipulating gene expression to get expression of the fetal hemoglobin gene HbF to partially compensate for the defective hemoglobin in the red cells (see Box 5.19).

Interestingly, there is a different kind of gain of function associated with this same sickle cell mutation that is considered to be dominant. Although people with two copies of HbS struggle with pain and illness, people with one copy of HbA and one copy of HbS are actually sometimes better off than people who only have the normal sequence. Specifically, people who are heterozygotes are better off if they live in areas

BOX 5.19

MANIPULATING GENE EXPRESSION TO TREAT SICKLE CELL ANEMIA

Hemoglobins are the proteins in the red blood cells that carry oxygen to the tissues and carry carbon dioxide back to the lungs to be breathed out of the body. HbS is a damaged version of the HbA hemoglobin present in adult red blood cells. By itself, the sickle cell form of hemoglobin (HbS) forms abnormal biochemical structures that cause the cell to sickle and become rigid. In heterozygotes, where HbA is also present, the abnormal hemoglobin structures that cause sickling don't take place. Fetal hemoglobin (HbF) is a form of hemoglobin that is mainly made before a baby is born. Expression of the HbF gene normally shuts down by the time a baby is born, but some expression may continue. HbF can provide a similar protection against sickling. Since the patient with the sickle cell anemia does not a have a gene present that can make HbA, you might think that manipulating gene expression would not help, since you can't turn on expression of something the patient does not have. However, since HbF is made by a different gene, it is available to be tapped for service when HbS is making the patient ill. By using a medication called hydroxyurea, doctors are able to turn on an increased expression of HbF. Presence of HbF reduces the amount of sickling and decreases the problems with bouts of pain caused by red cell breakage in the capillaries. It will be a long time before we know how beneficial this medicine is. It does not turn up expression of HbF in everyone who takes it, and it can have side effects. We will talk in later chapters about gene therapy in terms of actual changes to the DNA in the cell being treated, such as adding back a good copy of HbA into cells that cannot make HbA. Hydroxyurea, which has been in use since the mid-1990s, constitutes a different form of gene therapy that simply makes use of one of the patient's own endogenous genes to provide a substitute gene product that can take over at least some of the needed function. This is not an effective enough process to outright cure the patient, but it apparently does provide a level of remedy that can make a real difference for at least some patients. Unfortunately, not enough is known about the effects of this drug on children, so it is not available to them even though more than 10% of children with sickle cell anemia will have a stroke or other major problems before they are adults.

where malaria occurs. *The heterozygotes, with one normal allele, and one "sickle" allele, are less likely to die if infected by the parasite that causes malaria.* In fact, the frequency of the sickle cell mutation is higher in areas with endemic malaria, and lower in areas where malaria is rare or nonexistent. Thus the sickle cell mutation is a recessive gain of function in one sense (the disease called sickle cell anemia) but a dominant gain of function in another respect (the beneficial trait that causes resistance to malaria).

The incidence of this disease is the highest in African populations and approaches 1 in 25 births in some parts of equatorial Africa. (The incidence of sickle cell anemia among African Americans is approximately 1 in 500.) Under normal conditions, people who have one good copy of the gene are fine (although they may exhibit some symptoms at very high altitudes, where the oxygen pressure is low). Although we think of sickle cell anemia as affecting Africans and African Americans, sickle cell anemia can also be found among people who live in other parts of the world where malaria is present, including among some Mediterranean populations and in India. What we are looking at is a kind of trade-off between the optimal genotype for a malaria-free environment vs. the optimal genotype for an environment in which malaria is endemic. Because disease resistance occurs in the heterozygotes and is especially frequent in regions endemic for malaria, the individuals with sickle cell anemia pay the price for a mutation that benefits the population overall while harming them as individuals.

More information can be found by going to the New York Online Access to Health, which offers a long list of links to sources of information about sickle cell anemia and hydroxyurea treatments at www.noah-health.org/en/blood/ sicklecell/care/hydroxyurea.html.

In Summary

There would appear to be some standard correlations, that absent essentials (loss of function) usually will be recessive and monkey wrenches (gain of function) usually will be dominant simply because that is often what happens. In reality, there are no absolute correlations between the actual nature of mutations, in terms of their effect on gene function and their phenotype when a bad copy and a good copy are both present. The relationship between a given form of a gene and its phenotype depends on the nature of the encoded protein, its biological function, the cell type in which it acts, and the environmental factors that influence expression.

For these reasons, we prefer to couple the terms *dominant* and *recessive* with a separate description of mutations in terms of the gene's ability to synthesize functional or poisonous proteins. Thus we often couple terms together,

referring to a recessive loss-of-function allele or a dominant gain-of-function allele to give the combined information about how the mutation acts in a pedigree and what it actually does in terms of protein production. However, sometimes we do not have enough information to know which molecular mechanism is involved if we are dealing with a phenotype and a mode of inheritance for which the actual gene or biochemical pathway remains unknown.

Study Questions

1. What do we mean when we talk about a gain-of-function mutation?
2. What do we mean when we talk about a loss-of-function mutation?
3. What is a missense mutation?
4. What is a nonsense mutation?
5. List three different causes of mutations.
6. What is a silent mutation?
7. What is detected by the Ames test and how does the test work?
8. What kind of mutation causes lactose intolerance?
9. What is a sequence tagged site? Why did the development of sequence tagged sites allow for more rapid progress in the study of genes?
10. What is an *in silico* experiment?
11. Give two different definitions of the term polymorphism.
12. Give two different definitions of the term mutation.
13. Why would the Duchenne muscular dystrophy gene be considered the largest gene, and why might the titin gene be considered the largest gene?
14. What is a tandem repeat? What is the most common tandem repeat in the human genome?
15. What happens to the mutation rate in the Huntington disease gene in sperm as a man

grows older? How can this mutation rate be assayed?

16. What is anticipation? In some cases apparent anticipation is not real; what are two things that could lead us to think that anticipation is going on when it is not?

17. What is a transgenic animal?

18. How are Duchenne muscular dystrophy and Becker muscular dystrophy related?

19. What is the Tiresias complex?

20. What is the male biological clock?

Short Essays

1. How can methylation of DNA affect mutation, and why would this result in differences in male and female mutation rates? As you consider this read the article "Dear Old Dad" by Rivka L. Glaser and Ethylin Wang Jabs in *Science of Aging Knowledge Environment* 2004(3):re1 [DOI: 10.1126/sageke.2004.3.re1].

2. When we suspect mutation in the environment around us, it can be hard to sort out the real cause of the problem because the environments around us are so complex. How do animal models help us to understand which aspects of our environments we should be concerned about? As you consider this question read the article "Explaining frog deformities" by Andrew R. Blaustein and Pieter T. J. Johnson in *Scientific American*, February 2003.

3. Some data suggest that as men age the chance of having a child with autism increases. Because these autistic children have a substantially greater frequency of large deletions in their genomes, it has been suggested that these genetic changes play a role in the trait. How might this increased rate of deletions be present and yet not be the cause of the trait? As you consider this question read the article "Male biological clock possibly linked to autism, other disorders" by Charlotte Schubert in *Nature Medicine*, 2008;14:1170.

4. One of the characteristics that sets humans apart from animals is the ability to speak. One large family showing inheritance of a complex defect in the ability to speak lead researchers to conclude that mutations in the FOXP2 transcription factor gene play a role in determining who has the power of speech. How could the study of such mutations tell us about whether our distant ancestors had the ability to speak? As you consider this question read the article "Positive selection in the human genome: from genome scans to biological significance" by Joanna L. Kelley and Willie J. Swanson in *Annual Review of Genomics and Human Genetics*, 2008;9:143–60.

5. Sometimes those of us who live around smog and pollution might think we could protect our genomes and our descendants by going to live on a pristine tropical beach away from the contamination associated with high levels of technology. Why might this be trading one mutational problem for another? As you consider this question read the article "Evolution speeds up in the tropics" by Jeff Akst by going to TheScientist.com (posted June 24, 2009).

6. Copy number variation is just one of many mechanisms by which genes can contribute to our characteristics, but is it a major player in the game? Just how much copy number variation is present in your DNA? As you consider this question read "Too little, too much" by Melinda Wenner in *Scientific American*, June 2009.

Resource Project

Online Mendelian Inheritance in Man (OMIM) is a resource at the National Center for Biotechnology Information (NCBI) that offers brief overviews of human genetic traits and genes and a sampling of some of the mutations associated in the gene causing the trait. OMIM can be found at http://www.ncbi.nlm.nih.gov/omim/.

Go to OMIM and find the article on the ZEB1 gene. Write a paragraph about mutation type and phenotype that can result.

Suggested Reading

DVD

ABC News Nightline: Confronting a Genetic Legacy
The Code of Life: DNA, Information, and Mutation
When You Remember Me. (This movie presents the life of DMD patient Mike Mills.)

Articles and Chapters

"Environmental DNA damage may drive human mutation" by David Biello in *Scientific American*, May 2006.

"How trivial DNA changes can hurt health: Small changes to DNA that were once considered innocuous enough to be ignored are proving to be important in human diseases, evolution and biotechnology" by J. V. Chamary and Laurence D. Hurst in *Scientific American*, June 2009.

"There's something curious about paternal-age effects" by James F. Crow in *Science*, 2003;301: 606–7.

"Human mutation rate revealed: Next-generation sequencing provides the most accurate estimate to date" by Elie Dolgin in *Scientific American*, August 2009.

"The real cause of obesity: It's not gluttony. It's genetics. Why our moralizing misses the point" by Jeffrey Friedman, *Newsweek* Web Exclusive, September 10, 2009.

"Unfortunate drift" by Josie Glausiusz in *Discover Magazine*, June 1995.

"How can a genetic mutation cause muscle to turn into bone? A rare genetic disease leaves its victims debilitated by transforming soft tissue cells into bone cells, creating a strange second skeleton. A leading researcher explains how the disease works and what we can learn from it" by Katherine Harmon in *Scientific American*, December, 2009.

"Diversity revealed: From atoms to traits: Charles Darwin saw that random variations in organisms provide fodder for evolution. Modern scientists are revealing how that diversity arises from changes to DNA and can add up to complex creatures or even cultures" by David M. Kingsley in *Scientific American*, January 2009.

"Genomic rearrangements and sporadic disease" by James R. Lupski in *Nature Genetics* 2007;39:S43–7.

"'Methuselah' mutation linked to longer life study of long-lived Ashkenazi Jews may yield longevity genes galore" by J. R. Minkel, *Scientific American*, March 2008.

"The gene with two faces" by Lori Oliwenstein in *Discover Magazine*, May 1993.

"Neanderthal Man: Svante Paabo has probed the DNA of Egyptian mummies and extinct animals. Now he hopes to learn more about what makes us tick by decoding the DNA of our evolutionary cousins" by Steve Olson in *Smithsonian Magazine*, October 2006.

"What makes us human? Comparisons of the genomes of humans and chimpanzees are revealing those rare stretches of DNA that are ours alone" by Katherine S. Pollard in *Scientific American*, May 2009.

"Survival of the mutable" by Sarah Richardson in *Discover Magazine*, September 1994.

"The 2% difference: Now that scientists have decoded the chimpanzee genome, we know that 98 percent of our DNA is the same. So how can we be so different?" by Robert Sapolsky in *Discover Magazine*, April 2006.

"The structure of change" by Colin A. M. Semple and Martin S. Taylor in *Science*, 2009;323:347–8.

"An age of instability" by David Sinclair in *Science*, 2003;301:1859–1960.

"The migration history of humans: DNA study traces human origins across the continents: DNA furnishes an ever clearer picture of the multimillennial trek from Africa all the way to the tip of South America" by Gary Stix in *Scientific American*, July 2008.

"Sickle cell anemia, a molecular disease" by Bruno J. Strasser in *Science,* 1999;286:1488–90.

"Best in show: Scientists pursue selected gene mutations bred into dogs" by Carina Storrs in *Scientific American*, January 2010.

"Early to bed, early to rise: Scientists determine how gene behind sleep cycle works: A single amino acid in a particular protein can get you up long before dawn and into bed well before prime time" by Nikhil Swaminathan in *Scientific American*, January 2007.

Books

Gene Hunter: The Story of Neuropsychologist Nancy Wexler by Adele Glimm (2006, John Henry Press).

Living with Our Genes: Why They Matter More Than You Think by Dean H. Hamer and Peter Copeland (1999, Anchor).

The Legacy of Chernobyl by Zhores A. Medvedev (1992, W.W. Norton and Co.).

Physician to the Gene Pool: Genetic Lessons and Other Stories by James V. Neel (1994, John Wiley and Sons).

Abraham Lincoln's DNA and Other Adventures in Genetics by Philip Reilly (2002, Cold Spring Harbor Press).

Genome: The Autobiography of a Species in 23 Chapters by Matt Ridley (2005, Harper Perennial).

Madame Curie: A Biography by Eve Curie, translated by Vincent Sheean (2001, DeCapo Press).

The Seven Daughters of Eve: The Science That Reveals Our Genetic Ancestry by Bryan Sykes (2002, W.W. Norton and Co.).

Mapping Fate by Alice Wexler (1996, University of California Press).

The Woman Who Walked into the Sea: Huntington's and the Making of a Genetic Disease by Alice Wexler (2010, Yale University Press).

HOW CHROMOSOMES MOVE

Mitosis and Meiosis: How Cells Move Your Genes Around

The Human Genome. DOI: 10.1016/B978-0-12-333445-9.00006-X

THE READER'S COMPANION:
AS YOU READ, YOU SHOULD
CONSIDER

- How each daughter cell gets the right number of copies of each gene.

- Why there do not seem to be elaborate mechanisms for dividing other organelles into daughter cells.

- Which functions are taking place during the different steps of the cell cycle.

- How many chromosomes are present in a cell before and after the DNA is replicated.

- How a chromatid differs from a chromosome.

- How two homologous chromosomes are the same and how they differ from each other.

- What role the spindle apparatus plays in chromosome movement.

- What role kinetochores play.

- What holds sister chromatids together during meiosis I.

- What holds homologous chromosomes together during meiosis I.

- Why it is important that recombination takes place during meiosis.

- How meiosis II and mitosis are similar.

- How mitosis and meiosis differ.

- How homologous chromosomes manage to find each other and pair.

- Why aneuploid offspring often are not viable.

- Why most Down syndrome children are born to younger mothers, even though older mothers are at higher risk of having a Down syndrome child.

- Why some balanced translocations are benign while others are harmful.

- How the steps in meiosis bring about the rules of inheritance Mendel observed.

- Why sperm and eggs have half as many copies of each gene as somatic cells.

- Which of Mendel's rules is violated by linkage.

- What are the fundamental differences between male and female meiosis.

- How a pair of recombination events could look like no recombination event.

6.1 THE CELL CYCLE

Several years ago, a brief visit to the dermatologist left a precise surgical hole in Julia's thumb about the depth of a dime and half as wide. Over the course of several weeks, a pinkish mist of cells gradually spread inwards from the edges to fill the hole with solid skin and scar tissue. This migration of cells across the open space represented not just movement but cells growing and making new copies of themselves at a frantic pace. With each round of cell growth, the organelles and the genetic blueprint in those cells were being copied and passed along to the new daughter cells with a level of speed, efficiency, and precision that human industry would be hard pressed to match.

Have you had a rug burn lately? Or perhaps cut yourself? Have you ever wondered about the processes that go on as such damage is repaired? You spend your whole life replacing and repairing losses due to erosion and injury. In fact, cell types such as skin cells divide many times as we age in an effort to keep up with the rate at which we are losing cells. We have all repeatedly used this process of cell division to turn a small number of cells into a larger number of cells. It is how we developed into the large, complex animals we are from the single-celled zygotes that were created many years ago when that sperm met that egg and we came into being.

It seems a simple enough thing to imagine how a skin cell would duplicate itself as it joins the rush of cell growth that will fill in a damaged area. The cell is basically a sack full of organelles, the little biological engines and factories that run different functions within the cell. If this sack gets bigger while making extra copies of everything inside of it, there will be enough extras to make up two complete cells identical to the original cell. Once enough of the cell's innards have been duplicated, the cell divides down the middle to make two new cells.

It appears that cells do not need to take any specific, elaborate steps to ensure equal division of most organelles when the cell divides. There are thousands of copies of some organelles and millions of copies of others, and all of them are extensively dispersed throughout the cell. So simply pinching off the cell in the middle automatically results in many copies of each organelle on each half of that dividing line. If one daughter cell gets a few more copies of an organelle and the other daughter cell gets slightly fewer copies, it does not matter. The cell has all of the information and machinery it needs to make up for any small deficits that might happen if there is not an exactly equal division of resources at cytokinesis.

Passing the genetic blueprint along to new cells during cell growth is not nearly as simple as passing along the other organelles. The cell does not have vast quantities of copies of each chromosome. It only has two copies. Simply pinching the cell in half at cytokinesis offers no guarantee that the cell will end up with exactly one copy of each chromosome on each side of the dividing line. *Thus the cell needs a very precise mechanism for allocating exactly one copy of each chromosome into each daughter cell and ensuring that none of the daughter cells gets two copies or zero copies instead.* Having the wrong number of copies of all or part of the blueprint can cause major health problems.

Fortunately, we can use the microscope as we did in Chapter 2 to look at the movement of the chromosomes as the cell copies its contents and then splits into two daughter cells. We can start with cheek cells, white blood cells, or other sources of cells. The important thing is that we want to be looking at actively dividing cells so that we can watch the transfer of chromosomes from parent cell to daughter cells.

The series of processes leading up to cell division takes place in a carefully orchestrated, tightly controlled sequence of events called the cell cycle (Figure 6.1). The steps of the cell cycle include G1, a stage of active gene expression and resource production, S, a stage of DNA synthesis, G2, a stage in which the cell prepares for division, and mitosis, the process by which the replicated chromosomes are allocated into the daughter cells.

Through most of the cell's life the DNA molecules are loosely entwined with each other in the cell nucleus, going about the gentle business of running various aspects of metabolism and growth through transcription of genes to produce functional gene products. During this time, the chromosomes are not visible as separate entities; rather, the nucleus looks like an old Brillo pad. Only once the cell starts the process of mitosis do we begin to see distinct structures within the nucleus. So let's take a look at mitosis and see how it works.

6.2 MITOSIS

Mitosis Ensures that Two Daughter Cells Have the Same Genotype

Let us start our examination of mitosis, the process of getting the right chromosomes to end up in the right copy numbers in the right cells, by looking at a simplified case. Let's imagine a very simple fictitious critter by the name of *Organisma hypothetica*, otherwise referred to as *O. hypothetica*. This imaginary beastie consists of a small number of cells whose genes are arranged on only a single pair of homologous

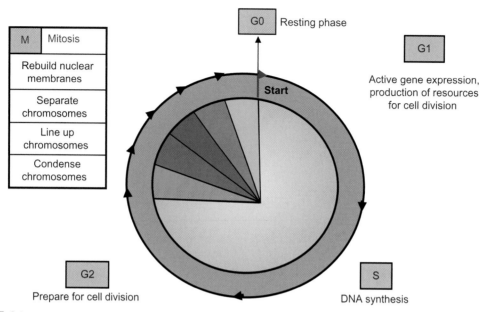

FIGURE 6.1   The cell cycle. Every dividing cell goes through the series of steps shown here. It starts with G1, the period at the beginning of the cell cycle when the cell accumulates resources needed for the next round of cell division. During the next step, S, DNA synthesis copies the chromosomes. During G2, the cell finishes off any remaining metabolic processes needed for cell division. Interphase consists of G1 plus S plus G2, a period during which the cell looks pretty much the same under the microscope. The microscopic view starts to change during M phase, or mitosis, when the chromosomes condense (prophase), line up (metaphase), move to the separate ends of the dividing cell (anaphase), and then are packaged back into a nucleus as the cell prepares to divide (telophase) in an orderly manner. Cytokinesis divides the cell into the two new daughter cells. Cells that are growing slowly spend a lot of time in G1. This pie chart shows an average representation of amount of time in the cell cycle spent in each of these stages. It also shows that mitosis is a very brief part of the cell cycle. If cells are truly inactive and not dividing, they go into a metabolic resting state called G0 instead of going into G1. *Take-home message: During interphase, when the nucleus looks like a Brillo pad, the cell makes copies of everything and gets ready for cell division. During the visibly distinct stages of mitosis, the cell carts the chromosomes around to where they should be (a process we can see under the microscope), and cytokinesis completes the separation into two cells.*

(which is to say, essentially identical) chromosomes (Figure 6.2). As you may recall, each of these chromosomes consists of one very long piece of double-stranded DNA. Now suppose that one cell in this organism needs to divide in order to form some necessary structure consisting of two or more cells.

When a chromosome is sitting in a non-dividing cell, or in the G1 phase of a cell that has not yet undergone DNA replication, it is composed of only one long, double-stranded DNA molecule. Once the cell cycle starts, the cell gets to S phase, the DNA synthesis phase. At this

point, it replicates the one, double-stranded DNA molecule in each of the two chromosomes. Each replicated chromosome is a complex structure consisting of two complete copies of the DNA molecule held together along their lengths by a complex of proteins called cohesins (Figure 6.3).

Each complete copy of the whole DNA molecule in this replicated chromosome is called a *sister chromatid* (Figure 6.4). Thus each replicated chromosome found in dividing cells contains two identical linear arrays of genes running in parallel, in exactly the same order along the length of the sister chromatids.

Once the copying, or replication, of the DNA is complete, the chromosomes begin to condense and become visible under the microscope as distinct entities. The stage is called prophase.

INTERPHASE
G1

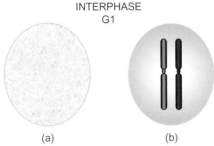

(a) (b)

FIGURE 6.2   A very simplified diagram of the nucleus of the original fictitious *O. hypothetica* cell that we are going to follow through mitosis. In a real cell, if we looked at the nucleus it would look indistinct, like the cell on the left, but since we know the pair of chromosomes is in there, we show a picture of the pair of chromosomes so that you can begin to follow what happens. For the sake of simplicity, we are leaving out the other organelles. To help keep track of this pair of chromosomes as we go, we are showing the chromosome that came from the father in blue and the chromosome that came from the mother in pink. These are *homologous* chromosomes, with the same genes in the same order arranged along the length of the chromosome. The dot in the middle of each of the chromosomes denotes the centromere, a structure that plays a critical role in chromosome movement that we will discuss later.

It is easy to remember this term if you think of "pro" as meaning "before," as in before the chromosomes start moving around within the cell. As prophase continues, the cell begins to assemble a scaffold with two poles called a spindle apparatus around the nucleus. This structure is comprised of protein assemblies

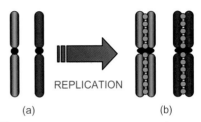

(a) (b)

FIGURE 6.3   Mitotic interphase – S phase copies the chromosomes. In a resting cell that is not actively growing and dividing, a chromosome is a long, single chain of complementary double-stranded DNA. Although the chromosomes and chromatids we show here display a constriction called a centromere, we usually can't see this structure, or even individual sister chromatids, when we look at interphase cells. Usually, we look at cells which are just entering mitosis (and thus have replicated their chromosomes) in order to see the replicated sister chromatids shown on the right, which actually contains one copy of the chromosome, with the two chromatids joined at the centromere so that the cell can move them around together.

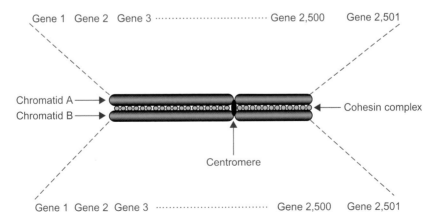

FIGURE 6.4   The replicated chromosome. Diagram of a replicated chromosome before cell division. Sister chromatids A and B are copies of each other that are highly similar, but not identical. Thousands of genes are arranged along the DNA in the chromosome rather like beads on a string, one after another. The order of the genes along the two chromatids is the same.

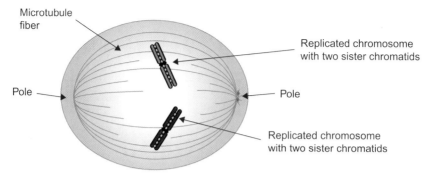

FIGURE 6.5   Nuclear envelope breakdown at the end of mitotic prophase – the nucleus opens and chromosomes attach to spindle fibers. Fragments of the nuclear membrane appear as fragments of dashed lines. Replicated chromosomes have condensed into a form that makes them easier to visualize under a microscope. A protein scaffolding called a spindle apparatus forms threads that run between the two poles of the cell and attach to the chromosomes.

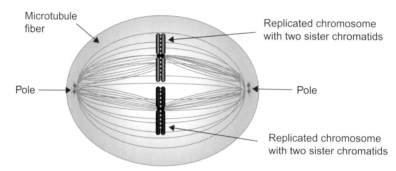

FIGURE 6.6   Mitotic metaphase – the chromosomes are pulled to the center of the cell in a schematic diagram of chromosomes on the spindle at metaphase in our fictitious cell, with only one pair of chromosomes. In a real human cell, there would be 46 chromosomes, each of which is attached to the microtubule fibers of the spindle, and the spindle apparatus would contain a vastly greater number of the threadlike microtubule fibers. Notice that fibers of the spindle attach to the chromosomes at one end and the opposite poles of the cell at the other end. Note that the spindle apparatus attaches to two sides of the same centromere.

called microtubules that will facilitate the process of chromosome movement. The end of prophase is signaled by the breakdown of the nuclear membrane that surrounds the chromosomes (Figure 6.5).

As the membrane breaks down, the centromeres of each chromosome attach to microtubules emanating from each pole, such that the two sister chromatids are attached to opposite poles at their centromeres (Figure 6.6). These microtubules, which are the protein "train tracks" along which the centromeres can move chromosomes, run from the centromeres to the ends or poles of the spindle. Complex protein structures, known as kinetochores, which are assembled at each sister centromere, actually connect the sister centromere of each chromosome to microtubule fibers. Each replicated chromosome ends up bound to its own set of tracks within the spindle apparatus, such that microtubules connect the centromere of one sister chromatid to one pole while other microtubules connect the centromere of other sister chromatid to the opposite pole.

By the time the cell cycle advances to the next stage, called *metaphase*, the chromosomes have moved to the center of the cell, midway between the poles, and lined up on a theoretical "plate" that is the cross-section through the center of the spindle. Again, kinetochores at the centromeres of each sister chromatid are connected by microtubule fibers to poles at both ends of the cell, such that one sister chromatid of each chromosome is oriented toward each pole (Figure 6.6). As a result of these attachments to the poles at opposite ends of the cell, chromosomes have lined up along the equator or midpoint of the cell (also known as the metaphase plate). In this case, we can think of meta- in metaphase as meaning "between" or "among" because it takes place right in the thick of things, after the chromosomes have been copied and before the cell divides. You might think of metaphase as the "middle stage" because of when it happens, or the "middle place" because of where it happens, at the metaphase plate in the middle of the cell.

The next step in cell division is known as *anaphase* (Figure 6.7). This step depends on two events. First the cohesin complex that has connected the sister chromatids along their entire length is destroyed and the sister chromatids are now free to separate. Second, the protein complexes known as kinetochores bound to the centromeres contain motor proteins that move the chromosome along the spindle fibers toward the pole during cell division. All of a sudden, the two sister chromatids completely split, right at their connection point at the centromere, and the centromere motors then move the separated sister chromatids rapidly to opposite poles of the cell by pulling them along the tracks of the spindle fibers. We are just now beginning to figure out how this terribly complex process works, and many but not all of the

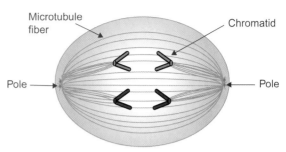

FIGURE 6.7 Mitotic anaphase – the sister chromatids of each chromosome split apart and move to opposite poles of the cell. The same cell seen in Figure 6.5 has now advanced to anaphase, in which the motors in the centromeres move the separated sister chromatids along the tracks of the spindle apparatus towards the opposite poles of the cell.

proteins involved have been identified. The key concept to understand here is that there are proteins at the centromere that function as motors that pull the chromatids along the spindle fiber tracks toward the opposite poles of the cell.

The phase of the cell cycle that occurs once the chromatids have reached the poles of the spindle is called *telophase*. At this point, the membrane around the nucleus reforms and we begin to see where the cell will split into two parts (Figure 6.8).

After telophase, actual cell division, called *cytokinesis*, occurs (Figure 6.9). During metaphase, each of the pair of replicated chromosomes had two chromatids (Figure 6.10), which means that the cell momentarily had four copies of each gene instead of the two copies normally present in a resting cell. Now, after cytokinesis, there are two daughter cells whose genotype and DNA content are identical to the original cell.

Remember, when we started out with our fictitious cell, we said that the blue chromosome came from one parent and the red chromosome came from the other. Notice that each cell has ended up with one blue and one red chromosome, not two red chromosomes in one cell and

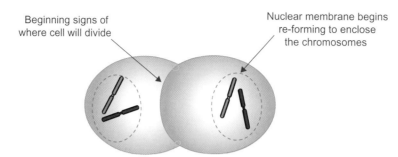

FIGURE 6.8   Mitotic telophase – the nuclear membrane begins to reform and the cell prepares to divide. The copies of each chromosome have been pulled to opposite sides of the cell by the motor apparatus and are now being set off from the surrounding cytoplasm as the nuclear membrane begins to form again.

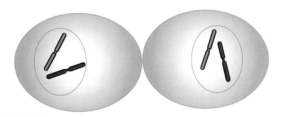

FIGURE 6.9   Cytokinesis – physical separation into two cells. The separation into two separate cells is completed, and the nucleus has completely reformed. The result is two daughter cells that are identical to the original cell, with identical copies of the pair of chromosomes that the original cell started out with.

two blue chromosomes in the other cell. As we will discuss later, it is important that each new daughter cell gets one copy of the pair from mom and the other copy of the pair from dad.

It is this process of mitosis that allows individuals to develop from a single-celled zygote (the product of sperm and egg fusion) to a complex organism with "gazillions" of cells, all of which are genetically the same. When there are more chromosomes, the process can be more complicated (Figure 6.11).

In a human cell, there are 46 chromosomes lined up in the center of the cell at metaphase. Each of these 46 chromosomes is attached to the spindle apparatus and needs to have the centromere motors pull the two halves of the

chromosome apart and carry them to opposite ends of the cell. In fact, the gathering of replicated chromosomes at the metaphase plate during mitosis in a human cell is a terribly crowded and complex event aimed not at passing along one pair of chromosomes but rather at seeing to it that a copy of each and every one of the 46 chromosomes ends up in each of the daughter cells at the end of mitosis.

The basic pattern of events in the human cell cycle is the same as in our hypothetical organism. In Figure 6.12 the whole series of steps in mitosis is shown in photographs of real cells. Green and red dyes show the locations of proteins, such as those that form the spindle apparatus, and blue dyes show where the chromosomes are.

So the process of getting the right number of copies of the genetic blueprint into the daughter cells doesn't seem that tough. During the cell cycle, the cell copies everything in it, including the chromosomes. Replicated chromosomes stay attached to each other while they line up at the center of the cell and become attached to "tracks" that connect to the poles at opposite ends of the cell. Motor proteins in the centromeres separate the replicated chromosome back into two single chromosomes that get pulled to the opposite poles of the cell along the protein tracks. The key to this successful allocation of

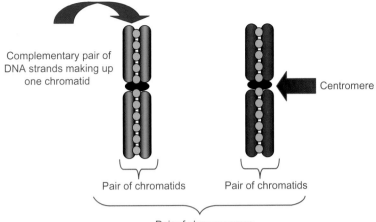

Complementary pair of DNA strands making up one chromatid

Centromere

Pair of chromatids

Pair of chromatids

Pair of chromosomes

FIGURE 6.10 Pairs of pairs of pairs. When we look at the chromosomes in an actively dividing cell, we end up with a lot of pairs of things. First, we find that we have pairs of chromosomes. Second, each of these chromosomes passes through a stage right after it is replicated but before the cell divides when the chromosome has a pair of sister chromatids. Finally, each chromatid consists of a pair of complementary single strands joined into a double-stranded structure. It is important to keep track of whether we are talking about pairs of strands, pairs of chromosomes, or pairs of chromatids within the replicated chromosome structure of a dividing cell.

copies into the new cells is the replicated chromosome in which the duplicated copies are kept locked together and moved around as a unit until the cell is ready to send them to opposite ends of the cell.

The system for mitotic allocation of chromosomes sounds simple, and it works with surprising efficiency to get the right number of chromosomes into the right daughter cells over and over and over again. But what happens when the two sister chromatids fail to segregate properly into the two daughter cells? In some cases the two sister chromatids both go to the same pole at anaphase I and thus end up in the same daughter cell after cytokinesis. The result is a daughter cell that inherits an extra copy of a given chromosome (thus ending up with *three* copies of that chromosome!) and a daughter cell that now has only one copy of that chromosome. This process is known as *mitotic nondisjunction*. In other cases a single chromatid gets "lost" and fails to make it to its pole, while its sister proceeds normally to the pole. This

process is known as mitotic loss and results in one of the two daughter cells missing a copy of that chromosome. Fortunately both mitotic nondisjunction and mitotic loss are very rare events that most often lead to cell death. There is unfortunately an exception to this rule in terms of tumor development that we will discuss in Chapter 10.

6.3 GAMETOGENESIS: WHAT IS MEIOSIS TRYING TO ACCOMPLISH?

In the first part of this chapter we focused on mitosis, a process that ensures the production of two daughter cells with identical genotypes (and thus an identical number of chromosomes). But sexual reproduction requires that gametes (sperm or eggs) carry only a copy of each pair of homologous chromosomes – in other words, although the somatic cells of our bodies each have 46 chromosomes, sperm or egg cells must carry only 23 chromosomes. In

Metaphase

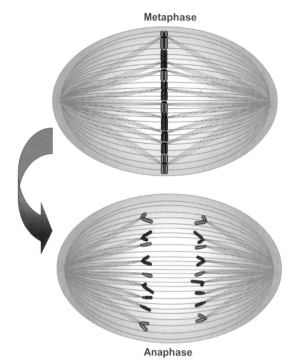

Anaphase

FIGURE 6.11 Mitosis in complex organisms. When many chromosomes are present in the cell, correct segregation into the separate cells still happens correctly because each chromosome gets moved around individually. Each replicated chromosome gets aligned at the metaphase plate and attached to the spindle apparatus separately. So, even though there are many chromosomes involved, the cell handles each one as an individual problem. Here we see the cell handling eight chromosomes (four pairs of homologous chromosomes, with a blue copy and a red copy of each chromosome) as it advances from metaphase to anaphase.

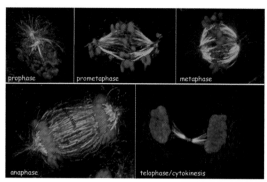

FIGURE 6.12 The whole mitotic process. These beautiful pictures show the whole process, from prophase, prometaphase, through metaphase, anaphase, telophase, and cytokinesis. DNA is labeled in blue, and protein machinery involved in putting the chromosomes through their paces is labeled in green and red. (Courtesy of William C. Earnshaw, University of Edinburgh.)

addition, the reduction in chromosome number required to make a sperm or an egg must be precise in the sense that while each of our somatic cells has two copies of chromosome 1, two copies of chromosome 2, and so on, the gametes must carry only one copy of each chromosome pair. It is this process of reducing of chromosome number by one-half that underlies the laws of Mendelian segregation, allowing an Aa heterozygote to make gametes that carry either the A or a allele – but not both.

The process that executes this reduction in chromosome number is called *meiosis*, and consists of two divisions (see Figure 6.13). The first division, which is known as meiosis I, results in the segregation of homologous chromosomes to opposite poles of the spindle. After meiotic prophase, during which homologs find and physically associate with their partners, pairs of homologs line up on the meiosis I spindle such that both sister centromeres of each chromosome are pointed to the same pole – but the centromeres of the two homologs are directed towards opposite poles. This stage is referred to as metaphase I. At anaphase I the homologs segregate to opposite poles of the spindle. Cytokinesis results in two daughter cells each with only one of the two members of the original pair of homologous chromosomes.

In humans the result of meiosis I is two daughter cells each of which has 23 chromosomes. It is critical to note that what segregates at meiosis I is whole chromosomes – not sister chromatids! When chromosome 1 separates from its partner it does not separate its two sister chromatids! It is precisely this absence of sister chromatid separation that necessitates the

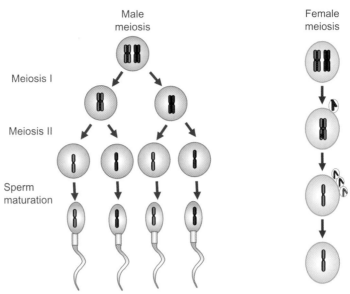

FIGURE 6.13 Gametogenesis produces four sperm from a single meiotic event in a male, but only one egg from a single meiotic event in a female. Note that this figure does not display recombination, which we will deal with later in this chapter.

mitosis-like second meiotic division, known as meiosis II. In a human cell that has just finished meiosis, the individual chromosomes now line up on a newly built spindle, with their sister chromatids pointed toward opposite poles, just as they do during a mitotic division. This is known as metaphase II. When you think of it, meiosis II is really just like any other mitosis, just with half the number of chromosomes.

The process of meiosis in humans is diagrammed schematically in Figure 6.13, which shows a cell with a pair of homologous chromosomes. This figure shows both the separation of the chromosomes during the two meiotic divisions (seen for spermatogenesis on the left). It also shows a fundamental difference between meiosis in males and females. *In males all four products of meiosis become sperm. But in females the cell divisions are grossly unequal, resulting in one very large cell that keeps virtually all the cytoplasm and three tiny cells, called polar bodies, that are discarded and then die.*

As we continue our discussion it is important to focus on two questions. First, what is a meiotic cell trying to do in terms of the chromosomes? The answer is that meiosis functions to precisely reduce the chromosome number by one-half. The second question is: what is the cell trying to do in terms of building two very different types of cells, a sperm and an egg? Sperm cells are short-lived and consist basically of a propulsion system that allows them to swim and a DNA "payload". An egg, on the other hand, needs to have all of the resources and information required for a zygote to become a rather complex embryo capable of attaching itself to the endometrium of the mother's uterus. A single conception event usually involves hundreds of millions of sperm donated by the father (only one of which succeeds in fertilization!) and a single egg provided by the mother. These are very different cells and not surprisingly they have very different histories.

Meiosis Is Executed Quite Differently in Human Males and Females

Given the importance of this process, it is surprising that it takes place in such different fashions and at different times in men and women, but it does, and these differences are truly impressive.

The most notable sex difference is in the timing of meiosis. In human females, meiosis begins during fetal development. Most if not all of the oocytes (eggs) that a human female will possess in her lifetime are produced while she is still *in utero* (in the uterus of her mother). These oocytes all begin the meiotic process during fetal development but arrest at the end of pachytene (the period of early meiosis during which chromosomes are observed to be fully paired along their length and are known to have completed meiotic recombination). Thus all of the meiotic recombination a human female will ever do is completed before she is born. These arrested oocytes remain quiescent until the girl enters puberty. At that point, a few oocytes are allowed to begin the maturation process during each menstrual cycle. Usually only a single oocyte is ovulated per cycle. It proceeds through the first meiotic division to metaphase of the second division, at which time it "arrests." Because completion of the second meiotic division is triggered only by fertilization, the number of completed meioses experienced by a human female roughly equals the number of conceptions.

Male meiosis begins at puberty and continues uninterrupted throughout the life of the male. Male meiotic cells, known as spermatocytes, are continuously reproducing cells known as stem cells. Thus, unlike oogenesis, in which all the oocytes exist at birth, spermatocytes are constantly produced throughout the life of the male. Once a spermatocyte initiates the meiotic process, it takes about 64 days to produce mature sperm. (Compare this with the process of oogenesis that must span decades!)

Thus, in human males, the meiotic process is basically free running, with cells usually progressing through the meiotic process in an uninterrupted fashion.

These differences in biology of oogenesis and spermatogenesis result in some rather large differences in the number of meiotic cells and of the number of *gametes* produced by the two sexes. Each female carries some 2 to 3 million oocytes at birth, but usually less than 400 of these oocytes will eventually mature during her life. However, the production of spermatocytes and the subsequent process of male meiosis occurs at a rate sufficient to produce the roughly 350 million sperm present in each ejaculate (approximately 1 trillion sperm during the life of the average male). The most important numerical difference is this: each female meiosis produces only a single oocyte; the remaining products of meiosis become nonfunctional cells called *polar bodies*. However, each male meiosis produces four functional sperm.

The actual molecular mechanisms that ensure meiotic segregation appear to be different, as well. In human males, the meiotic spindle is organized by cytoplasmic structures called *centrosomes*. The chromosomes then attach to the developing spindle. In females, the chromosomes themselves bind to the microtubules and build the spindle from the inside out without the assistance of centrosomes. Moreover, whereas human female meiosis includes frequent preprogrammed stops and selection appears to act at multiple points in the process, male meiosis appears to run uninterrupted once initiated. However, there do appear to be multiple checkpoints or control points in male meiosis that allow a spermatocyte that has made errors in meiosis to abort the meiotic process. Whether such checkpoints exist in female meiosis is a hotly debated issue.

It may be surprising to realize that meiosis is so different in the two sexes, but try to think about what the organism needs to accomplish.

A sperm and an egg are very different cells. A sperm is basically a genetic torpedo. It has a payload (23 chromosomes), a motor, and a rudder. Its function is to survive for a day or so and to swim to an egg. Once the content of the sperm nucleus (called a pronucleus) is delivered to the cytoplasm of the egg, the rest of the content of the sperm cell is destroyed. An egg, however, must possess all the supplies and determinants required to support embryonic development until the embryo can attach to the endometrium of the uterus and access the mother's blood supply. These two roles call for very different cellular machinery, and the process of human reproduction requires that a vast excess of sperm be produced for every egg, since the probability of any one sperm finding the one egg is very low!

6.4 MEIOSIS IN DETAIL

Passing Genes Between Generations

In 1834 a man with a blinding eye disease, a very early-onset form of glaucoma that was untreatable in the early years of the nineteenth century, passed glaucoma along to one of his two daughters. His affected daughter in her turn bore nine children. By the turn of the twenty-first century, she had more than 700 descendants, including more than 70 who inherited juvenile-onset glaucoma. This form of glaucoma is much more treatable than it was eight generations ago. The gene that causes it is now known, and it is even possible to test for mutations in this gene so that children at high risk can be identified and begin frequent eye exams to ensure that treatment will begin at the earliest possible moment to prevent vision loss. But the question remains, as each at-risk child is born into this family: what is it that decided that some of them would inherit the gene that causes the disease while others did not?

Traits get passed from one generation to the next when *that* sperm meets *that* egg and a zygote is formed. If we look at the many descendants of the young woman with juvenile-onset glaucoma, we find that even knowing which gene has a defect does not tell us the mechanism by which some of her children inherited a defective copy while others inherited an undamaged copy. Each affected member of the juvenile-onset glaucoma family has two copies of the genetic blueprint, but each of them makes sperm or eggs that have only one copy of the blueprint, that is to say, only one copy of each chromosome. So how does this happen? How does only one of the two copies of the blueprint get transferred into any one sperm or egg? Clearly, the processes of mitosis that are designed to preserve the full number of chromosomes won't work here.

To see what happens to the chromosomes during the creation of germline cells, let's return to our fictitious friend *O. hypothetica* with its one pair of chromosomes (Figure 6.14). It has gotten the urge to mate. Now it needs to make a gamete. In order to do things the Mendelian way, it needs to produce a sperm or egg with only one copy of each chromosome pair or, in this particular case, one chromosome. As with mitosis and cell division, getting the right number of chromosomes into the gamete is going to be something difficult that cell is going to have to actively orchestrate. The process that the cell will use to accomplish this is called *meiosis*.

Meiosis Made Simple

Meiosis actually encompasses two cell division events (cleverly called the first meiotic cell division and second meiotic cell division!). The purpose here is simple: get single copies of the blueprint into each germline cell. To do this, the cell carries out one round of DNA replication (just like in mitosis) but then carries out two rounds of cell division instead of the one cell division found in mitosis. (And yes, as the math wizards can tell us, one round

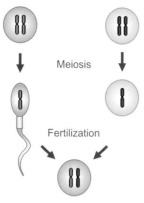

FIGURE 6.14 Meiosis made simple. This oversimplification shows how two *O. hypothetica* produce offspring. Papa *hypothetica* and Mama *hypothetica* each give a chromosome to the sperm and egg that fuse to make Baby *hypothetica*. Note that this figure does not display recombination, which we will deal with later in this chapter.

of chromosome replication accompanied by two cell divisions will in fact cut the chromosome number in half.) Each parental *O. hypothetica* has two copies of the chromosome, but a sperm or egg produced by *O. hypothetica* will have but one chromosome. Thus, when the sperm and egg come together, the new organism will once again have two chromosomes (see Figure 6.14).

Of course, it is all much more complex than just replicate–divide–divide, because once again we have to be sure that things end up in the right place at the end of this. So first, let's take a pictorial overview of meiosis (Figure 6.15) in terms of where the chromosomes are and how they get moved around and a description of those steps (Box 6.1). Once we have seen how the chromosomes get moved around, then we can move on to discuss more details about some of the critical steps, especially steps 2 and 3, first meiotic prophase and first meiotic metaphase.

Remember, the two chromosomes that *O. hypothetica* starts out with are homologous chromosomes; that is, they are effectively the same chromosome, with the same genes in the same

order along the whole chromosome. To keep track of the two copies separately, we will color them red and blue.

Recombination Commits Paired Chromosomes to Segregate from Their Homologs

Let's take a closer look at several steps in meiosis. We will only discuss details for a few of the critical steps. One of the most important steps in meiosis is recombination (Step 2, first meiotic prophase). As indicated in Figure 6.15 and Box 6.1, the homologous pairs of replicated chromosomes find their partner during prophase. During this step, homologous chromosomes pair along their entire length. This is a step not seen in mitotic cells. In the mitotic cells, the replicated chromosomes stay apart from each other and get moved around as independent units. However, as a result of homologous pairing, in meiosis the homologous chromosomes become associated.

Such paired homologs lock together by recombination, or crossing-over, a process that facilitates the exchange of large regions of homologous DNA between any two non-sister chromatids (Figure 6.16). The pairing and recombination of two homologous chromosomes creates a structure known as a bivalent. Bivalents are the basic functional until that facilitates proper segregation at the first meiotic division.

If we look through a microscope that provides sufficiently high magnification, we can sometimes see the details of the chiasma, or point of crossing over (Figure 6.17). While the chiasma structures are in place, the two homologous chromosomes are held tightly together by the sister chromatid cohesion on both sides of the chiasma. The cell has to release that cohesion from the site of the chiasma to the end of the chromosome (the telomere) before the chromosomes can separate and move to opposite poles of the cell.

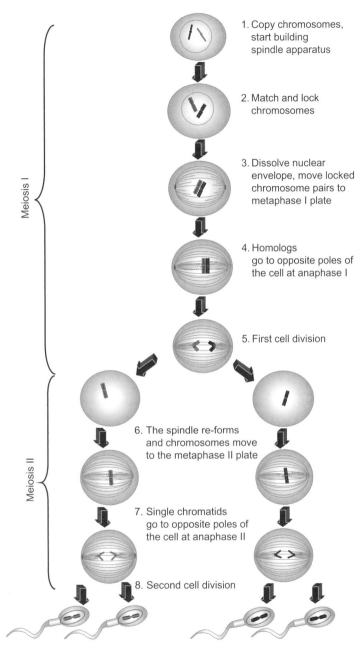

1. Copy chromosomes, start building spindle apparatus

2. Match and lock chromosomes

3. Dissolve nuclear envelope, move locked chromosome pairs to metaphase I plate

4. Homologs go to opposite poles of the cell at anaphase I

5. First cell division

6. The spindle re-forms and chromosomes move to the metaphase II plate

7. Single chromatids go to opposite poles of the cell at anaphase II

8. Second cell division

Meiosis I

Meiosis II

FIGURE 6.15 Meiosis starts with cells containing two copies of each chromosome. After two rounds of meiosis, the four resulting sperm each contain only one copy of each chromosome. Production of an egg similarly reduces the copy number from two to one, but as we saw in Figure 6.13, the product of female meiosis is a single egg with one copy of each chromosome plus three tiny discarded cells that cannot go on to be fertilized or form a zygote. Thus, in male meiosis one precursor cell produces four sperm but in a female one precursor cell produces one egg. Note that this figure ignores recombination, which we will deal with later in this chapter.

BOX 6.1

THE STAGES OF MEIOSIS

Meiosis is the process used to create the germ cells – sperm and eggs. Two homologous chromosomes replicate to become four copies that eventually separate to end up as four single chromosomes in the four separate sperm that are created. (In contrast, only one of the four cells at the end of oogenesis becomes a viable egg.) This list of stages from Figure 6.15 has been greatly simplified. For instance, first meiotic prophase alone is usually divided into five stages with unlikely-sounding names such as zygotene and diakinesis, but we have left out much of that. You don't have to struggle with distinguishing things such as different levels of condensation of the chromosomes to be able to follow the critical steps we present here and in Figure 6.15. In fact, if you compare this list to the description we gave when we talked about mitosis, you will see that we have even condensed some of the steps. For instance, we have grouped the telophase step when the nuclear envelope reforms together with the cytokinesis step in which the cells divide. This streamlined version of meiosis shows the essential steps that get the chromosomes replicated and then reduced in number so that each germ cell holds one unreplicated copy of the chromosome instead of two.

Meiosis I

- *Step 1.* During interphase, the cell copied everything, including copying the DNA to produce the replicated chromosomes.
- *Step 2.* Each replicated chromosome finds its homolog and recombination locks the chromosomes together.

- *Step 3.* After the nuclear envelope breaks down, the matched, locked-together chromosome complex – called a *bivalent* – moves to the metaphase plate, where the spindle apparatus attaches the centromere of one replicated chromosome to one pole of the cell, and the centromere of the other replicated chromosome to the other pole of the cell.
- *Step 4.* The two homologous chromosomes that make up each bivalent get pulled to opposite ends of the spindle. Chromatids in the replicated chromosome stay together because the sister centromeres were only attached to one pole of the spindle, not both.
- *Step 5.* Cytokinesis divides the cell into two cells, each with one bivalent.

Meiosis II

- *Step 6.* In each cell, replicated chromosomes move to the metaphase plate. The spindle apparatus now attaches to the centromeres of both sister chromatids so that two sister centromeres are attached to opposite poles of the spindle.
- *Step 7.* As the motors pull towards the opposite ends of the cell, the two separated sister chromatids move towards the opposite poles of the cell.
- *Step 8.* The nuclear membrane reforms and cytokinesis separates the cell into two separate cells, each one containing a single unreplicated chromosome.

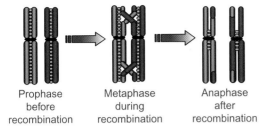

Prophase
before
recombination

Metaphase
during
recombination

Anaphase
after
recombination

FIGURE 6.16 Recombination during meiosis I. During prophase replication creates two replicated chromosomes. During metaphase, the two replicated chromosomes become linked into a structure called a bivalent, with the connection between the two chromosomes being recombination events between the two chromosomes. During anaphase the two chromosomes become separate again so that they can move apart, with each separated chromosome now carrying some DNA from the other copy. This recombination event is a process of exact exchange in which any region that is "given away" by one chromosome gets replaced by "taking back" exactly that same region from the other chromosome. Note that at the beginning of anaphase, sister chromatid cohesion is released along the arms, except near the centromeres, allowing the homologs to separate from each other.

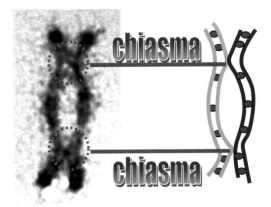

FIGURE 6.17 This picture, and the diagram next to it, show the structure that exists as chromosomes pair and exchange material. The chiasma at the point of exchange separates groups of genes above it from groups of genes below it when it results in trading of DNA between the two chromosomes. (After M. Petronczki, *et al.*, *Cell*, 2003; 112(4):423–40. Photo courtesy of Jasna Puizina. © 2003, with permission from Elsevier.)

We tend to think of meiotic handling of chromosomes during prophase as consisting of three functional steps:

Match them → Lock them → Move them

Pairing   Recombination   To opposite spindle
                                    poles

This process of hooking the chromosomes together has a purpose of much greater importance than trading DNA: these recombination events serve to link homologous chromosomes together to ensure that homologs are separated into two different daughter cells by the end of the first meiotic division – which is to say that they must move to opposite poles of the meiotic spindle before the cell divides. Think of it this way: pairing does not take place to allow recombination; rather, recombination takes place as a by-product of the pairing process that holds the chromosomes together at a critical point when

the cell needs to handle them as a single unit while moving them around.

Those who study human populations may sometimes be seduced by the view that the critical point of recombination is to generate diversity among the progeny. It is an especially attractive view because the existence of such diversity lets us carry out genetic studies. However, those who actually study meiosis know that cells don't lock chromosomes together to achieve recombination and diversity. Cells use the mechanisms that produce recombination as an engineering process to hold chromosomes together so that they will end up going where they are supposed to. When such exchange or recombination events don't take place, the homologous chromosomes may fail to go to the opposite poles of the cell at the first meiotic division. In such cases, the chromosomes don't end up where they are supposed to be by the end of meiosis. So the point of recombination is not diversity, however beneficial that side effect may be. The point of

First meiotic metaphase   Second meiotic metaphase   Mitotic metaphase

FIGURE 6.18   Comparison of first meiotic metaphase, second meiotic metaphase, and mitotic metaphase. On the left, where the centromeres are not attached to both poles at once, notice that the crossovers hold the bivalents together until the cell is ready to move the replicated chromosomes to the opposite poles of the cell. In the middle and on the right, attachment of one centromere to both poles is all that is needed to allow the cell to move things to where they should be in a mitotic division. Note that the gross structure of the second meiotic metaphase and the mitotic metaphase look the same, but when you look closer you see that the second meiotic metaphase chromosomes each have some material from the other chromosome, while the mitotic metaphase chromosomes have not recombined with each other.

recombination is getting the right chromosomes to show up at the right place at the right time.

Centromeres Mediate Chromosome Movement

When we look at Step 3 ("move them" or the first meiotic metaphase) in detail, we realize one of the ways in which the process of meiosis is critically different from the steps of mitosis. Whereas mitosis moves individual duplicated chromosomes to the metaphase plate, meiosis I moves bivalents, the locked complex of two duplicated chromosomes, to the plate. At this point in meiosis, the centromeres of the two sister chromatids do not attach to opposite poles, as they do in mitosis. Rather, in meiosis each chromosome has a centromere that is attached to only one pole by microtubule fibers; the two sister chromatids of that chromosome are both attached to the same pole!

Thus, at this point in the meiotic cycle, the centromere is doing something fundamentally different from what it does during mitosis: namely, the two sister centromeres of the mitotic centromere attach to two opposite poles, which, following centromere separation, leads to the movement of two sister centromeres to

opposite poles (Figure 6.18). However, at the first meiotic division the centromere of each replicated chromosome attaches to only a single pole, so the meiotic centromere travels intact to just one of the two poles.

Another critical event in meiosis happens at Step 4 (first meiotic anaphase) when the two replicated chromosomes that comprise each bivalent separate and move to opposite poles of the spindle (Figures 6.15 and 6.19). This is the crucial meiotic event. Movement of the two homologous replicated chromosomes to opposite poles explains Mendel's observation that only one copy of a given pair of alleles will be included in a gamete – the other allele just went to the opposite pole at anaphase I and will go into a separate cell at the end of meiosis. Or, as we see in Figure 6.12, the pair of blue chromatids went to one pole and the pair of red chromatids went to the other. The cells then proceed to meiosis II, which uses processes resembling mitosis to separate the two sister chromatids into separate cells, which each end up with one copy of the chromosome in place of the two copies the cell started with.

If we look carefully at the replicated chromosomes that are being moved to the opposite poles during the first meiotic division, we find

FIGURE 6.19 Chromosome movement during anaphase. Two homologous chromosomes that have exchanged DNA during recombination in meiosis I move to opposite poles of the cell. Each centromere is attached to one pole of the cell and remains intact as it moves through meiosis I. Cohesin attachment along much of the length of the sister chromatids has been released but cohesion attachment of sister chromatids remains near the centromere.

that each chromosome has recombined (traded material between the two locked chromosomes) and consists of some DNA obtained from its homolog. Also, we find that the cohesin proteins that held the sister chromatids together along the arms of the chromosomes are gone. *However, cohesion is maintained around the centromeres.*

So meiosis, when we reduce it to its simplest elements, amounts to this:

- *Replicate the chromosomes, so each chromosome consists of two chromatids.*
- Pair the replicated homologs.
- Allow the paired chromosomes to recombine, thus locking them together.
- Pull the two homologous chromosomes (each still possessing two sister chromatids) to opposite poles at meiosis I.
- Divide into two daughter cells; each has half the original chromosome number.
- Now, without any more replication, *line the chromosomes up on a new spindle.*
- *Split the sister chromatids, with one chromatid going to each pole.*
- *Complete cell division.*

We can compare this to a similarly reduced version of mitosis, which amounts to:

- Replicate the chromosomes, so each chromosome consists of two chromatids.
- Line the chromosomes up on the metaphase plate of a spindle.
- Split the sister chromatids, with one chromatid going to each pole.
- Complete cell division.

As you can see, the italicized steps for meiosis look an awful lot like the steps for mitosis. When we delve into the details, there are some differences there, but at the level of understanding how the cell moves the chromosomes around, it is quite striking that meiosis II operates much like mitosis. One of the key points here is that there is no round of DNA duplication after meiosis I and before meiosis II.

We can also see some key differences between mitosis and meiosis. Chromosomes do not pair or recombine during mitosis, or the second meiotic division that resembles mitosis; chromosomes only pair and recombine during the first meiotic division. Mitosis produces two identical daughter cells, each with two copies of each homologous chromosome; meiosis produces four gametes, each with one copy of each homologous chromosome. Mitosis takes place whenever cells divide, especially in cells that comprise the skin, the bone marrow, and the inside of the gut; meiosis only takes place in ovaries and testicles, with the objective of producing sperm or eggs.

6.5 MECHANISMS OF CHROMOSOME PAIRING IN MEIOSIS

The first step in meiosis, perhaps the most crucial step in meiosis, is the pairing of homologous chromosomes. In many, if not most, organisms the ends of chromosomes (called telomeres) are clustered in early meiosis and

FIGURE 6.20   Pairing of chromosomes during meiosis. A protein connected to each replicated chromosome connects to a protein complex in the nuclear membrane. This protein complex moves through the nuclear membrane, carrying the chromosomes along with the proteins as they move. This reduces the pairing problem from a search through a large three-dimensional space to a much smaller search along a two-dimensional surface. (After Hawley, R. S. and Gilliland, W. D., "Homologue pairing: Getting it right," *Nature Cell Biology*, 2009;11:917–18.)

that pairing appears to begin near the telomeres, which has led to a model in which pairing is initiated at sites near the telomeres and then the chromosomes simply pair by *zippering* along their lengths. While models that propose internal pairing sites (*buttons*) or a general "stickiness" along the chromosome arms that allows a gene-by-gene search for homologous partners (*Velcro*) have also been proposed, strong evidence for the zippering-in-from-the-ends model has been obtained now in several organisms.

A well-supported model for pairing in the worm *C. elegans* is shown in Figure 6.20. A geneticist named Abby Dernburg at the University of California at Berkeley and her colleagues have recently shown that specialized sites (known as pairing centers), which are located near the telomeres, bind a set of proteins that link the pairing centers to the inside of the nuclear envelope. Other proteins then connect the pairing center complex to a set of motors and "cables" in the cytoplasm that move the pairing centers around the nuclear envelop. Consider this: by linking the pairing centers to the nuclear envelope, the problem of a given pairing center finding its homolog is reduced from a 3-D problem (a search anywhere within the entire volume of the nucleus) to a much easier two-dimensional problem (a search that is limited to the inner surface of the nucleus). Once two homologous pairing centers collide

their pairing is stabilized – a mechanism that presumably involves interactions between the pairing center binding proteins and direct DNA–DNA interactions involving sequences at and near the pairing center. The movement of the pairing centers now tests the "strength" of homology by continuing to pull on the individual pairing centers, thus de-stabilizing inappropriate pairings.

Once strong pairings have occurred near the ends, the two homologs then "zip-up" along their length, forming an elaborate structure known as the *synaptonemal complex* between them. The function of the synaptonemal complex remains a bit mysterious, but it clearly plays a role in facilitating meiotic crossing over and may in some cases serve to stabilize pairings.

As we noted above, this mechanism is gaining strong support in a variety of organisms. However, there must also exist other mechanisms for pairing as well. We know, for example, that circular, or ring, chromosomes can occasionally be found in many organisms, including humans. These chromosomes pair and recombine with linear homologs, despite the lack of telomeric sequences. The mechanism by which these chromosomes pair remains unknown – but given the redundancy that seems to be characteristic of most biological mechanisms, it would not surprise us in the least if other mechanisms of pairing exist as well. Indeed, there is good evidence in Drosophila (fruit flies) that pairing is not mediated by telomere-associated sequences.

It may seem odd that after all these years we are still wrestling a problem as fundamental as how homologs can recognize each other and pair. It is a fascinating puzzle – a biological component of that much broader question of how does anything discriminate between self and non-self. And by presenting models derived from recent work in model organisms such as worms and flies and yeast, we run a real risk of giving an incomplete picture of what happens in humans. But we think this discussion is justified by our desire to tell you some of the most fundamental questions in biology remain unanswered – that's what makes it fun!

6.6 THE CHROMOSOMAL BASIS OF HEREDITY

And so we arrive at a junction of the different ideas we have been discussing. The genes Mendel was talking about, pieces of information connected to phenotypic characteristics, are in fact the bits of DNA sequence along the chromosomes, and meiosis is the mechanism that causes only one copy of a gene to be passed along because only one copy of the chromosome carrying that gene gets passed along.

So how does knowing that genes are being carried along on the chromosomes and segregated into separate germ cells during meiosis help us understand genetic and phenotypic variation between people, even people within the same family? If the A allele is carried by one homolog and the a allele by the other, and only one of the two homologs goes into the gamete, then any one gamete can only carry the allele present on that homolog, either A or a, but not both. Moreover, as we will show in the next section, two or more pairs of homologous chromosomes each assort into the daughter cells separately, with the

segregation of one pair of homologs taking place without any influence from the segregation of the other pair of homologs!

Doing Meiosis with Two Pairs of Chromosomes

Figure 6.21 shows what happens when two pairs of chromosomes segregate in our new fictitious friend, O. complexica. This more complex species has two pairs of chromosomes, one pair of metacentric chromosomes (having the centromere near the middle) and one pair of telocentric chromosomes (having the centromere near the end). If we color the chromosomes from his mom red and the chromosomes from his dad blue, we can keep track of what happens to the two forms of chromosomes and we can also keep track of their movement.

Figure 6.21 shows that there are two different kinds of results that come out of meiosis in O. complexica. One possible result is shown in the left-hand panel, where we can see that the chromosomes that came in from the mother both end up going into the same sperm after the first meiotic metaphase to produce a chromosome combination like that donated by the organism's mother. Chromosomes that came in from the father both end up going into the same sperm.

However, if we look at the right-hand panel in Figure 6.21, we see that sometimes the chromosomes that came in together leave in different gametes. For each "blue–blue" sperm the organism makes, we expect it to also make a "red–red" sperm and two "blue–red" sperm. So, what does this mean? It means that the cell does not care whether chromosomes that came in together go out together. However, it does care that one copy of each chromosome pair goes to each pole. Thus one sperm does not get all metacentric chromosomes while the other sperm gets only acrocentric chromosomes; each sperm always gets one acrocentric chromosome and one metacentric chromosome.

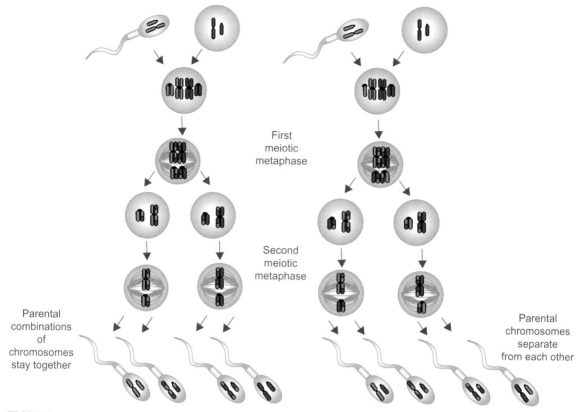

First
meiotic
metaphase

Second
meiotic
metaphase

Parental
combinations
of
chromosomes
stay together

Parental
chromosomes
separate
from each other

FIGURE 6.21   There are two different possible outcomes from meiosis in the same individual, in this case a male. Here we show only the key steps needed to trace where the different copies of chromosomes are going so we have not included recombination in this figure. This man makes more than one set of gametes in a lifetime; sometimes the outcome will be that shown on the right, where the parental chromosomes segregate apart from each other (blue plus red), and sometimes it will be what is shown on the left, where they stay together (blue–blue or red–red combinations). Notice that the decision to keep parental combinations together or separate them occurs at first meiotic metaphase.

Thinking about Meiosis in Terms of Genes

We can now consider the meiotic process in terms of two or more pairs of genes and in terms of organisms like us that have more than two chromosomes. Minimally there are two cases we need to consider: (1) when two genes are located on different pairs of chromosomes, and (2) when both pairs of genes are located on the same pair of homologous chromosomes. Because the case in which the genes to be considered fall on different chromosomes turns out to be both simpler and more common, we will consider it first.

Gene Pairs Located on Different Chromosomes Segregate at Random

Mendel asserted that the alleles of one gene will assort independently of the alleles of another gene. That is, an individual of the genotype AaBb, where A and a are alleles of one gene and B and b are alleles of a *different* gene, will produce four types of gametes (AB, Ab, aB, and ab) with equal frequency (Figure 6.22). By now, we hope you know why we do not end up with genotype combinations like AA in one sperm and Bb in the other.

Note that a gamete carrying the A allele is as likely to carry the b allele as it is to carry the

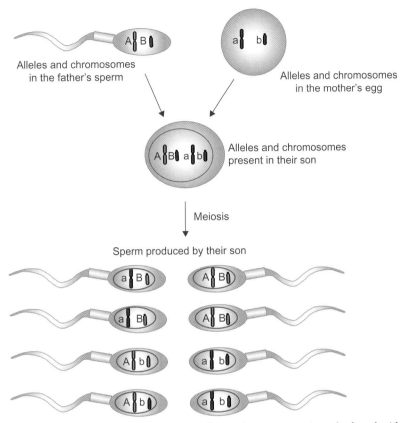

FIGURE 6.22   Independent assortment of genes on two different chromosomes. A can be found with either B or b, and B can be found with either A or a. Thus alleles that were present in a particular combination in the parents may transmit together to the next generation or may become separated, one passing to one progeny and the other passing to a different progeny. To better visualize the point we are making about independent assortment we do not display recombination.

B allele. The same is true for gametes carrying the a allele; about half of them will carry B and half will carry b. Mendel stated that the two gene pairs segregate independently such that there is no preference for a gamete to carry a particular combination of alleles. He referred to this rule of segregation as *independent assortment*.

As shown in Figure 6.22, independent assortment results from the fact that two bivalents will orient at random on the metaphase plate with respect to each other, so half the time the blue chromosomes are connected to the same pole and go to the same end of the cell, and half the time the blue chromosomes are connected to opposite ends of the cell and go to separate germ cells in the end.

Recombination Between Pairs of Genes Located on the Same Chromosome

Although you might expect alleles of genes on the same chromosome to travel together in moving between generations, alleles of two different genes that come in located on the same

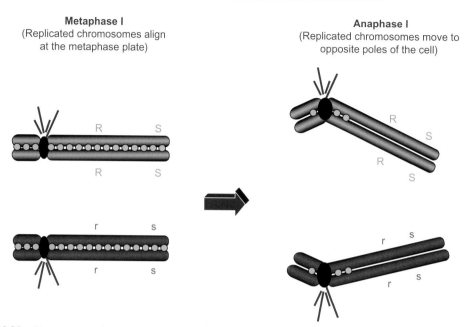

Metaphase I
(Replicated chromosomes align
at the metaphase plate)

Anaphase I
(Replicated chromosomes move to
opposite poles of the cell)

FIGURE 6.23   Two genes on the same chromosome at metaphase and anaphase I. If no recombination event happened, we would expect two genes that came in on the same chromosome to go out on the same chromosome no matter how far apart the two genes along the chromosome. What actually happens is shown in Figure 6.24.

chromosome can actually leave on different chromosomes. Let's see how this happens.

As shown in Figure 6.23, the rule of independent assortment does not apply when two genes, R and S, are located on the same chromosome. Indeed, in the simplest case, the previous pictures of meiosis we have looked at would suggest that an individual of the genotype RrSs, in which R and S alleles are on one homolog and the r and s alleles are on the other, would only produce RS or rs gametes. This exception to Mendel's rule of independent assortment is called *linkage*. This idea, linkage of things located on the same chromosome, makes intuitive sense because the two genes are located on the same physical entity that is one chromosome, but the situation is more complicated than what we see in Figure 6.23.

As shown in Figure 6.24, the rearranged combination of Rs or rS in the gametes can

only be produced when crossing over occurs in the region of the chromosome that is located between the two genes.

Recombination, or crossing over, events are, however, relatively frequent during human meiosis. (Note there are two synonyms for *crossing over*, namely *recombination* and *exchange*. All three will be used interchangeably.) There is usually at least one such recombination event for each pair of homologous chromosomes. In the case of large chromosomes, recombination may be more frequent. This is especially true for large chromosomes with the centromere near the middle, in which recombination events will likely occur on both arms of the bivalent.

Because recombination can occur at most sites along the chromosome, the probability that a recombination event will occur between two genes is dependent on the distance between those two genes on the chromosome. Thus, if

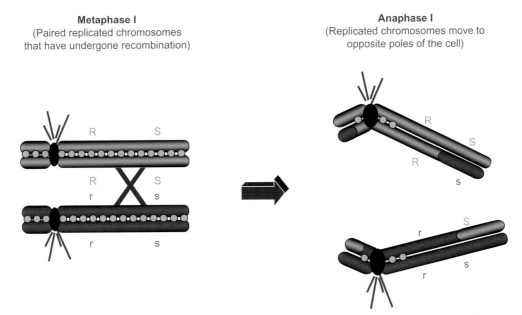

Metaphase I
(Paired replicated chromosomes
that have undergone recombination)

Anaphase I
(Replicated chromosomes move to
opposite poles of the cell)

FIGURE 6.24 Two genes on the same chromosome, at metaphase and anaphase I, with a crossover. This view shows only the chromosomes and leaves out the cell and its structures. Notice that the blue chromosome now contains a region of DNA with an S allele that used to be the red chromosome, and the red chromosome now contains a region of DNA with an s allele that used to be on the blue chromosome.

two genes lie at opposite ends of a chromosome, the probability of a recombination event occurring between them is high. If the genes are very close to each other, the chance that the recombination event will fall in between them is smaller than the chance that it will fall outside of the area containing the pair of genes. In fact, an approximation (a big approximation) would be this: if the length of DNA on the chromosome outside of the pair of markers is about 10 times as long as the length of DNA between the markers, the chance of the recombination event falling outside the markers is about 10 times the chance of the recombination event falling between the markers.

Of course, the situation becomes more complex as we start trying to follow more and more genes. In looking at just three genes on the same chromosome, Figure 6.25 shows exchange events involving a bivalent heterozygous for mutants at the E, F, and G genes: such that alleles E, F, and G are on one homolog and alleles e, f, and g are on the other. In this case all the exchange events fell between the E and G genes. Within that area between E and G, there were five crossovers that fell between E and F, but only one of the exchanges fell between F and G.

It is sometimes postulated that the frequency of exchange events between any two genes, or markers, that lie on the same arm of a given chromosome is approximately proportional to the physical distance between them. That said, this statement is a gross over-simplification. Crossing-over is most frequent near the middle of the arms and less frequent near the base or the tip of the arm. It NEVER occurs in the heterochromatin that surrounds the centromere. The chromosomal arms have some "hot spots" for recombination where recombination is seen more frequently than in surrounding regions. But for regions out in the middle of the

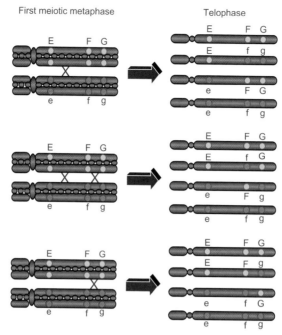

FIGURE 6.25   Recombination at different positions on a chromosome. X marks the point of recombination in each of these three meiotic metaphase events. The location of the recombination events in the bivalent can be correlated directly with the combinations of alleles present on the chromosomes that result from meiosis.

chromosomal arm, we can draw some general conclusions that correlate physical distance and map length. In Chapter 11, when we look at genetic mapping, we will discuss the correlations between physical and genetic maps.

Thus several different factors contribute towards a child having different combinations of genetic information than were present in the child's parents and grandparents. First, each parent passed only half of their genetic information along to their child. Second, through independent assortment of chromosomes, alleles carried on those different chromosomes can pass independently down through multiple generations so that alleles of two different genes that were present in someone's grandmother may no longer be present together in the same germ cell that produces the grandchild. Third, even when specific alleles of two genes are located on the same chromosome together in

the grandmother, recombination can exchange material between chromosomes in such a way that a different combination of alleles will be present on that chromosome in the grandchild. In Chapter 7, we will see that transmission of the X and Y chromosomes between generations leads to an unusual mode of inheritance and poses special problems in gene dosage for male and female organisms.

6.7 ANEUPLOIDY: WHEN TOO MUCH OR TOO LITTLE COUNTS

As we have talked about meiosis, we have emphasized the importance of getting the right numbers of chromosomes into the sperm and eggs that will be used to produce the next generation of human beings. Each sperm and each

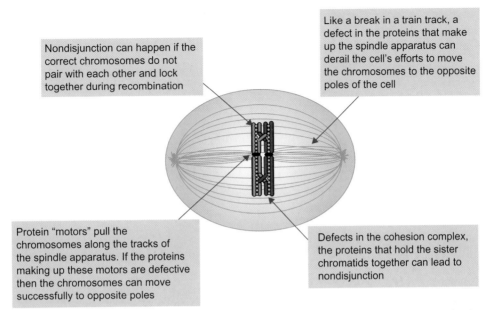

Nondisjunction can happen if the correct chromosomes do not pair with each other and lock together during recombination

Like a break in a train track, a defect in the proteins that make up the spindle apparatus can derail the cell's efforts to move the chromosomes to the opposite poles of the cell

Protein "motors" pull the chromosomes along the tracks of the spindle apparatus. If the proteins making up these motors are defective then the chromosomes can move successfully to opposite poles

Defects in the cohesion complex, the proteins that hold the sister chromatids together can lead to nondisjunction

FIGURE 6.26   Nondisjunction can result when necessary steps in meiosis fail, which sometimes may be due to defects in proteins that carry out those functions or make up the structures used to carry out the functions.

egg must end up with 23 chromosomes (and they have to be the right 23 chromosomes) if the resulting child is to have exactly the right number of copies of each of the genes in the human genome. Unfortunately, the cell machinery does not always succeed in its goal of getting all of the right chromosomes to where they are supposed to be by the end of meiosis, and missing or extra copies of chromosomes can mean illness or even death to a zygote produced by a sperm or egg with the wrong number of chromosomes.

The failure of two homologous chromosomes to *segregate* (separate from each other into the daughter cells after the first meiotic division) properly is called *nondisjunction*, and it can result from defects in any of several different kinds of structures and functions in the cell (Figure 6.26). Nondisjunction can occur either because two homologs failed to pair and/or stay together as they move to the metaphase plate at meiosis I, or because of a failure of the cell to properly move the segregating

chromosomes along the meiotic spindle (those track-like structures made of microtubules on which the chromosomes pull themselves to the poles) after the pairing and recombination steps. Indeed, considerable evidence exists that much of human nondisjunction may be due to failures of the processes that move chromosomes to opposite poles of the dividing cell at meiosis I. These failures or errors can include defects in the proteins that hold the sister chromatids tightly together during the first meiotic division, in the protein motors on the chromosomes that move them to opposite poles of the cell, or in the structural integrity of the spindle apparatus along which the chromosomes move.

Regardless of which structures and functions go wrong during meiosis, the result of nondisjunction is that the resulting sperm or eggs turn out to be *aneuploid* (i.e., having the wrong number of chromosomes). When an aneuploid sperm or egg is involved in a fertilization event, the resulting zygote is also aneuploid, and human biology is remarkably intolerant of aneuploidy. This is especially true for *monosomies*, zygotes with only a single copy of a given chromosome. With the exception of the X or Y chromosome (which we talk about more in Chapter 7), monosomy is simply not compatible with life and leads to early spontaneous miscarriage. We are not aware of a baby ever being delivered alive that carried one or three entire copies of large chromosomes (monosomies or trisomies) such as chromosome 1, chromosome 2, or chromosome 3. It's not that such zygotes don't arise; they do and then are lost, sometimes so early that the mother may not be sure whether she was pregnant. With the exception of four cases described below, most human trisomies do not survive long after conception.

More dramatic cases of aneuploidy, such as full triploidy (three copies of every chromosome per cell), occur as well. Again, these are not compatible with early fetal development. Since meiosis in humans is really pretty sloppy, these errors are fairly common among human conceptions, and autosomal nondisjunction occurs at a reasonably high frequency (perhaps as many as 40% of human conceptions are aneuploid). As such, aneuploidy has to be considered perhaps the most common cause of death in human beings. But it is one we are not usually aware of because many of these conceptions are lost very early in pregnancy.

The lethality of most types of aneuploid conceptions shows just how critical proper gene copy number is for correct development of a complex organism. While an organism might tolerate changing the dosage of genes encoding some kinds of enzymes, there may be serious deleterious effects from changing the dosage of genes whose products regulate the expression of other genes, carry out cell-to-cell communication, or serve as a structural component of a complex protein structure. Although even those changes might be tolerable in some cases if only a single gene were affected, you have to realize that each human chromosome carries hundreds

TABLE 6.1 Viable human aneuploidies

Aneuploidy	Alternative names
XO	X monosomy, Turner syndrome
Trisomy 13	Trisomy 13, Patau syndrome
Trisomy 18	Trisomy 18, Edward syndrome
Trisomy 21	Trisomy 21, Down syndrome
XXY	Klinefelter syndrome
XYY	
XXX, XXXX, XXXXX	

or thousands of genes. The additive effect of increasing the dosage of many genes is usually death.

However, a few types of trisomic zygotes are capable of survival, at least sometimes. These are *trisomy 21* (Down syndrome), *trisomy 18* (Edward syndrome), *trisomy 13* (Patau syndrome), and trisomy of the sex chromosomes (Table 6.1). Most instances of trisomy 18 and trisomy 13 are not viable and do not make it to live birth. Far more of the trisomy 21 pregnancies succeed. At least one factor that may make extra copies of chromosomes 21 and 18 compatible with survival is that they carry a relatively smaller number of genes than do the larger chromosomes in the human complement. There are some very different factors that allow survival with an incorrect number of X chromosomes (see Chapter 7).

Down Syndrome, or Trisomy for Chromosome 21

His name was Earl, and when Scott met him in high school they were both freshmen in the same physical education class. Earl had Down syndrome, a disorder caused by an imprecise segregation of chromosomes into the egg from which he arose. Like many, but not all, kids with Down syndrome, Earl was developmentally delayed and had been that way since birth. His intellect had stopped somewhere

around that of a five-year-old, but his body never got the message. Because of his limitations and because of the facial features that are characteristic of Down syndrome, Earl became the butt of an awful lot of high school humor. Kids couldn't resist making fun of the way he walked, ran, or talked. For four years Scott spent one hour a day in class with Earl. For various reasons, they became friends. Earl never did figure out why people made fun of him, but he knew that they did. Once, one of the other students tripped him in the hallway during break. The humiliation he felt seemed to hurt worse than the bloody lip. During Scott's junior year of high school the March of Dimes held a public lecture on the basis of birth defects. For reasons that have long faded into a mist of high school memories, Scott made his father drive him to that lecture at a nearby college. There he learned for the first time about genes and chromosomes, but mostly about Earl and about himself. Scott developed a passion for understanding how heredity works and how our genes make us what
we are. In many ways this book is a child of that obsession.

Down syndrome, or trisomy 21, is one of the most studied forms of aneuploidy. According to the March of Dimes, about one in every 800 babies born in the United States has Down syndrome. It is perhaps the best-known genetic defect, partly because it is the single most common cause of mental retardation among individuals outside of institutions, and partly because of the very distinctive characteristic appearance, including slanting, or epicanthic, eyes and small, frequently low-set noses (Box 6.2). Babies with Down syndrome grow slowly and have poor muscle tone. They often have rather short fingers and short, broad hands. They have a wide skull that is somewhat flatter than usual at the back, and the irises of the eyes often have obvious spots. In many cases, the mouth appears to remain partially open due to a protruding tongue.

BOX 6.2

PEOPLE WITH THE SAME DISORDER CAN BE QUITE DIFFERENT

Not all Down syndrome individuals will have all of the characteristics listed in this chapter. Whether you are considering the information in this chapter or in the rest of the book, whatever descriptors we use for Klinefelter syndrome, Turner syndrome, Down syndrome, or any other human disorder, not all features apply to *all* the people affected by that disorder. People are unique: these disorders can manifest themselves quite differently from one person to the next. Keep in mind that, even if a discussion of someone with Down syndrome focuses on chromosome 21, that person's overall characteristics are affected by differences on all of the other chromosomes too. Some individuals with Down syndrome are born with heart defects, some are not. Many suffer from substantial cognitive deficits, sometimes substantial enough to warrant institutional care, but in contrast, we have heard about at least one specific young woman with Down syndrome who is attending college. When a disorder is discussed, we try to give you a general description of the common features of that disorder, things that are found much more commonly in the disorder than in the general population. We know we are making generalizations. We know there will be exceptions, but it is the best we can do. We hope that you will carry this caution about variability away with you along with whatever generalizations you encounter here.

Perhaps the most commonly known aspect of Down syndrome is mental retardation. Their Intelligence Quotient (IQ) normally ranges from 25 to 50 (compared to an average IQ of 100 in individuals who do not have Down syndrome); however, some children do show higher levels of mental function, with some individuals with Down syndrome having near-normal IQs and the ability to read and write at high school or college levels. There is serious controversy, and some increasing degree of optimism, regarding just how much children with Down syndrome can be expected to achieve. Clearly, some children with Down syndrome greatly exceed our expectations and grow up to be happy and reasonably self-reliant adults, but many are severely limited. Growing evidence shows that certain types of early educational intervention, especially computer-assisted teaching, may be of real help to children with Down syndrome. Moreover, in these times, many adults with Down syndrome may be expected to live either semi-independently or independently and often are able to enter the work force. In some cases, such individuals seem to do better in so-called "sheltered workshops," but other individuals are able to find work in various aspects of the public and private sector.

Half of the children born with Down syndrome are born with severe heart malformations. These and other life-threatening conditions are so severe that some of these children die before age five. However, for those children who survive the fifth year of life, the average life expectancy is 50 years. Even so, these individuals are at high risk for leukemia and for a degenerative brain disorder similar to Alzheimer disease. Men with Down syndrome are usually sterile, but the women are fertile; from the few scattered reports available, it appears that half of their children are born with Down syndrome. On one hand, this result makes good sense – half of the eggs produced by such a woman should carry two copies of chromosome 21. However, given that some 80% of Down syndrome fetuses spontaneously miscarry, we have to wonder why the final result should be a 1:1 ratio.

Although we can imagine models for why the trisomy 21 children of a living trisomy 21 individual might have a higher survival rate than trisomy 21 conceptions by the general population, any explanations are still theoretical. In general terms, we might expect that there are elements of the genetic composition (particular alleles of genes on chromosome 21 or elsewhere in the genetic background) that might lead to the viability of a girl born with Down syndrome who survives to adulthood to reproduce. Given the rate at which trisomy 21 conceptions are lost, we might hypothesize that those alleles that kept her alive are not present in everyone in the population. Thus we can imagine that the trisomy 21 mother who is alive because of her particular genetic composition would have a greater chance than the general population of contributing alleles conducive to trisomy 21 viability, because we know she possesses those viability alleles to pass them along.

Most Cases of Down Syndrome Are Due to Nondisjunction in the Mother

Most often, a baby with Down syndrome is found to have three copies of chromosome 21 if chromosomes in their cells are examined via traditional karyotyping to classify the chromosomes present by size, centromere position, and banding pattern (Figure 6.27). Most cases of trisomy 21 are due to nondisjunction at the first meiotic division in the child's mother. We know this because we have developed several methods for determining which chromosome came from which parent, one of which makes use of subtle differences in banding patterns for chromosome 21 that can sometimes be seen with some staining techniques used in karyotyping.

Let's start with chromosomes marked with banding differences that let us track all four copies of chromosome 21 separately, two maternal copies and two paternal copies (Figure 6.28).

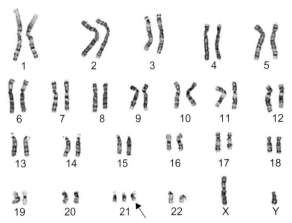

FIGURE 6.29   What we would see if nondisjunction of chromosome 21 in the mother happened at meiosis I. (Note that the egg has two copies of chromosome 21 instead of one, that both of the mother's banding polymorphisms are represented in the egg, and that three different banding polymorphisms are present in the child.) In cases in which banding polymorphisms can be distinguished, the most common outcome for trisomy 21 is the finding that both copies of chromosome 21 from the mother are present along with one copy from the father, meaning that nondisjunction happened at meiosis I in the mother. To better visualize the point we are making about recombination we do not display recombination, which is a separate issue.

FIGURE 6.27   Karyotype of an individual with trisomy 21 shows the presence of an extra copy of one of the smallest chromosomes that shows the shape and banding characteristics of chromosome 21. (Courtesy of the Clinical Cytogenetics Laboratory, University of Michigan, Ann Arbor, MI, Diane Roulston, PhD, Director.)

FIGURE 6.28   Cartoon of differences in banding pattern between individual copies of chromosome 21 that would let us tell whether the extra chromosome came from the mother or from the father. For purposes of illustration, we are showing banding patterns that are clear enough to allow distinction between all four parental chromosomes, but in real-life karyotyping situations, it won't always be possible to distinguish all four chromosomes.

If we can visibly distinguish the four chromosomes, we can track where and how the nondisjunction took place. In fact, we have other nonmicroscopic techniques in our genetic bag of tricks to distinguish maternal and paternal copies of chromosomes being passed along to the next generation, but this use of visible differences in the chromosomes is the technology that makes it easy for us to show you how conclusions can be drawn about where the duplicated chromosome came from.

Clearly, the consequences of nondisjunction at meiosis I in the mother are quite different than the consequences of nondisjunction at meiosis II. In the case of meiosis I nondisjunction, both maternal polymorphisms are present in the cells of the child with trisomy (Figure 6.29). If the problem had resulted from nondisjunction at meiosis I in the father, we would still see three different banding patterns among the three copies of chromosome 21, but two of them would come from the father.

If nondisjunction happens during the second stage of meiosis, the child with trisomy will end up possessing two copies of the same maternal chromosomes, that is, both copies that come from the mother would look the same (Figure 6.30). Nondisjunction at meiosis II results when the two sister chromatids split and then go to the same poles as opposed to going to opposite poles as they would normally do. This results in the child having three copies of chromosome 21 that demonstrate only two different banding polymorphisms between them. If you look at the different banding polymorphisms, you can see that nondisjunction at meiosis II in the

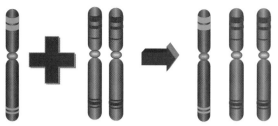

FIGURE 6.30   Consequences of nondisjunction of chromosome 21 in the mother at meiosis II. (Note that only one of the mother's banding polymorphisms is represented in the egg, but it is there in two copies. Also notice that the child has three copies of chromosome 21 but only two different banding polymorphisms.) This outcome is observed less often than the outcome in Figure 6.29. To better visualize the point we are making about aneuploidy we do not display recombination, which is a separate issue.

father would also have given only two banding polymorphisms among the three copies in the child, but it would have been one of the father's polymorphisms that would be present twice. Using these techniques, or other techniques that let us tell the chromosomes apart and track where the third chromosome came from, we can show that the extra copy of chromosome 21 almost always comes from the mother.

Why would the extra chromosome tend to come from the mother? The answer to that question is currently in rather hot dispute, but the best guess is that in male meiotic cells the failure of two autosomes to properly pair and segregate results in the cessation of meiosis and, indeed, in cell death. There appears to be a checkpoint in male meiosis that asks whether all of the chromosomes are properly paired and ready to segregate from their partners. If the answer to that question is "no," then the meiotic cell may be doomed and the potentially aneuploid sperm are never produced. Such checkpoints apparently do not exist in most female meiotic cells (oocytes). In oocytes the cell seems committed to completing meiosis despite whatever failures may occur. Thus, although the checkpoint system in sperm is not foolproof and some aneuploid sperm do get through, it does work efficiently enough to result in a substantially

reduced frequency of this kind of nondisjunction in sperm as compared to eggs. So it is not that nondisjunction fails to happen in males, but rather that the sperm cells in which it has happened are unlikely to survive to fertilize an egg.

Trisomies for Chromosomes 13 and 18

Excluding rare exceptions, only two other human trisomies involving autosomes have been reported among live births. These are trisomy 13 (sometimes called Patau syndrome) and trisomy 18 (sometimes called Edward syndrome). These usually have a characteristic set of congenital problems, including a high frequency of severe cardiac and neurological problems. Trisomy 18 is observed at a very low frequency; the March of Dimes says that it is found in about 1 out of 5000 live births. The incidence at conception is much higher, but most of these embryos are miscarried spontaneously. Trisomy 13 is even less frequent; according to the March of Dimes it occurs in about 1 in 16,000 live births. Like Down syndrome, the incidence of these trisomies increases dramatically with advancing maternal age.

Trisomies for other chromosomes are not viable, but, as noted above, they do indeed occur at the point of conception, and their frequencies increase with advancing maternal age. However, all of these other aneuploid conceptions lead to spontaneous miscarriages.

6.8 UNIPARENTAL DISOMY

Uniparental Isodisomy "Uncovers" Mutations

Arthur was born with severe early vision loss from a blinding eye disease called Leber's congenital amaurosis, or LCA, and he was found to be homozygous for a mutation in the RPE65 gene. RPE65 is located on chromosome 1 and is known to cause LCA. His father, Leonard, is a carrier for the same RPE65 mutation found in

Arthur. Because LCA causes a recessive form of vision loss, we expect that testing of his mother, Mallory, will show that she is also a carrier for the same RPE65 mutation. Researchers were very surprised to find that Mallory harbors no RPE65 mutation.

The next surprise came when additional genetic markers on chromosome 1 were tested in all four members of the family. Arthur turns out to be homozygous for every genetic marker on chromosome 1. A detailed analysis shows that while Arthur received chromosome 1 alleles from his father, he did not get a copy of chromosome 1 from his mother. Arthur's brother Gerald clearly received chromosome 1 alleles from both his mother and his father. While one possibility might have been that Mallory was not Arthur's mother, testing of markers on the other chromosomes show that Mallory really is Arthur's mom.

Does this mean that Arthur received a copy of chromosome 1 from his father but no copy of chromosome 1 from his mother? Everything we know about aneuploidy tells us that no one survives having only one copy of chromosome 1. What is going on? The answer is actually fairly simple, although the mechanism that caused it is not. Arthur indeed has two copies of chromosome 1. They are identical to each other and they are both identical to one of Leonard's copies of chromosome 1 (Figure 6.31).

Uniparental disomy is the term for having two copies of a chromosome where both copies come from one parent and no copy comes from the other parent. In this case, the two copies are identical copies, the product of replication of one of Leonard's chromosomes. The two copies are not the two homologs that were present in Leonard. We add the prefix "iso," meaning same or equal, to arrive at the term uniparental isodisomy to label the special circumstance in which the two chromosomes not only come from one parent but are exactly the same chromosome. As we will see in the next section, it is also possible to add "hetero," meaning other,

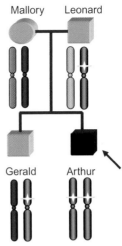

FIGURE 6.31 Uniparental isodisomy can uncover a recessive mutation in someone who has only one carrier parent. The father is a carrier for a causative mutation in the RPE65 gene on chromosome 1, but the mother does not have a causative defect in RPE65. Their son Gerald is a carrier who received a maternal chromosome with no mutation and a paternal chromosome with the causative mutation. The other son, Arthur, is affected with LCA. Arthur received two copies of the paternal chromosome with the mutation and no copies of the maternal chromosome. Uniparental = comes from only one parent. Disomy = two copies of a chromosome. To better visualize the point we are making about uniparental disomy we do not display recombination, which is a separate issue.

to create the term uniparental heterodisomy, which describes a state in which the two chromosomes from one parent are the two different homologs that were present in the parent.

How did Arthur end up with two copies from his father and no copies from his mother? Early in this chapter we discussed someone with an extra copy of a chromosome (trisomy) or a missing copy of a chromosome (monosomy). In each case a gamete with a normal single copy of a chromosome came together with an aneuploid gamete that had ended up with two copies of a chromosome as a result of a meiotic error. When the chromosome involved is one of the largest, such as chromosome 1, such aneuploidy is not viable. However, there are situations such as Arthur's in which meiosis creates an aneuploid gamete that produces a viable zygote with the right number of chromosomes as the result of a second error after fertilization. In some senses we might think of this as two wrongs making a right, but as we will see there are often consequences, so the outcome is often not entirely right.

Zygote Rescue

A possible mechanism by which Arthur could have ended up with two identical copies of chromosome 1 from his father and none from his mother involves one meiotic error, followed by a mitotic error known as zygote rescue. During meiosis in Arthur's mother, a normal egg was generated with one copy of chromosome 1 (Figure 6.32). In Arthur's father, a rare error in

male meiosis happened at meiosis II, when the cell containing a single replicated copy of chromosome 1 failed to correctly separate the two sister chromatids into the daughter cells. The resulting sperm that received both copies of chromosome 1 fertilized a normal egg to create a nonviable zygote that is trisomic for chromosome 1.

The zygote rescue step that follows is a mitotic error that discards the extra copy of chromosome 1 from the zygote during the first rounds of cell division after fertilization. This event can throw out any one of the three chromosomes to restore correct chromosome copy number. If it throws out one of the paternal copies, then a viable normal zygote is created with one maternal copy of chromosome 1 and one paternal copy of chromosome 1. If the cell

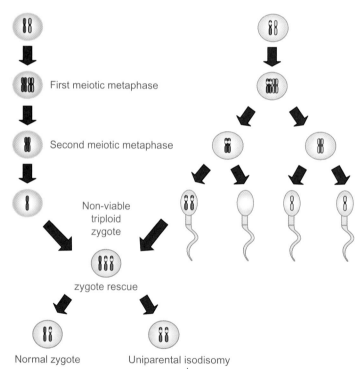

First meiotic metaphase

Second meiotic metaphase

Non-viable triploid zygote

zygote rescue

Normal zygote   Uniparental isodisomy

FIGURE 6.32   Uniparental isodisomy resulting when a mitotic error rescues a trisomic zygote. After an error at meiosis II creates a sperm with two identical copies of chromosome 1 that fertilizes a normal egg, a trisomic zygote results. This trisomic zygote would normally die, but if an error in mitosis during one of the first rounds of cell division kicks out one of the copies of chromosome 1 and restores a normal copy number, the zygote becomes viable and the cells of the resulting embryo are descended from the rescued cell. To better visualize the point we are making about uniparental disomy we do not display recombination, which is a separate issue.

throws out the maternal copy then the correct chromosome copy number is restored, but both copies originate from the father. If the chance of being thrown out is about equal for the three chromosomes, then about two-thirds of the time one of the father's copies would be tossed, leading to a normal diparental disomy. In one-third of the cases the lost extra copy would be the maternal copy, causing paternal uniparental isodisomy as shown in Arthur's family.

Arthur's example, which comes from a real family, involves a rare case of paternal isodisomy (inheritance of both copies from the father). It is actually more common that cases of uniparental isodisomy are maternal in origin because aneuploidy happens more often during the creation of eggs than sperm.

In Arthur's case, the effect on phenotype was the result of "uncovering" a mutation that under normal circumstances would have been covered for by a normal copy coming from the non-carrier parent. If two RPE65 mutation carriers married we could predict that each child would have a 25% chance of ending up with LCA. Where one parent is a carrier and the other is not, the chance of the child having the trait is proportional to the rate at which new mutations at the RPE65 locus might happen in the germline of the non-carrier parent and is also proportional to the rate at which uniparental isodisomy for chromosome 1 occurs. So when only one parent is a carrier, the chance of the child having the trait is small but that chance does not drop to zero.

Uniparental Disomy in the Absence of Mutations

Caitlyn is tall, slim, and very smart. Her younger sister Rosalyn is short, overweight, and mildly mentally retarded. All through childhood development their behaviors were quite different. Caitlyn was a light eater with a lot of self-discipline and flexible reactions to the environment around her. As Rosalyn grew she showed increasing problems with food cravings, compulsive behaviors, and tantrums. In this family of tall, slim, smart, calm people, there is no family memory of anyone who resembles Rosalyn. The family knew little about genetics so the term Prader–Willi syndrome meant little to them, but they all presumed that Rosalyn must have some kind of mutation. Some relatives suggested that Rosalyn's mother might have been exposed to something during pregnancy. When Caitlyn took a genetics class in college she learned about Prader–Willi syndrome and found out that there was a chance that Rosalyn's characteristics might have nothing to do with mutations or environmental insults. Rosalyn does turn out to have Prader–Willi syndrome, which GeneClinics tells us is found in about 1 in 10,000 to 1 in 30,000 individuals in the population. Like about a quarter of all Prader–Willi cases, Rosalyn turns out to have received two copies of chromosome 15 from her mother, but no copy of chromosome 15 from her father.

Prader–Willi results when the individual possesses only maternally derived copies of 15q11.2-q13, a region known as the Prader–Willi Critical Region (PWCR). Rosalyn has maternal heterodisomy because she received both of her mother's homologs of chromosome 15 but none from her father (Figure 6.33). Since she has both of her mother's homologs of chromosome 15, she is heterozygous for anything that is heterozygous in her mother who does not have Prader–Willi, so we do not expect that she has become homozygous for a mutation that is heterozygous in her mother.

Because the key issue here is the parent of origin of the DNA, and not the specific sequence, we face a novel concept – a genetic disease that is not the result of a mutation. Maternal disomy 15 causes Prader–Willi even if there are no mutations to be uncovered, so the mechanism here is unrelated to the mutuation-based cause of Arthur's problems. This suggests some completely different mechanism than the mutation-based mechanisms presented in Chapter 5.

In another family, a child named Oliver is born with Prader–Willi syndrome, yet he does not have uniparental disomy. Instead, a small interstitial deletion of the Prader–Willi Critical Region left Oliver with only a maternal copy of the DNA in this region (Figure 6.34). Thus the only copies of

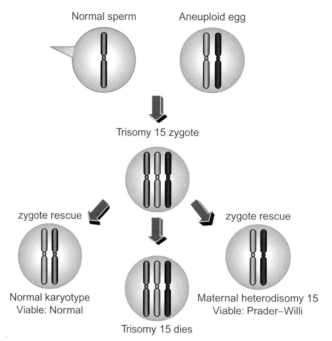

FIGURE 6.33 Prader–Willi results from combination of a disomic egg with a normal sperm. If the aneuploid egg resulted from an error in meiosis I, then the result will be either a dead trisomic zygote, a viable normal zygote with one copy of 15 from each parent, or a viable heterodisomy 15 as shown here. If the error happened at meiosis II, the result would be maternal isodisomy resulting from the same kinds of mechanisms shown in Figure 6.35. The individual with maternal heterodisomy is heterozygous so Prader–Willi cannot be the result of the kind of uncovering of a recessive mutation shown in Figure 6.32. To better visualize the point we are making about uniparental disomy we do not display recombination, which is a separate issue.

DNA that he carries from the Prader–Willi Critical Region carry a maternal methylation pattern and any gene expression from the region will follow only the pattern that can result from the maternal methylation pattern. If there are points in development, or cell types, that require the male methylation imprint, the cell is in trouble because it has no copies with the male imprinting pattern.

In the case of the deletions, we would consider Prader–Willi syndrome to be a contiguous gene syndrome resulting from effects on several genes located next to each other within one chromosomal region. The trait happens when the individual is missing the paternal copies of SNRPN, NDN, and possibly additional adjoining genes. This loss can be accomplished through deletion of the paternal copy of the relevant genes, through maternal isodisomy, or through balanced translocations with endpoints in the Prader–Willi Critical Region.

The Effects of Paternal vs. Maternal Imprinting

Other individuals with uniparental disomy for chromosome 15 do not have Prader–Willi syndrome. Instead they have Angelman syndrome, which has a very different phenotype.

Hailey was born with Angelman syndrome. She began showing developmental delays by 6 months of age. As she grew she never developed the ability to speak but did learn some limited sign language. She did not learn to walk until she was four years old, and as she learned she had problems with balance and she

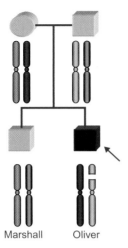

Marshall   Oliver

FIGURE 6.34 Prader–Willi resulting from a small interstitial deletion that includes the Prader–Willi Critical Region on chromosome 15. This deletion arose in the germline of the father, leaving Oliver with only one copy of each of the genes in this region of chromosome 15. Since there is only one copy, and that copy comes from the mother, that means that Oliver has only the maternal Prader–Willi allele. Uniparental = comes from only one parent. Disomy = two copies of a chromosome. To better visualize the point we are making about uniparental disomy we do not display recombination, which is a separate issue.

tended to walk with her feet far apart, with her arms upraised from the elbows. When excited she flaps her hands. At age two and a half she started having seizures and often needs less sleep than other kids her age. One of the most striking things about her is her sociable personality, sunny disposition, and big wide grin.

Angelman syndrome results from lack of a paternal copy of a region of chromosome 15 that is sometimes called the Angelman syndrome–Prader–Willi region. Although about 11% of Angelman cases have mutations in the UBE3A gene, most of the rest appear to result from a lack of a paternal copy in the Angelman syndrome. About 7% have uniparental disomy, 3% have a small interstitial deletion of less than 300 base pairs of the maternal copy of the region, and most cases have a large deletion of 5–7 million base pairs of maternal sequences spanning the same region involved in Prader–Willi

syndrome. Most cases that do not involve mutations in UBE3A all represent cases caused by the presence of imprinting (methylation) patterns typical of paternally derived chromosomes and the absence of imprinting patterns typical of maternally derived chromosomes.

Mechanism of Parent of Origin Effects

Mechanistically, what might uniparental heterodisomy, uniparental heterodisomy, and a small interstitial deletion have in common with each other? In the case of Prader–Willi syndrome, the individual with the trait has received DNA from the Prader–Willi Critical Region from their mother but no DNA from that region from their father. Why does it matter that a particular copy of a gene came from the mother rather than the father? The critical issue here is that *imprinting*, the pattern of methylation that regulates expression, changes as the chromosomes pass through formation of the germline. The pattern of methylation that is laid down in the sperm is different from the pattern of methylation laid down in the eggs. The result is that the amount of RNA transcribed from a maternally derived copy may be more than that transcribed by a paternally derived copy. Or methylation might also potentially change timing of expression during development or the geography of where in the body the gene is expressed.

Remember that there are two different ways to end up with no product being produced by a gene. One is to have a mutation that results in the cell failing to make the gene product or making a gene product that does not work. The other way to accomplish this is regulatory, by preventing the cell from transcribing the gene or changing when or where the gene is transcribed through imprinting, changes in methylation of DNA and histones. As you will recall from Chapter 3, important factors affecting transcription include the sequences of the regulatory regions (promoter and enhancer) and the state of methylation of those sequences. Two different individuals who have exactly the same sequence in a gene and its surrounding regions can show differences in expression of that gene depending on the state of methylation of regulatory sequences for that gene. A phenotype that is sensitive to different effects depending on parent of origin effects results from the imprinted genes being expressed where, when, and at the levels dictated by the pattern of methylation in the regulatory regions of those genes. If the methylation pattern for maternal and paternal imprinting patterns for a particular gene were different, then the effects in someone bearing only the maternal methylation pattern would be expected to be different from those in someone bearing only the paternal methylation pattern. Thus, for Prader–Willi and Angelman syndromes, even though the initial mechanisms that set up the situation seem very different – aneuploidy vs. deletion – the direct cause of the trait is altered gene expression that results from having an imprinting pattern specific to one parent of origin but not the other.

6.9 PARTIAL ANEUPLOIDIES

Partial Aneuploidy through Translocation of a Chromosomal Segment

Sometimes a broken piece of a chromosome becomes attached to another chromosome through a process called *translocation*. The extra piece of chromosomal material gets carried along through meiosis when the chromosome that it is attached to goes through normal pairing and segregation.

This type of breakage event can sometimes be reciprocal, with the chromosome that receives the broken piece of DNA giving back a piece of DNA in return. The result is that no DNA is lost. It is just moved around through trading of material between two non-homologous chromosomes. In cases where the exchange is reciprocal

and no DNA is gained or lost, we call it a balanced translocation. In such cases, apparently healthy people might go through their whole lives without knowing their cells hold such a balanced translocation if the points at which the chromosomes broke did not disrupt any genes or if the defect is recessive and requires that both copies be charged to cause the trait; after all, they still have the right number of copies of every gene. A healthy individual might first find out that they carry a balanced translocation when a doctor recommends that they be karyotyped in response to having a child with certain kinds of birth defects or when they develop fertility problems that include spontaneous miscarriages.

If the translocation break point falls within a gene, then the translocation may be a causative mutation leading to a phenotype. Sometimes there may be a phenotypic effect even if a gene is not disrupted if a gene is moved from an actively transcribed, metabolically active region into a region such as a heterochromatic region in which there is little or no transcription. If the translocation break point does not interrupt a gene or move it into a drastically different regulatory environment, then there might be no phenotypic effect at all since the individual still has the normal number of copies of every gene in the genome.

Translocations and Down Syndrome

Most Down syndrome cases that involve a translocation possess an isochromosome. This isochromosome is a duplicated chromosome 21 that contains two copies of the long arm of chromosome 21 (21q) attached on opposite sides of the same centromere. The short arm of chromosome 21 is dispensable because the only genes it carries are duplicated on the short arms of four other chromosomes. These highly repeated genes encode the ribosomal RNAs. The 21q-centromere-21q chromosome is, however, duplicated for a large number of single copy genes on 21q and it is possession of three copies of these genes that causes the Down syndrome phenotype.

In about 5% of Down syndrome cases the child has a translocation that has traded material between two non-homologous chromosomes. Let's consider what happens if a piece of chromosome 21 breaks off and sticks onto chromosome 13 (Figure 6.35). Each of the germ cells produced by meiosis will have a normal copy of chromosome 21. One of the daughter cells will get the normal copy of chromosome 13 and the other will get the translocated copy of 13 that has some genes from chromosome 21. The result of meiosis will be some normal germ cells and some cells in which the translocated part of chromosome 21 is aneuploid. After fertilization, the resulting zygote will have three copies of the translocated part of chromosome 21 and two copies of the rest of chromosome 21. The resulting child will have some or all of the Down syndrome characteristics, depending on whether the translocated region includes some or all of the genes involved in Down syndrome.

In looking at the children who could have been produced by one normal parent and one parent with a balanced translocation involving chromosomes 13 and 21 (Figure 6.35), we find that the four children they produced represent the four different genotypic combinations that could have resulted:

1. Lauri – Balanced translocation between chromosomes 13 and 21.
2. Jesse – Normal karyotype and phenotype.
3. Kaylee – Unbalanced translocation with a shortage of 13q and 21p along with an excess of 13p and 21q.
4. Larry – Unbalanced translocation with extra copies of 13q and 21p who have a shortage of sequences from 13p and 21q.

Because the translocation did not disrupt a gene, Lauri and Jesse would both have been born healthy. If no genetic testing were done, no one would be able to tell whether either of them carries a balanced translocation. Having their karyotype determined would tell them that Lauri carries a balanced translocation, and with every pregnancy she will face about a 50% risk that that child will face major problems comparable those of Kaylee and Larry. Genetic testing would reveal that Jesse has a normal karyotype and has no more risk than the general population of producing children with birth defects.

Although Kaylee and Larry both have unbalanced translocations involving the same chromosomes and break points, we expect that their phenotypes will be different since these children will differ in which pieces of 13 and 21 are present in excess and which pieces are underrepresented. Kaylee would have some features of Down syndrome due to some of the extra genes from chromosome 21 that include the

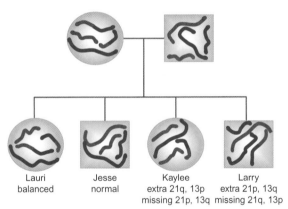

FIGURE 6.35 Children of a normal individual and an individual with a balanced translocation between chromosomes 21 and 13. Lauri has a balanced translocation and normal phenotype. Jesse has a normal karyotype and phenotype. Kaylee has an extra copy of some chromosomal regions and is missing a copy of other chromosomal regions. Larry has extra copies of the regions missing from Kaylee, and is missing a copy of regions that are extra in Kaylee. So we expect that Kaylee and Larry will each have an abnormal phenotype, but that the phenotypes will be substantially different. There is a possibility that either Kaylee or Larry would be lost very early in the pregnancy. Note that this figure does not deal with a separate issue, recombination events during meiosis.

Down Syndrome Critical Region, as well as additional major birth defects and health problems not typical of Down syndrome due to the missing genes. Larry would have had no characteristics of Down syndrome, since his extra sequences do not include the Down Syndrome Critical Region. Larry would also have different birth defects from Kaylee's, and there is a strong chance that he would have died before he was six months old because of the extra chromosome 13 sequences present.

Why should triplication of a chromosomal region be bad? The answer lies in the complex interactions between various genes in our genomes and of the proteins they encode. Many of the genes that produce proteins that act to regulate other genes are highly dosage-sensitive, and can often cause disease if present in either excess or reduced copy number. Other genes make proteins that are structural components of cellular organelles, and for many of these genes the cell is not tolerant of shortages or excesses. We still aren't sure exactly which triplicate genes are the real culprits in producing the various components of Down syndrome, but much progress is being made in the study of the roles of genes from this Down syndrome "critical region." Some individual genes have been implicated in specific features of Down syndrome. For instance the APP gene, which makes the amyloid precursor protein found in Alzheimer disease, might also be involved in the senile plaques seen in the brains of some of the oldest individuals with Down syndrome. Other specific genes from the 1.6 million base Down Syndrome Critical Region have been implicated in other typical Down syndrome characteristics, and many of the genes within this region show altered levels of transcripts and gene products in individuals with Down syndrome.

Could we cure the disease if we could answer the questions of which are the critical genes and why an extra copy is a problem? We just don't know. Our guess, and it is only a guess, is *maybe*. On one hand, even if we knew exactly what was wrong for any given component of the syndrome, we are unlikely to be able to fix the whole array of problems. We base this pessimistic prediction on the evidence that some of the most profound problems arising from the extra copy of chromosome 21 arise during development before birth, in structures such as the lungs and heart. Once that damage is done, we may simply have to rely on traditional medical and surgical processes for help. However, some of the Down syndrome problems, such as leukemia and Alzheimer diseases, develop after birth, so there may be a chance for prevention or to improve medical intervention if enough is understood about the roles of the particular genes and gene products in these later developments of the disorder. It remains to be seen whether the study of triplicate chromosome 21 genes will give us any capability to intervene in post-birth developmental processes that might affect IQ and other capabilities. And we expect that over time processes for prenatal intervention may even begin to offer the possibility of intervening to prevent or ameliorate some of the developmental defects that are put in place before birth. The key to it all will be a thorough understanding of how the various genes in the Down Syndrome Critical Region play a role in the different traits that make up Down syndrome.

6.10 THE FEMALE BIOLOGICAL CLOCK

Not only is Down syndrome normally a consequence of nondisjunction in the mother, but the frequency of Down syndrome births increases dramatically with advancing maternal age (Figure 6.36). Down syndrome children are found about once in every 800 live births. However, if we look at what we know about the ages of the mothers giving birth to these children, we find that the risk varies greatly

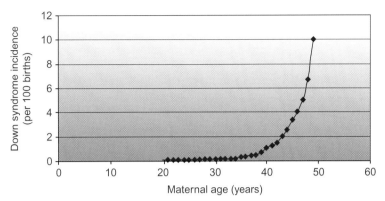

FIGURE 6.36 Maternal age and Down syndrome. According to the National Down Syndrome Society risk associated with maternal age is less than a tenth of a percent in very young adults but may get to be as high as 1 in 10 by age 49.

according to the age of the mother. According to the National Down Syndrome Society, about 1/1500 women under age 25 will give birth to a Down syndrome baby. On the other hand, by age 35 the risk increases to about 1/350, and by age 40 it may be as high as 1/100. By the mid-forties the risk is about 1/30, and by age 49 the risk is about 1 in 10, more than a hundred fold increase in risk over a period of less than 20 years! If you are a college-aged woman reading this, you have probably started to do some math. What is the trade-off of time spent on more education, career advancement, and trial relationships versus the risks evident in Figure 6.36? The biological clock, after all, is not something invented by the stand-up comics who joke about it. We should note that there is not strong evidence of age-dependent increases for the frequency of nondisjunction in men, although this issue is hard to sort out because of strong correlations between the ages of the two parents. Perhaps lack of evidence for age-associated risk of nondisjunction in men is partly for the reasons described in the previous section, such as the checkpoint during sperm development. However, as we see in the section on mutations, there are different age-associated risks for older men contemplating parenthood. Thus, there are actually important biological clocks ticking for both of the sexes.

Still, only about a quarter of children with Down syndrome are born to mothers over age 25. Think about that for a minute. Yes, the risk that any one woman will have a child with Down syndrome is much higher for older mothers. This is offset against the fact that far more women are bearing their children in their twenties than in their forties.

The mechanistic basis for the maternal effect is still unknown, but it is likely to be a consequence of events that occur during the long delay between prophase, when meiosis initiates, and the first meiotic division in human oocytes. Recall that human oocytes begin meiosis long before the female is born. Then the process stops in the late meiotic prophase, after pairing and recombination have taken place, but before the chromosomes have separated into the daughter cells. The chromosomes then stay that way for years. Beginning some months before birth, the chromosomes pair and then they just hang out, connected and poised to go their separate ways. They stay in this state of suspended animation until the girl hits puberty some 10–13 years later. Then several oocytes are given permission to restart meiosis each month, but usually only one of these is actually allowed to complete the first meiotic division! Realize then that the egg ovulated by a 45-year-old woman has been stalled part way through meiosis for 45 years, with the chromosomes sitting there paired and recombined but not segregated into separate cells. One can imagine that quite a lot could go wrong with the DNA supported on its complex scaffolding of proteins during that period of time, and apparently often does. However, just what actually does go wrong remains a bit of a mystery.

We can nonetheless offer a possible insight derived from recent studies of age-dependent female nondisjunction in humans and mice. In humans the data suggest that the nondisjunctional events that occur in mothers under 29 years of age are often the consequences of either a failure of the two 21st chromosomes to recombine at all or of recombination events that are in the wrong position (far too close to the end of the chromosome). But nondisjunction in older mothers seems to involve pairs of 21st chromosomes that have undergone what seem to be near normal levels of crossing over and have the usual distribution of crossing over. In other words, it is not just that nondisjunction becomes far more frequent as women age, but rather that the nondisjunctional events that occur in younger mothers have a different etiology than the nondisjunctional events that occur in older mothers. It is easy to understand why failed exchange might cause nondisjunction – after all, exchange is usually critical to ensure proper segregation. Without exchange, the homologs might often fail to separate properly. But why should properly recombined bivalents be nondisjoining in older mothers? An answer may lie in the finding by Patricia Hunt and her collaborators that mice heterozygous for a deletion of a meiosis-specific cohesin component display a dramatic and age-dependent increase in nondisjunction. Although 98% of the chromosomes were observed as bivalents in the eggs of one-month-old females, this number dropped to 35% in two-month-old mothers and to zero in six-month-old females. As mothers age, there were far fewer bivalents and a huge increase in both single homologs and single chromatids! Single homologs and single chromatids are often observed in oocytes from older human females as well. These data raise the fascinating possibility that the maternal age effect may be explain by defects in cohesin maintenance. We have said before that one of the most important features of meiosis is the careful control of sister chromatid cohesion by the control of cohesin. Thus, it may not be surprising that an age-dependent failure of cohesin may have such a dramatic effect on the frequency of failed meiotic segregation. Only time, and good experiments, will determine the validity of this model, but it does sort of tie this chapter together and allow us to end here.

APPENDIX 6.1

FAILED MEIOTIC SEGREGATION (NONDISJUNCTION) AS PROOF OF THE CHROMOSOME THEORY OF HEREDITY

Up until now we have blithely assumed that genes are located on chromosomes, which actually took our intellectual ancestors some effort to prove. Although the proof is one of the more elegant examples of genetics, it unfortunately did not involve the study of human subjects. Indeed the experimental organism was the common fruit fly, *Drosophila melanogaster*. Although we are loathe to deviate from our central focus, namely ourselves, this proof is critical to ensuring that the relationship between Mendelian inheritance and meiotic chromosome behavior

APPENDIX 6.1   (cont'd)

Cross A1

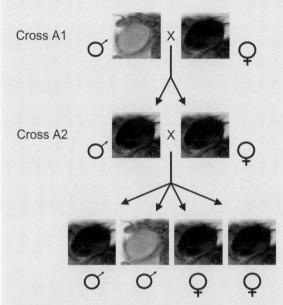

Cross A2

FIGURE 6.37   CROSS A1: White-eyed males were crossed to red-eyed females. This cross produced only red-eyed sons and daughters. CROSS A2: Red-eyed sons were crossed with red-eyed daughters. All the daughters were red-eyed, one-half of the sons were red-eyed, and one-half of the sons were white-eyed.

clear that the behavior of chromosomes during meiosis paralleled and explained the behavior of genes. But this was only correlative evidence. What was needed was proof.

The proof came from a rather egregious exception to Mendel's laws. Morgan found a mutation in fruit flies called white (*w*), which caused the eye color to be pure white rather than the brick red color that is characteristic of normal, or wild type, flies. As shown in Figure 6.37, crosses involving pure-breeding red and white stocks (the white stocks were homozygous for the white mutation) displayed some rather odd results. We will denote the allele that produces a white eye color allele as w, and the allele that produces a red eye color as W.

One could imagine that CROSS A1 was really ww males crossed to WW females, and that W is dominant. If you make that assumption, then CROSS A2 is a cross of Ww males to Ww females. So far, so good. One then expects one-fourth of the progeny to be white (ww) and three-fourths to be red (WW or Ww, often abbreviated W_). That is exactly what you see. However, note that all of the white-eyed progeny of CROSS A2 are males. You cannot wriggle out of this by saying that the white-eyed trait can only be expressed in males because there are white-eyed females in your original breeding stock. Something is terribly wrong. As shown in Figure 6.38, matters get worse when we do the cross backwards. Note that CROSS B1 produces 50% white-eyed sons and 50% red-eyed daughters. Nothing in Mendel's laws predicts this!

To figure this out, Morgan's student Calvin Bridges had to hypothesize that sons did not receive a copy of the white gene from their father. They instead carried only a single copy of the W gene (either the w or W allele), which they inherited from their mother. He further argued

is really understood. Moreover, in addition to its pedagogical importance, this inquiry, which centers on how chromosomes determine sex in flies, also serves as a useful transition to our next major topic: how chromosomes determine sex in humans. With that apology, we turn to the work of Calvin Bridges and his mentor Thomas Hunt Morgan in second decade of the 20th century.

By the first decade of this century, there were two lines of evidence that genes were carried by chromosomes. First, Boveri had shown that sea urchin embryos that were missing one or more chromosomes developed abnormally and that the pathology of that improper development differed according to which chromosomes were missing from the genome. Second, it was

APPENDIX 6.1　*(cont'd)*

Cross B1

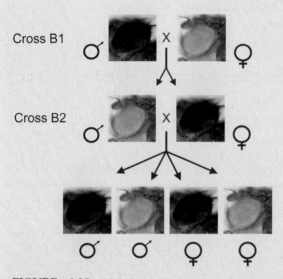

Cross B2

FIGURE 6.38 CROSS B1: Red-eyed males were crossed to white-eyed females. This cross produced only white-eyed sons and red-eyed daughters. CROSS B2: White sons were crossed to red daughters. One-half of the daughters were red-eyed, one-half of the daughters were white-eyed, one-half of the sons were red-eyed, and one-half of the sons were white-eyed.

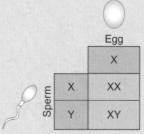

FIGURE 6.39　The fusion of Y-bearing sperm with X-bearing eggs.

that females carried two copies of the W gene; thus females receive a copy of the W gene from both mom and dad. Please read this paragraph several times until you are sure, really sure, that you understand it. This is crucial.

To explain this rather curious exception to Mendel's laws, Bridges turned to a difference between the chromosomes of male and female flies. Flies have four pairs of chromosomes numbered 1 through 4. Chromosomes 2, 3, and 4 are identical in both sexes, but the first pair looked different in males and females (recall that there is also a pair of sex chromosomes in humans). In males there is an acrocentric chromosome (known as the X chromosome), that pairs with a metacentric chromosome (called the

Y chromosome). In females there are only two X chromosomes. In males the X and Y chromosomes segregate away from each other at meiosis 1. Thus males produce either X- or Y-bearing sperm. Females produce only X-bearing eggs. The fusion of Y-bearing sperm with X-bearing eggs makes sons whereas the fusion of X-bearing sperm with X-bearing eggs makes daughters. Note that sons get their X chromosome from their mother and their Y chromosome from their father (Figure 6.39).

Bridges reasoned (correctly) that the sex in flies was determined solely by the number of X chromosomes – indeed, he proposed that males are males because they carry a single X chromosome and females are female because they carry two X chromosomes. (Bridges also proposed, again correctly, that the Y chromosome's only function in flies was to ensure male fertility, not to determine sex.) By placing the W gene on the X chromosome, Bridges found a way to explain the fact that males carried only one copy of the W gene that they received from their mother. We can now re-diagram CROSS A and CROSS B in a sensible fashion in Figure 6.40.

This was a truly elegant piece of science. Bridges had correlated an usual pattern of inheritance with an unusual pattern of chromosome segregation, but it was still a correlation.

APPENDIX 6.1   (*cont'd*)

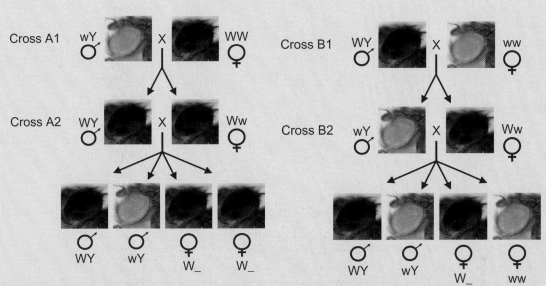

FIGURE 6.40   CROSS A1: White-eyed males (wY) were crossed to red-eyed (WW) females. This cross produced only red-eyed (WY) sons and red-eyed (Ww) daughters. CROSS A2: Red-eyed sons (WY) were crossed to red-eyed daughters (Ww). All daughters were red-eyed (W_), one-half of the sons were red-eyed (WY), and one-half of the sons were white-eyed (wY). Similarly, in CROSS B1, red-eyed (WY) males were crossed to white-eyed (ww) females. This cross produced only white-eyed (wY) sons and red-eyed (Ww) daughters. CROSS B1, red-eyed (WY) males were crossed to white-eyed (ww) females. This cross produced only white-eyed (wY) sons and red-eyed (Ww) daughters. CROSS B2: White-eyed (wY) sons were crossed to red-eyed (Ww) daughters. One-half of the daughters were red-eyed (W_), one-half of the daughters were white-eyed (ww), one-half of the sons were red-eyed (WY), and one-half of the sons were white-eyed (wY).

Nothing more, nothing less. To prove his hypothesis he would need to demonstrate that errors in chromosome segregation caused errors in gene transmission. Fortunately, the experiment he was already doing provided him with exactly the exceptional progeny that such an experiment needed.

Failed Meiotic Segregation (Nondisjunction) as Proof of the Chromosome Theory of Heredity

The proof that genes map on chromosomes came from rare exceptional progeny that came out of CROSS B1: red-eyed (WY) males crossed to white-eyed (ww) females. At low frequency (~1 in 1000), Bridges found white-eyed daughters and red-eyed sons. Bridges realized, and cytological studies confirmed, that these white-eyed females carried two X chromosomes and a Y chromosome (wwY) and that the males carried but a single X chromosome (W).

As shown in Figures 6.41 and 6.42, Bridges realized that such flies could arise if two X chromosomes had failed to segregate from each other at the first meiotic division in the mother. Assuming a normal meiosis II, such an event produces both eggs with two X chromosomes (diplo-X) and eggs

APPENDIX 6.1   *(cont'd)*

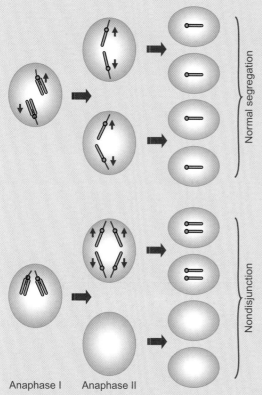

Anaphase I   Anaphase II

FIGURE 6.41   A diagram of X chromosomal nondisjunction in a Drosophila.

with no X chromosomes (nullo-X). Fertilization of a diplo-X (ww) egg with a Y-bearing sperm creates white-eyed (wwY) exceptional females, whereas fertilization of the nullo-X egg with an X-bearing sperm creates the exceptional red-eyed males (W). (Fertilization of diplo-X eggs with X-bearing sperm creates XXX zygotes that are lethal. Similarly, zygotes arising from the fertilization of nullo-X eggs by Y-bearing sperm die because they lack an X chromosome.)

In this experiment, Bridges showed that meiotic nondisjunction of the X chromosomes in the mother had resulted in nondisjunction of the carried X chromosome (i.e., a female had passed on both copies of the w allele to her XXY daughter). This was proof.

Bridges had also shown that sex in flies was determined by the number of X chromosomes an individual carried. Two X chromosomes made a female, whether or not they carried a Y chromosome. Similarly, individuals with just one X chromosome were male, whether or not they carried a Y chromosome. This finding that chromosomes, and the genes that they carried, could determine something as important as the sex of an individual convinced even the most

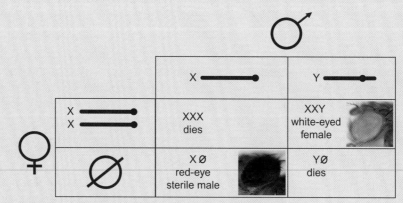

FIGURE 6.42   Bridges' explanation of Drosophila non-disjunction.

APPENDIX 6.1 (cont'd)

skeptical critics of the correctness of the chromosome theory and of the central importance of genetics.

Bridges' paper on nondisjunction was the first paper published in the prestigious journal *Genetics*. The reference is easy to remember: Bridges (1916) *Genetics* Volume 1, page 1. In a real sense this paper began the science of modern genetics. (Indeed the issues raised in this paper still form the basis of much of Scott's last 30 years of research. When pushed too far by his friends to explain "just what the heck I do all the time in that laboratory," he simply states that he is solving problems that were at the center of scientific interest in 1929.)

Study Questions

1. What is a balanced translocation?
2. What are five stages of activity that take place during the cell cycle?
3. What are sister chromatids?
4. What is a spindle apparatus and what purpose does it serve during mitosis?
5. Where are the chromosomes located during metaphase of mitosis?
6. Towards what cellular structure do the chromosomes move during mitotic anaphase?
7. As a cell moves through the stages of meiosis, what step has taken place during interphase?
8. What does cytokinesis accomplish?
9. What are the three basic events that occur during the first meiotic division?
10. In prophase the paired homologs include four chromatids. How many of these chromatids participate in a single recombination event?
11. When does female meiotic recombination take place?
12. List two differences between male meiosis and female meiosis.
13. What triggers the second meiotic division during meiosis in females?
14. What is the purpose of recombination in meiosis?

15. Based on the following diagram of a recombination event that takes place during meiosis, what are the genotypes of sperm that could result?

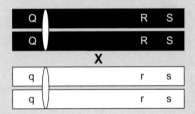

16. Sometimes a chromosome resulting from meiosis will appear to not have undergone recombination when it actually has. How would we detect that recombination event?
17. What is the difference between uniparental isodisomy and uniparental heterodisomy?
18. Why might someone with a balanced translocation expect to produce both normal children and children with at least two other phenotypes?
19. What does the checkpoint in male meiosis check for and what does it do if the cell fails the checkpoint test?
20. Why might we think that a child with Prader–Willi could have a first cousin with Angelman syndrome?

Short Essays

1. A male donkey and a female horse are different species. They can breed but their offspring, called mules, are sterile. What is different about a mule and why is a mule sterile? As you consider this question read "Ask a geneticist: Why can't mules breed?" by Monica Rodriguez at The Tech, June 20, 2007. http://www.thetech.org/genetics/ask.php?id=225.

2. In this chapter we talk about the involvement of genes and proteins in the process of meiosis, but for every gene-based process we have to ask ourselves: are there environmental factors that play a role? Do we all carry the same age-associated risks of aneuploidy, or can we limit our risk through limiting environmental exposures? As you consider this question read "Scrambling eggs in plastic bottles" by D. Warburton and R. S. Hawley in *PLoS Genetics* 2007;3:e6.

3. Meiosis leads to the formation of four sperm in males, but only one egg in females. The other three products of female meiosis form polar bodies. Are polar bodies simply discarded material or do they actually play a role in what happens to the egg? As you consider this question read "The good egg" by Stephen S. Hall in *Discover Magazine*, May, 2004.

4. We talked about the role of imprinting in Prader–Willi and Angelman syndrome. Since only some of our genes are imprinted, the question is: does imprinting play a role in any common traits that emerge later in life? As you consider this question read "Genetics, environmental factors and the emerging role of epigenetics in neurodegenerative diseases" by Lucia Migliore and Fabio Coppede in *Mutation Research*, 2009;667:82–97.

Resource Project

The March of Dimes offers information on a wide variety of traits that are seen in children.

Go to the March of Dimes website and look up Down syndrome. Write a paragraph about features of Down syndrome not mentioned in this chapter and explain why individuals with Down syndrome can or cannot have children.

Suggested Reading

Video

NOVA: *Ghost in Your Genes* (WGBH Boston, 2008).

Articles and Chapters

"Sizing up the cell" by Bruce A. Edgar and Kerry J. Kim in *Science*, 2009;325:158–9.

"To err (meiotically) is human: The genesis of human aneuploidy" by Terry Hassold and Patricia Hunt in *Nature Reviews Genetics* 2001;2:280–91.

"Heterochromatin: A rapidly evolving species barrier" by Stacie E. Hughes and R. Scott Hawley in *PLoS Biology*, e1000233.

"Sex matters in meiosis," Patricia A. Hunt and Terry J. Hassold in *Science*, 2002;296:2181–3.

"Pushing and pulling: Microtubules mediate meiotic pairing and synapsis" by Sue L. Jaspersen and R. Scott Hawley in *Cell*, 2009;139:861–3. "Chromosome 15: Sex" in *Genome: The Autobiography of a Species in 23 Chapters* by Matt Ridley (1999, Harper Collins).

"A DNA damage checkpoint meets the cell cycle engine" by Ted Weinert in *Science*, 1997;277:1450–1.

Books

You Will Dream New Dreams, edited by Stanley Klein and Kim Schive (2001, Kensington Publishing).

Adults with Down Syndrome by Siegfried M. Pueschel (2006, Brookes Publishing Company).

The Odd Couple: How the X and Y Chromosomes Break the Rules

THE READER'S COMPANION: AS YOU READ, YOU SHOULD CONSIDER

- How meiosis for the sex chromosomes differs from meiosis for autosomes.

- What allows the X and Y chromosomes to pair.
- What functions the PAR regions play outside of meiosis.

(Continued)

- How females cope with having more copies of X chromosome genes than males.

- What the presence of a Barr body in a cell indicates.

- What role the structural RNA XIST plays in mediating X inactivation.

- How X-inactivation facilitates the survival of individuals with sex chromosome aneuploidies.

- How a heterozygous female can end up expressing the mutant phenotype.

- Why maintenance of Y chromosome integrity would be a challenge.

- Why some genes are present in several copies on the Y chromosome.

- How segregation of X and Y chromosomes explains X-linked inheritance.

- Why males do not pass X-linked traits to their sons.

- Why X-linked traits are rare in females.

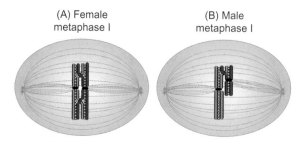

(A) Female metaphase I   (B) Male metaphase I

FIGURE 7.1 (A) During meiosis I in a female, the two copies of the X chromosome pair along their lengths based on sequence homology found all along the length of the chromosomes. (B) During meiosis I in a male, only the very small regions of the X and Y that have the same genes can pair, and other regions that contain genes specific to just the X or just the Y cannot participate in aligning the chromosomes or holding them together at the metaphase plate.

7.1 PASSING THE X AND Y CHROMOSOMES BETWEEN GENERATIONS

In previous chapters, we have talked about the fact that females have two X chromosomes and males have an X and a Y. Unlike the flies we discussed in Chapter 6, sex in humans is solely determined by the presence (male) or absence (female) of the Y chromosome.

During the formation of eggs in a female, the two X chromosomes are handled in much the same way the autosomes are handled, through pairing of the homologs (Figure 7.1A). But during formation of sperm, the cell is faced with a kind of genetic asymmetry that creates

a problem: how to get two different chromosomes, the X and the Y, to go through meiosis when neither of them has a homologous copy with which to pair.

As happens with the autosomes, both pairing and recombination are needed to get the X and Y safely through meiosis. Surprisingly, during formation of sperm, the cell's meiotic machinery treats the X and Y chromosomes as if they were a homologous pair, even though the simplest visual examination demonstrates that they are significantly different from each other. During meiosis I, the replicated X and the replicated Y come together to form a bivalent that lines up on the metaphase plate (Figure 7.1B) along with the bivalents that are autosomal (replicated versions of any chromosome other than the X or Y). At the end of meiosis I, we find the X in one daughter cell and the Y in the other daughter cell. At the end of meiosis II, there are four sperm, two containing the X and two containing the Y. How does this happen? What allows the cell to treat the X and the Y as if they were homologous structures?

Clearly, there is not enough DNA on the Y chromosome for it to contain genes homologous to all of the genes on the X. The X and the Y each contain a region near the tip of the shorter

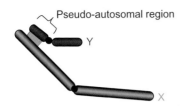

FIGURE 7.2   There is more than one region on the X and Y chromosomes that share genes, but there is a main region near the tip of the short arm of the X called PAR1 (pseudo-autosomal region 1) that is homologous to a region near the tip of the short arm of the Y chromosome. Crossing over between the X and Y PAR1 regions at each male meiosis is essential for proper X–Y segregation during male meiosis. The failure of crossing over between the PAR1 regions often leads to nondisjunction.

chromosomal arm called the *pseudo-autosomal region (PAR)*. The PAR contains a group of several dozen genes that exist as homologous copies on the X and the Y. This region of homology lets the X and Y pair, as if they were autosomes (Figure 7.2). DNA in this region can recombine, exchanging material between the X and the Y. Because of the exchanges, the chromosomes are locked together during male meiosis, but these exchanges are normally strictly contained within the pseudo-autosomal region. It is important to avoid exchange of material from non-homologous regions of the X and Y since such exchange events would put Y-specific DNA onto the X chromosome, and vice versa, in areas that do not contain copies of the same genes. There is also a second PAR region at the tip of the long arm of both the X and Y chromosome (Xq and Yq). This region also undergoes recombination, albeit at a much lower frequency (2–3% of meioses undergo exchange in this region).

The occurrence of exchanges within the PAR is critical for proper segregation of the X and Y chromosomes. Cases in which such exchanges fail to occur account for more than two-thirds of instances in which the X and Y nondisjoin at meiosis I. Moreover, loss or deletion of the PAR is associated with male sterility.

7.2 HOW HUMANS COPE WITH THE DIFFERENCE IN NUMBER OF SEX CHROMOSOMES BETWEEN MALES AND FEMALES

In our discussion of the effects of aneuploidy in Chapter 6 we repeatedly made the point that proper human development requires exactly two copies of every autosome. No autosomal monosomies (just one copy) are viable and most trisomies (three copies) are lethal as well. But half of all human beings are monosomic for the X chromosome, and trisomic XXX women are born at frequencies of one in a thousand, often with little phenotypic consequence. Such women are physically indistinguishable from their XX counterparts, and usually have no medical problems – though they may occasionally display learning difficulties or delayed speech as children. Clearly, either monosomy for the X in XY males or trisomy for the X chromosome in XXX females is viable.

This is true despite the fact that there are thousands of genes on the X chromosome that are not present on the Y. Although we can imagine that the amount of gene product might actually need to be different for the two sexes for some genes specific to sex, a large number of genes on the X chromosome have nothing to do with maleness or femaleness. So how does the human body cope with a two-fold difference in the copy number of the vast majority of X chromosomal genes between the sexes?

The first part of the answer lies in the fact that during early embryogenesis the cells of female embryos inactivate one of the two X chromosomes in every cell. This discovery was made by a mouse geneticist named Mary Lyon, and the process of X inactivation is sometimes referred to as Lyonization. There are a few X chromosome genes, including those few X genes that have counterparts on the Y, that escape inactivation, but the vast majority of genes on the inactivated copy of the X chromosome are shut down. This inactivation event is random. Thus, for each cell in the early embryo there is an equal probability of inactivating the X derived from the mother or the X derived from the father. However, once that decision is made it is permanent; all of the mitotic descendants of that cell will inactivate that X chromosome. As discussed below, the only exception to this permanence is the female germ cell – oocytes reactivate the inactive X chromosome!

The second part of the answer involves regulation of transcription. While we might have imagined that genes on the X would all be genes for which the cell requires only half as much gene product as would be needed from an autosomal gene, in fact the cell has just as much need for transcription and gene product amount from the X as from the autosome. The cell solves the problem of having half the number of copies it needs for X chromosome genes through up-regulation of transcription of X chromosomal genes by approximately two-fold in both male and female cells. Thus the genes on both the single X, in male cells, and the active X chromosome, in female somatic cells, is increased approximately two-fold relative to their autosomal counterparts. The result is that for both sexes one X chromosome serves the cells as well as two autosomal homologs.

It is worth noting that this rather complex system of dosage compensation is not the only way nature has found to solve the problem. In the fruit fly *Drosophila melanogaster* females do not inactive the X so they can simply use the copies they have at normal levels of expression, and the males simply double the transcriptional activity of the X chromosome in males – a seemingly simpler solution. This stands in sharp contrast to what we see in humans where the female first inactivates one copy of the X and then up-regulates the genes on the active copy to compensate for that inactivation! There are quite a few other mechanisms that exist to solve the same problem in other organisms. Why evolve so many different mechanisms to solve the same problem? We have said it before,

and will probably say it again: evolution is not Michelangelo and the Sistine Chapel. It does not act to make things simple or pretty. It is a teenager with a broken car and no money. In each case evolution selects for whatever solution works to solve the problem, however inelegant.

7.3 HOW X INACTIVATION WORKS

The inactivation of the X chromosome is the result of both modification of the bases of the DNA (usually by methylation) and changes in the proteins called histones that bind to the DNA. This combination of chemical modifications to both the DNA and the histone proteins that bind the DNA render most of the genes on the X chromosome inactive, that is to say, not actively used by the cell. These changes in chromatin structure have a morphological consequence, namely the condensation of the X into a very compact, densely staining mass known as a Barr body. The Barr body is easily observed as a dark mass in an interphase cell, or as a bright blue spot when modern dyes are used (Figure 7.3). Both XX women and XXY men display a single Barr body. But the cells of XXX women display two Barr bodies! Thus it is not the case that female cells simply inactivate one X chromosome, but rather that *in either sex only one X chromosome is kept active. Any extra copies of the X beyond that first one are inactivated, forming a Barr body in a cell with two X chromosomes, or Barr bodies in cells with more than two X chromosomes.*

A detailed discussion of the molecular mechanism of X inactivation lies beyond the scope of this text. Suffice it to say that the X chromosome possesses a site referred to as the *X inactivation center*, or *XIC*, and that once the decision to initiate X inactivation is made, inactivation spreads out from the XIC site in both directions (Figure 7.4). The first step in inactivation is transcription of a gene in the XIC known as XIST (see Figure 7.41).

The XIST gene is the only gene whose expression is limited to the inactive X chromosomes. Perhaps not surprisingly, mutations that cause damage to the XIST gene create an X chromosome that cannot be inactivated. This always forces the other copy of the X chromosome to undergo inactivation. XIST encodes a very large RNA molecule that is *not* translated into protein – but rather spreads out from the XIC, coating the soon-to-be inactivated X chromosome. (XIST does not bind to the other X – the X that will remain active.) The binding of XIST RNA to the X chromosome makes that copy of the X chromosome a target for the DNA and protein modifications that lead to its inactivation.

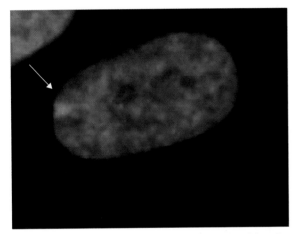

FIGURE 7.3 A Barr body is seen in the cells of females and XXY males. In black and white photos a Barr body classically appears as a small dark spot along the edge of the nuclear membrane, but modern staining techniques let us see it as a bright blue spot along the edge of the blue nucleus. The name Barr body is used in honor of the scientist Murray Barr, who discovered it. (Reproduced with permission. Photo of cell from Rego *et al. Journal of Cell Science*, 2008;121:1119–27.)

7.4 SKEWED X INACTIVATION – WHEN MOST CELLS INACTIVATE THE SAME X

Because X inactivation occurs randomly and at an early stage in fetal development, it is possible for a tissue or organ to be comprised of cells that only have active paternal copies of the X or that only have active maternal copies of the X. Thus, while we expect that on average about half the paternal copies will be inactivated, and about half the maternal copies will be inactivated, each inactivation event is random and not influenced by a tally of how many of the other active chromosomes are maternal or paternal. Thus, in rare cases we may see an

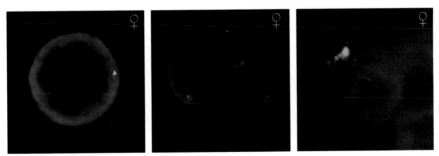

FIGURE 7.4 XIST is seen on X chromosomes of a female. In these panels, red marks the X chromosome and green marks XIST. On the left, at the eight-cell stage of development no green XIST is evident. Soon after, at a stage called the morula stage, as the cells of the embryo begin forming into a ball, the first signs of XIST appear (middle panel). By the blastocyst stage, when there are about 70 to 100 cells, a "cloud" of XIST begins to appear as the inactivation process spreads along the chromosome (right hand panel). (From van den Berg, I. M., Laven, J. S., Stevens, M., *et al.*, "X chromosome inactivation is initiated in human preimplantation embryos", *American Journal of Human Genetics*, 2009;84(6):771–9. © 2009, with permission from Elsevier.)

X-linked trait manifested in a female carrier who has inactivated the good copy of the X in most or all of the cells relevant to that trait.

There are two cases, both as rare as they are fascinating, in which all or the majority of cells in a female inactivate the same X. In each of these cases all, or nearly all, of the cells in a female have had one copy of the X inactivated while leaving the other X active in virtually all of her cells. This creates a situation much like what we see in males with only one copy of the X – there are no sequences from a second copy of the X chromosome to compensate for any defects present on the homolog of the X that is being used.

Skewed Inactivation in Monozygotic Twins

The first such case is the curious situation sometimes seen in monozygotic twins. There are multiple cases where two sisters, identical twins derived from the same fertilization event, show preferential inactivation of one X in the first twin and preferential inactivation of the opposite X in the remaining twin. A fascinating example of this phenomenon is shown in Figure 7.5, which shows the pedigree for a set of monozygotic twins which are heterozygotes for a mutation that causes red–green colorblindness. Curiously, one of these twins was colorblind, even though she produced sons with both normal and abnormal color vision. Indeed, molecular analysis reveals that in the colorblind twin most somatic cells have inactivated the X carrying the normal allele for color vision (the maternal X), while in the twin with normal color vision it was the paternal X (the one carrying the mutation) that was preferentially inactivated! The normal allele of the colorblindness gene was inactivated in the colorblind mother and yet activated in her son, allowing him normal vision. This indicates that the pattern of X inactivation in the mother was not still present in her son.

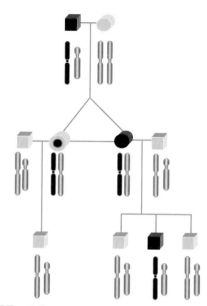

FIGURE 7.5 Phenotypic discrepancy in identical twins due to differential X-inactivation. In the affected twin almost all of the active X chromosomes are the copies from the father that carry the color vision defect. In the unaffected twin, who is an obligate carrier, almost all of the active copies of the X come from the mother. Note that the affected twin cannot be explained by a de novo mutation on the second copy of the X since she has unaffected sons. (After Jorgensen, A. L., Philip, J., Raskind, W. H., Matsushita, M., Christensen, B., Dreyer, V., and Motulsky, A. G., *American Journal of Human Genetics*, 1992;51:291–8.)

This is not surprising. Mothers inactivate half their X chromosomes and yet in cases where a mother has many sons, some with one of her Xs and some with the other X, all of the sons have active X chromosomes. A more dramatic example of this phenomenon may be found in the famous Dionne quintuples. Though all five of these women were derived from the same embryo, which was heterozygous for a mutation that can cause colorblindness, two of the twins were colorblind, while the others had normal color vision.

The literature carries many examples of pairs of monozygotic twins in which one of a pair of twins was affected by a disorder caused by a loss-of-function mutation on the X chromosome (for other examples, muscular dystrophy, fragile X syndrome, and hemophilia) while the other twin was not. It would appear that what happens is this: the embryo presumably splits into two separate embryos *after* X inactivation had already occurred and did so in such a fashion that those Xs which had inactivated one X were portioned into one embryo while those that had inactivated the other X were partitioned into the other daughter cell. The curious thing is that we can find no report of anyone observing a pair of heterozygous monozygotic twin girls both of whom expressed the mutant trait! Perhaps it is just that this would require three very rare events to all come together – the inheritance of a mutation that causes a rare recessive trait, the unlikely event of all of the cells in an embryo inactivating the same copy of the X chromosome, and the low probability of twinning. Nonetheless, the numerous cases of twins who demonstrate differential X inactivation has led some investigators to speculate that there may be a causal connection between twinning and skewed X inactivation. The question remains open.

Skewed X Inactivation in Females Heterozygous for Chromosome Rearrangements

Recall that a translocation results from the breakage of two non-homologous chromosomes (in this case, the X and an autosome) and subsequent re-healing by sticking the broken pieces back together incorrectly so that the broken end of one chromosome now caps the broken end of the other chromosome, and vice versa. Thus these females carry a normal X chromosome, a normal autosome, and the two rearranged chromosomes that resulted from the translocation. When the breakage events that created the translocation occur within a gene or genes, they can disrupt those genes and result in a loss-of-function mutation. RNA polymerase,

the enzyme that carries out transcription, can do many neat tricks, but it can't jump between chromosomes. By splitting a gene into two parts and moving one part to a new chromosome, you have killed that gene. The RNA polymerase molecule simply has no way to leap to another site in the genome to complete the transcription of this gene.

For some of the rearrangements, the women with balanced X-autosome translocations that disrupt the DMD gene end up affected, which at first seems counterintuitive. After all, they have a normal X chromosome in addition to the X involved in the translocation. The problem lies in the fact that all of the cells in the bodies of these females arose entirely from embryonic cells in which the normal X chromosome was inactivated (Figure 7.6). Why does that happen? It all comes down to a problem in gene dosage that kills off embryonic cells expressing the wrong number of copies of key genes.

In a normal female, half the cells will inactivate one copy of the X and half the cells will inactive the other copy. In a girl with a translocated X, inactivation of the translocated X causes a problem. Even though the balanced translocation had originally left her with the right number of copies of the genes on the two copies of the autosome, the translocated fragment lacking the XIC will fail to inactivate its X chromosomal genes, leading to too much X chromosomal gene expression. In addition, on the translocated chromosome bearing the XIC, inactivation *may* spread into adjacent autosomal regions, inactivating genes that need to be present in two copies.

Thus, all of the cells that inactivated the translocated X will die during embryogenesis because of both X and autosomal gene dosage effects. The embryo is then constructed from the remaining cells, all of which have the activated translocated X that can still express the attached autosomal genes. So females with a translocated X will all be born with the normal X inactivated because they have to keep the right gene

Live cells in the process of inactivating one of the two X chromosomes

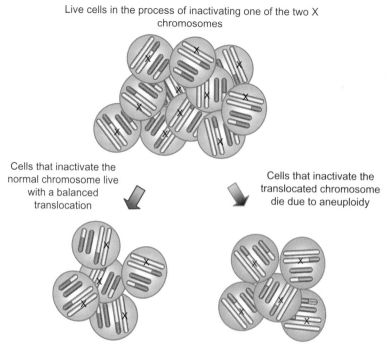

Cells that inactivate the normal chromosome live with a balanced translocation

Cells that inactivate the translocated chromosome die due to aneuploidy

FIGURE 7.6   An embryo that starts out randomly inactivating the normal X in some cells and a translocated X in other cells ends up with all cells having the normal X inactivated because autosomal gene dosage problems kill the cells that inactivate the translocated X. If the translocation breaks the DMD gene, then each cell in this female embryo will have only one activated X and it will always have the broken copy of the gene, leading to a girl with DMD. If you examine the cells on the left you will see cells that have two copies of each gene on a pink chromosome and one copy of each gene on the X, exactly as a normal cell would. On the right, examination of the chromosomes shows us a full complement of X chromosome genes plus additional copies of X chromosome genes on the small translocated chromosome, as well as a shortage of the autosomal genes on the inactivated long translocation chromosome.

dosage number for the cells to stay alive (Figure 7.6). If the translocation disrupted the DMD gene, they will have DMD because the only copy of the X chromosome from which they can transcribe the DMD gene carries a disrupted DMD gene. The lack of the DMD gene product, dystrophin, then causes DMD. Even if the translocation falls to one side of the DMD gene so that the DMD gene is not disrupted, this female may still have DMD if she was a carrier for a DMD-causative mutation located on the translocated copy of the X. The copy number problems will still force the issue by killing any cells that inactivate the translocated copy of the X, leaving only the translocated copy with the

DMD mutation in active form. Thus she is left with no extra copy of the X to compensate for the DMD-causative mutation.

So the fact that only the cells with the translocated chromosome are active in these cases might seem to suggest that there is selective inactivation of the normal chromosome, but that is not the case. The inactivation is just as random in these cases as others. The point at which the situation becomes unusual is not at the step in which inactivation takes place, but rather in the selective process that follows by which cells with a normal X chromosome die and leave only cells with an active translocated X from which to form the embryo.

7.5 GENES THAT ESCAPE X-INACTIVATION

While it would make writing this chapter simpler if we could just say that "all but one X chromosomes are inactivated in cells bearing more than one X," there are, as always, exceptions. A fraction of genes on the X chromosomes, primarily those genes that are located in the PAR regions, fail to be inactivated – nor is their transcription up-regulated by two-fold. There are several dozen genes in PAR1 (on Yp) and a half dozen genes in PAR2 (on Yq). All genes in PAR1 escape inactivation as do most of the genes in PAR2. Thus genes within the PAR regions, especially those within PAR1, will be present and expressed in two doses in *both* males and females. Such genes then become prime candidates for the genes whose presence in only one copy might underlie the phenotypes exhibited by XO Turner syndrome women. One strong candidate for such a gene is the SHOX gene located in PAR1. SHOX encodes a transcription factor critical for skeletal development and heterozygosity for mutations in, or deletions of, this gene are associated with short stature. Indeed, several lines of evidence strongly suggest that the presence of only a single copy of SHOX1 accounts for the short stature that is commonly seen in X0 women with Turner syndrome (see below).

Interestingly, very few of the genes located in PAR1 are present in the mouse – only three of the 24 known genes in the human PAR1 region are conserved in the mouse and these have diverged considerably. Perhaps the absence of such genes in the mouse PAR region accounts for the fact that X0 mice are both phenotypically normal and even fully fertile!

7.6 REACTIVATION OF THE INACTIVE X CHROMOSOME IN THE FEMALE GERMLINE

Following the specification of the germline, the inactive X chromosome is actually reactivated in primordial germ cells. The presence of two X chromosomes is required for the development of the female germline, and the absence of two functional X chromosomes results in female sterility. The reactivation of the X chromosome is associated with the loss of methylation, the absence of XIST RNA coating the X chromosome, and apparently with loss of transcriptional up-regulation as well.

One of the reasons for reactivation is probably straightforward: the inactivated X needs to be returned to a state where it is competent to pair and recombine with its homolog. The presence of a second active chromosome is clearly essential in humans, because the absence of the second X chromosome in XO Turner syndrome females causes rapid death (atresia) of oocytes during fetal development. There may, however, be other reasons as well, such as preventing the transmission of an inactivated X chromosome to the zygote.

7.7 X CHROMOSOME INACTIVATION DURING MALE MEIOSIS

During spermatogenesis the X chromosome in males is inactivated by a mechanism that is still not well understood. Although the XIST transcript is present at this time, it is not required. Male mice carrying null mutations in the XIST gene can still undergo X chromosome silencing during meiotic prophase. The inactivation of the X plays a critical role in meiotic progression, as part of a process used by meiotic cells to cope with unpaired regions of chromosomes. Although the X appears to be at least partially reactivated after meiosis and during sperm development, there is some evidence that the presence of XIST may leave an "imprint" that can in some cases predispose the paternally derived X chromosome to be inactivated. This phenomenon has not been observed in humans, but is seen in marsupials and in

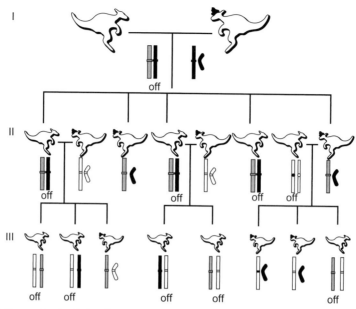

FIGURE 7.7   Imprinting made simple in a kangaroo family. In the kangaroo the copy of the X chromosome that is inherited from the father is inactivated (marked "off") and the copy of the X that is inherited from the mother is activated, even if the chromosome had been inactivated in the previous generations. Notice that the black copy of the X chromosome that came from Grandfather Kangaroo is inactivated in all of his daughters, but then activated again in generation III since only the daughters in generation II possess this copy of the X to pass along to three. The copy found in a male is always activated, whatever its status had been in his mother. This is not how X inactivation is determined in humans, where X inactivation is random, without regard to the parent of origin.

the extra-embryonic tissues in some other mammals.

Consider the process of X inactivation in female kangaroos (Figure 7.7). In many ways this process is strikingly similar to the events that occur in human females. The critical exception is that the inactive X in the cells of female kangaroos is always the X chromosome that they received from their father. When the X chromosome passes through the kangaroo male germline, it ends up inactivated in his daughters. This means that the genes on that copy of the X chromosome are not transcribed to make RNA and gene products from that chromosome are not present to affect the organism's characteristics. However, the switch gets reset as it passes through the germline of the next generation, and genes on the X chromosome that were not use in

one generation because they came from a male may be used in the next generation when they get passed along by a female. Thus, if you look at the X chromosome that the grandfather kangaroo contributes to the kangaroo family shown in Figure 7.7, it is easy to see that his copy of the X chromosome is turned off in all of his daughters and "on" in all of his grandchildren that receive his copy of the X. For his great-grandchildren who inherit an X from him, the activation status of that X will depend on whether it is being passed along by one of his male descendants or one of his female descendants.

The marking of the paternal X chromosome for inactivation is also observed in the extra-embryonic membranes of mice. Thus in all the cells of a kangaroo female and some of the cells of mouse embryos the choice of which X is inactivated at the 50 cell stage of embryogenesis can recognize some "mark" left on the X chromosome (perhaps the presence of XIST placed on the X during male meiosis?) that is donated to the embryo by its father! The nature of this mark remains obscure. Perhaps some proteins loaded on to the newly rebuilt paternal chromosomes predispose the paternal X to inactivation. Or perhaps the DNA is modified in some way, perhaps by adding a small chemical group (such as a methyl residue) to the X chromosomal DNA during spermatogenesis. Other modifications, such as the binding of structural RNA molecules to the imprinted X chromosomes, have also been proposed.

7.8 X INACTIVATION AND THE PHENOTYPES OF SEX CHROMOSOME ANEUPLOIDY

The lethality of most autosomal aneuploidies stands in stark contrast to the viability of most sex chromosome aneuploidies. This may not be surprising given that mechanisms already exist for dealing with the dosage differences between human males and human females. In fact, the mechanisms that allow this difference in chromosomal dosage also allow the survival of individuals with various sex chromosome aneuploidies. Nonetheless, most of those with a sex chromosome aneuploid constitution do show some differences from people with a normal chromosome complement. The most common examples of sex chromosome aneuploidies are Klinefelter syndrome (XXY males) and Turner syndrome, denoted X0 because Turner syndrome has only one sex chromosome, which is an X.

Variability of Turner Syndrome

Turner syndrome is a relatively common genetic condition that affects 1/2000 to 1/2500 females. Women with Turner syndrome are normally sterile with ovaries that appear as a rudimentary streak. This sterility almost certainly reflects the requirement for two active X chromosomes in the female germline (see Section 7.6). They are also shorter than average and may show immature development of the breasts. As discussed in section 7.5, the shortness of stature likely reflects the presence of only one copy of the SHOX gene. The reduced breast development may well be a straightforward consequence of the atrophy of the ovaries and thus lower levels of estrogen (see Chapter 8). Some Turner syndrome females *may* show some webbing of the neck as well as some medical difficulties. As we will discuss below, some Turner syndrome females may also display a distinctive set of behavioral phenotypes.

Surprisingly, the vast majority of all X0 conceptions are miscarried spontaneously *in utero*. Thus possessing only one X chromosome is almost always lethal to the zygote. Perhaps this high degree of fetal loss reflects a stronger requirement for two doses of those X chromosomal genes than might have been anticipated given the viability of at least some Turner syndrome females.

The reason for the survival of the rare Turner female and the phenotypic variability among such live-born females may be attributed to what scientists call *mosaicism*, the presence in the body of cells of more than one genetic type. Some live-born cases of Turner syndrome may be due to the fact that these surviving girls are not composed solely of X0 cells: they are composed of both X0 and XX cells. The loss of a single X chromosome, during mitosis (rather than meiosis), in one cell out of several cells present very early in zygotic development may produce a combination of both XX and X0 cells. Thus the resulting individual possesses both XX and X0 cells. As long as an XX karyotype is present in those cells that *absolutely require* two X chromosomes, or perhaps two copies of the PAR region, the individual will survive and it will not matter if an X chromosome is missing from some other cell types that do not require an X. Those cells that do not require two Xs will be able to survive as either XX or X0 cells. The more X0 cells the individual possesses, the more severely affected the individual will be, whereas the more XX cells the individual possesses, the more normal the individual will be. If different Turner females each have a different fraction of X0 cells in their bodies, it makes sense that the phenotype would be so variable.

Klinefelter Syndrome

XXY Klinefelter males occur at a surprisingly high frequency of 1/500 to 1/1000 live births. Like XX females, XXY males undergo X inactivation during early embryonic development. Thus, half of their somatic cells have inactivated one of their two Xs and the other half of their cells have inactivated the other X. Klinefelter males are sterile due to testicular atrophy at puberty, and some of these males exhibit some degree of external feminization, such as breasts and hips. For a rather complex set of reasons, XXY males usually show reduced levels of testosterone, beginning around puberty. In many of these men, there is almost an even amount of estrogen and testosterone in the individual's system. It has been found that some Klinefelter boys and men also have social problems and learning difficulties. Given the inactivation of the second X, the somatic phenotypes observed in Klinefelter males is surprising. The most reasonable explanations for these defects may lie in those genes on the X that escape inactivation. Perhaps the extra copies of these genes (along with the testosterone deficiency caused by testicular atrophy) provide a reasonable explanation for some of the observed phenotype.

Other Sex Chromosome Aneuploidies

Two other types of genotypic abnormalities of the sex chromosomes, which fail to have names, are XYY males and XXX females. Although XYY men do not exhibit any characteristic set of abnormal phenotypes, they are often taller than average males. Many XYY boys have learning disabilities, and some fraction of both XYY boys and men may have behavioral problems. Although the great majority of XYY men lead normal lives, the frequency of XYY men is increased in various kinds of prison populations, especially among inmates greater than six feet in height. The frequency of XYY men is 1/1000 among newborn males but may be as high as five times that in general prison populations; it has even been reported to be as high as ten times that frequency in one juvenile prison population. More strikingly, if attention is restricted to male prisoners over six feet tall, the frequency of XYY males has been estimated to be as high as 10–20%.

However, the proposal that extra testosterone in XYY men leads to aggressive behaviors that send them to prison is not supported by the evidence. Most XYY men are never sent to prison, and those XYY men who are sent to prison are not always, or even usually, incarcerated for violent offenses. Rather, they are more often jailed for repetitive violations of probation agreements, possession of stolen property, writing bad checks, etc. The simple conclusion is that having an extra Y chromosome may predispose some men to get into trouble with the law, but it does not make them more violent. Indeed, it may be the case that the tendency of XYY males to end up incarcerated may be less reflective of an influence of an extra Y chromosome on criminal aggression, and more the result of learning difficulties created by the extra X chromosome. Indeed, in any instance in which we find that a particular genotype is more common in a particular population, we need to be very careful in determining what the precise phenotype is that actually correlates with the genotype in question.

XXX females are not associated with any specific abnormal phenotype. Indeed, these females are most often detected only because they were karyotyped for some unrelated reason. The vast majority of them are fully phenotypically normal; however, there may be a decrease in fertility and presence of some learning disabilities has been observed during childhood. An XXX female inactivates two X chromosomes. Thus, like a normal cell (XY or XX), the XXX cell has one functioning X chromosome and the phenotype is normal or near-normal. In contrast, a cell carrying three copies of chromosome 21 has no way to simply inactivate the extra copies and there are major phenotypic consequences!

7.9 THE STRUCTURE OF THE HUMAN Y CHROMOSOME

The Y chromosome is often perceived as little more than a small chromosome that carries the male-determining SRY gene. The presence (in XY males) or absence (in XX females) of the SRY gene on the short arm (Yp) of the Y chromosome is what initiates the regulatory processes that eventually determine whether the individual will be male or female (see Chapter 8 for a description of this gene and its role in mediating sex determination).

Structure of the Y Chromosome

As we discussed at the beginning of this chapter, although the PAR region at the tip of Yp does recombine with the X chromosome, the majority of the Y chromosome (known as the *Male Specific Y* or *MSY* region) does not recombine with the X chromosome. Approximately two-thirds of the MSY region is comprised of heterochromatin, which you may recall is transcriptionally inactive. However, in recent years David Page and his collaborators at MIT have identified seven regions within the euchromatic portion of MSY that contain nine families of Y chromosome-specific protein-coding genes. By the term "families of Y chromosome-specific protein coding genes," we mean that each gene is present in more than one copy,

with the total number of copies per gene family ranging from 2 to 35 copies. These genes are specifically expressed in the testis and their products are required for spermatogenesis. These seven regions, which contain the vast majority of the genes located within the MSY region, are known as the ampliconic regions.

Within these ampliconic regions, these nine families of genes are embedded in a complex of three kinds of repeat structures:

1. Large palindromic regions. Most of the ampliconic region on the long arm of the Y chromosome (Yq) is made up of eight giant palindromic regions so large that in total they make up some 25% of the euchromatic MSY. Palindromic regions contain mirror-image inverted repeated sequences, taking a form such as 1 2 3 4 5 6 6 5 4 3 2 1. Some of these sequences are so long that we call them "giant palindromic regions"; in fact, one of these regions is 2.9 million base pairs long! The vast majority of genes in the MSY sit within these palindromes.

2. Inverted repeats in which the repeats are physically separated. The ampliconic regions also contain five sets of widely-spaced inverted repeats, sequences that might be represented as 1 2 3 4 5 6 7 8 9 10 11 6 5 4 3 2 1, such that the inverted repeats are separated by other DNA sequences.

3. Non-tandem direct repeats. The euchromatic MSY contains a number of long direct repeats, taking a form such as 1 2 3 4 5 6 7 8 9 10 1 2 3 4 5.

The presence of these repetitive and palindromic families of sequences around the critical Y-specific gene allows for an important process of self-repair that helps to maintain the integrity of the Y chromosome even though it never pairs with another Y chromosome during meiosis. This self-repair process involves a type of genetic recombination called *gene conversion* that allows one copy of a gene to replace the sequence of a homologous gene with its own sequence.

Imagine that one of three copies of gene X was mutated such that the Y chromosome carries two "good" copies of gene A and one mutant "bad" copy of gene A (denoted A*). Frequent gene conversion events within this family might well *convert* the non-functional A* copy back to a normal A sequence. Yes, the reverse could happen, the A* could convert one or both "good" copies of gene A to non-functional A* alleles. But realize that if gene A is really essential for male fertility, that A*-only Y chromosome goes no further. It cannot support spermatogenesis. Thus there is very strong selection against males carrying deleterious mutations on the Y chromosome – they are sterile and that Y chromosome goes no further. Thus gene conversion usually allows newly arising mutants within these repeated gene families to be converted (corrected?) back to the wild type sequence, and in those rare cases where the mutation begins to increase in copy number, selection prevents its transmission.

The Y chromosome clearly has a powerful mechanism for self-preservation. There are scientists who frequently make headlines in the newspaper by suggesting that the Y chromosome is decaying and will simply disappear altogether in 50,000 years. But frankly, the Y appears to be protected on two levels – its ability to repair a new mutation to match the more common sequence presence among the other copies of that gene, and its tendency to eliminate any cell in which the mutant genotype gets past the self-repair mechanism to take over the multiple copies of that gene. So we end up thinking that such bleak predictions for the future of the Y chromosome may be just a bit premature.

Recombination among Palindromic Repeats and Male Infertility

Unfortunately, this method of self-preservation does come with a cost – these repeated structures, and the occurrence of rare recombination

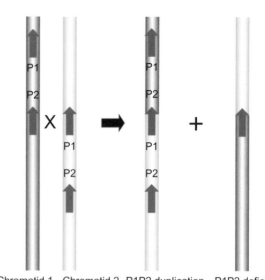

Chromatid 1   Chromatid 2   P1P2 duplication   P1P2 deficit

FIGURE 7.8   Recombination between repeats on sister chromatids can lead to structural aberrations in the Y chromosome present in the sperm that result from the meiosis in which the recombination occurred.

events between and within them, creates an "Achilles Heel" for the Y chromosome. A fraction of these intra-chromosomal gene conversion events are associated with reciprocal recombination or crossing over either within one chromatid of the Y chromosome or between homologous sequences on the two sister chromatids. Recombination between repeats can lead to chromosome aberrations such as deletions, inversions, and even Y chromosomes carrying two centromeres. As just one example, two of the larger palindromes (denoted P1 and P2) are flanked by a pair of direct repeats. Crossovers between these repeats, either as intra-chromosomal "loop-out" exchanges or unequal exchanges between sister chromatids, result in Y chromosomes that have lost both palindromic arrays (Figure 7.8). The occurrence of such deletions is a serious problem if the resulting deletion includes a family of genes known as the DAZ genes which are essential at multiple points in spermatogenesis. Moreover, although such crossover events occur in less than

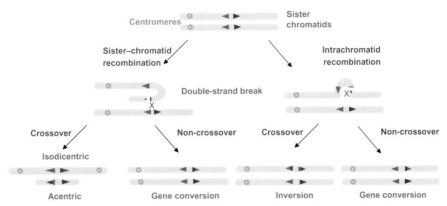

FIGURE 7.9 Crossing over between palindromic repeats on sister chromatids can lead to aberrant structures such as chromosomes with inverted regions or a Y chromosome with two centromeres at opposite ends of the chromosome, called isodicentric (After Hall, H. E. and Hawley, R. S., "The hows and Ys of genome integrity", *Cell*, 2009;138:830–2.)

1 in 1000 meioses, they are common enough that the resulting deletions may account for as many as one in 10 of all infertile males. While it is probably of very little comfort to males rendered sterile by the occurrence of such a deletion event in his father, it is worth noting that selection prevents the further propagation of that deleted Y chromosome. Thus loss of the testis-specific genes, or perhaps even a significant reduction in their copy number, is expected to have immediate effects on male fertility, with the expectation that infertility will end the lineage of a defective Y chromosome within the same male in which it was created.

As we noted above, there is also a low, but measurable risk of crossing over between palindromic repeats on the sister chromatids. These exchanges can produce a Y chromosome containing two centromeres which we would call an isodicentric chromosome (Figure 7.9). As you might imagine from our discussion of the role of centromeres in Chapter 6, the presence of two separate centromeres on the same chromosome might cause all kinds of chaos.

Suppose the two centromeres on the same sister chromatid were to orient toward the opposite pole at either meiosis or mitosis. If both the centromeres remain active, such a chromosome may be snapped apart during chromosome segregation (Figure 7.10). Moreover, by virtue of the recombination events that created them, these double-centromere chromosomes will have deletions for some genes on one chromosome arm and duplications for some genes on the other chromosome arm. The consequences of such deletions can lead to intersex conditions (if SRY is duplicated or lost) or disorders that mimic Turner syndrome! Since these situations often involve infertility, we see that once again the structure of the Y leads to a failure to propagate aberrant Y chromosomes into subsequent generations.

Thus there are many ways in which the Y chromosome can end up damaged, including both mutation and structural damage. But there are "backup" sequences from which the cell can correct errors in Y chromosome sequence, and structurally erroneous Y chromosomes tend to lead to male infertility, blocking that copy of the Y from continuing into future generations. When we consider the many ways in which Y chromosome errors are prevented from turning up in future generations, we find ourselves impressed with the self-preservation abilities of the Y chromosome.

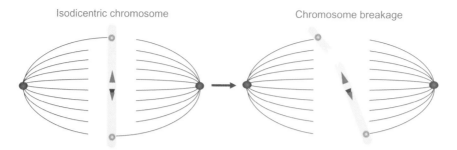

FIGURE 7.10 When a Y chromosome has been generated that has two centromeres, the aberrant chromosome ends up being broken as it passes through meiosis resulting in versions of the Y chromosome that have only one centromere but that have missing or extra copies of some genes, depending on where the break occurred. (After Hall, H. E. and Hawley, R. S., "The hows and Ys of genome integrity", *Cell*, 2009;138:830–2.)

7.10 X-LINKED RECESSIVE INHERITANCE

X-linked traits are caused by mutations in genes located on the X chromosomes. Most of these traits are inherited in an X-linked recessive manner. Even though these traits run in families, the mode of inheritance does not fit the simple Mendelian rules of inheritance that we talked about in Chapter 1. They are most often seen in males, but occasionally occur in females.

One of the most famous examples of X-linked inheritance is the family of Queen Victoria of the United Kingdom. In her family, hemophilia turns up in ten men among her children, grandchildren, and great-grandchildren (Figure 7.11). The form of hemophilia present in this family is called hemophilia A. According to the National Hemophilia Foundation this is the most common form of hemophilia. About 1 in 5000 males born in the US has hemophilia A and about 30% of them represent new mutations rather than inherited forms of the disease. Hemophilia A is the result of defects in the Factor VIII gene, which is located near the bottom of the long arm of the X chromosome, and people with hemophilia have problems with inadequate blood clotting that causes problems with healing, or even sometimes survival, when injuries occur.

X-linked recessive inheritance happens when a gene is located on the X chromosome and a normal copy of the gene can compensate for the presence of a defective copy of the gene. The Factor VIII gene does not have a homolog on the Y chromosome, so men with one copy of the X chromosome have one copy of the Factor VIII gene and their phenotype will reflect the genotype of that one copy. Women with two copies of the X chromosome have two copies of the Factor VIII gene. Their phenotype will be normal if they have either two normal copies or one normal and one mutant copy of the Factor VIII gene; in the very, very rare instances when a woman has a defective Factor VIII gene on both copies of the X chromosome, she ends up with hemophilia.

In Queen Victoria's family the mode of inheritance tells us clearly that the Factor VIII gene must be located on the X chromosome, but some of the other genes that encode clotting factors are located on the autosomes so hemophilia does not always show X-linked inheritance.

If we look at Figure 7.12 we can see the X and Y chromosomes segregating in one branch

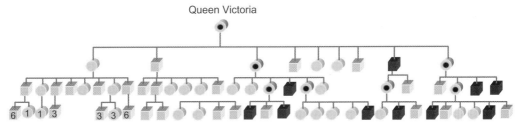

FIGURE 7.11   Hemophilia A in the royal families of Europe. This figure shows members of the first three generations of the descendants of Queen Victoria. Her children married into many of the royal families of Europe. Her descendants include British royalty King Edward VII, King George V, Kind Edward VIII, and King George VI, father of the current Queen Elizabeth. Other descendants shown here include royalty and nobility of many other European countries, including Wilhelm II, the last German Emperor and King of Prussia, Alexis, the last heir to the Russian throne, and Infante Juan, Count of Barcelona, father to the current King Juan Carlos of Spain. Queen Victoria's descendants include a son, three grandsons, and six great-grandsons with hemophilia (red squares), as well as many female descendants who are known or possible carriers (red circles). We expect that some of the white circles represent unidentified female carriers of the mutation while others represent non-carriers. Note that only individuals directly descended from Queen Victoria are displayed and that only the first three generations of her descendants are shown. To save space, limited information is provided for the branch of the family on the left in which no hemophilia has been observed; numbers on the left-most branch of the family indicate numbers of offspring of the indicated sex.

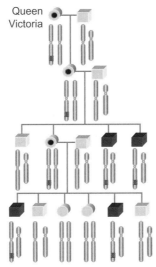

FIGURE 7.12   One branch of Queen Victoria's family showing the genotypes that go with the hemophilia phenotypes. Large gray chromosome – normal X. Small gray chromosome – normal Y. Gray chromosome with red band – copy of the X with the hemophilia mutation. We know that Queen Victoria and her daughter and granddaughter are carriers because they each have descendants with hemophilia. The two great-granddaughters shown in the youngest generation are presumed to not be carriers because no hemophilia turned up among multiple additional generations of their descendants, but we cannot know for sure since by random chance sometimes a woman who is a carrier produces only unaffected children. This happens when, by chance, each meiosis that results in a pregnancy happens to transmit the normal copy of the chromosome. Since each new meiosis is a new random draw from the two available copies, the chance that the next meiosis will use the normal copy is not affected by whether or not the previous pregnancies had used chromosomes with the normal or mutated versions of the gene.

of Queen Victoria's pedigree, with normal or mutant Factor VIII genotypes marked on each copy of the X chromosome. If we just look at the Factor VIII genotype for each individual we can predict their phenotype. Box 7.1 summarizes some features of X-linked recessive.

Why is Queen Victoria considered to be the origin of the hemophilia mutation that is segregating in this family? She had no family history of the trait, so it has been presumed that the mutation had arisen anew in herself or in the germline of her father. But when we consider that a mutation can pass from one carrier female to another without being detected, why do we not suspect that the mutation was passed to her through a line of carriers among her maternal ancestors? One of the most telling pieces of evidence is that Queen Victoria had lots of male relatives scattered through the preceding generations of her maternal lineage who all showed no signs of hemophilia. Does this mean that the mutation must necessarily have originated with her? No, there remains a small possibility that a mutation could have segregated through many generations of the female line of the family without ever happening to segregate into one of the men in the

---

BOX 7.1

HOW TO RECOGNIZE X-LINKED RECESSIVE INHERITANCE

Underlying concepts:

- Females have two copies of an X-linked gene; presence of defects in both copies results in the trait, but a second good copy can protect against a defect on the first copy, as for autosomal recessive inheritance.
- Males have only one copy of an X-linked gene; there is no second copy to cover for the defect present on the only copy of the X.

Rules for what we will see in an X-linked recessive pedigree:

- Affected males cannot pass the trait along to their sons.
- Most affected individuals are male.
- Affected females occur, but at a greatly reduced rate.
- Marriage of a carrier female with an affected male will mimic autosomal dominant inheritance by producing about 50% affected males and about 50% affected females.
- Affected males will inherit the gene through the maternal lineage.

- Female carriers will have about 50% affected sons and 50% carrier daughters.
- Daughters of affected males are obligate carriers.
- Because the defect can pass through multiple generations of unaffected carrier females, the trait may sometimes turn up in a male with no apparent family history, making it hard to tell whether the case represents autosomal recessive, X-linked recessive or complex inheritance.
- For some traits there is a carrier state in the heterozygous females, some detectable difference from the normal phenotype that may be very minor or difficult to detect.
- Heterozygous females can sometimes turn out to be affected in cases of selective inactivation of the X carrying the normal copy.
- Risk to one child is not affected by how many prior births in the sibship have the trait.
- Risk can be modified by environmental factors, mosaicism, errors in meiosis, de novo mutation, non-penetrance or age-related penetrance.

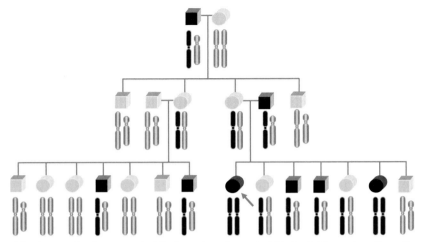

FIGURE 7.13   Inheritance of red–green colorblindness in a theoretical kindred. The proband's mother is a carrier and her father has the trait. She and her sisters each had to receive a defective copy of the color opsin gene from their father, but only some of them also received a defective copy from their mother. The branch of the family on the left looks like a classical X-linked recessive family, but the branch on the right shows us an alternative pattern for inheritance that we also have to watch for when we suspect X-linked recessive inheritance.

family, but it is most probable that the mutation arose in her or in the germline of a closely related ancestor.

As a general rule, when looking at a newly identified carrier with affected offspring and no family history, we have to balance the evidence for a new mutation against the probability that a female line of carriers could have been passing the mutation silently along through some generations without male offspring and other generations in which the meiosis did not happen to allocate an affected allele to any of the boys. For autosomal dominant inheritance it is easy to tell when a new mutation turns up in a family, but often we end up not being able to tell for sure unless we can do genetic testing that identifies the first point at which the mutation appears in the family with the prior generation lacking the mutation.

Sometimes a woman turns up with an X-linked trait. How can this happen? In the case of the X-linked recessive trait red–green colorblindness about 8% of men are red–green colorblind and about 0.6% of the women have the trait. We most often see women with this trait if their mother is a carrier and their father has the trait. In Figure 7.13 we see an example of a red–green colorblind man who married a woman who is a carrier for red–green colorblindness. Note that all of her daughters inherited a defective copy from their father, and half of her daughters received a defective copy from their mother. Thus the daughters experience the same 50% risk as their brothers based on the chance that they would inherit either a good copy or a defective copy of the color opsin gene from their mother.

7.11 X-LINKED DOMINANT INHERITANCE

This section is intended to deal with the inheritance of dominant X-linked mutations. In theory this should be an easy thing for us to do. After all, the rules for dominant inheritance predict that getting one copy of the X chromosome, one with the mutation, would result in having the trait. It seems like it should be a very simple situation, but actually it is not.

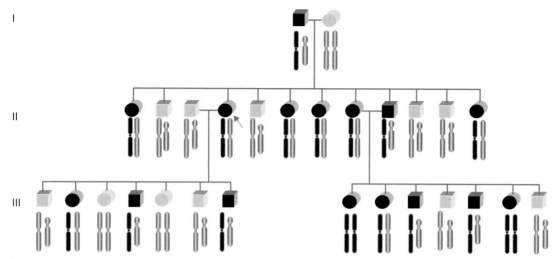

FIGURE 7.14   X-linked dominant inheritance pedigree showing X and Y chromosomes. Dark chromosomes contain the defect. Whether or not someone is affected (dark symbol) is a direct consequence of whether or not that individual received at least one dark chromosome.

Let's take a look at a pedigree for an imaginary trait that we will call pedigica imperfecta (PM) which causes an inability to complete the writing of chapters in textbooks. As shown in Figure 7.14, when a male in generation I, carrying the dominant PM defect, mates with an unaffected (pm/pm) female, all of their daughters develop writer's block while their sons are blissfully unaffected. This is true because daughters always receive the father's X and sons always receive their (in this case wild type) X from their mother. Conversely, as shown in generation II (on the left), when a male carrying the wild type pm allele mates with an affected but heterozygous (PM/pm) female, half of both their sons and daughters become unable to just finish writing chapter 7 and move on to chapter 8. If we look at the branch of the family on the right, we see that the heterozygous (PM/pm) daughter who marries a man with the dominant PM genotype, all of their daughters and half of their sons will be affected. Why are all of the daughters affected in the branch on the right but not in the branch on the left? The daughters on the left can only get a

PM-bearing X chromosome from their mother, and she is heterozygous PM/pm, while the daughters on the right all get a PM-bearing X from their father, whether or not they receive a PM-bearing X from their mother. Since the PM/PM and PM/pm genotypes are both affected, all girls in generation III on the right will be affected.

In generation III we find two girls who have the PM/PM genotype. Might we then predict a more severe phenotype? Might such an individual be one who would be unable to ever get started writing in the first place? For many human traits that are classified as dominant (but certainly not all), we may see a difference in severity between the heterozygotes and those who are homozygous for the causative allele. Clearly for the PM trait the worst that we expect is some issues relative to writing, but in fact sometimes the phenotype of the homozygote is not just different in severity but may even involve some different features not seen in the heterozygote. Nonetheless we predict that lethality in PM/PM individuals is unlikely to be an issue (even if the editors of PM individuals

---

BOX 7.2
_____

HOW TO RECOGNIZE X-LINKED DOMINANT INHERITANCE

Underlying concepts:

- Females have two copies of an X-linked gene; presence of defects in just one copy can cause the trait.
- Males have only one copy of an X-linked gene, and if that copy has the defect then they will have the trait.

Rules for what we will see in an X-linked dominant pedigree:

- Affected males cannot pass the trait along to their sons, and they did not inherit it from their father.
- Both males and females are affected.
- Marriage of an unaffected female with an affected male will lead to no affected sons and all affected daughters.
- Marriage of a carrier female to an unaffected male will lead to about 50% affected males but no affected daughters.

- Marriage of a carrier female with an affected male will lead to about 50% affected males, but all daughters will be affected.
- Affected males will inherit the gene through the maternal lineage.
- Daughters of affected males are affected.
- Because the trait will turn up any time one defective copy is present, inheritance can be seen continuously across multiple generations.
- Heterozygous females can sometimes turn out to be unaffected in cases of selective inactivation of the X carrying the defective copy.
- Risk to one child is not affected by how many prior births in the sibship have the trait.
- Risk can be modified by environmental factors, mosaicism, errors in meiosis, de novo mutation, non-penetrance or age-related penetrance.

---

sometimes want to throttle them) since the PM males survive and live healthy (if somewhat word-free) lives. But for some traits this difference will be so significant that we might miss detecting the difference in severity altogether if the homozygous mutant state is lethal during early embryogenesis, and thus we do not even see a live birth.

The problem with moving from such a hypothetical case to a bona fide human disease is that there are less than ten diseases that show this type of inheritance and all of them make poor examples, each for a different reason. For several of these diseases, such as Rett syndrome, CHILD syndrome, and Aicardi syndrome, the disease is lethal prior to birth in males. For others, such as Fragile-X syndrome

and Lugan–Fyrns syndrome, the phenotype is quite heterogeneous in females as a consequence of differences in which fraction of the inactivation X chromosomes are those that bear the mutation. The higher fraction of cells that inactivate the mutation-bearing X, the less affected the female. So in many of these cases the phenotype is highly variable, and only some of that variability has to do with differential X-inactivation. So we offer our list of rules for classifying X-linked dominant inheritance (Box 7.2), but keep in mind that most traits that should be classified in this way may well show exceptions to those rules, a lot of variability, and perhaps even some missing genotypes if the trait is lethal in homozygotes. In fact, for most of the lists of rules for classifying mode of inheritance, there will always be exceptions to be found because so many traits are subject to environmental effects and differences in how the trait manifests when present on different genetic backgrounds. We also face ascertainment issues – how many times do we fail to see some genotypic combination and start building models that include the idea that one of the genotypes is dying? In some cases, the real issue is that the phenotype associated with the homozygote is so different that we are not even looking at those individuals. And in some cases the answer is even simpler than that – patients with that genotype are not coming to our clinic, perhaps because they are the most easily treated cases so they stay home at their local medical centers and do not make their way to the big academic medical centers where the studies are taking place!

In talking about X-linked inheritance, we chose color vision, Rhett syndrome, and DMD to make it clear that many of the traits that turn up with sex-linked patterns of inheritance don't actually have anything to do with sexual traits. In fact, many traits encoded on the X chromosome, such as hemophilia, muscular dystrophy, mental retardation, have nothing to do with determining sex. But of course, the X and Y chromosomes are the fundamental determinants of whether we end up male or female. So it's not surprising to find that the X and the Y contain genes critical to the determination of whether we turn out to be male or female, a topic that we will take up again in the next chapter.

Study Questions

1. Why would we expect the body to tolerate extra copies of the X chromosome more easily than it tolerates most of the autosomal aneuploidies?
2. Autosomal monosomies are universally lethal, so why is it actually not so surprising that X0 is one of the aneuploid genotypes that survive?
3. What is sex linkage?
4. What is a Barr body?
5. How many Barr bodies are observed in XY men?
6. How many Barr bodies are observed in XX women or XXY men?
7. What does it mean if there are three Barr bodies in a cell?
8. Fill in the symbols on the pedigree below to indicate affected individuals so that the pedigree shows an X-linked recessive pattern of inheritance.

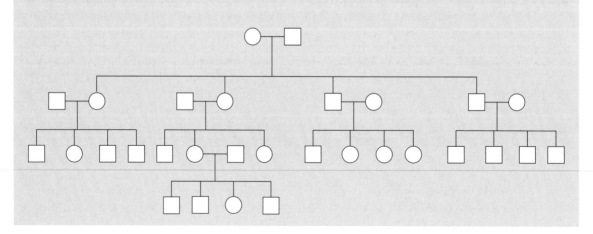

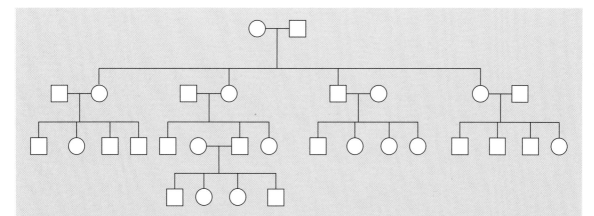

9. Fill in the symbols on the pedigree above to indicate affected individuals so that the pedigree shows an X-linked dominant pattern of inheritance .

10. What is the pseudoautosomal region?

11. What purpose does it serve?

12. If homology and exchange are critical for ensuring meiotic segregation, how can chromosomes that are as different as the X and Y chromosomes segregate from each other with the needed precision?

13. Every autosome is required in two copies for survival. Why can males survive with but one X chromosome?

14. Given that one chromosome is inactivated in XX females, and that males survive just fine with but a single X chromosome, why are X0 females phenotypically abnormal? In fact, the vast majority of X0 conceptions do not make it to term.

15. In a human female embryo, does XIST start showing up on the X chromosome before or after the eight-cell stage?

16. A woman is a carrier for a DMD mutation. She has inactivated almost all of the mutated chromosomes. Does this affect the chance that she could pass DMD along to her sons? Why or why not?

17. Men have one X and women have one active X. Does this mean that expression of X chromosome genes happens about half the level of autosomal genes?

18. Why is it hard to tell whether a trait is really X-linked dominant?

19. How can a woman end up with X-linked colorblindness if there are no translocations or biased X inactivation involved?

20. What are the features of a pseudoautosomal region?

Short Essays

1. When recombination between repeats creates an aberrant Y chromosome with two centromeres, it turns out that the instability of this structure increases as the distance between the centromeres increases. Why might this be the case, and why might the sex chromosomes be the only chromosomes in the human complement with which such a phenomenon can be observed and studied? As you consider these questions please review Chapter 6 and also read "The hows and Ys of genome integrity" by Heather E. Hall and R. Scott Hawley in *Cell*, 2009;138:830–2.

2. TSIX is an anti-sense sequence transcribed from the opposite strand from the same region that encodes XIST. Why might we expect to see TSIX transcribed from one copy of the X chromosome while XIST is transcribed from the other copy

of the X chromosome? What is the difference in the roles of XIST and TSIX? As you consider this question please read "New twists in X-chromosome inactivation" by Jennifer Erwin and Jeannie Lee in *Current Opinion in Cell Biology*, 2008;20:349–55.

3. As we think about the functional importance of the PAR there are two interesting questions that arise. What can comparison across species tell us about how important it is that particular specific genes are the ones located within that region? As you consider the answer to this question please read "The human pseudoautosomal region (PAR): Origin, function and future" by A. Helena Mangs and Brian J. Morris in *Current Genomics*, 2007;8:129–36.

4. Rett syndrome is an X-linked dominant trait, but some women turn out to be carriers instead of affected individuals. How can this happen? As you consider this question, please read "X chromosome inactivation in clinical practice" by Karen Helene Orstavik in *Human Genetics*, 2009;126:363–73.

Resource Project

There are small differences in names of genes that help to clue us in when we are looking at information on a human gene in the literature. Go to the Human Genome Nomenclature Committee online and look up XIST and XIC. Then go to the Mouse Genome Database and look up the same genes. Write a paragraph on the difference between the human and mouse gene symbols. Predict what the mouse gene symbol would be for the hypothetical PM gene that was described in this chapter and describe how you arrived at it.

Suggested Reading

Articles and Chapters
"Fragile X syndrome" by Kathryn B. Garber, Jeannie Visootsak, and Stephen T. Warren

in *European Journal of Human Genetics*, 2008;16:666–72.

"The gift of observation: An interview with Mary Lyon", interview by Jane Gitschier in *PLoS Genetics*, 2010;6:e1000813.

"The hows and Ys of genome integrity" by Heather E. Hall and R. Scott Hawley in *Cell*, 2009;138:830–2.

"The human Y chromosome: Rumors of its death have been greatly exaggerated" by R. Scott Hawley in *Cell*, 2003;113:825–8.

"Different patterns of X inactivation in MZ twins discordant for red–green color-vision deficiency" by A. L. Jorgensen, J. Philip, W. H. Raskind, M. Matsushitat, B. Christensen, V. Dreyer, and A. G. Motulskyt in *American Journal of Human Genetics*, 1992;51:291–8.

"X chromosome inactivation in clinical practice" by Karen Helene Orstavik in *Human Genetics*, 2009;126:363–73.

"Cytogenetics fmilial nonrandom inactivation linked to the X inactivation centre in heterozygotes manifesting haemophilia A" by Maria Patrizia Bicocchi, Barbara R. Migeon, Mirella Pasino, Tiziana Lanza, Federico Bottini, Elio Boeri, Angelo C. Molinari, Fabio Corsolini, Cristina Morerio and Maura Alquila in *European Journal of Human Genetics*, 2005;13:635–40.

"Infertility" in *Is It In Your Genes?* by Philip R. Reilly, pages 13–16 (2004, Cold Spring Harbor Laboratory Press).

"Cytogenetics of infertile men" by E. van Assche and colleagues, in *Human Reproduction*, 1996;11(suppl. 4):1–26.

"Rett syndrome: New clinical and molecular insights" by Sarah L. Williamson and John Christodoulou in *European Journal of Human Genetics* 2006;14:896–903.

HOW GENES CONTRIBUTE TO COMPLEX TRAITS

Sex Determination: How Genes Determine a Developmental Choice

The Human Genome. DOI: 10.1016/B978-0-12-333445-9.00008-3

THE READER'S COMPANION:
AS YOU READ, YOU SHOULD
CONSIDER

- How many separate traits make up the characteristic that we label with the single term "sex".

- How many different ways we can classify what someone's sex is.

- What the X and Y chromosomes have to do with sex determination.

- What the rest of the genome has to do with sex determination.

- Which initial developmental events determine what someone's sex will be.

- What the "default" sex is in the absence of a directed signal to become a particular sex.

- What happens if we move the SRY gene off the Y chromosome onto the X chromosome.

- What happens if we eliminate the SOX9 signal that is downstream of SRY.

- How loss of the androgen receptor can result in external sex reversal.

- How other hormones can play a role in sex determination.

- What evidence there is that sex, gender, and orientation are separate phenomena.

- What defines someone as male or female.

- What evidence there is for a biological basis for gender identification.

- What evidence is there that there might be a "gay gene" on the X chromosome.

- How self-identification for a trait can affect a study of that trait.

8.1 SEX AS A COMPLEX DEVELOPMENTAL CHARACTERISTIC

With her full bust, small waist, broad hips, delicate facial features, and female genitalia, the Parisian fashion model fitted a very classical view of femininity. She was a beautiful woman who was about to be married, a seemingly non-controversial move for a woman at the beginning of the twentieth century. When she sought to have some "tumors" removed before the wedding and the doctors found them to be undescended testicles, they reclassified her as a male and informed her that her sexual attraction to men ("such as her fiancé?" we find ourselves musing) made her a male homosexual.

A gonadal definition of sex such as this has been only one of a variety of evolving medical views of intersex individuals, who show some characteristics of each sex or intermediate development of external sexual anatomy. Most individuals fall into one of two categories, clearly male or clearly female with complete consistency of genetic, gonadal, anatomic, and psychological aspects of sex within any one individual; however there clearly is a complex gradient that runs from male to female occupied by many different varieties of people who do not fall neatly into one of the two standard sexual definitions.

Whenever a child is born, it seems that there would be a simple answer to the question "Is it a boy or a girl?" However, for a surprising number of people in this world, the answer is unclear, or the answer may even change over the course of a lifetime as new information comes to light or as medical views change. As we see in the example above of a woman who suddenly found herself being told that she was a man, sometimes efforts to answer the question are perplexing because the answer may be different depending on what aspect of the person you ask about. More details about the Parisian model and others with intersex phenotypes can be found in the writings of Alice

Domurat Dreger, who offers many insights into the sexual complexity of people who occupy the gradient in the middle between the conventionally defined male and female. We in the field of genetics find that many of the cases that hold sociological and historical interest for Dr Dreger also offer potential insights into the role of genes in the determination of different aspects of sex. In this chapter, we will tell you about some of the genes that determine whether we will look or feel female, including the gene that we know about today that might have led the doctors to tell this surprised woman that she was "really" a homosexual male, and we will talk about some of what is known (or mostly not known) about underlying genetic contributions to gender identity and sexual orientation.

Defining Features of a Complex Developmental Phenotype

As biologists, we spend much of our time trying to understand how the cells of the developing organism make choices, such as whether or not to become a nerve cell or a muscle cell. As geneticists, we find ourselves asking about how genes play a role in this process. One of perhaps the most fascinating of such processes is the pathway of sexual differentiation – in other words: how does the embryo make the developmental "choice" between producing a male fetus or a female fetus, and thus between producing a man or a woman?

In this chapter, we are going to talk about some of the key genes that control sexual differentiation, by which we are referring to several different things: gonadal sex (whether you have ovaries or testes), somatic sex (whether you have male or female body characteristics; Table 8.1), sex role (gender) identification, and sexual orientation.

In talking about sexual characteristics we use the term "primary" to refer to those characteristics directly related to reproduction, which is to say the reproductive organs and associated tissues. We use the term "secondary" to refer to other sexually dimorphic characteristics that are not centrally involved in reproduction, such as issues of body shape, muscle strength, and hair distribution. Many of the secondary sexual characteristics might be thought of as those that are involved in attracting mates, surviving to reproduce or protecting and caring for offspring. You will notice that many of the secondary sexual characteristics are those that develop after birth as the individual matures, while the gonadal and primary characteristics are those the individual is born with. While we can now explain the development of primary and secondary sexual characteristics solely in terms of increasingly well-understood biological pathways – a genetic understanding of sex role identification and sexual orientation proves far more elusive.

We are fascinated by the complexity of sexual differentiation, both in terms of the well-studied biology of building female or male bodies and in terms of the complexity of creating masculine or feminine psyche and determining sexual orientation. Here we find that what most people think of as a rather simple binary process ("It's a boy" or "It's a girl") is actually replete with complexity, and occasionally some ambiguity. From the biological perspective, it is curious that the key "genetic switch" in sex determination acts only once at a very early point in fetal development and solely determines whether structures called indifferent gonads (which are possessed by all fetuses) become testes or ovaries. From that point on it is the hormones produced by the gonads that determine biological sex; these hormones act by mediating an extremely complex series of events that activates genes required to build male structures or by allowing the developmental of female structures. As sexual developmental continues, these hormone-mediated processes continue to govern the physical aspects of sexual differentiation. You may not be surprised, or perhaps you will be, by our

TABLE 8.1   Different classifications of biological sexual characteristics

	Conventionally male	Conventionally female
GENETIC		
Karyotypic	XY	XX
Genotypic	SRY positive	SRY negative
GONADAL		
Gonadal	Testes	Ovaries
SOMATIC		
Primary somatic	Penis, scrotum	Vagina, cervix, uterus, fallopian tubes, clítoris
Secondary somatic	Face and body hair Narrower hip structure Greater upper body strength Ability to rapidly gain muscle mass	Breasts Little face and body hair Broader hip structure Less upper body strength Less ability to add muscle mass Increased body fat Menstrual cycle

discussion of the phenotypes exhibited by mutations in genes that produce or recognize these hormones or whose activity is controlled by hormones. Such mutations often result in people in which the sex is "reversed" (for example XY females such as the Parisian model) or even ambiguous.

From a behavioral and psychological perspective, sexual differentiation is even more intriguing. To what degree is biological sex tied to gender identification? There are XY males who identify themselves psychologically as women and XX individuals who identify themselves as men. What processes facilitate the usual congruence of biological sex and gender identity, and what events (genetic or otherwise) create non-congruence? We can ask similar questions about sexual orientation. Much has been said about the existence of so-called "gay genes" – but to what degree is sexual orientation genetically determined, and if so how is it determined?

Thus, a discussion of the genetics of "sex" provides us an opportunity both to widely explore questions within our "comfort zone", such as: how does the SRY gene act as an initial genetic switch that controls development of the indifferent gonad, and how do hormones act to control gene expression during development? What are the phenotypes of mutations in sex-determining genes and what do those phenotypes tell us about how these processes work?

But perhaps more importantly, such a discussion also allows us to leave that same comfort zone as we try to understand the effects of genotype and experience on complex processes such as gender identification and sexual orientation. Sexual identification (gender) and sexual orientation define our sex roles and our choices in sexual partners. They are independent phenomena determined separately from whatever determines our gonadal and somatic sexual characteristics. Later in this chapter, we will talk about how sex, gender, and orientation are related to each other. Ultimately, in actual practice, sexual categories usually end up being defined socially and not biologically, and we will discuss how subjective classification of phenotypes confounds our ability to sort out the genetic underpinnings of a phenotype.

The Biological Foundations of Sex Determination

The question we want to explore revolves around the degree to which each of these components of sex in human beings is genetically determined. Just how do our genes determine our sex, and to what extent do genes determine our sexual behaviors? To begin with, we will consider in detail some of the peculiar properties of the sex chromosomes in humans. Then we will address the more controversial issues of the role of genes in establishing sex roles or sexual orientation. But let's begin at the beginning by defining the various levels of biological classification of the different components of human sexual differentiation shown in Table 8.1.

Karyotypic Definition of Sex

It is well established that biological sex in humans is determined by the presence or absence of a Y chromosome – the XY individuals are male because they carry a Y chromosome and XX individuals are female because they do not have an X chromosome. As we will discuss below, we know that the critical factor in determination of biological sex is the presence or absence of a Y and not the number of X chromosomes below, since XXY and X0 individuals who arise by nondisjunction are male and female respectively.

Genotypic Definition of Sex

The most straightforward way to evaluate what sex someone will turn out to be is to evaluate the genotype of one gene on the Y chromosome, which is known as *SRY* (*Sex-determining Region on the Y*). This gene, which is also sometimes referred to as *TDF* (*Testis Determining Factor*), determines whether a developing embryo will makes testes or ovaries, the first step in sexual development. As we will discuss below, many other genes are also involved in

sex determination, and there are rare individuals in which presence or absence of SRY is not the primary switch that determines sex, but for most individuals, SRY is "it."

The Gonadal Definition of Sex

If your *gonadal sex* is male, you have testes. If your gonadal sex is female, you have ovaries. Some who write about sexual characteristics include gonadal sex characteristics in with the primary sexual characteristics while others list them separately.

Primary Sexual Characteristics

Primary sexual characteristics are those characteristics other than the gonads that are directly required for reproduction. Male primary somatic sexual characteristics are the penis and the scrotum, all of which allow a male to make and deliver sperm. Female primary sexual characteristics are the vagina, uterus, fallopian tubes, clitoris, cervix, and the ability to bear children.

Secondary Sexual Characteristics

Secondary sexual characteristics are those sexually dimorphic characteristics that are not directly involved in reproduction. For males, secondary characteristics include facial and chest hair, increased body hair, pelvic build (lack of rounded hips), upper body muscular build, and the ability to generate muscle mass at a faster rate than the female. For females, secondary sex characteristics include relative lack of body hair, thicker hair on the head (in some cases), rounded hips/figure, a decreased ability to generate muscle mass at a fast rate, decreased upper body strength, breasts, ability to nurse children, a menstrual cycle, and increased body fat composition. There are, of course, exceptions to any efforts to use a list of features to classify people into the conventionally defined sex categories. For instance, not all women succeed in breastfeeding their infants, even if they otherwise fit the conventional definition of female.

Having defined biological sex in terms of chromosomes, genes, hormones, and organs, let's begin dissecting the process of sexual differentiation.

8.2 WHAT DO THE X AND Y CHROMOSOMES HAVE TO DO WITH SEX?

As we noted in Chapter 7, a genetically normal human being contains 23 pairs of chromosomes, the 22 pairs called autosomes and one pair called sex chromosomes (the X and the Y). A normal female possesses two X chromosomes, whereas a normal male possesses one X and one Y chromosome. The finding of XXY males and X0 females convinced geneticists that sex in humans was determined solely by the presence or absence of a Y chromosome (Table 8.2), and that it was not determined by the number of copies of the X chromosome. This is especially interesting because, as you will recall from the Appendix in Chapter 6, some organisms, like Drosophilia, do determine sex based on numbers of copies of the X chromosome.

Although XXY individuals are clearly male, they exhibit a trait known as Klinefelter syndrome whose most common characteristics or symptoms are listed in Table 8.2. This table also describes the most common characteristics of X0 Turner syndrome women in terms of secondary sexual characteristics. We provide these phenotypes a bit warily because not all XXY males or X0 females will exhibit all of these characteristics, and those that do show them may exhibit them to varying degrees. The one certain and common characteristic of both traits is sterility. The cells in the germline are less forgiving of changes in X chromosome dosage than are somatic cells. Thus XXY male germline cells die at puberty and

TABLE 8.2   Features of Klinefelter and Turner syndromes

	Klinefelter syndrome	Turner syndrome
GENETIC		
Karyotypic	XXY	X0
Genotypic	SRY positive	SRY negative
GONADAL		
Gonadal	Testes often of reduced size after puberty Reduced levels of testosterone No sperm produced	Ovaries (greatly reduced in size due to loss of oocytes)
SOMATIC		
Primary somatic	Penis, scrotum	Vagina, cervix, uterus, fallopian tubes, clítoris
Secondary somatic	May show varying degrees of somatic "feminization," including breast development (gynecomastia) and a feminine pattern of hip development, depending on whether or not therapeutic intervention with testosterone is used to ameliorate feminization	May display reduced stature May display less breast development

X0 female germline cells die during fetal development.

The fact that XXY individuals are male and X0 individuals are female offers compelling argument that sex in humans is determined by the presence or absence of the Y chromosome, and not by the number of copies of the X chromosome. However, variant Y chromosomes have been found that were missing quite a bit of material but were still capable of determining maleness. All that seemed to matter in terms of being able to determine maleness was a small region on the short arm of the Y chromosome. These data demonstrated clearly that it is not simply the presence of the Y chromosome that creates a normal male but rather a small amount of genetic material now known to be a single gene located on the Y chromosome called the *SRY* gene. The product of the SRY gene promotes the body to develop male genitals.

8.3 SRY ON THE Y: THE GENETIC DETERMINANT OF MALE SEXUAL DIFFERENTIATION

SRY Initiates Male Sexual Differentiation

Several lines of evidence argue that the SRY gene is both necessary and sufficient by itself to initiate male sexual differentiation. First, XX human beings are occasionally found in which a piece of the Y chromosome bearing the SRY gene has been appended (or translocated) onto the tip of one of the two X chromosomes, creating a chromosome that we will call X(SRY+). Suppose a sperm bearing that translocated X(SRY+) chromosome fertilizes an X-bearing egg? Such XX(SRY+) individuals will develop as a male but will suffer from testicular atrophy or, more simply put, small testes and thus sterility (Figure 8.1). (Why sterility? This is because

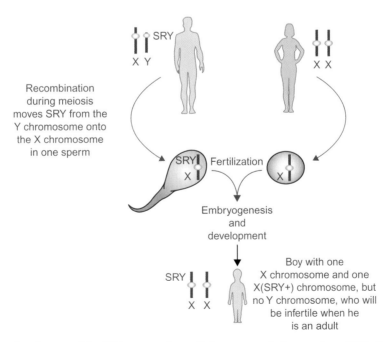

FIGURE 8.1   Transfer of a copy of the SRY gene creates an X chromosome that can make an XX individual be male.

it is not possible to have two Xs present in the male germline and still be fertile; the mere presence of another X chromosome acts almost like a poison to the germ cells at puberty and kills them during meiosis.) Hence, this individual is unable to produce healthy and happy living sperm. Nevertheless, regardless of the two XXs present, this individual is a male! He has male gonads, he has male genitals, and the rest of his primary and secondary sexual characteristics are male.

The second line of evidence that SRY causes an individual to become male comes from the finding that several XY *females* differ from normal males only by mutation of one base pair within the SRY gene (Figure 8.2). These women possess

A girl who inherited a defective Y chromosome that lacks the ability to make SRY protein will grow up to be an infertile woman

X Y (no SRY)

FIGURE 8.2   A defect in the SRY gene results in a girl with a male karyotype.

a normal or near-normal outward appearance, a cervix, a uterus, and normal vagina. However, because oocytes require two functional X chromosomes, oocyte death occurs during fetal development and, as a result, the ovaries are rather small and such women are sterile.

Finally, to prove that the SRY gene alone is responsible for male gonadal sex, researchers used some rather clever tricks of DNA manipulation to insert a mouse SRY gene, and just the SRY gene, into the genomes of XX mouse embryos. (This experiment works to answer this question because mice and humans determine sex in exactly the same way.) These XX mouse embryos, which would have become female if they had not had a copy of the SRY gene added, developed into healthy but sterile male mice without having any of the sequences of the Y chromosome except SRY (Figure 8.3).

Why do we need to ask this question using the mouse if we already know that an XX human becomes male if the SRY gene has transferred to the X chromosome? In fact, when genes move between chromosomes by natural mechanisms, the process involves moving not only the gene itself but also adjoining

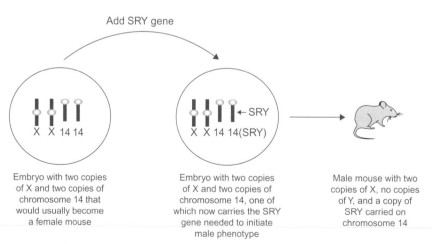

Add SRY gene

X X 14 14

X X 14 14(SRY)   ← SRY

Embryo with two copies of X and two copies of chromosome 14 that would usually become a female mouse

Embryo with two copies of X and two copies of chromosome 14, one of which now carries the SRY gene needed to initiate male phenotype

Male mouse with two copies of X, no copies of Y, and a copy of SRY carried on chromosome 14

FIGURE 8.3   Addition of the SRY gene to an XX mouse embryo results in an XX with SRY associated with one of the chromosomes that does not have to be either the X or Y chromosome. Here we show it having become part of chromosome 14, but it could have joined with any of the chromosomes.

sequences. So we cannot be sure that the phenotype is not due to something in the adjoining regions of sequence rather than the gene itself. When we add the SRY gene into the XX mice we are able to control exactly what sequence is introduced and to be sure that SRY is the only new thing that has been added.

So the primary determinant of gonadal sex determination is the SRY gene, but what does the SRY gene do, and how does it do it? We will explore that question next.

The SRY Gene Causes the Indifferent Gonads to Develop as Testes

When you were first conceived, you began life with a pair of indifferent gonads. The term *indifferent gonads* is self-explanatory: the fetus' organs are literally "indifferent" to becoming either ovaries or testes. They are equally willing to become one or the other, depending on whether the fetus' germ cells do or do not carry the SRY gene. The presence of the SRY gene during the seventh to eighth week of fetal development gives the instruction to the indifferent gonad that it should develop into a male gonad. Note that the SRY gene acts only during this brief moment in development and is inactive the rest of the time. Moreover, it acts only in a certain specific subset of the cells in the indifferent gonad. The expression of SRY in those cells is, however, sufficient to induce the indifferent gonads to become testes, which is the step that initiates all of the rest of the subsequent male development processes.

The SRY gene turns on for a brief period of time in a minor fraction of fetal cells and is then done for the rest of that individual's lifetime and not heard from again until the next generation. We are reminded of the lines of Shakespeare's *Macbeth*:

Life's but a walking shadow, a poor player
That struts and frets his hour upon the stage,
And then is heard no more.

Macbeth concludes that this moment on the stage produces a tale "signifying nothing," but in fact, SRY's brief turn on the biological stage to determine who among us will be male is truly significant. We see a manifestation of this significance every time someone's first question about a newborn baby turns out to be, "Is it a boy or a girl?"

SRY performs its function of inducing testicular development by enhancing the transcription of an autosomal gene called SOX9. The protein produced by SOX9 functions as a transcription factor that regulates a large battery of hundreds of genes required to induce male development, including genes required for the formation of a set of testicular cells referred to as Sertoli cells. The Sertoli cells facilitate the proper development of male germ cells both early in development and later on during active spermatogenesis. By facilitating the commitment of cells within the indifferent gonad to become Sertoli cells, both SOX9, and the genes it regulates, stimulate testicular development.

Given the critical role of SOX9 as the primary target of SRY, it is perhaps not surprising that XY individuals who are *heterozygous* for mutations in the SOX9 gene display sex reversal, which is to say that the sex they display is the opposite of that predicted by their karyotypic sex. In this case, they are XY females. In the absence of two good copies of the SOX9 gene, the cells in the indifferent gonad simply cannot produce sufficient SOX9 to induce maleness, even in the presence of functional SRY. (That SOX9 performs additional functions in the soma is shown by the fact that both XY and XX individuals heterozygous for a loss-of-function mutation in SOX9 also manifest a skeletal malformation syndrome called campomelic dysplasia – *the same gene often has multiple functions in a variety of cell types during development*.) Similarly, XX individuals that carry an extra copy of SOX9 often develop as males! Thus, SRY does its job by up-regulating the expression of SOX9, which in turn regulates the genes required for testicular development,

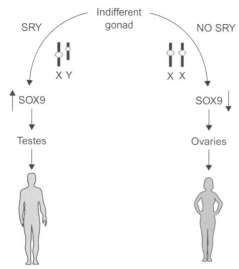

FIGURE 8.4   The default state for the indifferent gonad that receives no signal is to become an ovary. Only if the SRY signal is received will it become a testis.

and as we shall see below, other functions as well.

Without SRY, the Indifferent Gonad Becomes an Ovary

So we arrive at a very important concept: if the SRY gene is not expressed, the cells of the indifferent gonad will follow a separate path and the indifferent gonad will develop as an ovary. It used to be thought that ovarian development was a default state, the state that occurs if no signal is received from SRY via the elevation in the expression of SOX9 (Figure 8.4). However, it is now clear that there are a number of genes that function to repress SOX9, genes that are required for proper ovarian development. In mice, mutations in these genes cause partial masculinization of the developing ovary in XX females.

In humans, a recessive mutation in one of these genes downstream of SOX9, called RSPO1 (R-spondin1) is sufficient to create XX males. An RSPO1 mutation has turned out to be the underlying genetic lesion responsible for four

XX male brothers in a consanguineous family in Italy (Figure 8.5). In each of these cases, the indifferent gonads become testes even in the absence of SRY! (We might also note, at least in mice, over-expression of at least one of these downstream genes causes ovarian development in XY animals.)

Thus it becomes clear that becoming an ovary isn't just a default state. Rather it requires the repression of some basal level of expression of SOX9, just as becoming testes requires the over-expression of SOX9.

8.4 THE ROLE OF HORMONES IN EARLY DEVELOPMENT

Gonads Dictate the Next Step in Development of Somatic Sexual Characteristics

Unlike gonadal sex, somatic sex (the sexual characteristics of the body) is independent of the presence or absence of the SRY gene and the Y chromosome. It is determined by the hormones that are produced by the developing gonads. You began life with two sets of reproductive "plumbing": the Müllerian ducts destined to become the female reproductive tract: uterus, primitive fallopian tubes, ovaries; and the Wolffian ducts destined to become the male reproductive tract: vas deferens, epididymus, prostate gland, seminal vesicles (Figure 8.6). You also possess a small bud of tissue called a genital tubercle that will form either a penis or a clitoris. This is to say that, the indifferent gonad starts out as neither and then proceeds depending on whether or not it gets a signal. Nature's first choice for the plumbing is to make both and then get rid of the one that it is not going to use!

As noted above, in males the expression of the SRY gene, and the concomitant up-regulation of SOX9 during the eighth week of development, causes the indifferent gonads to

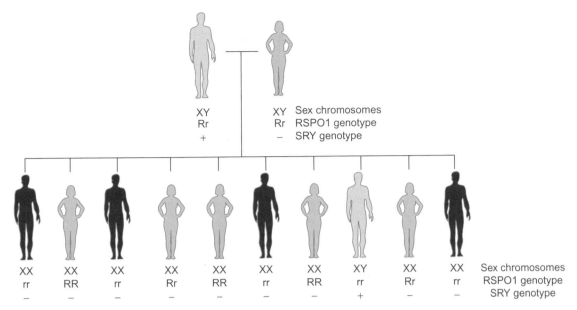

FIGURE 8.5 One branch of a large Italian family with four cases of sex reversal. Men shown with dark blue icon have two X chromosomes, no Y chromosome, and a defect that eliminates activity from the gene product made on both copies of the RSPO1 gene. Note that the only individual in the family with SRY is the male with the Y chromosome, and that all of the other males possess two X chromosomes that do not have SRY. (After Parma, P., Radi, O., Vidal, V., *et al.*, *Nature Genetics*, 2006;38:1304–9.)

become testes and begin secreting androgen (testosterone) and the Müllerian inhibitory factor (MIF), which is sometimes referred to as the Anti-Müllerian Hormone. The expression of MIF, which is a direct consequence of the up-regulation of SOX9, causes the regression of the Müllerian ducts. In females, the indifferent gonads become ovaries and produce estrogen. During the 13th week of development, the Wolffian ducts degenerate (due to the absence of high levels of testosterone) and the Müllerian ducts develop. The relative levels of estrogen or testosterone also determine the development of the primary sexual sex characteristics. The high levels of testosterone produced by the testes cause the genital tubercle to develop into a penis, and a scrotum is formed. In the absence of testosterone, the same tissues will form a clitoris and a vagina. Notice that, once again,

as with the gonadal differentiation, this takes place in response to a lack of signal.

Somatic sex manifestations can be altered in ways that are not the result of the infant's genes. Some developmental events are influenced by the uterine environment. As we will discuss in detail below, both testosterone and estrogen are secreted by the adrenal cortex in both sexes. Thus, if the mother has an otherwise benign adrenal tumor during pregnancy, her daughter might be born with masculinized genitalia.

Congenital Adrenal Hyperplasia and Ambiguous Genitalia

The cells of the adrenal cortex (a part of your adrenal gland) also produce low levels of both estrogen and testosterone. Sometimes

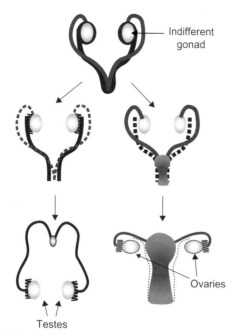

FIGURE 8.6 In the embryonic male, the indifferent gonads develop into testes, and the Wolffian ducts (blue) develop into tissues of the male reproductive tract while the Müllerian ducts (pink) shrink away. The testes descend to a position in the scrotum at the surface of the body where they are protected from the high temperatures within the body. In the embryonic female, the indifferent gonads develop into ovaries, and the Müllerian ducts (pink) develop into tissues of the female reproductive system while the Wolffian ducts (blue) shrink away.

as a consequence of overactivity of the adrenal gland during development or of a defect in hormone synthesis, high levels of either estrogen or testosterone can be produced by the adrenal cortex. Congenital adrenal hyperplasia is not an intersex condition for males, but it is for females. Thus a developing female fetus could be exposed to high levels of both testosterone (from the adrenal cortex) and estrogen (from the ovaries). The result is a mixture or confusion of developmental processes, resulting in a newborn whose genitals seem to be "a little of both," or in some cases may even appear to be clearly male. These cases of ambiguous genitalia are

quite disturbing to some parents and physicians, who may rapidly push to make the child's situation unambiguous at a very early age, before the child begins experiencing a wide array of sex-specific social interactions. Often surgeons seek to treat the situation as promptly as possible with plastic surgery, but some other medical specialties tend to prefer waiting before deciding about intervention. Treatment of these infants requires genetic evaluation to determine the sex chromosome composition and the presence or absence of ovaries and testes, surgical evaluation to determine the treatment most likely to produce a functional adult, and psychological evaluation and counseling of the parents.

The frequency of such congenital adrenal hyperplasia births (approximately 1 in 10,000) requires that we mention this trait. We also mention ambiguous genitalia (which can be as frequent as 1 in 2000) because it vividly makes our point that genitals and other external features of sex are determined by hormonal messengers and not by SRY gene on the Y chromosome. We are also well aware of the use of surgery and hormone treatment in the sexual reassignment of adult transsexual patients. Both of these cases should focus attention on the fact that the only step in primary or secondary sexual differentiation that is controlled by the SRY gene is the choice of testes or ovaries. The rest is determined through environmental events and a secondary set of steps directed by the hormones produced by testes or ovaries and other glands.

Hormones Carry Signals to Other Cells in the Body

As puberty begins, testosterone or estrogen will also determine the development of secondary sexual characteristics. The high levels of testosterone flowing through a male's body are responsible for his physically masculine appearance, whereas the high levels of estrogen flowing through a female's body are responsible for her physically feminine appearance.

A general definition of the hormone is a chemical messenger that is produced by one cell type and released into the bloodstream and received by a target cell with the intention of altering this target cell's pattern of gene expression. Testosterone and estrogen are steroid hormones. Steroid hormones include testosterone and estrogen. Testosterone is excreted from the testes and the adrenal cortex in the male, whereas estrogen is excreted by the ovaries and the adrenal cortex in the female. Actually, both sexes produce both hormones. However, there is much more testosterone than estrogen in males and much more estrogen than testosterone in females.

When sex hormones are excreted into the bloodstream, they circulate until they encounter the target cell where they are needed to carry out their purpose, which is telling the target cell to alter its pattern of gene expression. These target cells have receptors that wait for the needed hormone to float on by. When the receptors detect the presence of the hormone, they bind to the hormone and carry it through the nuclear membrane and into the awaiting nucleus. Once inside the nucleus, the hormone and the receptor complex bind to DNA regulatory elements and promote gene expression. The protein products of these testosterone- or estrogen-induced genes allow the cells and organs to execute sexual differentiation.

8.5 ANDROGEN RECEPTOR ON THE X: ANOTHER STEP IN THE SEXUAL DIFFERENTIATION PATHWAY

Mutations in the Gene that Encodes the Androgen Receptor

Imagine what would happen if a steroid hormone receptor in your body was not there or was not functional. Your hormones would continue to flow throughout your body, but when they arrived at the target cell, there would be no place in the cell for them to dock. If they don't dock with their receptor, the target cell cannot tell that the hormones are there and thus does not know that it needs to change which genes it is expressing and to change the levels of expression of some of the genes it is already using. In the case of sexual development, one of the key receptors is the *androgen receptor gene* (AR gene). It is encoded by a gene on the X chromosome, and loss-of-function alleles of the AR gene are referred to as *AIS mutations*. Because these mutations prevent the production of functional testosterone receptor, the phenotype of XY individuals with the AIS mutation is the result of a pattern of gene expression that has not been altered by signals from testosterone. The result is a trait known as *complete androgen insensitivity syndrome* (CAIS), sometimes also known as *testicular feminization* (TFM). CAIS is seen in approximately 1 in 20,000 live births.

In XY embryos with a CAIS mutation, the indifferent gonads receive the SRY signal and develop as testes while the Müllerian ducts regress in the presence of MIF. However, the cells of this embryo cannot sense the testosterone that is running around the body looking for androgen receptors. Instead, the somatic cells respond to the normal, low level of estrogen secreted by the adrenal cortex of both sexes, *and the embryo develops along a female pathway* (Figure 8.7). Consequently, the child at birth appears as a perfectly normal female. However, her vagina ends in a blind duct. The CAIS female has no cervix, uterus, or fallopian tubes. Instead of fallopian tubes, there are two fully developed but un-descended testes producing testosterone. These females are externally normal throughout childhood, puberty, and adult development, with the exception of a scarcity of underarm and pubic hair. Obviously, they will neither menstruate nor be able to bear children.

Given that such women are often detected as children or teenage girls, this is a serious issue in terms of how much information should be

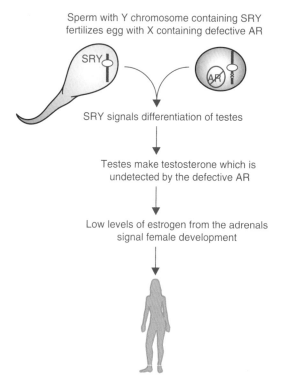

Sperm with Y chromosome containing SRY
fertilizes egg with X containing defective AR

SRY signals differentiation of testes

Testes make testosterone which is
undetected by the defective AR

Low levels of estrogen from the adrenals
signal female development

FIGURE 8.7   CAIS syndrome. Individuals with a defective androgen receptor have an XY karyotype. Their gonads are undescended testes, and their sexual anatomy is female, except they have no ovaries, cervix, or uterus. The phenotype can be either complete or partial, with the latter resulting in some sexual ambiguity.

provided during diagnosis and counseling, how it should be provided, and who should receive the information. Some girls were not told that they had the AIS mutation, even though their doctors and parents knew. In such cases, the news can come as a shock if discovered later as an adult. However, there can also be serious health repercussions to not knowing if you have an CAIS mutation, so withholding such information can be dangerous.

As noted above, the CAIS female also possesses a fully developed set of testes that are located internally above where the scrotum would normally be. These testes reside inside the body, existing at a higher temperature inside the body than would normally exist for testes that have descended into the scrotum. It is recommended that such a female have these testes removed as a young adult because of an increased risk of testicular cancer that can develop later in life as a result of the elevated temperature.

CAIS females are often considered quite attractive by contemporary standards, and they are often taller than the average woman. The health implications of CAIS are infertility, osteoporosis, and risk of testicular cancer. The cancer risk can be addressed by removing the testicles and use of woman hormone therapy can help reduce her risks of osteoporosis. Psychologically, these women are as stable and happy (or not) as women with two X chromosomes who are coping with fertility issues. They can be expected to live perfectly happy, normal lives and, when they so choose, become parents of adopted children or stepchildren.

Do CAIS women (with XY karyotype and no functional androgen receptor) have the same characteristics as SRY-negative women (with XY karyotype and no functional SRY protein)? No. Recall that the SRY-negative woman, in the absence of the SRY signal, has produced female gonads, which provide an estrogen-dominated hormonal environment. Although they are infertile because their meiotic processes needed two copies of the X chromosome, they have a full set of female anatomy. The CAIS woman, with an active SRY signal, has male gonads, only as much estrogen as the adrenal glands can supply, and no ovaries or uterus. Thus a SRY woman would be harder to identify without genetic testing, and a CAIS woman has cancer risk to deal with in addition to infertility.

So what we see from this step-by-step walk through of the first several steps in sex determination in humans is the requirement for at least four elements:

- a Y chromosomal signal;
- a sensing mechanism in the indifferent gonad to respond to the Y chromosomal signal;

- a hormonal signal produced by the gonads (androgens or estrogens);
- a set of sensors, androgen, and estrogen receptors, in the somatic tissues responding to the secondary signal coming from the gonads.

In fact, there are other genes involved in sex determination that can affect a variety of the secondary steps that take place in different cell types and tissues in different portions of the anatomy as the primary and secondary somatic characteristics emerge.

8.6 GENETICS OF GENDER IDENTIFICATION

Genetic, gonadal, and somatic sex are consistent for most human beings. Similarly, an individual's sex is most often consistent with their *gender* (how individuals identify with male and female sex roles) and their *sexual orientation* (attraction to same, different, or both sexes); however, different combinations of sex, gender, and orientation can occur. Examples of the disconnect that can occur among these three traits can be found most noticeably among homosexuals attracted to individuals of the same sex and transgendered or transsexual individuals who grow up feeling as if they are trapped in a body of the wrong sex. The situation is further complicated by the existence of intersex individuals who have some or all of the physical characteristics of both sexes, some individuals who are bisexual (are attracted to both sexes), some individuals who self-identify with different gender roles at different times or under different circumstances, and some individuals who grow up to decide that they are a different gender than the gender they were raised as.

The fact that sex, gender, and orientation can occur in different combinations suggests that these three traits could have some different underlying determinants, whether genetic or environmental, just as some key determinants of

gonadal and somatic sex are distinct. Although a number of the key elements leading to sex determination have been identified and turn out to be genetic, there is noticeably less known about the genetic components of gender or orientation, and some of what has been found is considered controversial.

Mechanisms of Gender Identification

The controversial issue of whether gender is biological or acquired has been debated over the last century. For a long time, it was argued that the primary determinants of gender were environmental and that a child would acquire the sex roles with which they were raised.

As we have already mentioned, some children are born with ambiguous genitalia or who are intersex individuals with some biological properties of each sex. Depending on the exact condition of the infant, the treatment of those children has often included "sexual correction," that is to say, surgical revision of the child's sexual anatomy, sometimes to recreate their anatomy to more closely resemble the anatomy usually expected for their genetic and gonadal sex, but sometimes instead to arrive at an external sexual anatomy that is different from their chromosomal sex.

Part of the medical argument that such surgeries should be done, and done early, arises from reports that reassignment works well. Early reports suggested that, where the parents are comfortable with the outcome of early revisions, the children will usually identify properly with the genders they have been assigned.

However, things may not be that simple. Two different schools of thought have developed – that a child will take on the sex roles and gender identity with which he or she was raised, or that there are biologically inherent determinants of sex roles and gender identity that cannot be reprogrammed by raising a child as if he or she were the opposite sex (Box 8.1). Perhaps the truth lies somewhere in between. If we look to other types of studies in the scientific literature,

BOX 8.1

THE GIRL WHO WAS REALLY A BOY

A dramatic case that suggests a biological basis for gender is that of a male child who was "reassigned" as a female after irreversible genital damage during circumcision. During childhood, this child with a Y chromosome, who dressed in dresses and had a collection of "girl" toys such as dolls, was presented in the literature as evidence that surgical reassignment of sex would result in the child's successful acceptance of his newly assigned gender role. Although reports in the literature repeatedly presented the view of a normally adjusted little girl, his real patterns of play as a child showed evidence of a taste for the toys and activities of the boys around him. In fact, this was a child struggling with a gender identity that did not fit. As a teenager, when he was finally told his medical history, he rejected the female identity that had been assigned to him and re-embraced a male role in life. He took a male name and chose to live as a man. Because of cases like his, a number of workers believe that gender identification is biologically inborn and cannot simply be assigned (or reassigned) based on how the child is raised or what their external genitalia look like.

we find some evidence of both genetic and environmental components of gender identification. Twin studies of gender identity suggest that there is a strong genetic component to gender identity, but that genetics cannot account for all of the determinants of gender. In mice when animals of opposite sex develop together in the same womb, siblings may acquire sex-specific behaviors of the opposite sex, something possibly explained by exposure of one embryo to hormones being produced by another embryo sharing the same uterine environment. Such effects have been observed for organisms that given birth to litters in which adjacent pups have a shared blood supply, such as rats and rabits. When asking similar questions in human opposite sex twins, who do not share placenta or blood supply, it is hard to separate the biological influences from the social environmental influences, so matters in humans are far less clear and still under investigation.

Thus, although much on the subject remains confused, the overall picture we find is one of both genetic and environmental effects on gender identity. While the lack of a simple answer complicates efforts to make decisions about sex reassignment surgeries or to understand the processes that produce transgendered and transsexual individuals, it is perhaps not surprising if the real answer on such a complex subject is not a simple answer. Overall, it is rather surprising how little is known about biological determinants of gender in humans.

8.7 GENETICS OF SEXUAL ORIENTATION

We will now turn our attention from the development of sexual or gender identities to the development of sexual orientation, another topic where not nearly enough is known about the real underlying determinants. We can only apologize in advance if our treatment of this topic (or anything else in this chapter) in any way fails to be adequately sensitive to the broad array of perspectives on such controversial topics.

In a 1994 *Scientific American* article two major workers in this area, named Dean Hamer and

Simon LeVay said "Most men are sexually attracted to women, most women to men. To many people, this seems only the natural order of things, the appropriate manifestation of biological instinct, reinforced by education, religion, and law. Yet a significant minority of men and women, estimates range from 1 to 5%, are attracted exclusively to members of their own sex." This statement raises some fascinating questions. First, just how is sexual attraction or orientation determined? Is it biological? Are there genes that direct males to be attracted to females and vice versa? Second, if sexual orientation is biologically programmed, how are we to understand the etiology of cases in which men choose men as lovers or women choose women? Could such people reflect genetic variation in genes for sexual orientation? If such genes and such variation do exist, what are those genes and what do they do? These questions will be our focus for the remainder of this chapter.

The Genetics of Sexual Orientation: Population Studies

Sexual orientation is defined by the sex to which a given individual is sexually attracted. When, as is usually the case, a person is attracted to an individual of the opposite biological sex, that individual is referred to as *heterosexual*. In the case in which people are attracted to others of the same sex, they are referred to as *homosexual*. In some cases, in which an individual is attracted to both sexes, the term *bisexual* is used. Terms used in popular culture seem to keep changing, but in recent years, common parlance in the United States often refers to homosexuals as "gay" men and "lesbian" women.

Before the early 1990s, there were two lines of evidence to suggest that male homosexuality might be genetic. The first line of evidence came from studies of heritability, the measure of how often the trait is concordant or discordant in identical twins vs. fraternal twins. In the case

of homosexuality, such heritability estimates are suggestive of an important role of genes in determining the phenotype (Box 8.2). For both gay males and lesbians, their homosexual orientation is found in more than half of their identical twins, compared to one-sixth (lesbians) or one-quarter (gay males) of their fraternal twins, and about one-eighth of their non-twin siblings. Genetically identical individuals are more likely to be concordant than genetically different individuals if some aspect of the trait is genetic, so the fact that the identical twins show a much higher concordance for being gay or lesbian suggests a substantial genetic contribution to the trait. Also notice that brothers of gays tend to also be gay more frequently than expected than the 1–5% rate estimated for the American male population, another piece of information that helps support the view that there are genetic factors contributing to the gay or lesbian phenotype.

However, these data also suggest that the determination of sexual orientation cannot be wholly genetic. If the gay or lesbian phenotypes were completely genetic, we might expect the concordance of identical twins to be 1.0, as it is for traits such as color blindness or cystic fibrosis. Clearly, genotype alone cannot account for those 50% of cases in which the twins were discordant.

How are we to explain these data? One explanation is that there may be genotypes that predispose individuals to a particular sexual orientation to others (sexual orientation genes, if you will) but that these genotypes interact with the environment. These environmental influences may include obvious things such as family values, peer pressure, societal responses, personal relationships and specific sexual experiences or religious influences, but environmental effects could also be nonsocietal and could *theoretically* include things such as medical events or nutrition. The allelic differences, if indeed they exist at all, appear to predispose rather than dictating a specific outcome.

BOX 8.2

SOME LIMITATIONS OF HERITABILITY STUDIES

There are real limitations to what you can tell from measures of heritability. Any estimate of heritability is only good for that one particular population at the time that estimate is made and might or might not offer insights into other populations. Some of this is due to differences in the genetic composition of different populations. However, some of it is due to differences in exposure to environmental factors influencing manifestation of the trait. In addition, the accuracy of this method is limited by the accuracy with which the researcher can validly and accurately score people for the trait in question. If you want to compare levels of protein in urine, it may be possible to make simple quantitative assessments of whether the values are the same in two individuals. If you want to know whether or not two individuals are concordant for a trait that you cannot directly measure, the amount of non-concordance in the test will be directly related to the chance that self-report of the trait is inaccurate for any one individual being questioned, either because the answer given is untrue or because the individual does not know the correct answer. Another factor that can confound heritability studies is something called age-related penetrance – the tendency for

the same trait to develop at different ages in different individuals – which can make it hard to tell whether lack of concordance indicates that twins don't share the trait or whether it means that one of them simply has not yet developed a trait that will appear later in their development. In the case of homosexuality, the social environment could vary for different individuals in ways that not only influence the willingness of the study subject to self-identify as a homosexual, but that also influence the age at which individuals admit to themselves that they are homosexual. So studies of heritability in homosexual populations may well be confounded by a variety of factors – differential environmental influences on the development of the trait, differences in accuracy of self-report, differences in age at which the individual realizes they have the trait, and differences in study participation rates for some individuals depending on their attitudes towards their status and towards surrounding social reactions to their status. We expect to get a much more accurate assessment of the role of genetics in homosexuality in a society that fully accepts it than in a society that is critical of it or seeks to suppress it.

An alternative explanation could be that the genetic components of homosexuality are even larger than they appear to be, with apparent cases of discordance representing underreporting of gay or lesbian status. Even in cases in which social pressures on the situation are not in evidence, self-reporting of medical status can often be inaccurate when self-reports are compared to medical records, so how much more of a problem could this be if there are societal or personal pressures against self-identifying as gay or lesbian?

The Genetics of Sexual Orientation: Family Studies

A different form of support for a genetic basis for male homosexuality comes from studies of families. There are many pedigrees in which male homosexuality appears to segregate in a predictable and sex-linked fashion through a given kindred. The pedigree shown in Figure 8.8 is an example. If you didn't know the phenotype under consideration, you might have glanced at

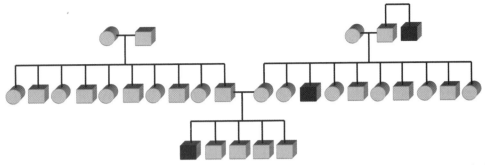

FIGURE 8.8   Family in which homosexuality appears to segregate as an X-linked recessive trait. The three dark squares represent three homosexual relatives. (After Hamer *et al.*, "A linkage between DNA markers on the X chromosome and male sexual orientation," *Science*, 1993;261:321–7.)

the pedigree, and wondered if it could be "sex-linked recessive inheritance." It was pedigrees such as this that caused Dean Hamer and colleagues at the National Institutes of Health (NIH) to begin a careful study of the genetics of male homosexuality in the early 1990s.

The initial subjects in Hamer's study were 76 self-identified gay men and their relatives over the age of 18, as well as 38 pairs of homosexual brothers and their relatives. The researchers recruited through an AIDS clinic, through local gay organizations in Washington, DC, and through advertisements in gay-oriented magazines and newsletters. Before we go any further, we need to think about this study group. This population consists of gay men open enough about their sexuality to agree both to participate in this study and involve their extended families. In other words, all of these men were fully "out," and functioning in the context of families that did not reject participation in such a study. (We are left with questions here about how this result differs from what they might have found if they had studied a population of homosexuals who are not open about their status.)

After evaluating the initial group, assessing the phenotype among relatives can actually be complicated. Hamer and colleagues used two methods to ascertain the phenotype: self-assessment and a set of psychological tests known as the

Kinsey scales (Figure 8.9). Amazingly, both the self-assessment and the assessment by the original members of the study group (the probands) were remarkably concordant. To quote the Hamer paper:

> All (69/69) of the relatives identified as definitely homosexual verified the initial assessment, as did most (27/30) of the relatives considered to be heterosexual; the only possible discrepancies were one individual who considered himself to be asexual and two subjects who declined to answer all the interview questions.—*Hamer, D. H., Hu, S., Magnuson, V. L., Nan Hu, and Pattatucci, A. M. L.*, Science, *1993;261:321–7*

Thus again quoting from the Hamer paper, "describing individuals as either homosexual or heterosexual, while undoubtedly over-simplistic, appears to represent a reliable categorization of the population under study."

What Hamer is saying is that in this population of men and their relatives, homosexuality or heterosexuality can be considered as a discrete pair of traits, such that each individual can be reliably classified as one or the other. Hamer and co-workers are backed up in this assertion by their data from the use of the Kinsey scales mentioned. Using these scales, people rate themselves on four aspects of their sexuality: self-identification, attraction, fantasy, and behavior. The ratings range from 0 for exclusively

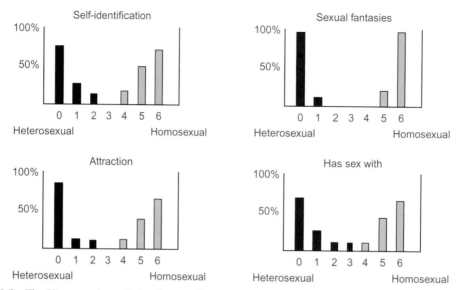

FIGURE 8.9   The Kinsey scale analysis of a population reported in Hamer *et al., Science,* 1993;261:321–7. Dark bars denote self-identified heterosexual men, while light bars denote self-identified homosexual men.

heterosexual to 6 for exclusively homosexual. Thus a man who has never had even a fleeting attraction to another man would rate himself a 0 on the attraction scale, whereas a man only attracted to other men would give himself a rating of 6. As shown in Figure 8.9, the graphs are bimodal for each of these four characteristics.

The graphs in Figure 8.9 might surprise you, because they suggest a discrete bimodality in human sexual orientation that is not consistent with many people's experience. Do Hamer's data really suggest that bisexuals do not exist? No, Hamer's data say only that such people don't exist in his study group, which is a highly selected and precisely defined study group, so these graphs represent only his study group and not any other populations.

Hamer's analysis yielded some fascinating conclusions. Hamer and co-workers confirmed previous studies by noting that a brother of a gay man had a 14% chance of being gay as compared with a 1–5% percent chance for males in the general population. Hamer and colleagues also noticed something even more interesting among more distant relatives of these gay men. Maternal uncles and sons of maternal aunts had a higher chance of being gay (7–8%) than expected for the general population, but no such effect was observed for paternal uncles or sons of paternal aunts. This is highly suggestive of an X-linked determinant for sexual orientation. Remember, paternal uncles or cousins on the father's side cannot share an X with the homosexual male in question, but maternal uncles or cousins can. (See the pedigree diagram in Figure 8.8.) Thus Hamer saw only a high frequency of concordance among relatives who could share an X chromosome, evidence again for sex linkage of a gene or genes.

In additional studies of this effect, Hamer's group further refined their study group to 38 families in which at least two sons were gay. They excluded any family in which the father was gay (ruling out any case of father-to-son transmission), or with more than one lesbian relative. It was hoped that, by excluding other causes of homosexuality, this population might be enriched for the X-linked form of homosexuality they were seeking to study. It sort of worked. Maternal uncles and sons of maternal aunts had a higher chance of being gay (10–13%), and again no such effect was observed in paternal uncles or sons of paternal aunts. However, the ratios were still lower than those expected for a simple Mendelian trait (50% for a maternal uncle and 25% for a son of a maternal aunt).

To sort this out, Hamer and colleagues fell back on what we call "Hawley's first law": *If a gene exists, you can map it; and, if you can map a gene, then it exists.* They now focused on only 40 pairs of brothers. Realize how important this was. No matter what other environmental conditions need to be met, or other genes need to be present to develop homosexuality, they must all be there in these males. If there really is an important gene on the X chromosome that determines homosexuality, and if the mother was heterozygous for that gene, then these brothers should share a specific region of one of the mother's X chromosomes, the region bearing the allele predisposing them to homosexuality.

Consider thinking about it this way: If a woman is heterozygous for the color-blindness allele (cb), and two of her sons are affected, it is because both inherited the cb allele from her. Because recombination is frequent on the human X, approximately five exchanges per bivalent on the long arm of the X alone, one doesn't expect the brothers to share the same alleles for all genes at other sites, but they should share the cb allele and other closely linked alleles as well. This is an example of a fundamental concept in gene mapping – the concept of allele sharing that assumes identity by descent, the situation in which two people have the same trait because they inherited the same causative allele(s) from the same ancestor.

Hamer and colleagues analyzed the inheritance of homosexuality in these pairs of brothers by studying a large number of genetic markers distributed at various points along the length of the X chromosome. By looking at markers that have two different alleles in the mother, it is then possible to ask whether both brothers received the same allele, or whether one brother received the first allele and the second brother received the other allele.

If there were only one gene for male homosexuality on the X chromosome and it was completely penetrant, we would expect to find every pair of gay brothers carrying the same allele of a marker located next to the "gay" gene. Because of recombination events along the X chromosome, we expect that they will not be identical for all of the markers on the X. The closer a marker is to the "gay" gene, the more often the brothers will share it. Markers that are farther away will be shared less often, and markers that are a long way away will seem to be randomly assorted in the brothers. If there were not "gay" genes on the X chromosome, we would expect to find that each marker on the X would present the same allele about 50% of the time, and a different allele about 50% of the time.

Also note that we do not expect that the allele shared by brothers in the first family would necessarily be the same allele shared by the brothers in the second family (Figure 8.8). For many genes, different mutations have arisen over time in the same gene leading to the same phenotype, so to do this experiment we look for allele-sharing between brothers based on identity by descent but we do not ask that the same allele be present in all of the families.

So what did Hamer find? For most of the X chromosome, the brothers were as likely to have two different alleles as they were to both share one of the two given maternal alleles. However, for one region, Xq28, near the tip of the long arm of the X, the two homosexual brothers shared the same alleles in 33 out of 40 cases. This finding is very highly significant and provides strong evidence for an important gene in this region. Although this is a very strong result, you do need to note that there were seven pairs of brothers who carried different alleles of the Xq28 region. Thus, even in this highly refined population, the Xq28 region cannot account for all cases of homosexuality. Nonetheless, the basic result is still indicative of some *correlation* between genotype at Xq28 and sexual orientation phenotype in a large fraction of these sibling pairs. In a more recent study, Hamer and colleagues have repeated this mapping and extended their studies to include heterosexual brothers of the two gay brothers initially studied. Not surprisingly, these heterosexual brothers carried the alleles in the Xq28 that were shared by their gay brothers much less often (22%) than would be expected by chance (50%).

We should point out that Hamer's data also suggest that whatever genes might be in Xq28 that affect sexual orientation in men, there is no evidence for that gene, or any other genes, affecting sexual orientation in women. Although the heritability of lesbianism is as high as it is for male homosexuality, very little is known about a genetic basis for lesbianism, or indeed if one exists at all.

The Findings Are Still Considered Controversial

So Hamer's data *suggest*, and only suggest, that there may be a gene, or genes, in region Xq28 of the X chromosome that affects sexual orientation. However, not everyone in the scientific community agrees with that suggestion. Many workers worry about the small sample size of the study group. Other workers are trying to repeat the results using different populations. More recently there have been additional findings pointing to possible loci on three of the autosomes, but all of these findings await confirmation by other studies. The final verdict is anything but "in." Finally, even if Hamer and his group are correct and some region of the X chromosome is determining sexual orientation in their population, it is not at all clear how

generalizable this result is to the general population. Even if Hamer is wrong about this particular gene, this seems like an approach that could eventually answer these questions.

What If the Results Mean Something Else?

If Hamer is wrong, does that mean there was something wrong with his study or that he has produced what we would call a statistical artifact? Is there some way his study could be right and yet not mean what he thinks it means? Consider this: they mapped something to Xq28 that is held in common between the brothers in the study. The brothers were selected because they had their sexual orientation in common. What else might they have in common? It is interesting to note that another similar study conducted in Toronto, in a different gay population, did not find the same thing. One thing that comes to mind is that gay brothers who are very "out" about their orientation, who are very public and outspoken about their status, might not only share their sexual orientation but perhaps also other personality characteristics, such as assertiveness or self-confidence or rebelliousness. Recruitment among a group of gay men who are circumspect about their status, maintain a low profile, or even hide their status might identify a group of men who share sexual orientation with the first population but do not share certain other personality characteristics with them. This sort of thing is a big risk in studies of this kind – that you select a set of study subjects on the basis of sharing a particular trait while not realizing that they also share other things that might actually be the basis for your findings. We do not know that this happened in the Hamer study and if there are differences between the two study populations that are responsible for the difference in findings, we do not know that those differences have anything to do with factors we have suggested. There might be a "gay" gene on Xq28, and the failure of a second study to reproduce the finding might really represent some other methodological difference between the studies, or perhaps a difference in the genetic backgrounds of the two populations. There might not be a "gay" gene on Xq28. It remains to be seen.

Beyond those caveats, suppose Hamer and friends are right. Just what kind of things might such a sexual orientation gene specify? How might it work? Simon LeVay has presented evidence for a structural difference between a small region in the brains of gay and heterosexual men thought to be involved in controlling sexual orientation. In rodents and primates it has been shown that hormonal influences on development of the brain are associated with sex-specific differences in behavior, but it remains to be seen whether these observations are related to LeVay's observations on human sexual orientation. In other studies, a different region of the brain has shown association with gender identity but not sexual orientation or sex. Might genes play a role in the formation of such structural differences? And might such structural differences play a role in the associated gender identity or sexual orientation? As interesting as these results are, at this time the scientific community is far from fully persuaded on this matter.

And so we close both this section and our formal discussion of sexual differentiation in humans. There are clearly proven roles for specific genes in determination of gonadal and somatic sex. Much less is known about other aspects of sex, but evidence suggesting that genetic factors contribute to gender identity and sexual orientation also suggest that it may not all be genetic. And we are left with the interesting question to consider: are the environmental components of this equation affecting gender and orientation, or are they simply affecting whether or not people act on or admit such differences to the world around them (or to the scientists doing the study!)?

As we go on to talk about other aspects of human genetics, we hope you will keep in mind some of the lessons of this chapter:

One line of evidence, such as family studies, can help validate findings from other types of studies, such as twin studies or population-based studies. There can be both genetic and environmental effects on a trait, and some of those environmental traits can be societal rather than the usual environmental effects we think of, such as diet or exposure to toxic substances. You might get different answers depending on what population you look at or how you assay for the existence of the trait. And a single trait may actually be made up of multiple independent traits, with different underlying causes of separate aspects of things that we think of as a being a single trait.

Reprise

And so we return to the Parisian model from the start of this chapter (Figure 8.10). It has been hypothesized that she was a case of androgen insensitivity with undescended testes in a body that is externally female but lacks the reproductive machinery present in most females (although we cannot know specifically that she had an AIS mutation, since no tissue samples were available to test). In 1909, when the doctors said, "you are male," medicine seemed ill equipped to cope with someone who did not fit neatly into one of the two sexual niches, male and female. As we gain further understanding of the underlying genetics and biology of sexual development, we expect this to be one of the topics that will push policy makers and society to arrive at reasonable reactions to situations that do not fit our preconceptions.

Consider how someone would feel about finding out that they had the karyotype and gonads of a male even though they seemed to be anatomically female. How would you feel if you found out that your karyotype or gonads were not those expected for the sex that you appear to be? If you discovered that the person you were married to had the same set of sex chromosomes you have, even though their

FIGURE 8.10 A Parisian fashion model who was hypothesized to have represented a case of androgen insensitivity because she had normal external female sexual anatomy but had undescended testes. More information about this fashion model and the ways in which the medical establishment handled her case can be found in Alice Domurat Dreger's *Hermaphrodites and the Medical Invention of Sex* (1998, Harvard University Press.) (Courtesy of Alice Domurat Dreger, used by permission. This photo previously appeared in *Diseases of Women* by Lawson Tait [1879, William Wood & Company], p. 22.)

dilemma that faced her, but Dr Dreger tells us that historical records do not indicate whether the planned marriage took place.

Clearly, much of our society feels a strong urge to fit people into known classifications, to be able to react to them as a man or a woman and not as someone somewhere in between. Much of our society seems to want people to be congruent, to have different aspects of their sexual being all match according to conventional definitions of male and female, even though, as we now see, different aspects of the sexual phenotype seem to be the very separable result of different genes and biological processes. However, wanting the world to be neat and tidy does not make it so. We have argued before that diversity is one of the greatest gifts ever granted humanity, and that applies not only to issues of race or culture but also to sex. Sexual diversity offers us lessons that can grant us increased understanding of ourselves and our sexuality if we can learn what that sexuality consists of and realize that some of the things we feel are carved in stone are actually variables with real biological underpinnings.

anatomy is that of the opposite sex? What would you think of laws declaring your union homosexual and your marriage invalid under the law? Would it change your perceptions of yourself and how you fit into the world? What would you think of the Parisian model's situation if she had simply said, "that's ridiculous," and went ahead with marrying her fiancé? Alternatively, how would you feel if the model had compliantly responded to the doctors by taking up male attire and going in search of a woman to marry? Consider why you have the reactions you have and what your justifications are, and think about how you would answer the same questions if you knew you were considering an individual with a mutation in a gene controlling gender or sexual orientation, instead of somatic sex. We were quite interested in knowing how the Parisian model dealt with the

Study Questions

1. What is the characteristic that determines gonadal sex of a male? Of a female?
2. What are the primary sex characteristics of a male? What are the secondary sex characteristics of a male?
3. What are the primary sex characteristics of a female? What are the secondary sex characteristics of a female?
4. How is sex different from gender?
5. What is the evidence that homosexuality in men may be genetically determined?
6. What biological process is determined by the presence or absence of the SRY gene?
7. What is the indifferent gonad?

8. What controls the development of the Müllerian ducts?

9. How do we *know* that the entire Y is not required to determine sex?

10. Why are Klinefelter syndrome males sterile? Why are Turner syndrome females sterile?

11. Why do XY women with CAIS have testicles? Why don't they have other male sexual characteristics? Why don't they have a cervix, uterus or fallopian tubes?

12. Why don't XY women with a mutation in the SRY gene have testicles? Why do they have a cervix, uterus or fallopian tubes?

13. What role does SOX9 play in sex determination?

14. What causes levels of SOX9 to change?

15. If male sex can be determined by the simple step of expressing SRY, why can sex reversal result from mutations in RSPO1 or the androgen receptor?

16. If there were no "gay gene" on Xq28, what else could have caused the Xq28 mapping result?

17. Why might the Parisian model be considered male, and why might she be considered female?

18. Is there evidence of a biological basis for gender?

19. How can age-related penetrance complicate interpretation of heritability studies?

20. Why might the female sex be considered the default sex?

Short Essays

1. One of the arguments for a role of the X chromosome in male sexual orientation is the observed trend towards family history tracing through the maternal lineage. If the "gay" locus on Xq28 does not turn out to be real, then how else might we imagine that the X chromosome could be involved in this trait? As you consider this question please read "Extreme skewing of X chromosome inactivation in mothers of homosexual men" by Sven Bocklandt, Steve Horvath, Eric Vilain, and Dean Hamer in *Human Genetics*, 2006;118:691–4.

2. For many years, the birth of an intersex child led to doctors' recommendations that parents immediately arrange for surgery to create a sexually unambiguous situation. Policies regarding whether or not to carry out corrective surgery have changed over time. What are some of the factors that have caused the field to reconsider the earlier policies? As you consider this question please read "Developmental endocrine influences on gender identity: implications for management of disorders of sex development" by William Byne in *The Mount Sinai Journal of Medicine*, 2006;73:950–9.

3. We list gonadal, primary, secondary, and psychological classifications of sex, but sexual dimorphism extends far outside of the features that play a role in mate selection and reproduction. What is the medical significance of sexual dimorphism and how should an understanding of this affect the way we study medicine and select medical treatments? As you consider this question please read "Sex-specific genetic architecture of human disease" by Carole Ober, Dagan A. Loisel, and Yoav Gilad in *Nature Reviews Genetics*, 2008;9:911–22.

4. We tend to think of sexual selection as applying to aspects of sexual attractiveness that can be easily identified such as beauty and strength. Is there evidence for selection of mates based on features we do not consciously perceive? As you consider this question please read "Is mate choice in humans MHC-dependent?" by Raphaëlle Chaix, Chen Cao, and Peter Donnelly in *PLoS Genetics*, 2008;4(9):e1000184.

Resource Project

SRY, SOX9, and the androgen receptor (AR) each carry out specific, critical steps in sex determination during embryogenesis. Are their functions limited to sex determination? Search on the term Genecards to get to the Genecards homepage to look up each of these genes, and scroll down the page to where the bar graph shows the levels of expression of the gene in various adult tissues. Notice the normalized intensity scale on the left that tells us how high the level of expression is for the displayed gene, and compare levels between the different genes. Write a paragraph on how specific or broadly distributed gene expression is for each of these genes, and speculate on what that might tell us about specificity of action of each gene.

Suggested Reading

Video

NOVA: *The Miracle of Life* (1996)

Articles and Chapters

"Sex differences in brain and behavior: Hormones versus genes" by S. Bocklandt and E. Vilain in *Advances in Genetics*, 2007;59:245–66.

"Choosing sex" by Blanche Capel in *The Scientist*, 2009;23:36.

"The mysteries of sexual identity: The germ cell's perspective" by Judith Kimble and David C. Page in *Science*, 2007;316:400–1.

"Sex: The greatest lottery on earth" by Nick Lane in *Life Ascending: The Ten Great Inventions of Evolution* (2009, W. W. Norton, pp. 118–43).

"A difference in hypothalamic structure between heterosexual and homosexual men" by Simon LeVay in *Science*, 1991;253:1034–7.

"Evidence for a biological influence in male homosexuality" by Simon Levay and Dean Hamer in *Scientific American*, 1994;270: 44–9.

"Male or female? The answer depends on when you ask" by H. K. Salz in *PLoS Biology*, 2007;5:e335.

"Sex determination and SRY: Down to a wink and a nudge" by R. Sekido and R. Lovell-Badge in *Trends in Genetics*, 2009;25:19–29d.

Books

Hermaphrodites and the Medical Invention of Sex by Alice Domurat Dreger (1998, Elsevier).

She's Not There by Jennifer Finney Boylan (2003, Broadway Books).

Brain Gender by Melissa Hines (2005, Oxford University Press).

Christine Jorgensen: A Personal Autobiography by Christine Jorgensen (2000, Cleis Press).

Straight Science: Homosexuality, Evolution and Adaptation by Jim McKnight (1997, Routledge).

Conundrum: The Evolution of Homosexuality by N. J. Peters (2006, Authorhouse).

Genetic Mutiny and Gender in The Red Queen: Sex and the Evolution of Human Nature by Matt Ridley (Perennial, 2003).

Whipping Girl by Julia Serano (2007, Seal Press).

Complexity: How Traits Can Result from Combinations of Factors

The Human Genome. DOI: 10.1016/B978-0-12-333445-9.00009-5

THE READER'S COMPANION: AS YOU READ, YOU SHOULD CONSIDER

- How defects in two different genes combine to cause a trait.

- What digenic diallelic and digenic triallelic inheritance are like.

- What a quantitative trait is.

- How a threshold trait differs from a quantitative trait.

- The rules for recognizing multifactorial inheritance.

- Why phenocopies pose a problem for genetic studies.

- Why it might be a problem to not recognize that osteogenesis imperfecta is an inherited trait.

- What kinds of situations might run in families without being inherited.

- How studying twins tells us something about the role of genetics in a trait.

- Why identical twins might not be identical for some traits.

- Why cardiovascular disease is often multifactorial.

- Why it is hard to tell who has an inducible trait like malignant hyperthermia.

- What an iatrogenic trait is.

- The difference between identity by state and identity by descent.

- How variation in host defenses affects resisting and surviving infections.

- How someone else's genotype can affect your phenotype.

- Why some phenotypic modifiers can affect more than one kind of trait.

9.1 DIGENIC DIALLELIC INHERITANCE

"...five, four, three, two, one, here I come, ready or not!" Drew took his hands off his eyes and turned to peer into the darkness that was deepening across the grass and trees of the backyards where he and his friends were playing hide and seek. Light stabbed out into the darkness from the patio doors and kitchen windows, and the street lights cast a halo of light across the sidewalks. Somewhere in the adjoining yards five of his friends were hiding, waiting to be discovered. Before he went exploring into the edge of the woods and the large vegetable gardens, he headed straight for the front yards and the street lights. After poking around under several well-lit front porches he headed straight for a large bush across from a street light where he hauled out a protesting girl who stomped her foot. "It's not fair," she protested. "You always find me first!" He laughed as he brushed some leaves off her shoulders. "What do you expect, Jackie? You always hide by the street lights." She scowled and replied, "Of course I do. If I go into the back yard after dark I can't see anything." Drew laughed off this idea, and started backing towards the space between two of the houses. "Sure it's dark, but if you just wait your eyes get used to the dark." As he turned to run off into the back yard in search of the other kids Jackie shook her head and muttered, "No, they don't. My eyes never get used to the dark."

Why would Jackie have eyes that don't get used to the dark? Jackie is in the early stages of retinitis pigmentosa, and she does not see well in the dark because she is gradually losing the rod cells responsible for detecting faint light at night; she is also starting to lose daytime vision around the periphery. Some adults with retinitis pigmentosa have commented on having the same problem Jackie had, spending their childhood being the first child found in a game of hide-and-seek because everyone knew they would be hiding near the lights.

Retinitis pigmentosa is a form of retinal degeneration that begins with damage in the

FIGURE 9.1   Retinitis pigmentosa. On the left is a scene through the eyes of someone with normal vision. On the right is the same scene through the eyes of someone with retinitis pigmentosa. (Images from the National Eye Institute)

periphery of the retina, with degeneration of rod cells gradually progressing towards the center of the field of vision (Figure 9.1). Some people with retinitis pigmentosa retain vision at the center of focus (Figure 9.1) but others eventually lose all vision, arriving at a status we refer to as "no light perception."

Many dozens of loci play roles in retinitis pigmentosa. Many of the mechanisms that lead to retinitis pigmentosa involve processes that damage the rod cells in the retina. It takes the products of many different genes to carry on all of the essential functions of the cell to keep it healthy. Inactivating any of those essential genes can damage the cell. It is an important concept – there are many different ways to kill a cell, and genetic traits involving cell death often result from any of many different genetic defects, each affecting one of many genes important to the health of that cell. Most causes of retinitis pigmentosa show simple Mendelian inheritance. For instance, Jackie's disease is the result of a mutation in the gene that encodes rhodopsin, the photoreceptor protein in rod cells. A surprise turned up when one form of this trait turned out to offer the first human example of the simplest of the complex traits – digenic diallelic inheritance.

Most of our discussions have centered on simple Mendelian traits caused by defects in a single gene. There are a large number of different simple Mendelian traits. Some of them are important topics of study because they cause very severe phenotypes. For others, understanding them is important because they offer insights into fundamental developmental processes. And many of them cause traits that are quite common. An understanding of Mendelian traits not only offers us insights into specific diseases but has also offered many lessons in the basic processes by which inheritance works.

There are many other traits that do not follow the simple rules of Mendelian inheritance, including many of the greatest public health problems that are both common and severe; these are much more complex than the kinds of examples we have offered so far. Heart disease, Alzheimer disease, and age-related macular degeneration all affect large numbers of elderly individuals, and each of them is genetically complex (having more than one genetic component in an individual), genetically heterogeneous (not always having the same cause in different members of a population), phenotypically complex (showing variation in trait characteristics from one person to the next), and subject to important environmental influences. As we will see in this chapter, an understanding of the factors that affect the simple Mendelian traits provide as with

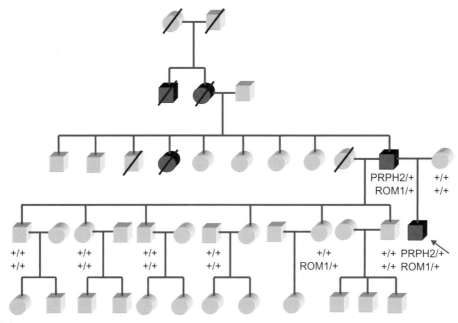

FIGURE 9.2 Digenic diallelic inheritance in one of three digenic diallelic families. This mode of inheritance is hard to classify based on looking at the family. We see affected individuals in four generations in a row, which suggests dominant inheritance, but we also see two large sibships with substantially less than 50% of the children affected, which is not consistent with dominant inheritance. (After Kajiwara, K., Berson, E. L., and Dryja, T. P., *Science*, 1994;264:1604–8.)

an important theoretical framework for understanding situations that are more complicated.

So let's start in with a discussion of *genetic complexity* – the presence of multiple genetic factors that come together in one individual to cause a trait. And let's start with the simplest of the complex cases, a form of inheritance that is called *digenic diallelic*, sometimes just called digenic, inheritance because it is caused by the simultaneous presence of mutations in each of two different genes.

The discovery of digenic diallelic inheritance in humans was the product of a large project studying the genetics of retinitis pigmentosa, although the concept had long been known in animal model systems such as the fruit fly. When researchers were designing this experiment, defects in the PRPH2 gene, which encodes the perpipherin protein, were already known to cause autosomal dominant retinitis

pigmentosa, and ROM1, which encodes the rod outer segment membrane protein, was considered a candidate gene for the same trait. PRPH2 and ROM1 encode proteins that can be found localized in the same region within the rod cell. A group of 232 unrelated individuals with retinitis pigmentosa were screened for mutations in PRPH2. Two individuals turned out to both have the same missense mutation that puts a leucine in place of a proline at position 185 of the PRPH2 protein sequence (L185P). When the ROM1 gene was screened for mutations, the two individuals with the PRPH2 mutations also turned out to have a ROM1 nonsense mutation. A study of three families with the PRPH2 L185P mutation showed a total of 19 individuals with retinitis pigmentosa, each with one copy of the PRPH2 L185P mutation and one copy of the ROM1 nonsense mutation (Figure 9.2). In these same three families they found 15 individuals with a single copy of the L185P defect and 10 individuals with a single copy of the ROM1 nonsense mutation, none of them with retinitis pigmentosa. So clearly having both mutations in the same individual causes retinitis pigmentosa, but either mutation alone does not cause disease.

The distribution of genotypes in these families tells us that ROM1 and PRPH2 are segregating independently of each other. This is not surprising when we look at where the two genes are located. PRPH2 is located on chromosome 6. ROM1 is on chromosome 11; thus many individuals ended up with only one or the other of the two mutations and no disease, but everyone who ended up with both of these mutations ended up with retinitis pigmentosa.

This digenic diallelic mode of inheritance is not a general property of the PRPH2 gene. Other PRPH2 mutations show simple autosomal dominant inheritance, and those other PRPH2 mutations can cause the disease without any ROM1 mutation being present. But when the PRPH2 mutation is the L185P mutation, it only causes disease if there is also a ROM1 mutation present.

Our understanding of this situation is helped if we know that the PRPH2 gene product and the ROM1 gene product interact in the rod cells in the retina of the eye (Figure 9.3). The cell clearly requires the presence of tretramers which are usually made up of two copies of the PRPH2 gene product and two copies of the ROM1 gene product.

Why does the ROM1 defect alone not cause retinitis pigmentosa? Because the normal PRPH2 protein product is capable of forming tetramers if ROM1 is not around, using two copies of the PRPH2 protein to substitute for the two copies of ROM1 that would normally be used (Figure 9.4B). Why does the L185P mutation not cause retinitis pigmentosa unless the ROM1 defect is also present? Because the L185P protein can form a tetramer when joined with the normal ROM1 protein (Figure 9.4C). Why does the combination of L185P PRPH2

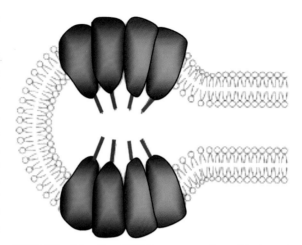

FIGURE 9.3 The PRHP2 gene product (blue) and ROM1 gene product (pink) come together to form tetramers anchored in the membrane of the outer segments of rod cells in the retina. (After R. G. Weleber, *Digital Journal of Ophthalmology*, 1999; 5(2), adapted from an illustration by Drs G. Clarke and R. McInnis.)

and ROM1 nonsense cause the disease? Because ROM1 is missing and L185P PRPH2 can only make dimers, not tetramers (Figure 9.4D); the resulting dimers cannot substitute for the tetramers. Thus the normal PRPH2 protein product can step in and take the place of missing ROM1 protein, but the defective PRPH2 protein cannot replace missing ROM1.

We told you about Jackie, who has retinitis pigmentosa. We have to wonder how many other situations like this will be found, in which one mutation in a gene can play a simple role in disease and yet a different mutation in that same gene is involved in a more complex form of the disease. There is a growing list of human traits that have been found to be digenic diallelic. Digenic diallelic inheritance is turning up in more and more cases involving independently segregating genes, including the cardiology phenotype called long QT syndrome, a form of combined deafness and blindness known as Usher syndrome, type II diabetes, and hereditary hemochromatosis. In some cases, the two different genes involved in the digenic

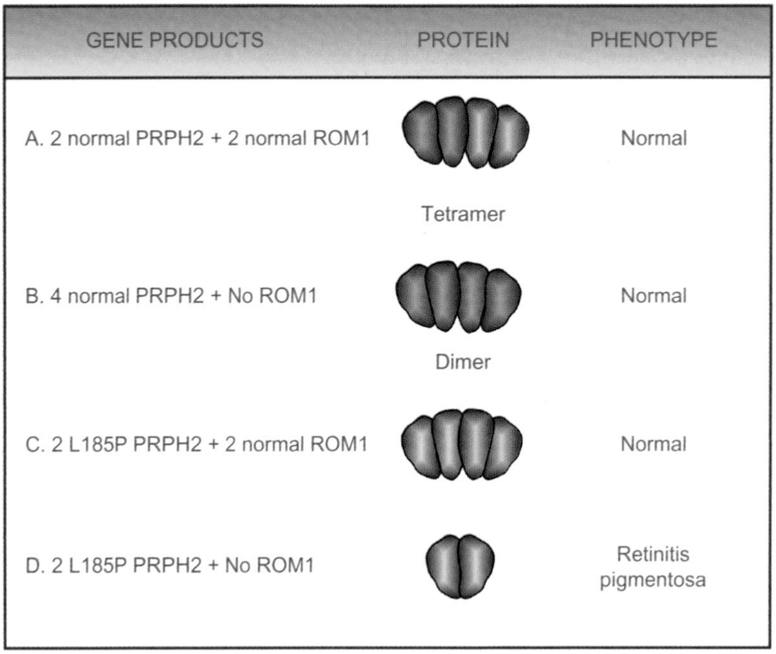

FIGURE 9.4   The rod cell requires tetramers in the outer segment membrane for cell health. (A) Normal tetramer protein complexes contain four proteins – two copies of the PRPH2 gene product and two copies of the ROM1 gene product. Disease results when the tetramer can't be formed. Some PRPH2 mutations can cause retinitis pigmentosa when only one copy is defective; the protein product of the autosomal dominant PRPH2 mutations cannot participate in tetramer formation and in the absence of functional PRPH2 protein ROM1 cannot make a tetramer by itself. (B) Normal PRPH2 can form tetramers if ROM1 is missing. (C) The Leu185Pro PRPH2 mutation seen in digenic diallelic inheritance, which substitutes a leucine in place of proline at position 185 of the PRPH2 sequence, does not cause disease if normal ROM1 is around because it can participate in tetramer formation as long as ROM1 is normal. (D) Leu185Pro PRPH2 cannot make a tetramer by itself in the absence of normal ROM1. In individuals who are heterozygous for both mutations, some normal PRPH2-ROM1 tetramers and some tetramers made entirely of normal PRPH2 are formed, but the cell is sensitive to dosage and the amount of tetramer formed is not enough. Normal PRPH2 gene product, dark blue; mutated version, lighter blue; ROM1, pink. (After A. F. Goldberg and R. S. Molday, *Proceeding of the National Academy of Sciences of the USA*, 1996;93:13726–30. © 1996 National Academy of Sciences, U.S.A.)

diallelic disease may be located near each other, so there would not be a lot of independent segregation of the alleles at the two genes; this has been seen in an early onset form of Parkinson disease where the affected individuals are heterozygous for a mutation in the PINK1 gene and heterozygous for a mutation in the DJ1 gene, both of which are located on chromosome 1p36. It is quite common that there will be some mutations or combinations of mutations that turn up in the digenic diallelic forms, while many of the genes that can play a role in digenic diallelic disease often have alternative alleles

that cause the same trait as a simple Mendelian trait involving (apparently) only that one gene.

9.2 DIGENIC TRIALLELIC INHERITANCE

Bardet–Biedl syndrome has a highly variable phenotype. It includes retinal degeneration, obesity, polydactyly, learning difficulties, kidney disease, genitourinary abnormalities, and may sometimes include a variety of other variable features such as diabetes or heart disease.

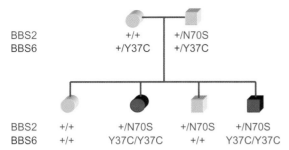

BBS2
BBS6

+/+
+/Y37C

+/N70S
+/Y37C

BBS2   +/+
BBS6   +/+

+/N70S
Y37C/Y37C

+/N70S
+/+

+/N70S
Y37C/Y37C

FIGURE 9.5 In a digenic triallelic Bardet–Biedl family we see that the only individuals who are affected are
those who have two defective alleles in one BBS gene and
one defect in another BBS gene. Some BBS genes participate
in digenic triallelic cause of disease, and some BBS genes
cause simple autosomal recessive disease. (After Katsanis
et al., *Science* 2001;293:2256.)

In 2001 six different genes had been reported
to cause autosomal recessive forms of Bardet–
Biedl syndrome. Then Nicholas Katasanis and
his colleagues discovered families in which
mutations in two different Bardet–Biedl genes
were present, but in this case the situation was
more complicated than digenic diallelic inheritance – instead of two mutations, one in each
gene, they found that each affected individual
had one mutant allele in the BBS2 gene plus
mutant alleles in both copies of the BBS6 gene
(Figure 9.5). They referred to this mode of
inheritance as digenic triallelic.

Since then the list of genes that can cause
Bardet–Biedl syndrome has expanded to include
more than a dozen genes. Some of these BBS
genes have not shown digenic triallelic inheritance. One mutation in BBS1 accounts for about
80% of Bardet–Biedl syndrome cases and no evidence has been found of BBS1 involvement in
digenic triallelic inheritance. Digenic triallelic
inheritance of Bardet–Biedl syndrome is very
interesting but so far it looks like most cases do
not involve digenic triallelic inheritance. As more
genes are found, and more combinations of genes
are tested, this picture might change, and the use
of animal models will assist in sorting out whether
some of the rare combinations of mutations

in more than one BBS gene might be responsible
for the disease phenotype.

Digenic triallelic inheritance has been found for
a variety of other traits, such as cortisone reduct-
ase deficiency. Long QT syndrome, which we presented above as an example of digenic diallelic
inheritance, can also be found in autosomal recessive and digenic triallelic forms. But as we will
discuss below, many other forms of cardiovascular disease show inheritance patterns that are even
more complex than digenic triallelic inheritance.

9.3 MULTIFACTORIAL INHERITANCE

Many of the most common public health problems would be considered multifactorial – being
caused by a combination of multiple different
factors. A prime example of multifactorial inheritance is cardiovascular disease. Consider all of
the events that take place in your cardiovascular system. If you have a pulse rate of 70 beats
per minute, your heart will beat more than a
hundred thousand times in the course of a day.
The average adult has about five liters of blood
and each heart beat serves to push that blood
through many miles of arteries, veins, and capillaries. Blood includes red blood cells to carry
oxygen, white cells that carry out functions of the
immune system, and a large number of different
serum proteins including growth factors, hormones, and immunoglobulins. Cardiovascular
disease can be a plumbing problem (blocked
blood vessels or damaged valves), an electrical
problem (aberrant timing of heart beats), or a
muscular problem (damage to the heart muscle).

Although there are rare instances of simple
Mendelian forms of heart disease, such as the ven-
tricular fibrillation caused by long QT syndrome,
most forms of cardiovascular disease result from
the additive effects of many different genes that
play a role in many different biological systems
(Table 9.1) combined with the effects of environmental factors such as diet, smoking, and exercise.

TABLE 9.1   Examples of biological systems involved in cardiovascular disease

Cholesterol levels
Triglycerides
Strength and flexibility of blood vessel walls
Formation of plaque
Adherence of plaque to vessel walls
Inflammatory processes
Heart muscle strength
Heart valves
Valves in blood vessels
Heart rhythms
Blood clotting

Many of the loci involved in cardiovascular disease have minor effects, increasing risk or decreasing risk without being a simple outright cause of disease. Genes whose products are involved in transport or processing of lipids can affect cholesterol levels. Reactive oxygen species can play a role in ischemic heart disease so genetic variants affecting levels of reactive oxygen species can affect cardiovascular risk. Channel genes, whose products transport ions into or out of cells, affect cardiac rhythms, as can many other kinds of genes. But in most people a combination of many of these factors is going on, with multiple genetic variants that each cause only a small variation in the system combining to increase or decrease any one person's risk.

If there are so many different factors adding up to the final phenotype, how can we figure out how much of the effect is due to genetics? By looking at twins, as we discussed in Chapter 8, we can ask about how much of the cause of a disease comes from genetic factors and how much comes from environmental factors (Table 9.2). We expect that a complex phenotype will be shared more often between identical twins, who share 100% of the same DNA, than between fraternal

TABLE 9.2   Concordance between twins in one study

	Monozygotic	Dizygotic
Migraine	34%	12%
Schizophrenia	46%	14%

twins, who on average share about 50% of the same DNA. Heritability is a measure of the extent to which genetic factors are responsible for variability in the trait. We decide how heritable a trait is by looking at how often identical twins are concordant (both share the same trait) and how often they are discordant (differ with regard to the presence of the trait).

Rules for Multifactorial Inheritance

As is the case for autosomal recessive and autosomal dominant traits, there are some rules that let you determine that a given trait is best explained by multifactorial inheritance.

1. Although the trait obviously runs in families, there is no distinctive pattern of inheritance (autosomal dominant, autosomal recessive, or sex-linked) within a single family. In other words, when nothing else makes any sense, start thinking about multifactorial inheritance. Nonetheless, a few rules are helpful in identifying a trait whose expression reflects multifactorial inheritance. These are:
   a. The risk to immediate family members of an affected individual is higher than for the general population.
   b. The risk for second-degree relatives (aunts, uncles, grandchildren, etc.) is lower than it is for first-degree relatives (parents, siblings, children), but it declines less rapidly for more remote relatives. This latter point is a hallmark of multifactorial inheritance and distinguishes it from autosomal recessive inheritance.
2. The risk is higher when more than one family member is affected. Traits reflecting multifactorial inheritance are

controlled by the number of deleterious or advantageous alleles segregating in the family. If there are several affected individuals in a family, the odds increase that a large number of deleterious or advantageous alleles are segregating within the family. In such a family the risk to subsequent children is increased when the parents are consanguineous.

3. The more severe the expression of the trait, the greater the risk of recurrence of the trait in relatives. The model for this is that the phenotype is presumed to be proportional to the number of deleterious alleles distributed over a large number of genes carried by that individual, with any one gene having a very small effect on the phenotype. An individual with many such alleles will, on average, pass on half of those alleles to their children and share half of them with their siblings. Imagine that some trait, such as cleft lip or cleft palate, was governed by alleles at 400 genes and that "normal" and "bad" alleles exist at all of those genes. Assume that any 50 such "bad" alleles distributed among 400 genes are sufficient to produce a phenotype and that the severity of the phenotype gets worse as the number of "bad" alleles increases. Thus an individual carrying 60 such "bad" alleles will be mildly to moderately affected. On average, he will pass only 30 "bad" alleles onto his children, which is usually too few to cause a problem. However, a more severely affected individual with, say, 150 "bad" alleles might be expected to produce children with 75 such alleles. Such offspring are very likely to be affected. Note that these are 400 hypothetical genes since many of the genetic factors affecting cleft lip or palate remain to be determined.

4. Because of the influence of the environment on such traits, identical (or monozygotic) twins need not always be concordant with respect to the trait. Indeed, the concordance can fall anywhere between 100% and the level of concordance observed between siblings. This may be especially true in cases where the number of alleles is just over the threshold at which characteristics of the trait can occur. Keep in mind that if the trait starts late in life, lack of concordance among younger twins could indicate variability in onset rather than genetic complexity or environmental influences.

If all of this seems less solid or more confusing than simple Mendelian inheritance, that is because it is. The understanding of polygenic traits and of genotype–environment interactions is becoming an increasingly interesting and profitable avenue of inquiry in human genetics, and we will spend more time on this topic in Chapter 12 when we look at how we find genes involved in multifactorial traits.

9.4 QUANTITATIVE TRAITS

Ira listened in frustration as his doctor explained that his blood pressure was borderline. Today it had registered 135 over 85. On the last visit it had been 120 over 73, and the visit before that it had been 142 over 90. "So, what does that mean?" Ira asked. "Do I have high blood pressure or not?" His doctor smiled ruefully and shook her head, saying, "It's not really that simple. Although I sometimes treat people at this blood pressure, I don't always. We know the higher values at which we clearly want to treat, and lower values that we never treat. But your values are moving around and you seem to fall in a range that we did not usually treat in the past. But as we are learning more about the effects of supposedly borderline blood pressures on long-term health, it has become clear that the pressures that should alarm us are lower than those that alarmed us in the past. There is no dramatic step from 130/75 to 131/76 that suddenly moves you from a healthy category into a disease category. And frankly, for some people the blood pressure is higher when they come running into a doctor's visit. So you need to start doing a blood pressure diary to record values every day. We can use that information when we talk about medication on your next visit." Ira walked out of the doctor's office shaking his head, wondering why the answer could not be some clear-cut yes or no. As we will discuss here, many human traits are quantitative traits that differ not by being present or absent, but rather by differing continuously over a range of values.

Many of the traits we have talked about so far have been *binary traits*, that is, traits with two states – people either have the trait or they do not. However, for many human traits, such as height, weight, or blood pressure, the variation is not binary but rather continuous across a broad continuum of values. Although some of those at extreme ends of the weight and height continuums may sometimes be suffering from pathological conditions, it is difficult to draw a simple line that lets us say that everyone above the line has the trait and everyone below the line does not. Moreover, many of these traits, even those that are strongly genetically determined, may also be affected by the environment. A real understanding of these traits requires a more sophisticated view of the mechanism by which a given genotype produces a given phenotype.

Many Traits Are Specified by Additive Effects of More than One Gene

A complication facing geneticists who are trying to find genes responsible for human traits is that the traits in which we are most interested, those posing the greatest public health risks to the most people, are often not simple

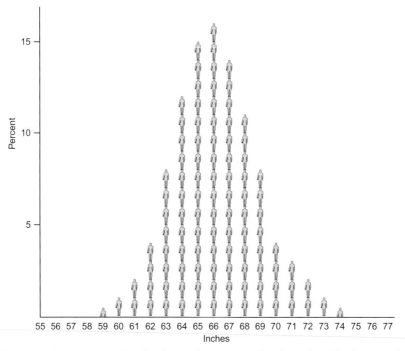

FIGURE 9.6   A histogram for variation of height of men. Over time, the height and weight charts used decades ago have gradually become outdated, raising questions about whether the different current distribution of height may represent some combination of altered nutrition in the population and changes in the overall composition of the population in this country, since we would see very different histograms if we plotted this information separately for Caucasian, Hispanic, African-American, and Asian men. (Adapted from Harrison, G. A., *et al.*, *Human Biology*, 2nd edition, 1977, Oxford University Press.)

yes/no traits. Rather, many of the major public health problems are complex *quantitative traits* that show some fairly continuous pattern of distribution of quantitative values. These quantitative, polygenic traits are the result of many separate effects that are added (and subtracted) together to arrive at the resulting phenotype.

Thus, for a trait such as height (Figure 9.6), a very large number of gene products (as well as additional environmental factors) go into determining one's height. One recent study found 20 different loci that influence human height and found that those 20 loci account for only about 3% of the variability in height in a population. Other studies have found additional loci but everything found so far seems to account for less than 10% of the variation in human height. Genes that affect height include genes encoding growth factors, developmental signals, hormones, proteases, and molecules of the extracellular matrix. Clearly, there are occasional Mendelian traits that affect height, such as shortness in cases of dwarfism or tallness in cases of Marfan syndrome, but most of the variation in human height results from a combination of dietary factors and the cumulative effects of sequence variants in different genes.

It has been estimated that perhaps 60–80% of the variation in human height is heritable.

9.5 ADDITIVE EFFECTS AND THRESHOLDS

Imagine that there was a trait that was quite variable in expression but not truly continuous. We think of cleft palate and other similar disorders such as *anencephaly* and *spina bifida* as being *threshold traits*, although some studies have suggested alternative models. A threshold trait is one in which individuals with the trait are thought to carry more than a certain threshold number of "deleterious" or "advantageous" alleles that are required to create a phenotype, as shown in Figure 9.7. Although the severity of this trait varies between affected individuals, it is not a continuous trait within the population. Babies are either born with a normal lip and palate or they are born with a cleft lip or palate, although the severity of the trait can vary greatly among those who have it. Cleft palate does seem to "run in families" and follows the rules for multifactorial inheritance that we presented above.

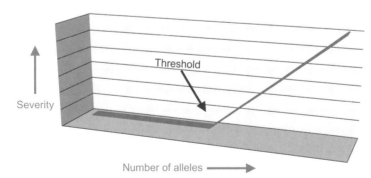

FIGURE 9.7 A schematic view of the threshold model, in which the key issue is how many total genes are affected. Thus different individuals with the trait may not all have defects in the same genes, but out of a set of many genes that can affect the trait, the ones who will be affected will be those with more "hits" on the set of relevant genes, and individuals who have some mutations in such genes will not be affected if the number of "hits" they carry is below the threshold level. For cleft palate, many of the genes have not yet been identified and we do not yet know how many genes are actually contributing to the trait, or how many are needed to pass the threshold. Might the number of alleles needed vary depending on the additional environmental factors that affect the trait?

Let's consider a model for how many genes might come together to result in cleft palate. Forming the top of the palate during embryogenesis requires that two masses of tissue in the head of the developing fetus arch up over the tongue and the mouth cavity and fuse properly to form the upper palate and the two sides of the upper lip. Those tissue movements would be influenced by many genes acting in concert. These movements also must occur in a very short window of time during fetal development. Clearly, some alleles might produce defective proteins that retard this process. As long as they don't retard it too much and the arch of the mouth is built before the window of time is closed, things will be fine. However, if more deleterious alleles are added to the genome of this fetus, progressively more impairment of movement is observed. Finally, as shown in Figure 9.7, in true "the one straw (or allele) that broke the camel's back" form, one too many proteins

is impaired, and the arch fails to be completed before the temporal window closes. The severity of the phenotype will then vary depending on how many of the deleterious alleles beyond the threshold are present in the fetus.

One of the more interesting consequences of proposing such thresholds is that it becomes easy to suggest that thresholds might be altered by environment, such that two identical genotypes might display different phenotypes in different environments. For spina bifida, this idea is supported by the observation that differences in prenatal nutrition affect the chance that a baby will be born with spina bifida.

9.6 IS IT GENETIC?

For most traits of medical importance, we end up wanting to know if the trait is hereditary (Box 9.1). One of the ways we tell whether

---

BOX 9.1

ONE OF SCOTT'S FAVORITE RULES

If a given trait is genetic you can map it AND if you can map it then it is genetic.

Among geneticists, a favorite example of nongenetic familial traits is the finding that in some families, attendance at Harvard Medical School runs in the family. We might hypothesize that a variety of genetic factors affect intelligence, specific talents, temperament, and other factors that contribute over the course of growing up to influence what someone will decide to do with their life or what they will be capable of doing with their life. But there are a great many other very obvious factors, such as geography, family wealth, family traditions, opportunities for relatives of alumni, and support or pressure from relatives that also contribute to decisions about

attending college. So is attendance at Harvard Medical School a genetic trait, and might we consider it to be genetically complex with strong environmental components? Or might we instead consider it an issue of sociology and economics outside of the geneticists' realm?

Some studies use information from pairs of sibling – or sib pairs, as they are known – or from twin studies (Box 9.2) to address such questions. If attending Harvard were genetic then we might expect to find greater concordance for attending Harvard for identical twins than for fraternal twins. If it is not genetic then we might expect that concordance for attending Harvard would be about the same for fraternal twins and identical twins.

BOX 9.2

TONGUE-ROLLING TWINS

A trait that has previously been reported as running in families is the ability to roll the tongue so that the sides of the tongue come up towards the roof of the mouth while the whole center of the tongue stays down. Tongue-rolling was studied in identical twins (who are genetically identical) and fraternal twins (who share on average 50% of their genetic information). It turns out that the fraternal twins are about as likely to share their tongue-rolling ability (that is, that they would both be able to or both be unable to roll their tongues) as are identical twins. So pairs of people with very different percentages of gene-sharing show the same frequency of trait-sharing. This suggests that the primary differences between people who can roll their tongues and ones who cannot may not be genetic.

a trait has genetic determinants is by studying the trait in families, looking for the kinds of patterns of inheritance we have talked about earlier. But we have to watch out for some pitfalls when we try to conclude things from what we see in families.

It may be harder to sort out whether a trait is genetic if it is recessive, especially if the trait is rare, because it may be hard to find families with multiple affected individuals. However, occasional large families can provide answers that we cannot get from pooling information from many small families. One of the other clues that a trait is recessive and inherited may come from the study of *consanguineous* families, that is, families in which a married couple shares ancestry (Figure 9.8).

Once a gene defect is identified, we can test many individuals with that disease to determine how many of them have that particular genetic defect. In some cases, such as cystic fibrosis, we find that everyone with the disease has a defect in the cystic fibrosis gene. In other cases, such as epilepsy, we may find that there are many different epilepsy genes and that only a tiny fraction of the population has a defect in any one of the genes.

Thus, although you may be able to look at your family and easily identify some simple autosomal dominant traits, for other traits you may need to consult a medical geneticist or a genetic counselor to find out if the item is hereditary, since any one family may not hold enough clues to provide the answer.

Our efforts to understand the genetic components of a trait are confounded by three key issues – phenocopies, penetrance, and expressivity. A *phenocopy* is a trait that has the same characteristics as the trait we are studying, and that can be mistaken for the trait we are studying, when it is actually a different trait with a different underlying cause. *Penetrance* is the extent to which an individual with the genotype ends up with the phenotype. *Expressivity* is the variation in the characteristics of the trait – such as severity or how early the trait starts. In the next sections we will discuss each of these three confounding factors and talk about some examples.

Genetic Classification Assists Risk Estimation

So far, the complex issues of interaction between genotype and environment are understood for only a small fraction of genetic diseases. This can make it difficult to sort out the genetic components of many traits that involve

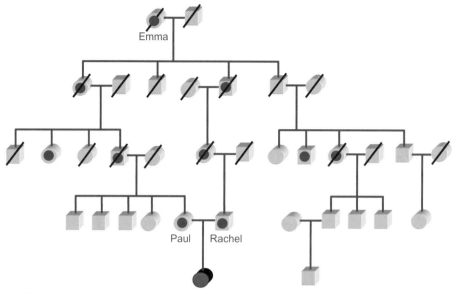

FIGURE 9.8   If Paul and Rachel each had parents who died young, they might not know a lot about their family tree and might not realize that they share a pair of great-grandparents. Even if they knew that they had shared ancestry, there is nothing in this family tree to tell them that they each inherited the same recessive deafness allele from their shared great-grandmother, Emma.

phenocopies or that may be conditional upon environmental effects, especially if there are many different genes and environmental factors that can affect the trait. Through the studies of geneticists, epidemiologists, and others, the complex interplay of genes and environment is gradually coming into focus. Will this eventually let us pinpoint who will become alcoholics if they take up drinking, or who it is that will develop lung cancer if they take up smoking? The goal is to end up being able to offer estimates of altered risk in the presence of environmental effects. Those risk estimates will tell us who has susceptibilities that should lead to more medical monitoring, different treatments, or advice on how we should change our diets or other behavior patterns. For many diseases that involve multiple different genetic components to susceptibility in addition to a complex array of environmental factors, it will continue being difficult to make predictions with any certainty about the medical fate of a specific individual even if we can make statistical statements of risk for groups of people.

Sorting out which effects are genetic and which effects are not will assist those who are hunting for the genes that affect susceptibility. It will also take us a long way towards a better understanding of many different risks in life. Risk estimation is now possible for many simple Mendelian traits, but we do not yet know enough for effective risk estimation based on genotyping for many of the most common complex traits that are major public health problems.

9.7 GENES AND ENVIRONMENT: INDUCIBLE TRAITS

Some traits are strongly affected by environmental factors. Height can be affected by the number of calories consumed at specific points in development during childhood. For some traits, disease severity can be affected by nutritional factors such as levels of key vitamins. In some cases a trait may be caused entirely by an environmental event without there being anything hereditary about the event. High levels of radiation can lead to cancer even if the individual did not inherit a cancer gene defect.

We use the term conditional trait to refer to those traits that occur only if some external eliciting event occurs. The quintessential conditional trait would be an infectious disease such as measles or influenza. Whatever the genetic factors that influence susceptibility to the infection, the disease will not happen in the absence of the infectious agent. A variety of other inducing conditions exist – dietary, pharmaceutical, meteorological, or even social. In fact, we might think in terms of many people being phenocopies of "normal" for a particular trait not because they lack the genetic component of that trait but rather because they have not been exposed to the necessary stimulus that turns a genetic susceptibility into a disease in reality. When an environmental influence is needed to bring about the manifestation of a trait that is not always present in an individual, we call it a *conditional trait*.

A profound example of a conditional trait is *malignant hyperthermia*, a reaction in which individuals exposed to general anesthetic or some other environmental conditions experience "explosive" elevation of body temperature. In operating theaters that are not prepared to cope with elevation of core body temperature this can be fatal, but doctors now know which anesthetics carry risk of malignant hyperthermia and there are precautions that can help save lives when it happens. There are many different causes of malignant hyperthermia. It has been suggested that there might be links to sudden infant death syndrome (SIDS) or heatstroke in some members of malignant hyperthermia families. Individuals who have never had general anesthesia do not know whether they carry a malignant hyperthermia defect, although clues to family history can come from looking at whether your relatives have received the high-risk categories of general anesthesia and whether they fared well. For most people, it might not seem like a great risk because the frequency of malignant hyperthermia has been reported to be about 1 in 50,000 cases of general anesthesia. However, the risk is substantially greater for those who have had a relative manifest malignant hyperthermia.

Malignant hyperthermia is an example of what is called a *pharmacogenetic disorder*, a disease state that occurs in response to a pharmaceutical agent (Box 9.3). For many pharmacogenetic traits, we expect that there will be multiple genes in the genome that will affect the combination of effectiveness and side effects that can result from use of any given drug. When dealing with a pharmacogenetic trait, it can be difficult to tell who is affected or at risk, since often there will be many family members who have never been exposed to the eliciting event.

Mad Cows and Cannibals

One group of conditional traits results from exposure to something that was originally thought to be an infectious agent, such as a virus or bacterium. This agent later turned out to be something much more perplexing, an infectious brain protein called a *prion* protein. A prion is a mutant form of a protein that not only takes on an aberrant structure, but also forces normal copies of the protein to take on that abnormal structure. Although they do not contain genetic information, they nonetheless get the host organism to produce more of the defective protein that causes damage to the brain. So prions turn the normal version of the protein into an altered disease-causing version of itself.

Kuru, a lethal human neurodegenerative disease called a *transmissible spongiform encephalopathy* (TSE), that was found to be transmitted between members of the Fore tribe in New Guinea via cannibalism. In cases of kuru, plaques and lesions are formed in the brain. These

---

BOX 9.3

PHARMACOGENETICS

The emerging field of *pharmacogenomics* involves the study of human genetic and genomic influences on variation in reactions to drugs. In some cases, such as *malignant hyperthermia*, the variability may take the form of an *iatrogenic*, or drug-induced, illness. In other cases, there may be a difference in response to the drug, as has been seen in the cases of some asthma sufferers who do not respond to an asthma medication because they have a mutation that changes the receptor protein that binds the drug. Often, the real array of variability in response to a drug is expected to be complex and involve differences in many different genes in the human genome. Thus, for example, we might expect the efficacy of a drug to be affected not only by the sequence of the proteins that is the target of the drug's action, but also by differences that affect proteins that will transport the drug into the cell, degrade and eliminate the drug, pump the drug out of the cell, target the drug to a particular place in the body, or carry out an immune reaction to the drug. Eventually, it is hoped that it will be possible to carry out genetic tests that will tell us whether one drug will work better than another for a particular person, or what side effects or risks there would be for that person if they take the drug.

---

plaques and lesions look like those found in other diseases, such as scrapie in sheep, mad cow disease in cattle, and similar transmissible spongiform encaphalopathies found in other animals. Such plaques and lesions even occur in some kinds of wild game consumed by humans.

The disease appears to originate from exposure to a misfolded form of the prion protein. Exposure to the externally supplied, misfolded prion protein coming from an infected person or animal apparently causes misfolding of endogenously produced protein so that the patient ends up with far more misfolded protein than he or she actually ingested.

Of great concern is the fact that prion infections can jump between species, as shown in the 1990s when a small number of people in Great Britain were reported to have developed transmissible spongiform encephalopathy from eating beef containing the "mad cow" version of the prion protein. We are left with questions about what genetic susceptibilities contributed to the production of disease in the rare few who developed transmissible spongiform encephalopathy out of a huge number of beef eaters. Questions are being asked about whether some of the increase in cases of Alzheimer disease might actually be undiagnosed transmissible spongiform encephalopathies rather than Alzheimer disease.

This conditional trait mimics a genetic disorder because the pathology of the disease resembles that of a human genetic disease called *Creutzfeldt–Jacob disease* (*CJD*). This hereditary neurodegenerative disorder presents with the same kinds of plaques and lesions found in the transmissible spongiform encephalopathies but is transmitted genetically within a family instead of via consumption of infected meat.

It is perhaps reasonable that kuru and mad cow disease, which are transmitted by prions, would resemble CJD since the CJD genetic defect is a mutation in the gene that encodes the endogenous protein that can misfold when the prion version of the protein is present. It has been pointed out that four out of six individuals who contracted a transmitted (rather than inherited) form of CJD all had a rare variant of the prion protein in which a valine is found at position 129 of the protein sequence. This suggests that genetic variants in the prion protein gene might also help determine whether someone is susceptible to the development of transmissible spongiform encephalopathy in people who are exposed to misfolded prion protein. This might also explain why the predicted mad cow epidemic has not materialized.

Diet

Environmental components of dietary sensitivities and allergies can also complicate our efforts to tell whether a particular item is genetic because the trait will only be manifested by people who have been exposed to the food or allergen. A trait called *favism* is characterized by a form of hemolytic anemia that happens after consumption of fava beans. Someone who has never consumed fava beans would not know whether they would be susceptible to favism or not. The prevalence of lactase persistence/lactose intolerance, discussed in Chapter 3, is hard to evaluate in countries where dairy products are not part of the diet. Sorting out the genetic components of allergies can be especially difficult because often people with allergies react to many different allergens, not just one, while being exposed to a vastly larger array of potential allergenic culprits. However, in some cases, such as penicillin allergy, the eliciting event (taking the antibiotic) and the unusual nature of the allergic reaction (a rash) make it easier to identify than some more generalized reactions to airborne allergens.

Age as an Eliciting Agent

While aging might not seem like an inducing agent, since it is not something coming from outside the body, it is something that can bring on the presence of a trait that was not present in someone younger. Consider this: why would someone who is born with the causative mutation(s) for a later onset disease, such as Huntington disease, not have that disease at birth? Why do they go through childhood and young adult life untouched by symptoms of this late onset disease? As the developmental process of aging proceeds, someone whose cells and organs were able to tolerate the presence of the mutation(s) is converted into someone whose cells and organs can no longer tolerate the defect. Then the disease process begins. In a world in which many things start going wrong with people as they age, we need to think of aging as one of the eliciting events, and to consider age-associated traits much as we consider conditional traits. *Age-dependent penetrance* is the term used to express the increased expression of a trait in older populations when compared with younger populations. Issues of aging and conditional expression of traits are genetically complex. This becomes complicated as we consider that genetic variation affects the rate at which we age, which in turn affects the rate at which aging causes us to express other traits caused by genetically variable susceptibilities.

9.8 GENES AND ENVIRONMENT: INFECTIOUS DISEASE

A specialized instance of an environmental factor is an infectious disease. There are a number of genes in the human body that affect our abilities to resist various kinds of infections, and the outcome of an infection can depend in part on how well our bodies are prepared to cope with the particular invader causing an illness (Box 9.4). Some of the mechanisms by which we protect ourselves seem fairly obvious. If an infectious organism has to attach to a particular protein or receptor to enter a cell, a human being may be protected if she is lacking that protein or if she makes a variant form of the protein that the "bug" doesn't recognize.

BOX 9.4

WHO WILL BE AROUND AFTER THE NEXT EPIDEMIC?

The processes that generate genetic diversity, while sometimes causing problems, can also provide advantages under the right circumstances. Diversity turns out to be of especially great importance in terms of responses to infectious disease. Who becomes infected and who survives in an epidemic might seem like a random event, akin to being struck by lightning, but there are actually a large number of genetic differences between individuals that affect their resistance to different diseases. Some of these differences actually involve molecules of the immune system but can also affect genes involved in many other host defense systems. The result is that people who survive a polio epidemic may be a different subset of the population than those who would have survived if it had been a smallpox epidemic instead. If we built a population of clones derived from one individual, they would share similar (though not necessarily identical) fates in the next epidemic. If it happened that the population had been built from an individual with good defensive mutations against that particular disease, they would fare well, but what is the chance they would all be similarly genetically prepared for the next

infection, and the next, and the one after that? This is a problem already faced in agriculture, where trends towards growing certain popular genetically identical strains puts crops at risk of an all-or-nothing fate depending on whether they are or are not resistant to the next pest or bug that comes through the region. Imagine the hazard to a human community, or humanity as a whole, if we all shared identical sequences in the genes that affect resistance to infectious diseases. Even if we could engineer it so that we all started out resistant to the known diseases, new diseases and new strains keep coming along. Influenza, and the need to keep getting new flu vaccinations each year to keep up with the constant trickle of new antigenic types, offers one of the strongest lessons in the rate at which infectious diseases can keep changing almost faster than our ability to cope with them. Our diversity, then, is one of our greatest protections; not in the sense of protecting any one individual, but rather in the sense of protecting populations overall so that there is someone left to carry on after the epidemic is over, and someone to take care of those who are ill while the epidemic is ongoing.

If the invading organism causes damage when a protease cleaves an important human protein, protection can come from a change in the protein sequence that eliminates the cleavage site.

Just one type of infectious organism involved in a single infection in a human being may be using quite a diverse arsenal of biochemical tricks to assist in establishing a connection with the host, facilitating invasion to arrive at its favorite target cells, diminishing host defenses (immune and otherwise), and causing damage to the cells and/or surrounding tissues. Every single point

of interaction of that infectious organism with the human body represents a possible point of susceptibility or resistance to the invader, depending on whether mutations in human genes have produced altered forms of the human proteins with which the infectious organism interacts. In many cases, even if someone has a sequence change in one of those key proteins, it may not be a change that affects the critical points of interaction between human and bacterial proteins. In a large population, with many different mutations having occurred over long periods of time, it is

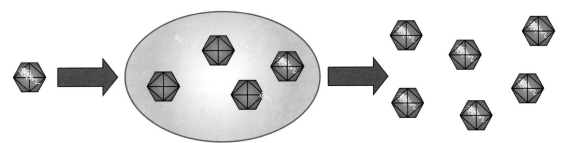

FIGURE 9.9   Generalized concept of a viral infection. A virus enters into the cell and uses the cell's enzymes to replicate the viral genome and uses the cell's ribosomes to make copies of the viral proteins needed to assemble a protein coat. The many copies of the virus are then released, either by killing the cell or by the cell releasing the viruses. There are many specific details that vary from one virus to another, from one cell type to another, but the fundamental problem for the virus and the host is that the virus has to invade the host cell and use its machinery to copy itself.

likely that some people will have different reactions to any given pathogenic factor in the bacterium's arsenal.

Infectious Organisms

Three major classes of organisms are responsible for many of the most common human infections – viruses, bacteria, and parasites. They have very different properties that influence how the human body reacts to the infection.

Viruses have the general structure of a genome (made of DNA or RNA) packaged in a protein coat; even though they can cause infections that result in production of many new copies of the virus, they need a host cell to provide the machinery that copies the viral genome and produces new copies of the viral protein coat (Figure 9.9). Viruses are tiny, measuring in the nanometer size range, but there is a lot of variation in the sizes of the different kinds of viruses. There are thousands of different kinds of viruses, including both bacterial viruses that infect bacterial cells and other viruses that infect humans or other multi-celled organisms. Viruses are sometimes not considered to be "alive" since they do not possess all of the structures and functions needed to replicate the virus, although we commonly refer to live virus

or killed virus when distinguishing between viruses that can cause an infection and viruses that have been rendered incapable of causing an infection.

Bacteria are cells that possess all of the machinery needed for replication of the bacterial genome and cellular structures (polymerases, ribosomes, etc.). Bacteria are tiny cells; one typical bacterium called *E. coli* is about 2 micrometers (or microns) across. Bacteria commonly each have a single circular chromosome, and may often have one or more copies of a small additional circular piece of DNA called a plasmid carrying a small number of additional genes. Many effective antibiotics have been developed that can target things about bacteria that are different from similar structures in a human cell; for instance, bacteria and human cells both have ribosomes, but bacterial ribosomes are different from human ribosomes in ways that make them targets for some types of antibiotics.

Parasites are a third kind of infectious organism. Parasites are much more similar to the cells of the human host. Cells like bacteria that lack a nucleus are considered prokaryotes, and the nucleated host cells are considered eukaryotes. Parasites are also eukaryotes and have a nucleus. Parasites can be especially difficult to treat because of their similarity to human cells.

Host Defenses

The human body offers many different *host defenses* against infectious diseases, and each of these defenses uses structures or processes involving proteins encoded by one or more genes in the human genome. In some cases, susceptibility to infection may be accounted for by some generalized reduction in effectiveness of the immune system, such as that seen in individuals with mutations in the adenosine deaminase gene who develop *severe combined immune deficiency* (SCID). General effects in efficiency of the immune system can result from mutations that affect the MHC-II genes whose products are responsible for "presenting" antigens to the immune system to initiate an immune response.

Mutations in some genes play a specific role in resistance to specific bacterial infections. People with type O blood are more likely to have more serious cases of cholera if they become infected with *Vibrio cholerae*. In some populations, sequence differences in the NRAPM1 gene are associated with predisposition to leprosy, tuberculosis and salmonella.

Similarly, mutations in specific genes can affect susceptibility to infection with specific types of viruses. Human immunodeficiency virus (HIV) was discovered in 1981, and in 1996 a human allele was reported that reduces susceptibility to HIV infection. Many strains of the human immunodeficiency virus (HIV) use the chemokine C-C receptor CCR5 as the point of viral entry into critical cells of the immune system including CD4+ T helper cells, an important cell type that becomes depleted during the disease process, with devastating consequences in most untreated individuals. However, rare individuals have mutations in the CCR5 gene, mutations that interfere with viral entrance into the cell. The CCR5 sequence variant CCR5-delta32 was first detected in an HIV-1 infected individual who showed slow progression of disease and two other individuals who did not become infected in spite of repeated exposures.

The CCR5-delta32 variant is rare and mostly found in individuals of European ancestry. Since then additional CCR5 variants have been identified that can affect HIV risk or long-term progression of disease. The understanding of the role of CCR5 in viral entry into the cell, and the observation that inactivating CCR5 can help protect against the HIV virus, have led to efforts to develop drugs that target CCR5 protein.

There are many genes whose variants affect susceptibility to the parasite malaria, such as genes that encode alpha-globin, beta-globin (remember sickle cell anemia?), glucose-6-phosphate dehydrogenase, nitric oxide synthase 2, the Duffy blood group, and others. We see dramatic differences in frequencies of the key variants of these genes in populations living in areas where malaria is endemic. It is not surprising to see an increase in frequency of alleles that confer a survival advantage, since the disadvantageous allele is being eliminated from the population one person at a time as each new death occurs. For many infectious agents, many different genes can be involved in determining susceptibility, and the result is often *mitigation* (reduction in the chance of becoming ill or reduction in severity of the resulting illness) rather than outright prevention of the disease. Thus even determining whether anyone became ill as a result of exposure to an infectious agent may not offer a clear answer to what that individual's genotype holds by way of resistance or susceptibility to that infectious agent. Some alleles that confer disease resistance increase in frequency in regions where the disease is endemic – chronically present in the population.

Many host responses involve genes that actually play other roles in the body. The IgE antibody helps fight certain types of parasitic worms. The fact that such worms are not present in North America does not eliminate IgE reactions in the human body. We notice its presence because IgE is responsible for allergic reactions to things like pollens.

9.9 PHENOCOPIES

Chloe sits quietly, one hand in her lap, the other holding onto the white cane that helps her navigate without tripping on things when she walks. She starts talking with tears in her eyes. "When I was growing up I didn't know my father so I didn't know that he had glaucoma and I should be getting my eyes examined. I lost a lot of vision before I was diagnosed and the doctor started doing things to help me. But my kids? My kids know what happened to me, but they are in denial. Each one is sure that he or she won't be the one to end up with it, and I can't get any of them to get their eyes examined. But if they wait until they have problems before they see a doctor, won't they have damage that can't be fixed?"

More than half of those who have glaucoma do not know that they have it, and will not know until after they have suffered irreversible damage. There are effective treatments available, but they can only prevent damage, not fix existing damage. So the hardest thing about this situation is that the damage that can't be reversed is damage that probably could have been prevented if an eye exam had led to a diagnosis or glaucoma. Is there a particular age at which you should start being examined for glaucoma? Your eye doctor can answer that question, based on general information about prevalence in your ancestral population as well as specific information about your own risk factors such as whether or not you have a family history of the disease.

According to the World Health Organization, more than 100 million people worldwide are believed to have glaucoma, and more than 5 million of them are blind because of it. Glaucoma is the leading cause of blindness among African-Americans and the second leading cause of blindness among North Americans of European ancestry. Among the millions of individuals in the United States who are affected with glaucoma, about half of them do not yet realize that they have the disease and many of them will not realize it until after

irreversible damage has occurred. Although increasing age is one of the risk factors for glaucoma, and most cases of glaucoma are found among those who are over forty years of age, there are some forms of glaucoma that can also be found less often in young adults and children, and one form of glaucoma even occurs in newborns.

In glaucoma, nerves that carry visual signals from the eye to the brain die in a characteristic pattern, usually slowly over a long period of time. This loss of nerves is accompanied by the loss of visual field (or area of vision) from the regions served by those nerves. Typically, visual field is first lost in local regions while vision remains excellent in surrounding regions. The local and arc-like regions of missing vision can gradually merge to form large regions of visual deficit if the disease remains untreated. Many people think that having glaucoma means having increased pressure inside the eye, because the most common form of glaucoma involves elevated intraocular pressure. However, elevated intraocular pressure is no longer part of the definition of the disease since the characteristic pattern of nerve cell death in glaucoma happens to people who have never shown any increase in the pressure inside their eyes.

There is a lot of evidence for genetic causes of glaucoma. As many as half of the patients being seen for glaucoma may have relatives who also have glaucoma. In some cases, autosomal dominant inheritance of glaucoma can be traced for three or four generations, with one family we know about showing transmission of glaucoma through eight generations in a row. With the mapping of more than a dozen different loci that can cause glaucoma and the identification of disease-causing mutations in more than a half dozen glaucoma genes, glaucoma now ends up being classified as a genetic disease.

There are also many cases of glaucoma in which a clearly non-genetic cause of the disease can be seen. Not only can some kinds of injury cause glaucoma, but it can also be caused by some medications or inflammatory situations and may sometimes be an unwanted consequence of some kinds of eye surgery. Although we can easily see how genes play important roles in the processes that lead to glaucoma, in some cases it is equally clear that the actual event that started the disease was an external non-genetic event.

The simple Mendelian forms of glaucoma are rare. The most common types of glaucoma appear to be multi-factorial. The combination of genetic complexity, environmental factors, genetic heterogeneity, and phenotypic heterogeneity make the process of finding glaucoma genes a real challenge.

Identity by Descent

When we use genetic information to map glaucoma genes we make the assumption that family members with the same trait share *identity by descent*. The term identity by descent refers to a situation in which the same underlying cause is producing the same trait among relatives because they have inherited the same genetic defect from a shared ancestor. The term *identity by state* refers to a situation in which two people have the same trait but we cannot tell whether the cause of that trait is the same. In some families where people share a trait, we assume identity by descent as the simplest explanation for what is going on, but sometimes someone in the family is a *phenocopy* whose trait is the result of something environmental or from a different underlying genetic cause. As a further complication, a case of *non-penetrance* shares the genetic defect with the affected relatives, but does not manifest the trait.

The four concepts – identity by descent, identity by state, phenocopy, and non-penetrance – are related because they all relate to the issue of

whether or not the trait we see in different family members is really the same trait. A phenocopy will be a case of identity by state, which can also be considered a false-positive (something that seems to be the trait we are studying when it is really something else that looks just like it), and a case of non-penetrance will be a false-negative (a case of someone who inherited the same genotype as the affected relatives but does not have the trait). Those cases that are identical by descent are the ones we want to be using in our genetic studies.

Phenocopies and Glaucoma

Phenocopies become a problem in glaucoma genetics studies in several ways. There are many different forms of glaucoma. When we see glaucoma running through a family, and think that glaucoma in that family results from a genetic defect, we expect that there is a chance that some family member affected with glaucoma might actually not have the familial form of the disease but rather glaucoma due to some outside, unrelated cause. In some cases, as in Figure 9.10, there may be clues to help us tell who the phenocopies might be, but in other cases, we may not have enough information to tell whether there are any phenocopies. Especially when families are large, and information about deceased relatives is coming from distant family members who might not have known them well, information can sometimes be incomplete without it even being obvious that anything is missing.

Identity by State

Identity by state (having the same trait without sharing the same underlying cause) can come about in several different ways. If someone has glaucoma because of something environmental, then we have some hope of being able to sort out the fact that that person does not have what the rest of the family has. But identity by state becomes a confounding factor when

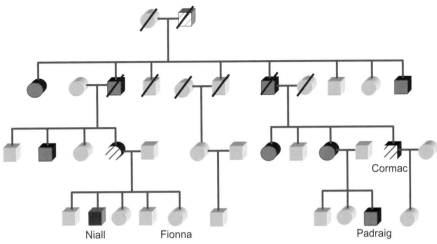

FIGURE 9.10 If Fionna's family history looked like this, what might we conclude about her chance of developing glaucoma, which depends partly on whether or not her mother was carrying the family's defective glaucoma gene? We presume that glaucoma in Padraig and Fionna's mother would be identical by descent. Fionna's mother ended up with glaucoma after years of using a prescribed corticosteroid that is known to cause a form of glaucoma called steroid glaucoma. Fionna's brother Niall moved to Australia, where he was rumored to have developed glaucoma, but the family has not been able to find out what kind of glaucoma he had. The family knows that Cormac has glaucoma but is he identical by descent or is he a phenocopy? He is a phenocopy, since they do not know that he has a completely different kind of glaucoma that resulted from a blow to the eye. Fionna's great-grandfather was rumored to have been blind but the cause of his blindness was unknown. So even though we know that Fionna has a strong family history of glaucoma, we have trouble estimating her risk because of individuals in the family whose glaucoma is not identical by descent.

someone with glaucoma in their family marries someone who also has glaucoma in their family. The possible input of more than one glaucoma genetic defect into the same family might seem an unlikely thing, but when we are dealing with such a common disease, we find that people with a family history of glaucoma often marry other people with a glaucoma family history.

Since most people do not develop glaucoma until late age, and some people who carry glaucoma genes pass away before their disease ever manifests itself, it is also possible for some people with a glaucoma gene defect to marry into a family without realizing that they have a family history or that they will later develop the disease themselves. Some people do not realize their relatives have glaucoma even when the affected relatives are currently quite alive and under active care for their vision problems. So it is possible

to have different genetic defects coming into the same family from different points in the family structure for several different reasons, and to have very similar medical courses even if the underlying causes include more than one gene and at least one unrecognized environmental factor.

What If We Don't Realize It Is Genetic? What If It Looks Genetic but It Isn't?

Serious situations can arise when a genetic defect exists but is not recognized as being genetic in origin, or when something is deemed genetic that is not. In the case of osteogenesis imperfecta, some parents have been jailed for child abuse, accused of causing broken bones through physical abuse, when in fact the child had a genetic defect that causes incredibly fragile bones that could not tolerate even the normal stresses of daily life without breaking. In other cases, it can be a big problem if neurological problems in multiple children in a family are dismissed as genetic, if what was really needed was to identify the source of lead exposure in the home environment that needs to be remediated.

9.10 GENOTYPIC COMPATABILITY: WHOSE GENOME MATTERS?

Now stretch your brain around this concept: in some cases, your phenotype might not be caused by your genes. Sometimes it might be caused by someone else's genes, or by a combination of your genes and someone else's.

A prime example of this is the case of a mother with *phenylketonuria* (*PKU*) who has two defective copies of the *phenylalanine hydroxylase* (*PAH*) gene that makes her unable to correctly metabolize phenylalanine (Figure 9.11). Even if she ends up healthy because she grew up eating a diet low in phenylalanine, her baby will be damaged by high phenylalanine levels if she does not keep her diet controlled during the pregnancy.

Logically, in addition to *maternal-fetal incompatibilities*, we can imagine *in utero* situations by which the genotype of one fraternal twin could lead to the production of a gene product that could affect the other twin who is not making that gene product. This would be especially important in the case of hormones or other exported proteins that could travel from one sib to the other by getting into the amniotic fluid. In studies of mice, it has been shown that male mice that undergo prenatal development surrounded by female siblings can come out with phenotypic differences from male mice that developed surrounded by other male mice. As with unusual genotypes leading a mother to expose her child to something the baby is not producing, so we can imagine human twins potentially influencing each other while still

in the womb. This should only apply to certain kinds of proteins, such as hormones that can leave the first baby and get into the amniotic fluid that contacts the second baby in that shared environment.

Modern technology leads to another phenotype that only happens in reaction to some combinations of genotypes of two different individuals – organ transplantation. Each of us has a set of proteins present on the surfaces of our cells that are recognized as being self or non-self. Each of us can muster immune reactions that are aimed at attacking foreign cells such as parasites and fungal infections. This type of immune response will not only attack infectious cells but can also mount an attack on cells from another human being. When someone who has signed up as an organ donor passes away, there are lists of people waiting in line to get a heart or a kidney or a variety of other organs. When someone receives an organ transplant, doctors have to be very careful about matching the donor and recipient tissues to avoid rejection of the transplanted cells by the new host.

In the case of one of the other blood type loci, the Rh locus, we see a phenomenon called *Rh incompatibility. At each Rh locus, we either do or do not make the Rh blood type factor that appears on the surfaces of blood cells.* When the baby has an Rh+ blood type and the mother has an Rh− blood type, the baby can develop anemia after birth as a result of exposure to maternal antibodies directed against the Rh proteins in the baby's blood. However, this can only happen after the mother has borne a previous baby with the Rh+ blood type.

How many other factors that affect a baby's development before birth are also affected by the maternal genotype? We can imagine that a variety of things, including hormone levels that affect whether the baby is delivered prematurely, and other factors affecting things such as nutrition and oxygenation could all affect the traits that will be observed in the new baby that do not

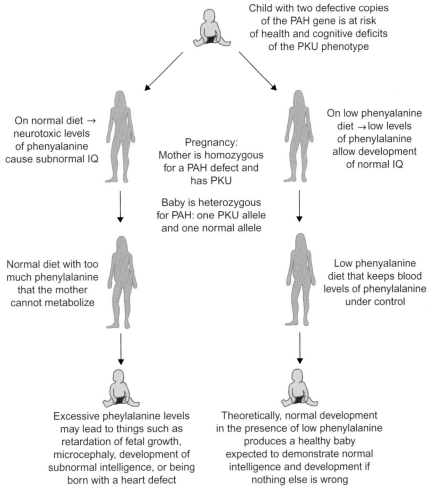

FIGURE 9.11   Deficiency in the enzyme phenylalanine hydroxylase can have devastating consequences not only on the person with the deficiency if they grow up eating a normal diet but also on their child if they do not use careful diet management during the pregnancy. This turns out to be a major problem because it is hard to stay on a diet that has just the right amount of phenylalanine, since damage can come from too little phenylalanine as well as too much.

depend on the new baby's genotype at all. So in this case the problem has underlying genetic and environmental factors but only some of those genetic factors are found in the baby since one of the critical gene-based elements is only present in the mother.

The most common situation in which this is a potential problem is in the case of blood transfusions. Any time we need a blood transfusion great care is taken to be sure that the blood we receive is the right blood type. The main piece of information that most of us know about our blood type indicates our phenotype (and genotype) for the ABO blood type genes on the long arm of chromosome 9 (Table 9.3). On each copy of the ABO locus, we have either an A allele, a B allele, or an O allele. The A allele makes a cell surface structure called A, the B allele makes a cell

TABLE 9.3   The ABO blood group

Recipient genotype	Recipient phenotype	Donor phenotype
AA or AO	A	A or O
BB or BO	B	B or O
AB	AB	AB, A, B, or O
O	O	O

surface structure called B, and the O allele makes neither A nor B. A we see in Table 9.3, the combinations of possible alleles present on the two copies can result in four different phenotypes (blood types), A, B, AB, or O. Thus the blood type (phenotype) depends on which combination of alleles someone has (A, B, and/or O alleles), but an individual's own genotype and phenotype does not produce the transplantation phenotype. Blood transfusion reactions occur only when the donor and recipient phenotypes are incompatible, and both donor and recipient blood types depend on the genotypes at the ABO blood type locus. So we cannot look at any one individual's genotype or blood type to tell whether or not they will have a reaction to a blood transfusion, but the information on both donor and recipient blood types is fully informative regarding whether or not there will be ABO incompatibility.

9.11 PHENOTYPIC HETEROGENEITY: ONE GENE, MANY TRAITS

One of the surprising revelations that emerged as molecular geneticists began identifying human genes was the finding that sometimes apparently different traits can be caused by mutations in the same gene. If you think back to our analogies regarding absent essentials and monkey wrenches, it makes sense that you might get a different effect from eliminating a gene product (and its corresponding function) than you would get by adding a new function for the same gene. Is it really that simple and straightforward, that knocking out a gene causes one trait, partially knocking it out causes another, and adding a new function to it causes the other trait? What kinds of variation can there be, and how much can we predict just from knowing which gene it is and which kind of mutation? In the long run, sorting out such issues will be important to be able to make the most use out of genetic testing information.

One of the most dramatic examples of the one gene–multiple phenotypes phenomenon is the androgen receptor (AR) gene. Back in Chapter 8 we noted that loss-of-function mutations in the AR gene cause CAIS, but the story turns out to be somewhat more complicated. If we look at the known types of mutations in the AR gene, we see not two but three phenotypes that are dramatically different:

- If the AR gene is knocked out, that is, if a mutation eliminates the gene product or prevents the AR protein from performing its function, the result is the CAIS phenotype described in Chapter 8 – external female anatomy, no uterus, and presence of undescended testes in an individual with both X and Y chromosomes who is apparently female but infertile.
- If a trinucleotide repeat sequence located within the AR gene is expanded from its normal copy number of about 21 copies up to 40 or 50 copies, in the same way the Huntington disease gene repeats are expanded, then the result is a neuromuscular trait called spinal bulbar atrophy. In this case, there are questions about whether the phenotype results from anything about AR function or loss thereof, or whether the interjection of the amino acid repeat is adding a new function that actively causes a set of events unrelated to the protein's normal function, as we see in Huntington disease.

- In some cases, an AR missense mutation turns up in cases of prostate cancer, with the altered receptor chronically signaling the cells to proliferate even when testosterone is not present. Since binding of testosterone to the androgen receptor signals proliferation of prostate cells (and prostrate cancer cells), some treatments for prostate cancer inhibit testosterone production to stop tumor growth. However, the "always on" mutations in the AR gene bypass this prohibition, allowing continued tumor growth in the absence of testosterone.

This finding of such disparate phenotypes associated with the same gene seems surprising but makes sense when we look at the underlying mechanisms in the three cases. We expect loss of the gene product to prevent hormone signaling, and we now know that loss of that hormone signal changes the pathway by which secondary sexual characteristics develop. Based on studies of other diseases, we might expect that a repeat expansion could cause a gain of function (such as making the protein sticky) that might bring about effects completely unrelated to whether or not the hormone is successfully transmitting its signal to the receptor. A missense mutation might be expected to alter how the protein folds and/or to change the binding properties of the protein, perhaps causing changes in the interaction of hormone and receptor but not eliminating the signaling process. More work is needed to understand these complex situations, but at least we have a framework for the formation of hypotheses to be used in designing new experiments.

9.12 GENOTYPIC AND PHENOTYPIC HETEROGENEITY

Dana hates going to movies because she cannot see well enough in the dark to navigate her way out of the darkened theater to get to the bathroom in the faint runway lighting used when the theater lights are turned down. She has no problems watching videos at home on her TV, where she can keep the lights turned up while she watches. Dana has a trait called congenital stationary night blindness. It is called congenital because she has always had it. It is called stationary because it does not change as she ages, which is good news if it means that her condition will not get any worse than it is now. It is called night blindness because it affects her ability to see things at night like the stars or to see in dark situations like in the theater. Her eyes work just fine for anything she wants them to do during the day or in a lighted room. Dana has a mutation in the gene that makes the rhodopsin protein. Dana's vision problems actually make sense when we consider rhodopsin's normal function – detecting faint light at night, such as starlight. But hold on – at the beginning of the chapter we said that Jackie had retinitis pigmentosa as the result of a rhodopsin mutation. But most people with rhodopsin mutations have retinitis pigmentosa, like Jackie at the beginning of this chapter. So why doesn't Dana have the same phenotype as Jackie if the same gene is causing the problem? They have different mutations in the same gene. Jackie's mutation affects where the rhodopsin protein travels to in the cell and results in a very unhealthy rod cell; Dana's mutation affects the ability of rhodopsin to correctly signal its detection of light but it does not do anything to make the rod cell ill. Dana, with night blindness, and Jackie, with retinitis pigmentosa, represent a situation that turns out to be rather common, a mix of genetic and phenotypic heterogeneity. What are some of the factors that affect the phenotype that results from any particular mutation, and are there theoretical considerations that let us look at a newly discovered mutation and predict what trait will result from the mutation?

More than 120 different loci have been mapped that can cause forms of retinal degeneration that involve the death of photoreceptor cells in the eye, and the genes corresponding to more than 80 of those loci have been identified. One of the genes in which mutations were found

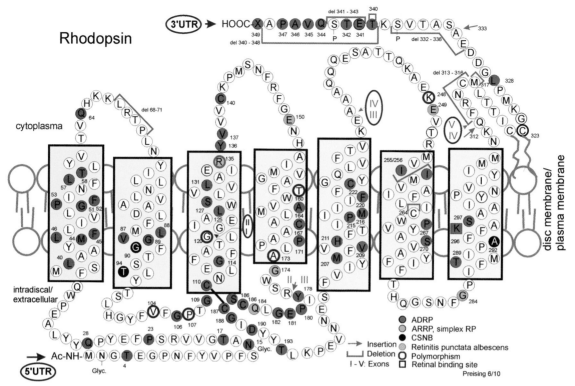

FIGURE 9.12 Mutations in rhodopsin are spread throughout the molecule. Both missense and nonsense mutations can cause either dominant or recessive retinitis pigmentosa, and some missense mutations that are not located anywhere near each other can cause a much milder disorder that involves night blindness without retinal degeneration or daytime blindness. (Image courtesy of Retina International Scientific Newsletter – http://www.retina-international.org/sci-news/rhomut.htm.)

that can cause retinitis pigmentosa, one of the main forms of retinal degeneration, is the gene that encodes *rhodopsin*, the protein that detects faint light at night. Mutations in rhodopsin are responsible for about a third of the retinal degeneration that is found in families with autosomal dominant inheritance of retinitis pigmentosa.

Retinitis pigmentosa is considered *genetically heterogeneous* because it can be caused by mutations in many different genes, including rhodopsin. Rhodopsin phenotypes are considered *phenotypically heterogeneous* because mutations in rhodopsin can cause multiple phenotypes, including retinitis pigmentosa and congenital stationary night blindness. Modes of inheritance of

retinitis pigmentosa include autosomal dominant, autosomal recessive, and sex-linked recessive inheritance, and different rhodopsin mutations can cause autosomal dominant disease in some people and autosomal recessive disease in others. There is currently no cure for retinitis pigmentosa, although as we will discuss in Chapter 14, there now exists an apparently successful form of gene therapy for another form of retinal degeneration that promises important possibilities for the treatment of many forms of retinal degeneration.

When we consider rhodopsin, we cannot simply look at the type of mutation and predict what the phenotype will be like or, frankly, even what the mode of inheritance will be.

Missense mutations have been found that are located throughout the gene affecting many different positions in the protein (Figure 9.12). Most of those missense mutations cause retinitis pigmentosa, which results in progressive death of the photoreceptor cells that make rhodopsin. However, several different missense mutations in rhodopsin instead result in a fairly simple form of night blindness, with any retinal degeneration being very minor and occurring very late in life, if at all.

The rhodopsin molecule had already been studied extensively before it was identified as a disease gene. It's a fascinating molecule that sits in the rod photoreceptors and catches photons of light from faint sources in the dark, such as starlight. Because the processes by which rhodopsin "catches" light and sends the signal along to the brain had been so well studied, it was possible to make some intelligent predictions regarding some specific amino acids that would be expected to cause disease if changed. For instance, it was not surprising to find that one of the disease-causing mutations changes amino acid 187, which had been shown to contribute to the correct folding and shape of the protein by forming a bond that holds two specific points in the molecule together. Another mutation was found that changes amino acid 296, the amino acid that was already known to bind to the vitamin A derivative that is essential to rhodopsin function.

However, for many of the amino acids, their importance, and their role in disease pathology only became evident once a mutation at that position was identified and the phenotype examined. For instance, one missense mutation at position 90 (Gly90Asp) had no particular known role until it was found to cause the kind of night blindness that Dana has.

Once this was discovered, extensive studies revealed its role in the process of detecting light. When the rhodopsin molecule is folded up into its three-dimensional structure (Figure 9.13), it contains a binding pocket in which the

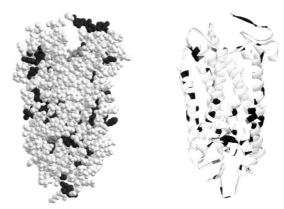

FIGURE 9.13 Three-dimensional view of rhodopsin showing locations of different mutations that cause autosomal dominant retinitis pigmentosa (black) or other forms of disease, including autosomal recessive retinitis pigmentosa and congenital stationary nightblindness (gray). On the left is a space-filling model showing the positions of the different atoms within the molecule. On the right is a ribbon model that gives a view of how the protein is folded into helical, pleated, and other structures to achieve the three-dimensional structure. These images were generated with Accelrys' Discovery Studio Viewer Pro software from the results of submission to SWISS-MODEL the Automated Protein Modeling Server. The resulting model was based upon the coordinates of PDB entries 1HZX, 1L9H, and 1F88, all of which are x-ray diffraction crystal structures of bovine rhodopsin. Positions of human mutations of the rhodopsin gene were taken from Retina International's mutation database. (Courtesy David M. Reed.)

vitamin A derivative sits. The glycine at position 90 is one of the amino acids that lines the surface of this binding pocket. Glycine is a neutral (uncharged) amino acid and asparagine has a negative charge, so a Gly90Asp mutation introduces an extra negative charge into the binding pocket in the vicinity of the vitamin A derivative. Rods normally function only in dark environments and shut down in daylight, perhaps a mechanism that protects them against high light levels that they are not designed to use. Adding a new negative charge into the binding pocket seems to cause a chronic low level of signaling from rhodopsin that makes the rods think they are in a high light level environment, so they shut down even at night just as they would at noon on a sunny day.

Sometimes when a new disease gene is identified, it turns out to encode a known protein such as rhodopsin, one of the globins, an immunoglobulin, or other well-characterized proteins that had been studied for years before cloning and sequencing came along. However, when many genes are first found to be disease genes, we do not yet know what role they play in normal cellular processes or disease pathology. This makes it even harder to figure out how a mutation is bringing about the trait being studied.

9.13 VARIABLE EXPRESSIVITY

Sharon has a juvenile onset form of glaucoma that was diagnosed many years ago. Half of her brothers and sisters have glaucoma. Each affected family member developed high pressure inside of the eye as the first sign of the disease. When researchers found the MYOC mutation that causes glaucoma in her family, they were not surprised to find that each family member with glaucoma has the mutation. They were surprised to find that Sharon's sister Shirley also has the mutation. So does Shirley have glaucoma? No. But she does have one of the key symptoms that other family members developed in the earliest stage of the disease – elevated pressure inside of the eye. Why would Shirley just be showing the first signs of the disease decades after all of her brothers and sisters had progressed to full blown cases of glaucoma? Join us now as we talk about variation in phenotype severity and onset.

People with the same disease do not always have exactly the same characteristics. We are not talking about different diseases or traits as we did above, we are simply talking about differences in how mild or severe the disease is, or how early or late in life it begins. We use the phrase *variable expressivity* to refer to the situation in which different individuals with the same disease show quantitative or qualitative

differences in the severity of the trait. Variable expressivity encompasses things like difference in severity of outcome, difference in rate of progression, or difference in age at diagnosis. Variable expressivity occurs for most diseases, although the variability can be large for some traits and small for others. Sometimes individuals with big differences in severity of disease are affected because they have different mutations in the same gene. You may recall from Chapter 5 that nonsense mutations and deletions or insertions that cause frameshifts can all eliminate the DMD gene product and cause the early severe disorder Duchenne's muscular dystrophy; missense mutations, and in-frame deletions that alter but do not eliminate the gene product cause the much less severe, later-onset Becker's muscular dystrophy.

Different Mutations Associated with Variation in Expressivity

We also see different kinds of causative mutations in the MYOC glaucoma gene, but the severe cases are the missense mutations instead of the nonsense mutations. Not only do we see severity differences between the mutations classes, but we also see differences between the different missense mutations. When we compare different mutations in the same category – missense – we see a correlation of the particular mutation with the average age at which disease starts in a family. For instance, when we look at the ages at which disease is diagnosed in six different glaucoma families (Table 9.4), we see that the Pro370Leu mutation (a proline in place of a leucine at position 370 in the protein sequence) has the earliest start. We see a lot of variation in when the disease starts for individuals with the same mutation, but there are clear overall trends, with some mutations causing earlier disease, on average, than the disease caused by other mutations.

An additional complication is seen when we look at some rare MYOC homozygotes. For one

TABLE 9.4   MYOC mutations show great variation in the age at which glaucoma is first observed in six families

MYOC mutation	Average age at diagnosis (yr)	Earliest age at diagnosis (yr)	Latest age at diagnosis (yr)
Glu323Lys	19	9	43
Gln368Stop	36	28	49
Pro370Leu	12	5	27
Thr377Met	38	34	44
Val426Phe	26	16	46
Ile477Asn	26	4	80

MYOC missense mutation that causes disease in heterozygotes, homozygosity for the disease allele results in a normal phenotype; a different missense mutation, when homozygous, causes disease that is even more severe than what we see in the heterozygotes! And yet each of these mutations causes disease when heterozygous. Even if the situation looks complex, the underlying concept makes sense: different mutations in a gene may produce a different phenotype because the two mutation types are having very different functional effects on the resulting protein. A harder concept is this: that there can be variable expressivity among people whose disease results from the same mutation.

9.14   PHENOTYPIC MODIFIERS

In fact, we now know that there is at least one modifier locus that can affect how early the disease starts when the causative mutation is a MYOC mutation. Some of the earlier onset MYOC glaucoma cases are individuals who are also heterozygous for a mutation in CYP1B1. Individuals heterozygous for the CYP1B1 mutation without a MYOC mutation do not have glaucoma. But those who are homozygous for the CYP1B1 mutation have a form of infantile glaucoma often observed at birth or soon after.

Sequence variants in genes that normally protect us against various forms of damage have the potential to either increase or decrease our ability to compensate for or protect against such damage. If part of the disease pathology involves oxidative stress, a normal biochemical process that goes on at some level in all of us, there are quite a number of genes whose products can generate or protect against oxidative by-products; allelic variants in such genes might be expected to affect a large number of traits influenced by oxidative stress. If the disease pathology leads the cell to die through the process called programmed cell death, we can point at a variety of genes that could play a differential role in the programmed cell death pathway, if mutated. These modulatory factors are acting on fundamental processes that could well apply to many traits and disease processes, so something that modulates one trait might very well be able to modulate many others that involve the same common damage mechanism. Mutations in the SOD2 (sodium oxide dismutase 2) gene have been found in individuals with kidney disease as a complication of diabetes, and researchers have concluded that oxidative damage may play a role in this secondary form of kidney disease.

Another important class of modulatory gene encodes a type of protein called a *chaperone*. These proteins were originally discovered as part of a category of proteins called heat shock proteins that are produced in response to the stress of high temperature inside the cell. In some cases, we see chaperones helping to "escort" proteins to where they need to go, an important role in the intracellular trafficking process. However, in some cases, what we see chaperones doing is helping newly synthesized proteins fold correctly. This is especially important in the case of mutant proteins that are having their effect if they are misfolded.

We can imagine that someone whose cells make more of a particular chaperone, or make a genetic variant of that chaperone that is more efficient at helping proteins fold, might end up with less misfolded copies of the mutant protein clogging up the cell. Although this might not completely prevent problems, it might very well slow down the initial development of the disease and the rate of subsequent progression.

For some mutations and some proteins, the amino acid substitution might cause a major change in the local chemical properties of the protein that is so severe that it can't be overcome by a chaperone. For other combinations of protein and mutation, the new amino acid might be only a bit different from the amino acid normally used at that position, so it would be much easier for a chaperone to refold it. In the latter situation, we would expect a slightly more efficient chaperone or a slightly higher level of the chaperone to have a beneficial effect on the disease pathology. Of course, it is easy to see that a mutation that makes a chaperone less effective at its job might make the disease pathology even worse. Because each chaperone interacts with many different proteins, we might expect that many different traits could potentially be modulated by sequence variants in the same chaperone. Thus the mutation causing the primary disease pathology might be something very rare, but the mutations in modulatory genes could potentially be much more common.

Cataracts offer an important example of a disease phenotype that results when a chaperone fails. The lens of the eye has to be clear for light to transmit to the back of the eye without the light bending or being blocked. The lens has a large, but non-renewing store of the chaperone alpha B crystallin. As long as alpha B crystallin can help other proteins in the lens stay folded correctly, the lens stays clear. Over the course of aging and environmental exposures such as UV light, the alpha B crystallin gets used up. When a protein in the lens misfolds and the chaperone is not available to help refold it, then misfolded proteins begin to accumulate with gradual clouding of the lens.

There are other generalized functions in the cell that can affect the disease phenotype. The apoptosis programmed cell death pathway acts downstream of many disease genes. A cell may survive longer if it is better able to resist the damage that leads to physiological dysfunction that causes the apoptotic cell death pathway to turn on. Where phosphorylation activates a particular protein, a mutation that slows down the rate at which proteins are phosphorylated would reduce the number of activated copies of the protein and potentially make the disease phenotype even worse.

There are many other ways to end up with modifiers. In the case of nail–patella syndrome, defects in the first copy of the LMX1B gene cause the disease, but there is some evidence to support the idea that disease severity might be affected by allelic variants in the second copy of the LMX1B gene. This effect might be expected where the cell is highly sensitive to the amount of the gene product or the level of activity of the gene product; allelic variants in the second copy can have small differences in expression level or activity (either upwards or downwards) without making that second copy into a disease gene with its own separate phenotype. So if knocking out the first copy of the gene takes the cell to 50% of its normal level of the gene product, a small decrease in efficiency of the second copy can drop that level further to 40% (with increased disease severity). Similarly a small increase can bring the level of gene product back up to 60% or 70% and ameliorate the severity.

9.15 BIOCHEMICAL PATHWAYS UNDERLYING COMPLEXITY

Amidst the various concepts originated by Mendel, we find that the "one gene–one trait" concept was an important one for helping simplify and organize the genetic models that were being developed. Although this was a very useful model that helped Mendel and others sort out some of the initial rules of heredity, it is an oversimplification. The more we know about the roles of individual genes in any given trait, the more complexity we find; and yet, looking at it from the molecular viewpoint, Mendel was fundamentally right. Any given protein in fact derives from one gene, or rather from the two copies of that one gene that reside in our cells. Whatever trait may be caused by loss or alteration of that protein corresponds with that one gene. This really shifts the problem backwards a step. Wouldn't we then say: "one protein–one trait?" The answer, we discover, is no. When we looked at the many diseases that can be caused by mutations in rhodopsin, and when we talked about mutations in modulatory genes, we were just looking at the tip of the complex genetic iceberg.

What do we mean by complex genetics, and how complex can it get if it is all based on this "one gene–one protein" variant of Mendel's original "one gene–one trait" concept? We use the same term – complex genetics – to cover two rather different situations: those in which multiple genes cause the trait or those in which a combination of genetic and nongenetic factors lead to the trait.

A trait is heterogenous when multiple people with the trait each have a mutation in a different gene. Let's start out by looking at the situation we started to explore when we talked about the many different rhodopsin mutations that can cause disease: the simple trait (only one thing is causing this trait) in a heterogenous population (different things cause the trait in different members of the population).

The Biochemical Assembly Line

A model for genes and traits that are simple in the individual but complex in the population is the *assembly line model* in which you can get the same effect – a car that you can't drive out of the factory – from messing up any of a large number of different steps in the assembly of the car (Figure 9.14).

In a *biochemical pathway*, multiple steps may be needed to reach the desired end point, with each step carried out by a different gene product. Whether a *metabolic defect* results in the failure to make something essential or from the accumulation of toxic levels of some material that the pathway normally eliminates, there may be multiple different ways to get that effect (Figure 9.15).

So in some cases, breaking a step in the pathway may have two consequences: failure to make things beyond the break point is one consequence, and accumulation of intermediates before the break point can be another consequence. Sometimes only one of those two items causes a problem. For instance, if the item that is not being made is something that you can also obtain from your diet, there may be no major consequence to your body's inability to make it as long as you are consuming it.

In biochemical pathways that convert something like a protein or fat or sugar into some other biochemical form needed by the body, accumulation of an unwanted intermediate may not always be a big problem. In some cases, the intermediate may be something readily excreted from the body. In other cases, the intermediate that is created may also participate in some other biochemical pathway that

FIGURE 9.14 An early twentieth-century Ford assembly line demonstrates the sequential nature of the events that lead to the completed assembly of a series of cars that have gone through the same assembly stages. The auto assembly line is a model for some complex genetic traits. A defect created at any of many steps along the assembly line can result in the same trait – a car whose engine won't run. In theory, a car that fails to turn on and makes no noise when the key is turned could be distinguished from a car in which the engine turns over and then stops. The idea that we can tell the difference between two cars that won't start presumes that we have the ability to distinguish subtleties beyond the fact that this car can't drive away from the factory because the engine doesn't run. Similarly, hits at any of a number of points in a biochemical pathway may cause related traits with the same main characteristics. Such traits might have additional features that can help distinguish the two situations, but if we do not understand enough about the causes of the disease, we may not be able to detect the subtle differences that would let us tell whether we are studying one disease or several different diseases with very different causes but the same final outcome. (Copyright © 2003 Detroit News. Used by permission, All rights reserved.)

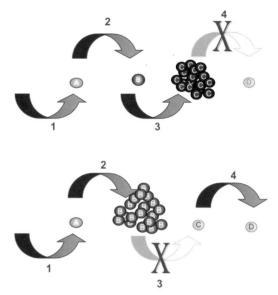

FIGURE 9.15 In a biochemical pathway A→B→C→D that requires four steps to make the item labeled D, knocking out one of the steps in the pathway by mutating the gene that makes one of the enzymes in the pathway can cause problems one of two different ways: by preventing the cell from making D, or by accumulating too much of an earlier intermediate metabolite in the pathway. In the upper panel, end product D is eliminated by knocking out enzyme 4 in the pathway (by mutating the gene that makes it), the enzyme that converts biochemical C into biochemical D. If the only way the cell can get rid of C is by turning it into D, a cell with this defect will accumulate an excess of C. In the second example, if the second gene in the pathway were knocked out, excess B cannot be converted to C, so excess B would accumulate, and both C and D would be missing. Note that in this case D is missing even though enzyme 4 is present, since enzyme 4 does not have the substrate that it would act on, which is C. If the only essential item in the pathway is D, if the body has a way to get rid of excesses of A, B, and C, and if the trait is caused by the absence of D, mutations at any point in this pathway would produce the same trait. However, if the trait is caused by accumulation of excess C, we might see very different effects from knocking out the fourth enzyme and accumulating C than we would get from knocking out the third enzyme and accumulating excess B. In real-life situations, we might see loss of D causing some aspects of a trait, with accumulation of B or C causing some of the differences between the two traits. If loss of D causes the primary characteristic for which the trait is known, hits on any of the four steps in this pathway would cause the trait.

is capable of eliminating it (Figure 9.16). Thus, even though both consequences can theoretically be there, often the problem may involve only one of them and it is not always obvious which one it will turn out to be.

A good example of this kind of assembly line phenomenon is the *retinoid cycle* that is responsible for processing and delivering the vitamin A derivative used by rhodopsin when it detects

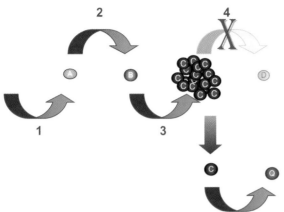

FIGURE 9.16 In a biochemical pathway that blocks the conversion of B to C, we may not see accumulation of B if there is a second pathway that uses excess B to make Q and R. For some pathways, this will solve the problem, especially if R is something that is excreted, and for other pathways this kind of diversion away from the main pathway will simply create a different problem if excess Q or R is toxic.

light (Figure 9.17). A variety of different diseases result that share the major feature of retinal degeneration but that can be distinguished by differences such as the age at which the disease begins, the appearance of the retina as viewed through the pupil of the eye, the rate of progression of the disease, the extent of vision loss that can result, which subtypes of photoreceptors are dying, and the types of electrical responses made by the retinal cells when responding to a light signal. So if someone has a mutation in certain genes of the retinoid cycle, they may not be able to convert vitamin A consumed in the diet all the way through to the final chemical form of it that the body uses. This has inspired researchers to begin working on ways to feed some intermediate form of vitamin A (or other forms of vitamin A that are normally not part of the cycle at all) to bypass the block in the cycle. This general principle may eventually provide answers for a number of metabolic disorders, if the details can be figured out: if the step that creates intermediate four is blocked,

then deliver intermediate four and it won't matter that you can't manufacture it yourself. For a variety of reasons, this will not always work, but in some cases, it might someday transform the lives of people with some retinal dystrophies involving the retinoid cycle.

Gene Families

Another level of complication is that some genes have close relatives, genes that can carry out the same or similar functions. Thus some traits may require alteration of more than one gene to get the effect. In these cases, the trait that could potentially result may virtually never be seen or might be incredibly rare because people who have a mutation in one of the genes would be protected by the fact that other members of that gene family or a related pathway are still functioning. In other cases, the other members of the gene family may carry out a similar function but not one that can substitute for the missing function. Sometimes, the genes may be functionally similar enough but may not be expressed in the same tissues or at the same stage in development. If you need to replace the activity of gene A in the kidney but its potential substitute gene B is only expressed in the brain, it cannot come to the rescue.

This raises questions about whether some therapies will eventually result from getting a gene to change where or when it is being expressed so that it can cover for a defect in a different gene. We already can see an example of this in treatments that compensate for the HbS sickle cell anemia defect by turning back on some expression of a fetal hemoglobin that can help compensate for the defect. Of course, there are many genes whose loss cannot be compensated by another gene simply because nothing else performs that function, even in a different cell type or at a different point in development.

Thus, if we try looking for THE gene that causes the primary pathology – cell death, toxic metabolites, absent essentials, inflammatory

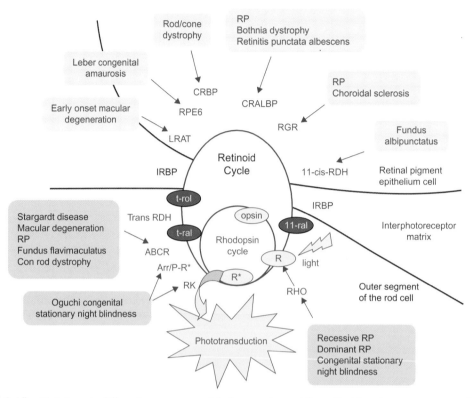

FIGURE 9.17   Mutations in different genes responsible for carrying out the retinoid cycle can cause more than a half-dozen different but related traits. Arrows point from the names of different retinal degenerative diseases (pink or blue backgrounds) to the names of genes that can cause those diseases, and that also carry out the different steps in the retinoid cycle (the process by which the different vitamin A derivates are formed that are needed to complete the cycle) and the rhodopsin cycle (the process by which a complex of rhodopsin plus a vitamin A derivative detect a photon of light and send the information that light hit the complex down the line via a process called *phototransduction* that leads to the transmission of a signal to the brain). Vitamin A derivatives are shown in yellow letters against a blue background, and different forms of rhodopsin are framed in yellow ovals. You don't have to understand any of the chemistry of the retinoid cycle to understand that mutations in the many different genes of the retinoid cycle can cause a related set of diseases by affecting different steps in the same pathway. RP stands for retinitis pigmentosa. (After McBee *et al.*, *Progress in Retinal and Eye Research*, 2001;20:469–529.)

reactions, susceptibility to infection, etc. – we would discover that we are actually looking for many different genes that could each lead to that main functional defect if they were mutated, but we may still be looking for only one gene causing the trait in any given individual. Sometimes, in science, we ask a simple question and get back a complex answer. If we did an especially good job of designing the experiment, or at least if we are flexible in our thinking, we may even recognize the complex answer as telling us something important and not just dismiss it because it is not the answer we expected.

9.16 BEHAVIORAL GENETICS

In 1997 in California a man was tried and convicted for a terrible crime, the murder of a young

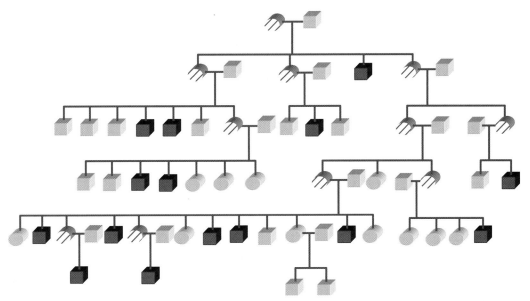

FIGURE 9.18 Pedigree exhibiting X-linked inheritance of a violent behavioral phenotype. (Adapted from Brunner, H. G., *et al.*, "X-linked borderline mental retardation with prominent behavioral disturbance," *American Journal of Human Genetics*, 1993;52:1032–9.)

girl. It was reported that he displayed a total lack of remorse or respect for our society and its laws. During the coverage of the trial a local television station aired a report in which someone who had known this man for quite some time claimed, "He was just wrong from the beginning, just a bad seed." More recently, another man on trial for murder is actually trying to use genetic inheritance of criminality in his family as a legal defense that implies he was predestined to do bad things. If there were such a thing as a "natural born killer," a genetically predestined "bad boy" or "bad girl" from birth, would such a person be missing some crucial gene product essential for building the parts of the human psyche that proscribe most of us from such behavior, genes that make empathy, caring, and guilt possible? Can the situation possibly be so simple? Do such genes even exist, or are they just part of a prejudice ingrained in our culture? What follows in this chapter is an attempt, however unsatisfying, to gain some insight into this complex issue.

Genetics of Violent Aggression in a Dutch Family

Figure 9.18 displays a pedigree for a family that was studied in the Netherlands. Indicated males in this family were often subject to seemingly unprovoked and uncontrolled violent outbursts. These aggressive episodes ran the gamut from ranting and shouting to exhibitionism and serious crimes, such as rape, arson, and assault with deadly weapons. (One of these men forced his sisters to undress at knife point. Another man raped his sister and then later, while incarcerated, attacked the warden with a pitchfork. A third member of this family attempted to run over his employer with a forklift.) All of the affected males were mildly retarded, with an average IQ of only 85 (100 is considered "normal"). Although there were no affected women, sisters of affected males frequently gave birth to affected sons.

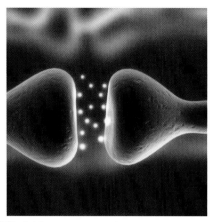

FIGURE 9.19   Neurons and neurotransmitters in synapses. Although signals within a neuron are carried electrically from one end of the cell to the other, a given neuron communicates with the next neuron in the sensory or motor pathway by releasing neurotransmitters into the small space between these cells, known as a synapse. The second neuron absorbs the neurotransmitters, triggering it to fire an electrical signal along its length. (Image from iStockphoto/Sebastian Kaulitzki. © All rights reserved.)

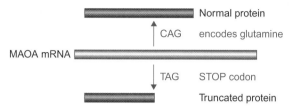

FIGURE 9.20   A point mutation in the MAOA gene introduces a stop codon that prematurely truncates the MAOA gene product in a group of male relatives who share a trait that includes aggressive behavior and intelligence below normal.

All of this evidence pointed strongly not only to a genetic basis for this trait but indeed to a sex-linked mutation in a gene on the X chromosome underlying this behavior pattern. To test this possibility, researchers set out to determine if they could find the gene responsible and determine what mutation caused this set of behaviors. The gene they found, called MAOA, encodes a protein known as monoamine oxidase A that is required to break down molecules known as neurotransmitters in the brain.

Neurotransmitters are small molecules that facilitate communication between cells, known as neurons, that comprise the nervous system (Figure 9.19). Obviously, the presence of neurotransmitters at the synapse needs to be very tightly controlled if the nerve cells are to function properly. They must be released rapidly by the stimulating cells and absorbed and/or degraded by the responding neuron. One of the enzymes required to break down some neurotransmitters is MAOA. A variety of biochemical studies on urine samples taken from the aggressive males in the Dutch family indicated markedly abnormal metabolism of neurotransmitters that are normally broken down by MAOA, including dopamine, epinephrine, and serotonin. Very high levels of these compounds were found in the urine of these males, consistent with an inability to break down these compounds.

As diagrammed in Figure 9.20, in a subsequent paper, affected males in this kindred were shown to carry a point mutation in the eighth exon of the MAOA gene that changes a glutamine codon to a stop or termination codon. This nonsense mutation was not found in the MAOA genes carried by unaffected brothers of affected males; obligate female carriers were also found to carry one normal and one mutant allele, although they were phenotypically normal. Thus there was a precise correlation between the violent aggressive behavior and the presence of the MAOA mutation.

Similar observations were made in a strain of mice that were genetically engineered to delete the MAOA gene. Mice homozygous for this mutation displayed highly increased levels of aggression and greatly increased levels of neurotransmitters. More critically, adding back a functional copy of the MAOA gene to the mouse genome restored both the ability to break down the neurotransmitter molecules and normal levels of aggression. More recently, studies in Macaque monkeys also confirm these findings.

How is the defect in MAOA correlated with the violent outbursts exhibited by these men? Realize that, among their many roles, neurotransmitters function as part of the body's "fight or flight" response to threats or danger. In most of us, as the levels of these neurotransmitters increase in our brain in response to various stresses, they in turn are broken down by MAOA. Thus most minor stimuli produce only transient increases in neurotransmitter levels. One can then imagine that the degradation of some neurotransmitters is greatly impaired in males lacking the MAOA enzyme, thus allowing levels of neurotransmitters to rise far in excess of a normal level. Indeed, in several cases, crimes committed by these males closely followed traumatic family events, such as the loss of a loved one. It is then possible that stressful events might overstress these males to the point that violent outbursts become more likely.

9.17 GENE EXPRESSION: ANOTHER LEVEL OF COMPLEXITY

How important is the genotype for MAOA (and the environment) in terms of violent aggression in the general population? Most workers now seem to accept the conclusion that this null mutation at the MAOA gene largely explains the violent phenotype seen in this family. However, this type of mutation at MAOA is extremely rare in the human population. The differences that do exist among most human beings result in altering the level of MAOA expression, not abolishing it. Attempts to correlate the polymorphism that was observed with violent aggressive behavior in the general population remained inconclusive for several years after the report of this family from the Netherlands.

A recent study in Australia and New Zealand by Caspi and colleagues reveals that variations in the level of MAOA expression do not simply

TABLE 9.5   Different outcomes for abused children with high and low MAOA genotypes

Genotype	Violent antisocial
High MAOA	40%
Low MAOA	80%

Derived from Caspi, A., McClay, J., Moffitt, T.E., *et al.*, "Role of genotype in the cycle of violence in maltreated children," *Science*, 2002;297:851–4.

create a predestined behavioral phenotype; rather, the MAOA genotype appears to mediate the effects of mistreatment of young boys, at least in terms of whether or not those children develop into adult men who exhibit antisocial problems, specifically: a disposition towards violent antisocial behavior, an antisocial personality disorder, or eventual conviction for a violent offense. They found that mistreated or abused male children with high levels of MAOA expression were much less likely to develop into adult men with antisocial violent personalities than were male children with low MAOA levels (Table 9.5). For example, among severely maltreated boys with low MAOA expression, greater than 80% of these children developed into adults with behavioral difficulties, while the fraction of mistreated children with high levels of MAOA expression that developed similar difficulties fell just above 40%. Because it was not asked in this study, we cannot know what other kinds of environmental difficulties might similarly confound a "low-MAOA" genotype. Sadly, similar differences were observed when the metric was the percentage of individuals of a given expression level who had been incarcerated for violent offenses by age 26.

The critical finding in this study was that the genotype for MAOA, by itself, did not predict violent or aggressive behavior unless something environmental was factored in. Differences were observed only when childhood abuse or mistreatment was superimposed on these differences. Thus the final phenotype (violent aggression or antisocial personality disorder) seems to result from the combination of two separate components, genotype (low MAOA levels) and environment (childhood abuse). The biochemistry and pharmacology of all of this makes good sense. It turns out that in both animals and humans there are data to suggest that early maltreatment or stress can alter neurotransmitter levels in a fashion that can last well into adult life. Perhaps higher levels of MAOA in a child may make that person more resilient, or more resistant to the stresses of abuse and maltreatment. Similarly, lower threshold levels of MAOA might sensitize a child to the same effects. The critical finding was that, although the MAOA genotype contributed to the behavioral phenotype, the environment in which the child was raised also played an important role in whether antisocial, violent behavior turned up in the men with the low MAOA genotype. Development of violent, antisocial behavior was not a foregone conclusion for these individuals but rather highly dependent on environment.

Given the well-documented correlations that have come out of two different studies in two different parts of the world, you might be wondering why we don't just start sequencing the MAOA gene from every serious criminal who will stand still long enough to let us draw blood. There are several reasons why such studies are both ethically difficult and scientifically incomplete.

- First, there is a serious problem of "informed consent" when doing this sort of research with inmates. The first principle of informed consent is that the individual participating in a study must be participating freely and of their own will, without being under pressure to participate. Can people in prison truly give free informed consent (imagine what type of pressures they might feel to agree even if they don't want to)?
- Second, what would be the legal status of this kind of genotype information? Suppose

we did find inmates bearing such a mutation. What effect would this have? Would we change anything about how we handle that individual's case? Would his legal status change? Before trial, would the finding of such an MAOA mutation constitute a legal defense? Might the Governor be more or less likely to pardon a condemned man if he thought the crimes were driven by his genes, and would that inclination be valid or not?
- Third, do we know enough to understand what the information is telling us? As with XYY males discussed Chapter 7, would finding a higher fraction of inmates with such a mutation really prove "cause and effect"? Here we have to be very concerned because a finding of co-occurrence does not always tell us what we think it is telling us, and our ability to interpret the answers we get is very much limited by our ability to frame the right question in the first place.
- Fourth, even if it were all true, if there really were a cause-and-effect relationship between MAOA genotype and behavior, do we know enough to know what should be done with the information? Do we know enough about how to design an environment in which to raise a low MAOA child so as to avoid his development of unwanted behavioral characteristics? If someone has already grown up in an abusive environment, do we know whether changes in his adult environment can help undo any of what developed during his childhood? If someone has a low MAOA genotype, are there other genetic factors modulating that effect? To put it differently: How do we account for the 20% of low MAOA cases raised in abusive environments who did not develop the predicted phenotype?

Still, the question in most of our minds is: Just what do we do if all, or even some, of this pans out? Would we, as a society, screen male babies or fetuses for this mutation? If we did such screening, what action would we take? Would bearing this mutation be a cause for termination of a pregnancy? Twenty students in a senior seminar class were once asked if they would choose to terminate a pregnancy if they knew that the male fetus carried the MAOA mutation that had been found in the family in the Netherlands. The answer was an overwhelming "yes." Stop and think what your answer might be, then ask yourself why. Then ask yourself whether we know enough to be basing such decisions on a genotype, and if not, what else do we need to know to make such decisions valid and fully informed?

So the question arises: How many other genetic influences might there be? Unfortunately, we cannot answer this question yet. Even more unfortunately, the issue has led to some rather careless speculation. Scott once listened in horror while a professor told a class in medical genetics that he thought that virtually all criminal behavior was genetic: that the likelihood of dying in a hail of bullets over a bad drug deal was as genetically influenced as other human traits, such as blood clotting and color vision. There is currently no hard data to support such an assertion, and in fact the Australia/New Zealand MAOA study raises very serious questions about this proposed genetic model of criminality. While there is clearly a major gene-based component to the MAOA story, there is also a major environmental component.

Still, this is an issue that is not going to go away. To lay our prejudices on the line, we suspect that, while much criminal or violent behavior may have roots in environmental causes, such as child abuse, hunger, drug addiction, and seemingly hopeless poverty, there will be more cases like the MAOA mutation that will render some people more susceptible to a poor outcome from being raised in such circumstances. Our society is going to have to find ways to cope with the crime and punishment of such individuals, and to be sure that decisions about how

to handle such cases are based on real knowledge about the cause-and-effect relationships, and real knowledge of the level of complexity involved. The existence of genotypes that render an individual susceptible to the development of violent or antisocial tendencies in response to environmental influences challenges the basic and cherished concepts of free will and individual responsibility. If some mutations turned out to apparently make crime essentially inevitable, how would we arrive at a truly just view of the punishment of such crimes? Perhaps in those cases we will refocus our interests as a society from punishment to treatment. Perhaps.

Genes Are Not Always Destiny

These issues bring us to the question: Are genes destiny? Will a particular genotype necessarily manifest a particular phenotype? Clearly, for a wide array of traits the answer is that genes play a major role in what is going on, and yet genes are not destiny. Just knowing a genotype does not tell you the final phenotype. So if we can gain better control over our environments can we make big improvements in our phenotypic outcome relative to the potential in our genomes? Yes, but in many cases we do not yet know enough and in other cases, such as the MAOA abuse story, the real solution is one that society knew about long before any human genes were ever found – if we can't find ways to stop all abuse and protect everyone, can we at least find better ways to protect vulnerable individuals (those with a predisposing genotype)?

Or are there other ways to go about intervening in some of these situations? Those of us who spend our lives looking for causative genes do so on the presumption that genes are not destiny, that we can find ways to intervene to alter the course of a theoretical phenotype to end up with a different final phenotype. And some of those interventions will be behavioral but many others will involve medications or gene therapy. In Chapter 14 we will talk about some efforts to modify the genotype in beneficial ways through gene therapy and gene-based therapies.

Genes are not always destiny. But until we learn more about what modulates the factors responsible for variation in the outcome of a particular genotype, looking at our genotypes may continue to feel a bit like looking into a crystal ball.

Study Questions

1. Why might the quality of the home environment be of greater importance to the growth and development of a child who is producing low levels of MAOA? Should screening for MAOA in children be instituted? Why or why not?

2. Why are studies of genetic factors that contribute to criminal behavior considered ethically difficult and scientifically problematic?

3. What does digenic diallelic mean and how does this complicate efforts to understand disease mechanism and develop new treatments?

4. How can mutations in the retinoid cycle lead to more than one disease?

5. What type of genetic disease is cardiovascular disease? How do different factors lead to different cardiac disease risk factors and outcomes?

6. What are the rules for determining if traits are due to multi-factorial inheritance?

7. What are the three major classes of organisms responsible for many of the most common human infections?

8. Can your phenotype be affected by someone else's genes? What are common examples of this phenomenon?

9. Two people have the same disease but with different levels of disease severity? What is a common reason for this occurrence?

10. If the genome plays such a critical role in regulating disease, why not just sequence everyone's genome and tailor diet, exercise, clinical treatment plans, etc. to prevent disease?

11. Why do we think that level of MAOA expression is not *solely* predictive of outcome?

12. Why do we think that the mode of inheritance in Figure 9.18 is X-linked?

13. Why might it be hard to classify individuals in a family like the one shown in Figure 9.18?

14. Why might some modifiers impact the severity of disease for only one disease while others have the potential to affect multiple severity of multiple different diseases?

15. If affected members of one family all show onset of glaucoma in their teens and affected members of another family consistently show onset in middle age, why do we think that the age variability is not due to a modifier gene?

16. What is the similarity between a threshold trait and a quantitative trait?

17. What is the difference between a threshold trait and a quantitative trait?

18. In the case of digenic diallelic inheritance of retinitis pigmentosa, why can normal PRPH2 protein yield a normal phenotype if ROM1 is missing but the L185P variant of PRPH2 cannot?

19. If two people with autosomal recessive retinitis pigmentosa have children together, why do we think there is a low chance that they will have a child with retinitis pigmentosa?

20. Why do we consider congenital stationary night blindness to be related to retinitis pigmentosa?

Short Essays

1. Height is 80% heritable in white males in the United States. It is only about 65% heritable in West African males. How is it possible for a trait to be more heritable in some populations

than in others? As you consider this question, please read "How much of height is genetic and how much is due to nutrition," by Chaio-Qiang Lai in *Scientific American*, December 11, 2006.

2. Aging is a very complex process influenced by a variety of genetic and environmental factors. Even though many factors are involved, it is possible to identify individual factors that contribute to such a complex process and evaluate how much influence that factor might have. Why do some people look older than others, and why do some people develop age-associated diseases sooner than others? What is an important factor that seems to play a role in the aging process, and why might we expect it to be important to aging? As you consider this question, please read "Researchers identify genetic variant linked to faster biological aging" by Katie Moisse in *Scientific American*, February 28, 2010.

3. The article by Erika Check Hayden highlights the difficulties inherent in using genetic tests to direct or manage patient care. Warfarin, and similar anti-coagulant drugs, are life-savers but they are not without some risk. If the prescribed dose is too low fatal clots can be formed, and if the dose is too high the patient can bleed internally. Setting the right warfarin level is an art that requires frequent monitoring of blood clotting rates, an expensive and aggravating process. The article notes that much of the patient-to-patient variation in the response to a given dose of anti-coagulant can be explained by specific genetic variants, but we know that some of the variation is also due to differences in dietary factors such as levels of vitamin K. Are tests for these genetic variants in common usage and what factors are affecting physician attitudes towards use of such tests? As you consider your answer, please read "Cardiovascular disease gets personal" by

Erika Check Hayden published online in *Nature news* and in *Nature*, 2009;460:940–1.

4. When sequence variants in the CCR5 gene were first shown in individuals who had resisted HIV infection in spite of HIV exposure, it started a campaign to identify the mechanism by which such variants could work and move towards the development of drugs that could act via similar pathways. We might expect that screening for CCR5 variants would tell us who is at risk for HIV and who is not, but the story turns out to be more complex than that. What other factors can affect whether someone is at risk of HIV infection and how does this complicate development of predictive genetic testing? As you consider this question, please read "Safety in numbers" by Cornelis J. M. Melief in *European Journal of Human Genetics*, 2005;13:795–6.

Resource Project

Go to the Center for Disease Control and look at the statistics on causes of death. How many of the top causes of death are genetic in origin, or might be affected by genetic factors? If you have questions about whether a trait is genetic you can check it on Online Mendelian Inheritance in Man (OMIM). Write a one-page paper on the genetics of the leading causes of death.

Suggested Reading

Articles and Chapters

"A new dynamic: With an eye toward host–pathogen interactions, can a Penn State center predict and prevent the next pandemic?" by Brendan Borrell in *The Scientist*, 2007;21:32.

"Why your DNA is not your destiny," by John Cloud, January 6, 2010, Time.com. at http://www.time.com/time/health/article/0,8599,1951968-1,00.html.

"Sports genes" by David Epstein in *Sports Illustrated* May 2010, p. 53.

"The real cause of obesity" by Jeffrey Friedman in *Newsweek*, September 10, 2009.

"Triggering addiction" by Markus Heilig in *The Scientist*, 2008;22:30.

"Returns an ancient disease" by Janeen Interlandi in *Newsweek*, December 8, 2009.

"Twin disorders" by Alison McCook in *The Scientist*, 2008;22:32.

"Adulthood: Heart diseases" in *Is It In Your Genes? The Influence of Genes on Common Disorders and Diseases That Affect You and Your Family* by Philip R. Reilly (2004, Cold Spring Harbor Laboratory Press, pp. 81–106).

"Rare history, common disease" by David Secko in *The Scientist*, 2008;22:38.

"Signaling neurons make neighbor cells 'want in.' Synapses are primed to strengthen (and thus enable learning) if a nearby one has just been stimulated" by Nikhil Swaminathan in *Scientific American*, December 2007.

"Eat your way to better DNA," by Kate Travis in *The Scientist*, 2006;20:50.

"A genetic diet by the numbers," by Kate Travis in *The Scientist*, 2006;20:54.

"The gene puzzle" by Carl Zimmer in *Newsweek*, June 27, 2009.

Books

Living with Our Genes: Why They Matter More Than You Think by Dean H. Hamer and Peter Copeland (1999, Anchor).

The Multiple-Hit Hypothesis: How Genes Play a Role in Cancer

- The different kinds of genes that can play a role in cancer.
- The different kinds of mutations that can play a role in cancer.
- Which cancer risk factors are associated with which kinds of cancers.
- What a tumor suppressor gene is.
- What a proto-oncogene is.
- Why defects in DNA repair can cause tumors in so many different cell types.
- What kinds of cells sit in G0 and what kinds of cells divide on a regular basis.
- How customized medications can target the causes of specific cancers.
- How PCR can be used to monitor recurrence of cancer.
- What FISH tells us about the state of the genome in advanced metastatic cancers.
- What the two-hit hypothesis is.
- What the multi-hit hypothesis is.

10.1 THE WAR ON CANCER

To dream … the impossible dream …
—**Don Quixote in Man of La Mancha**

Healthy young people aren't supposed to die. Even amidst the many dangers that arise from the exuberance and hazards of youth, the death of someone young is always a shock. And when the blow is delivered from some direction we never expected, were not waiting for, had never considered, when someone young is felled by an illness such as leukemia, we are left feeling stunned. It seems impossible to understand such an outcome, and we find ourselves asking, "How could this have happened?"

And the next question that comes to mind is, "What can be done so that this does not happen again?"

Brenda Knowles (Figure 10.1) was a graduate student in Scott's lab back in the late 1980s. She was bright and funny and totally unimpressed by Scott's supposed seniority. She was trained as a chemist and had begun graduate school doing biochemistry. However, Brenda had a strong connection to biology and the organisms that embody so much more complexity than simple biochemistry. Soon she found her way into a lab where there were organisms to work on, maybe just fruit flies, but organisms nonetheless.

She shared her time in Scott's lab with the usual array of characters that populate a "working lab." Science is a business that cherishes eccentricity, even encourages it. A healthy, growing lab will have its share of unusual characters. The basic foundation on which any new lab is started is unusual and novel ideas. Such ideas often come from and attract unusual and novel people.

In some ways, Brenda resembled the classic image of a young scholar. Her radio played classical music or National Public Radio, drowning out the competing styles of rock music from other desks or that much-ridiculed country music emanating from Scott's office. Her desk was neat and her ideas were equally

FIGURE 10.1 Brenda Brodeur Knowles (1962–1996.) (Photo courtesy of James Knowles.)

well organized. She was rigorous in her critical thinking and tenacious in her pursuit of answers to scientific questions. She wrote two papers from Scott's lab and went on to continue her scientific training by taking on a postdoctoral fellowship at Yale.

On her way to that fellowship, she married a handsome young doctor and they bought a beautiful little house in rural Connecticut. If you sense a fairy tale being told here, there's a reason: Brenda's life always seemed a bit of a fairy tale to Scott. This fairy tale was unusual only in the sense that Brenda was enough of a feminist to slay most of her own dragons.

That is, until Brenda got sick. Sometime in the early 1990s, Brenda acquired acute myeloid leukemia (AML). We'll talk more about leukemia later in this chapter. The disease results from a rather nasty genetic alteration that occurs in one of the stem cells that produce the circulating cells in our blood. The result is an instruction for the altered stem cell to divide repeatedly.

Leukemia was the ultimate dragon in Brenda's life, and she committed all of her resources to slaying it. She tried everything that was available, or even close to available. She suffered more than our words can convey. In the end, she lost the battle.

The battle she lost was just one battle in what the press often refers to as the "war on cancer." In 1969 a full-page ad in the *New York Times* urged President Nixon to begin a war on cancer, saying "We are so close to a cure for cancer. We lack only the will and the kind of money and comprehensive planning that went into putting a man on the moon." The war on cancer was proposed in 1969. Brenda lost her battle with cancer in 1996.

There have been too many such battles. For most of history the idea of a cure for cancer has seemed like an impossible dream. We daresay that there will not be a single reader of this book who does not know someone touched by cancer. After all, one in four of us will be directly affected by cancer in our lifetimes.

Fortunately there are now many cures, remissions, and cases in which the cancer is simply held in check for years at a time.

With impatient excitement, we watch advances in cancer treatment begin building on the results coming out of genetic studies of cancer. Breakthroughs in understanding of the molecular mechanisms of various forms of leukemia have led to breakthroughs in the development of new treatment approaches. Scientists have begun creating molecular "lances" aimed at slaying the monsters that are the various kinds of leukemia. Their molecular lances are drugs designed based on an understanding of what has gone wrong at the molecular level in the leukemia cells. How wonderful that these weapons against leukemia are emerging; how terrible that they come too late for Brenda.

Increasingly, we are seeing "magic bullets" emerge based on breakthroughs in our understanding of the underlying mechanisms of diseases caused by defects in genes. Some of the cures being developed use gene therapy, but we are going to see a lot of other pharmaceutical treatments emerge that will not use gene therapy even though they will be based on the information gained from the study of genes.

As we read news articles with stories about new advances in the war on cancer, we need to understand what cancer is, and how it is caused. Many different genes play roles in cancer, and this chapter will try to give you some of the basic concepts and talk about some of the types of genes involved in cancer.

10.2 CANCER AS A DEFECT IN REGULATION OF THE CELL CYCLE

Cancer is a disease that results when a single cell in your body overrides the normal controls of cell division. Most of the cells in your body are not supposed to divide anymore, or if they do divide, they are supposed to do so slowly and rarely under very tightly regulated control. There are some notable exceptions in terms of cell types that continue dividing frequently throughout your life, such as the cells that make up the inside of your intestines, the cells that comprise your bone marrow, the cells in the lowest layer of your skin, and your germ cells. The basic pathway by which each cell must divide is diagrammed in Figure 10.2. As we told you in Chapter 6, cell division begins in a resting or preparatory state called G1. Once the cell commits to cell division, it then moves into the S phase, where replication occurs, moves to another preparatory phase called G2, and then enters the mitotic division (M). But many cells

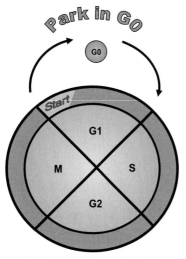

FIGURE 10.2   Resting in G0. As the cell finishes dividing, each of the new daughter cells has to decide whether to move back into the cell cycle, going through another round of synthesis and mitosis, or whether to park itself in G0. Cells can hang out in G0 indefinitely, carrying out some specialized metabolic functions but not proceeding towards cell division. Some genes make products that tell the cell to move into the cell cycle. Other genes make products that tell the cell to go park in G0. Some specialized cell types continue dividing throughout your lifetime, such as cells that line the gut, but many other cell types have either given up on dividing or will only divide in specialized circumstances, such as during repair of injury. This simplified version of the cell cycle figure that we showed in Figure 6.1 emphasizes the G0 pause that is the resting state for cells not trying to divide.

spend a lot of their time permanently parked in a sidetrack of the cell cycle called G0 (pronounced "gee zero"). Sitting quietly in this stage, your cells have permanently foregone the possibility of division in favor of a stable commitment to execute their particular function.

Sometimes, however, something inside a single cell overrides that inhibition in the absence of a need for wound healing. The cell loses that commitment to just "being" and begins to divide. This is the start of a tumor, the hallmark of the most awful word in our language: cancer. Usually, the daughter cells descended from one miscreant cell will divide slowly, staying together in a dense, well-defined, and tightly bordered mass that can usually be removed by a surgeon. Such tumors are referred to as *benign*.

Sometimes, however, the daughters of those cells change further and lose their inhibitions regarding the invasion of normal neighboring tissue and begin to spread throughout the organ or tissue in which they arose. Such tumors are described as *invasive*. In some cases these tumor cells, now committed to rapid unbridled cell division, find their way via the lymph nodes or bloodstream into other sites or tissues, establishing new sites or nodes of tumor formation. This movement to new sites is called *metastasis*. Invasive or metastatic tumors are referred to as *malignant*. Unfortunately, malignant tumors (and even some benign tumors) have the potential to kill people if their growth and their spread to new locations can't be stopped.

10.3 CANCER AS A GENETIC DISEASE

Accumulation of Mutations in Somatic Cells Can Cause Tumors

Tumors begin from single cells within our bodies. In the case of prostate cancer, a single cell in the prostate gland leaves the "G0 parking lot" that we described above and begins dividing repeatedly – something that none of the cells in the prostate gland of an adult male is supposed to do! There are actually two counterbalanced processes going on during the cell cycle: signals that tell the cell to move forward in the cell cycle (to grow and divide) and signals that tell the cell to stop (to pause before moving on to the next step in the cycle or to stop and wait indefinitely). So the simple little diagram of the cell cycle in Figure 10.1 actually represents a very complex combination of start signals and stop lights that advance the cell from one step to the next until the entire cycle has been completed, or until the cell is left paused in G0. *There are two ways that a cell can escape the normal controls of the cell cycle: by failing to stop when it should or by getting a go-ahead signal that it should not be getting.* Normally, this series of stop and go-ahead signals take place in a very tightly regulated series of events in which many of the signals are "on" only part of the time.

The stop signals and the go-ahead signals of the cell cycle are each given by proteins that regulate the cell cycle, and it is through changes in these genes that cancer comes into being. The ability of some cells to restart cell division after many decades in a non-dividing state develops as a consequence of mutations in three types of gene:

- First, there are *tumor suppressor genes* (or stop lights) whose protein products protect the cell against unwanted cell division. Mutations that inactivate both copies of a tumor suppressor gene can allow the cell to slip past a point in the cell cycle at which the cell should have stopped. As a consequence, the ability of the cell to stay in the resting state is compromised. In some senses, we can think of the two copies of a tumor suppressor gene as being a pair of guards protecting a step in the cell cycle and preventing it from advancing in an uncontrolled manner, with the presence of

even just one of the two copies of the gene being sufficient to keep that step protected and correctly regulated.

- Second, there are *tumor promoter genes* (or go-ahead signals) known as *proto-oncogenes*. Although these genes are normally silent during adult life, they play critical roles in promoting cell division during the early stages in development. Sometimes genetic events occur (described below) that activated these genes in non-dividing cells. These activated cell division genes, known as *oncogenes* or *tumor promoter* genes, turn on cell division by supplying a go-ahead signal on a continuous basis. This "go ahead and divide" signal, which should have been there only transiently as a tightly controlled event at a specific point in the cell cycle, forces the cell into continuous division. *Proto-oncogenes*, whose normal and highly regulated cellular functions involve supplying a go-ahead signal in the cell cycle can become oncogenes by mutation or translocation which turns them on constitutively.

- Third, *DNA repair genes* act as cancer genes by affecting the rate at which the other two categories of genes (especially the tumor suppressor genes) are mutated and begin causing unwanted cell proliferation. Mutations that impair certain DNA repair systems impair the ability of cells to properly replicate their DNA without making errors and/or diminish the ability of the cell to repair the DNA damage done by the environment. The loss of these repair proteins allows damage to the DNA to escape repair or to be repaired improperly. Thus there ends up being a drastically increased probability of a cell acquiring a cancer-causing mutation in a tumor suppressor gene or a tumor promoter gene. DNA repair genes can cause cancer by greatly increasing the rate of mutations in the other two classes of cancer genes.

An important risk factor for some but not all types of cancer is family history. For some types of cancer, especially those in the tumor suppressor genes or DNA repair genes, family history may be readily evident. However, for the tumor promoter genes, we do not expect to see a family history since the events that cause these cancers happen somatically in an individual after conception, and often do not take place until well into adult life.

10.4 CANCER AND THE ENVIRONMENT

As for many other genetic traits, cancer also has major environmental contributors. Smoking is a major risk factor for lung cancer; according to the National Cancer Institute 180,000 Americans die each year from tobacco-related cancers. Sun exposure can increase the risk of skin cancer, and limiting sun exposure through actions such as wearing sun screen and limiting tanning activities can help reduce your risk of skin cancer. Cancer risk can be increased through exposure to radiation from sources such as radon gas that can seep into our basements and go undetected for years if we do not have radon testing done. A high fat diet increases the risk of colon cancer. Aging, chemical exposures, some viruses, some hormones, and even simply being overweight can also increase risk of a variety of cancers.

Understanding these risk factors is very important since it lets public health services develop proactive programs to try to reduce risks through reduction of risk-associated behaviors. It is also turning out that there are a variety of actively beneficial steps we can take, such as eating broccoli, to reduce risk of many different kinds of cancers.

As we look at the list of things that increase risk of cancer, we see a lot of things that are known to bring about either changes in DNA sequence (radiation, sunlight), changes in gene expression (hormones), or changes in methylation that affect gene regulation (some chemicals and viruses). In addition there are some viruses that raise the risk of certain kinds of cancers, such as HIV and HPV. HPV infection is associated with the changes in the cell structure in the cervix that can lead to the development of cervical tumors. Fortunately, there is a vaccine (such as Gardasil™) that can protect women against HPV infection and thus greatly lower the risk of subsequent cervical cancer, as well as some other types of vaginal or vulvar cancers.

Even as scientists work towards the development of gene therapies and new medications that target aspects of cancer we are learning about in our labs, ways of affecting key environmental exposures remain some of the most important approaches to cancer prevention.

10.5 TUMOR SUPPRESSOR GENES AND THE TWO-HIT HYPOTHESIS

Hereditary Retinoblastoma: A Model for Understanding the Genetics of Tumor Formation

The first insights into tumor suppressor genes come from the study of families in which a strong predisposition to develop a specific type of cancer is passed along in a family through multiple generations. In such families, half of the children are at risk because they are born carrying one mutant allele in a gene whose normal function is to prevent improper cell division. Although every cell in the body has this mutation in one of the two alleles of the gene, those cells all go through cell division normally, under tightly regulated control. However, during the life of that child, at some point the body's normal low level of mutation knocks out the other normal allele in one or a few cells that need that particular suppressor gene to regulate their cell cycle. This leaves that cell and its mitotic descendants in a position to eventually give rise to a tumor. They have lost both copies of a gene that makes a critical guard protein, and they are now prone to inappropriate division in the cell that is missing both copies of the tumor suppressor gene.

An example of such a tumor suppressor gene is found in the inherited form of the ocular cancer retinoblastoma. *Retinoblastoma* is a cancer of the retinal cells of the eye that is most commonly diagnosed in young children. In certain families, retinoblastoma appears to be caused by a simple autosomal dominant mutation. These cases of inherited retinoblastoma make up about 40% of the total cases of retinoblastoma identified each year. Children with this inherited form of the disease usually end up with multiple tumors in both eyes. Other cases of retinoblastoma are sporadic (turn up in people with no family history of retinoblastoma). Inherited retinoblastoma is often manifested earlier than the sporadic form, and the sporadic form may turn up in only one eye, not both.

In families with inherited retinoblastoma, the causative gene maps to the long arm of chromosome 13. When a child inherits the mutant retinoblastoma gene, or RB gene, from only one parent, he or she is likely to develop tumors in both eyes at an early age. In this case, we can point at one mutation in just one gene and say that this is the hereditary defect responsible for mutation that causes the cell proliferation.

It is important to realize that the normal product of the RB gene plays a crucial role in the non-dividing cells in the retina. This RB protein is required to block cells from entering S phase and starting the mitotic cell cycle. If it is missing, cell proliferation begins and the cell starts on the path to tumorigenesis.

The Two-Hit Hypothesis

When we consider inherited forms of cancer, one of the first and most important questions to ask is: Why do only some retina cells in patients carrying an inherited RB defect form tumors? If every cell of this child carries the one copy of the RB mutation, why doesn't every cell initiate tumor formation? The normal copy of this allele, obtained from the unaffected parent, is sufficient to control cell division and prevent tumor formation in every cell that keeps an intact copy of the normal allele.

Then why are any tumors formed? Don't all the cells have the normal copy of RB that they inherited from the other parent? The retinal cells of heterozygotes only become competent to form tumors when they lose that normal allele of the RB gene; that is, they must carry two loss-of-function alleles of the RB gene to form a tumor. In these children the first hit was inherited and is present in every cell. The second hit happened in a single retinal cell at some later point in the child's development and is only present in cells that are descended from the cell in which that second hit took place (Figure 10.3).

This is the basis of the *two-hit* model of tumor formation formulated by Alfred Knudson. To form an eye tumor, cells have to knock out both good copies of the RB gene. Cells throughout the eyes and bodies of these RB heterozygous children already have one bad copy, but tumors will only occur in those cells that lose the other copy, as well. The inherited RB mutation is thus dominant in terms of pedigree analysis: most offspring that receive one copy of this mutation will develop tumors. However, at the cellular level we may think of it as classically recessive because both copies have to be missing to manifest the trait.

How can cells lose the normal allele of the RB gene? Mutations in any gene, including the normal allele of RB, occur at a very low frequency (approximately 1 per 100,000 or 1,000,000 cells per cell generation) during the process of DNA synthesis in each cell cycle. Many cycles of cell division are required to produce the millions of cells in the retina, called retinoblasts. At each division there was a very

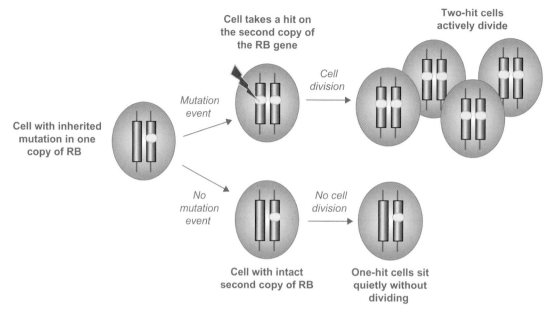

FIGURE 10.3   The two-hit model for tumor formation. The first hit, an inherited defect in a tumor suppressor gene in every cell in an individual's body, does not itself result in cancer. It is only when an additional mutation event (the second hit) takes out the second copy of the tumor suppressor gene in a single cell that loss of regulation of cell division occurs and the cell begins to divide in an uncontrolled fashion. It is unlikely but possible for defects to develop in both copies of a tumor suppressor gene in the same cell in an individual who did not start out in possession of a defective copy, but such events are very rare. In an individual who has inherited an RB copy with the first hit already present, most cells will retain their protective good copy and only rare cells will receive a second hit and become cancer cells. Second hits like this can happen independently more than once in the same individual bearing an inherited tumor suppressor defect. The two-hit hypothesis for tumor formation was proposed by Dr Alfred Knudson.

low risk of mutating the second normal allele of the RB gene. So even though the somatic mutation frequency is low, if we look at enough cells, we will see more than one independent case in which a cell loses the second allele. As a consequence, several cells in each retina will endure such mutations and thus be left without a functional RB gene.

Loss of Heterozygosity

This process of losing the remaining normal allele is referred to as *loss of heterozygosity (LOH)*. It is a common event at many sites in the genome in human tumors. It is important to realize that the normal product of the RB gene plays a crucial role in the non-dividing

cells in the retina. This protein, the RB protein, is required to block cells from starting the mitotic cell cycle. In its absence, cell proliferation begins and the cell starts on the path to tumorigenesis.

Basically, there are four classes of mechanisms that allow loss of the wild type copy. First, one could imagine a simple new somatic mutation that inactivates or destroys the wild type copy (Figure 10.4A). As we noted above, although the mutation rate is low, there are 10 trillion cells that make up a human being and thus such events will eventually occur – and once they do occur the resulting cell will begin the process of becoming a tumor.

The second general mechanism is the occurrence of a deletion that removes the wild type

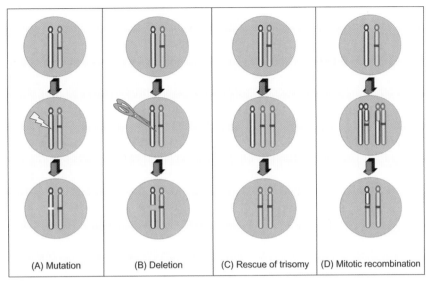

(A) Mutation   (B) Deletion   (C) Rescue of trisomy   (D) Mitotic recombination

FIGURE 10.4   Mechanisms of loss of heterozygosity. Starting with individuals who are heterozygous for a mutation on one copy of the retinoblastoma gene (red band), examples of ways in which the second copy can be lost include: (A) mutation of the second copy (yellow band), (B) a deletion that removes the second copy (white band), (C) a mitotic error leading to trisomy 13 in a cell that ends up with two copies of the chromosome with the defect, followed by a second mitotic error that restores the correct copy number. Loss of heterozygosity results when the copy that gets tossed out is the copy containing the normal version of the gene. (D) Mitotic recombination, which is not a normal step in mitosis, sometimes takes place and results in exchange of chromatids between the two copies of chromosome 13 below the position of the recombination event. After the mitotic recombination event the two chromosomes independently align on the mitotic metaphase plate. In half of such cells the chromatids carrying the two normal RB alleles (one on each chromosome) are pointed toward one pole and the two RB mutation-bearing chromatids are pointed to the other pole. Following the completion of cell division, one of the daughter cells ends up with both copies of chromosome 13 carrying the normal RB gene. More critically, the other daughter cell ends up with both copies of chromosome 13 (below the recombination point) carrying the mutant RB gene and thus has undergone loss of heterozygosity.

RB gene as well as some number of neighboring genes leaving the remainder of the chromosome intact (Figure 10.4B). The observation of such deletions was a major part of developing the loss-of-heterozygosity model for explaining hereditary retinoblastoma. When the normal copy is deleted, the only copy left in the cell is the defective copy. This one cell is no longer heterozygous but is now hemizygous for the mutated allele. Such deletions will be rare, but this event only needs to happen once among the hundreds of millions of cells that make up the retina of the eye.

The third mechanism involves complex errors in the mitotic divisions that lead to the formation of the retina (Figure 10.4C). Imagine that a mitotic error occurs that results in a trisomic cell with three copies of chromosome 13. In some cases, two copies of those three copies will be the mutant-bearing chromosome and one will be the chromosome carrying the wild type RB gene. Such cells will be quite unhealthy and there will be strong selection for a second mitotic event (mitotic loss) that causes the loss of one of those copies of chromosome 13. If that lost copy carried the wild type allele, the cell is now homozygous for the mutant RB gene and a tumor will result. This mechanism can be referred to as "reduplication and loss". (We should note that others often refer to the order of the duplication and loss events, and call the process "loss and reduplication."

In this case the cell initially loses the copy of the chromosome carrying the wild type allele and then duplicates the mutant copy. Although the result is the same, we think the probability of a cell carrying only one copy of chromosome 13 surviving for any length of time is quite remote.)

The fourth mechanism, known as mitotic recombination, involves recombination in response to various kinds of DNA damage at mitosis (Figure 10.4D). These recombination events occur between two non-sister chromatids of the two copies of chromosome 13 during the G2 phase of *mitosis* (again, note that we are *not* talking about the normal process of pairing during meiosis). The occurrence of recombination is then followed by a normal mitotic division, and can in some cases result in the creation of cells that are homozygous for all of the markers that occur between the mitotic recombination event and the telomere (for a full description of this process see the legend to Figure 10.4). Thus, a recombination event between the RB gene and the centromere in a cell that is heterozygous for a mutant RB gene can produce a daughter cell that is homozygous for that RB mutation!

If you have appropriate DNA markers on chromosome 13, or the sequence of the whole chromosome from the tumor cell, you could conceivably tell exactly which event led to formation of the tumor. A simple mutation will lead to LOH only at the RB gene, all other sites will remain heterozygous. A deletion will result in LOH for loci flanking the RB gene as well as at the RB locus itself. Reduplication and loss will cause LOH to occur along the length of the entire chromosome. Mitotic recombination will cause LOH from the point of the recombination to one tip of the chromosome. That said, as interesting as it may be to determine just how LOH for the original mutation occurred, the result is still the same – the cell is now either homozygous (carrying two copies of the original RB mutation) or hemizygous (as in the case of the deletion heterozygote) for the RB mutation, and the first step toward tumor progression has been taken.

If mutations are going on throughout our lifetimes, can't a cell sometimes have separate mutations hit both alleles of the same gene even though the person does not carry a mutated allele in all of their cells? In both inherited and sporadic retinoblastoma both copies of the gene are defective in the tumor cell. They differ only in whether it takes one mutational hit, or two hits both falling on the same cell, to bring about a functional defect. The sporadic form is quite rare (about 1/40,000) because it is so unlikely that both copies of a gene will be knocked out in the same cell, and two hits on different cells will not cause tumors because each of those two cells will still retain one good copy. Because it is so unlikely that this double hit will happen even once to a particular person, the chances become vanishingly small that someone with two normal copies of the RB gene will have a double hit happen independently in the same cell. Thus sporadic cases of RB characteristically present with a tumor in only one eye. The inherited form is found in only those who have inherited a first hit, but it is likely that they will experience a second hit more than once and end up with multiple tumors originating from different cells at different points in time. Thus individuals with inherited forms of RB are more likely to end up with tumors in both eyes.

10.6 CELL-TYPE SPECIFICITY OF TUMOR SUPPRESSOR GENE DEFECTS

Some children suffering from the inherited form of retinoblastoma are also at risk for some other kinds of cancers later in life, especially those children with retinoblastoma who are treated with radiation therapy who may sometimes develop tumors of the eyelids or elsewhere. However, these people are not at risk for all types of cancers. That is to say that, even though many of the proteins that protect and direct the cell cycle are expressed in many cell types, defects in a particular tumor suppressor gene will often be limited to causing cancer in a small number of cell types.

Other types of inherited cancers show similar relatively specific effects. The *Wilm's tumor* gene causes hereditary development of kidney tumors if a second hit causes the loss of the one normal allele that was passed along from the unaffected parent. Two *breast cancer* genes, BRCA1 and BRCA2, lose their ability to block uncontrolled cell division in individuals who lose both copies of either gene. Although BRCA1 and BRCA2 are primarily known for causing breast cancer, functional loss of either one can also predispose to other cancers, especially *ovarian* cancer, in some individuals, with the risk of ovarian cancer being lower for BRCA2 mutations than for BRCA1 mutations.

However, the mutational inactivation of some tumor suppressor genes leads to a broad array of different kinds of cancers. One such gene is p53, which produces one of the most important tumor suppressor proteins in our cells. This protein plays a crucial role both in preventing unwanted cell division and in regulating the response of the cell to DNA damage. More than 50 types of cancer have been shown to carry new mutations in the p53 gene; indeed, more that 70% of all colorectal cancers carry mutations in this gene. The same is true for many other kinds of tumors. So what happens if one inherits one defective copy of the p53 gene? The result is a hereditary disorder known as *Li–Fraumeni syndrome*, in which heterozygotes are at risk to develop a wide variety of different tumors in different tissues of the body, including tumors of the ovary and sarcomas; the type of tumor depends on which cell types experience a mutation in the second, normal copy of the gene.

10.7 THE MULTI-HIT HYPOTHESIS

Adenomatous Polyposis Coli

Single-gene traits like retinoblastoma are rare and represent only a small fraction of all cancers. This is a situation that you will often see in science. In many cells we can make huge advances in our understanding of fundamental processes through the study of the rare and exceptional situations. As important as some of these single-gene cancers have been to understanding the two-hit hypothesis, in fact most cancers are much more complex, starting later in life and progressing more slowly in part because they progress through more steps than just an initial "second hit."

Like the tumors just described, *adenomatous polyposis of the colon (APC)* is inherited as a simple autosomal dominant disorder. Affected individuals in these families already have the first hit on one copy of the APC gene on the long arm of chromosome 5. The first step in the formation of the tumor(s) is the loss of the normal (second) allele of the APC gene. Again, the protein product of the normal APC gene appears to be required for the control of cell division. However, in this case, the result of the second hit taking out the second good copy of the APC gene is not cancer but rather a precancerous polyp (Figure 10.5). The further progression of the small polyp into a malignant tumor can be divided into clear stages that can be distinguished by a pathologist. A study of these various stages reveals that the development of an invasive and metastatic tumor requires multiple new mutations at other places in the genome. Thus, although the formation of the early polyp appears to require only one mutation, additional mutations in other genes (most notably the gene encoding a protein called p53 discussed above) are required for that early tumor to become a dangerous malignancy (see Figure 10.5). Other kinds of cancers can also be found in individuals who inherit an APC gene

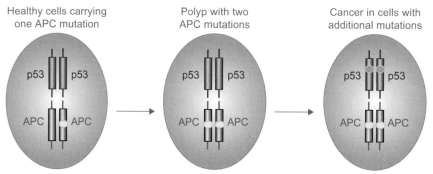

Healthy cells carrying one APC mutation

Polyp with two APC mutations

Cancer in cells with additional mutations

FIGURE 10.5　The more-than-two-hit hypothesis. The two-hit hypothesis deals with the initiating event that leads to the initial loss of control of cell division. In many cases, as an initial tumor cell divides and the tumor grows, additional mutation events affecting other genes in the genome can enhance cell growth and help confer other properties, such as invasiveness, that contribute to metastasis. In the case of the APC gene, the second hit results in growth of a polyp, but if the polyp is removed, those cells are not available to be hit with further mutations in other genes that could turn that polyp into a metastatic form of cancer. Since the probability of mutation is proportional to the number of cells, if there are fewer cells, there is less overall chance of a mutation. Thus a strategy used in APC is to screen for and remove polyps before they have a chance to acquire the additional mutations that will convert a polyp into metastatic cancer. Yellow marks mutations in APC; pink marks mutations in p53.

defect, such as stomach cancer, and other non-malignant features such as pigmented scarring of the retina can sometimes help in identifying individuals who are carrying an APC mutation.

How Do the Products of Tumor Suppressor Genes Act to Prevent Tumor Formation?

Some tumor suppressor proteins, such as the RB protein, act by shutting down the function of other proteins that activate steps in the cell cycle. The RB protein is a critical member of a suppressor complex that keeps a number of cell cycle proteins locked up in a multi-protein complex, and specifically regulated events are normally required to free a protein from this complex so that it can act. One of the key gene products that the RB "guard" protein locks up in this complex is a protein called E2F1, which is one of the proteins that signals the cell to move from G1 to S in the cell cycle (Figure 10.6).

Other tumor suppressor genes, such as BRCA1, appear to encode proteins that regulate the transcription of other genes, most

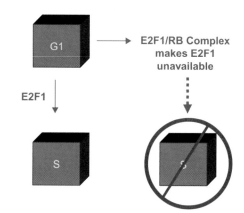

FIGURE 10.6　E2F1 is a critical key that unlocks the ability of the cell to enter S phase and proceed with cell division. RB is a tumor suppressor protein that blocks control of entry into S by helping to sequester E2F1 in a complex that contains a number of other proteins important to the cell cycle. The cell can release E2F1 from this complex when it needs it to act, and sequester it again when it needs to keep from proceeding into S phase. If there is even one good copy of the RB gene in the cell, there is enough RB protein being made to keep E2F1 under control. In a normal cell, E2F1 carries out its role in initiating S phase under very carefully controlled conditions, so that cell division only happens when it is supposed to. If there is no RB in the cell, E2F1 can escape suppression and activate entry into S phase and the subsequent events leading to cell division.

notably genes that control the cell cycle. Another such example of this phenomenon is found in *Von Hippel–Lindau disease*, or *VHL*, a hereditary predisposition to brain tumors. In this case the mutated gene is a known compon-ent of the enzyme complex that transcribes genes into mRNA.

10.8 THE ACTIVATION OF PROTO-ONCOGENES AND THE ROLE OF ONCOGENES IN PROMOTING CANCER

The End Run: Dominant Tumor-Promoting Mutations Push the Cell into the Division Cycle

Chromosome rearrangement and instability are hallmarks of tumor cells. Indeed, such rearrangement may play a crucial role in the initiation of some tumors. One of the best examples is *chronic myelogenous leukemia (CML)*. Greater than 90% of individuals with CML carry a specific translocation, referred to as the *Philadelphia chromosome*, involving chromosomes 9 and 22. This translocation is not present in normal cells of these patients. In this case a normally inactive gene (called *ABL*) on chromosome 9, which acts to promote cell division, is moved by translocation to fuse it to a gene called BCR from chromosome 22 (Figure 10.7). After translocation, the activity of the BCR–ABL fusion protein is now carried on continuously instead of happening transiently in response to specific signals that promote cell division. The activity of the BCR–ABL fusion protein triggers the activation of a regulatory cascade that promotes cell division. As a result, these white blood cells begin to constitutively enter the cell division cycle, repeating the cycle over and over instead of moving into G0 and parking until cell division is needed again.

In the case of *acute lymphoblastic leukemia (ALL)*, a translocation chromosome that looks like the Philadelphia chromosome when viewed under the microscope has actually created a different gene structure that includes a smaller portion of the BCR protein than occurs in the CML translocation.

Burkitt's lymphoma, another form of cancer, results from a similar translocation mechanism, in this case involving chromosomes 8 and 14, that perpetually turns on another tumor promoter gene called *MYC*. *The Metabolic and Molecular Bases of Inherited Disease* (Meltzer et al., 2001) lists dozens of different combinations of chromosomes that cause various forms of leukemia as the result of a translocation in a single cell bringing pieces of genes together in a way that turns on the activity of a cell cycle regulator, which normally should be expressed only under tightly controlled circumstances.

Another example of specific rearrangements that promote tumor development has been identified in prostate cancer (Figure 10.8). Recall that the prostate gland is a Wolffian duct derivative that depends on the presence of testosterone for its proliferation. Prostate cancer is often first identified by screening for a protein in the blood that is known as PSA (Prostate Specific Antigen) and whose level is correlated with the development of prostate tumors. Of tumors identified in such a fashion, approximately 50% are shown to carry gene fusions in which the 5' regulatory regions of genes whose levels are up-regulated in response to testosterone are fused with the coding sequences of a family of proto-oncogenes called ETS. The resulting gene fusions create true oncogenes in which the ETS transcription factors promote tumor development under the control of testosterone or a prostate gland-specific promoter. A gene fusion known as the TMPRS22:ERG fusion is the most common such rearrangement, but the TMPRSS22 promoter can also be found fused with other ETS family members. Prostate cancers carrying the TMPRS22:ERG fusion have a characteristic morphology and the overexpression of the ERG gene appears to play a role in making tumors more invasive.

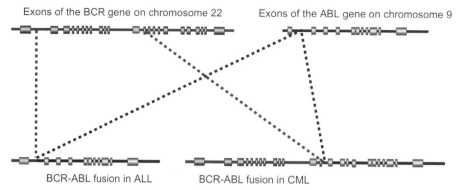

FIGURE 10.7   Translocations that connect the BCR gene on chromosome 22 with the ABL gene on chromosome 9 can cause either acute lymphoblastic leukemia (ALL) or chronic myelogenous leukemia (CML), depending on where the break-point occurs. (After Meltzer, P. et al., "Chromosome alterations in human solid tumors," in *The Metabolic and Molecular Bases of Inherited Disease*, Eighth Edition (New York, McGraw-Hill, 2001), p. 558.)

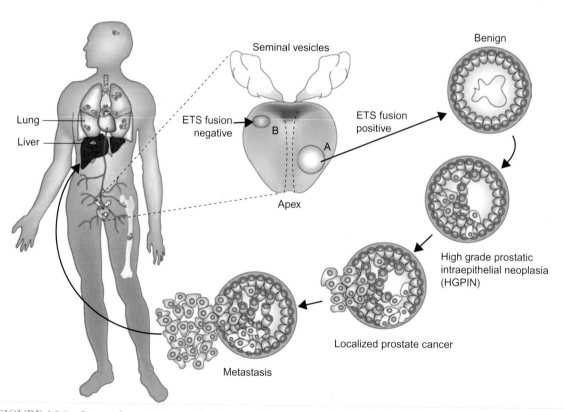

FIGURE 10.8   Stages of prostate tumor development. Some tumors that develop in the prostate develop an ETS fusion and these tumors progress towards localized and metastatic cancer. Other tumors do not develop an ETS fusion and are more likely to stay as a localized tumor. (From Kumar-Sinha, C., Tomlins, S. A. and Chinnaiyan, A. M., "Recurrent gene fusions in prostate cancer," *Nature Reviews Cancer*, 2008;8(7):497–511. Used by permission of the Nature Publishing Group. All rights reserved.)

As common as they are, such rearrangements are not the sole determinants of how aggressive (and thus lethal) a tumor might be. Loss-of-function mutations in a tumor suppressor gene known as PTEN, which normally functions to block rapid cell divisions, also appear to play a critical role in prostate tumor development. As we write these words in 2010, data are accumulating that tell us how both the genotype of a prostate tumor with respect to ETS gene family fusions that are highly expressed in prostate cells *and* the status of the two PTEN genes may well predict how fast or how slowly a tumor might grow. Such information may soon direct the course of treatment as patients make decisions between watchful-waiting, surgery to remove the prostate, and/or radiation and chemotherapy.

There is a crucial difference between such rearrangements and the inherited tumor promoting mutations discussed earlier. These rearrangements are found only in tumor cells. They are *not* present as a "first hit" throughout the body. Thus they are dominant on the cellular level (in the sense that hitting only one of the two copies can cause the problem even when a second "good" copy is present) but are not transmitted as a dominant disorder within a family because the mutation is not present in any cells of the body except the tumor cells.

10.9 DEFECTS IN DNA REPAIR

Tumors may often begin with mutations in genes that impair the ability of cells to repair their DNA and thus to fix either errors that occur during the replication process or DNA damage that occurs as a result of exposure to environmental mutagens. The relationship between defective DNA repair and carcinogenesis is well illustrated by a hereditary form of colon cancer called *hereditary non-polyposis colon cancer (HNPCC)*. Researchers have found two sets of families segregating for HNPCC. One

set of families allowed them to map a gene that predisposed individuals to HNPCC to chromosome 2, and the second set of families allowed them to map a second cancer-causing mutation to chromosome 5. Analysis of tumors from both families revealed an unusual genetic instability in the tumor cells. This instability was most easily manifested as the expansion or contraction of sequences in the human genome called *microsatellite repeats*, the short runs of repeated simple sequences such as dinucleotides and trinucleotides (e.g., CACACACACACACA) that are scattered throughout the genome (see Chapter 5 for a review). Although the number of repeats at each site was constant in normal cells of these individuals, cells in the tumor showed huge expansions of the repeat (e.g., CACACACACACACACACACA CA CACACACACACACA) or contractions of the repeat (such as CACACACACA). The investigators quickly found that HNPCC resulted from repair-deficient mutations in genes that encoded enzymes involved in a DNA repair process called mismatch repair (see Box 10.1).

In this case, the colon cancer-causing mutations on chromosomes 2 and 5 were mutations in one copy of the MSH2 gene or the MUTL gene. In the tumors themselves, the normal allele of these genes often appears to be deleted. This is another example of the "two-hit" model. Individuals inheriting a cancer-causing allele of one of these genes are at high risk for HNPCC. The "second hit" is the mutational "knock-out" of the normal copy of the MSH2 or MUTL gene in one or more of these cells, which creates a cell with no functional copies of these genes, a cell that can no longer accomplish mismatch repair. That cell is now going to experience a high frequency of new mutations every time it replicates its DNA.

Why should a defect in DNA repair cause cancer? If you have an error-prone system for replicating DNA, one that cannot repair the occasional errors made during replication or

---

BOX 10.1

MISMATCH REPAIR

Mismatch repair is one of several processes that repair the damage done to DNA both by exogenous agents (such as sunlight, radiation, or carcinogens in food, water, or smoke from tobacco) and by everyday life in the cell, such as errors resulting from mistakes that happen in the normal course of copying DNA. Because such errors during replication are not infrequent, our cells have a fairly efficient mismatch repair system for dealing with them. In the absence of a mismatch repair system, those errors that do occur during replication are not repaired but rather end up being incorporated into one of the two daughter strands follow-

ing the next round of replication. The resulting daughter strand will then carry a new mutation at the site of the replication error. Cells that cannot repair errors through mismatch correction will have greatly increased mutation rates. Please note that errors occur throughout the genome, not just at cancer genes or at microsatellite repeats. However, errors in mismatch repair are detectable through monitoring of changes in lengths of microsatellite repeats because errors seem to be more frequent where DNA polymerase has been trying to copy a length of repeated simple sequence in the absence of mismatch repair.

---

spontaneous damage to DNA, every round of replication is a potentially mutagenic event. Every round of replication gives you a chance to lose another tumor suppressor gene by mutation. Every round of replication makes the loss of control of cell division more inevitable.

If DNA repair defects really make a cell susceptible to cancer, shouldn't defects in other repair systems also make cells more cancer prone? Yes! Indeed, there are a number of human diseases in which individuals who are demonstrably repair defective are highly cancer prone, including *xeroderma pigmentosa*, *Bloom syndrome*, *ataxia telangiectasia*, and *Fanconi anemia*. Each of these disorders results from homozygosity (or compound heterozygosity) for inherited recessive mutations in genes required for various aspects of DNA repair. For example, patients with xeroderma pigmentosa are defective in various aspects of a process called excision repair, whereas children with ataxia telangiectasia are deficient in a process

that forces cells with unrepaired DNA damage to stop dividing until they can repair their DNA. In each of the disorders, mutations in genes whose protein products are required for DNA repair predispose their bearers to develop tumors. How? The best guess is that these repair-deficient disorders effectively raise the mutation rate and, in doing so, increase the probability of mutating the tumor-suppressor genes.

10.10 PERSONALIZED MEDICINE

For more than eight years, Scott's father fought a long, hard battle with prostate cancer. During the course of his treatment, the doctors considered and/or used each of the three primary modes of cancer treatment: surgery (to remove the dividing cells), irradiation (which preferentially kills dividing cells by fragmenting the chromosomes), and chemotherapeutic drugs (which act to block cell division, often by inhibiting DNA replication or mitotic spindle function). Because the cancer in question was prostate cancer, hormone-based treatments were also used to try to remove the testosterone signal that was giving a "go ahead and divide" signal to the prostate cells. Over time, each treatment stopped working, a reflection of the mutations ongoing in the multi-hit process by which initially benign tumors become malignant and increasingly aggressive, and tumor cells become more resistant to treatments. In addition, some of the side effects were seriously affecting Scott's father's quality of life and, at times, seemed worse than anything the cancer cells were doing.

Why would cancer treatments come with so many bad side effects when other medications we take, such as antibiotics, have so little effect beyond curing the malady at hand? The answer is that cancer treatments are almost the only therapies we experience that are aimed at killing actively dividing human cells, and we walk a fine line between killing the cancer cells and killing other cells in our bodies that we need to keep around. So in the vicinity of the tumor, the cancer treatment may do an excellent job of distinguishing between the actively dividing tumor cells and the non-dividing cells in the surrounding tissue, but in other parts of the body these same treatments have deleterious consequences when they kill the normal cells that are also dividing. For example, blocking the normal active division of white blood cell precursors in the bone marrow leads to a compromised immune system and susceptibility to infections, whereas impairing the normal ongoing division of gastrointestinal cells can produce debilitating nausea and other complications. If we just had a pharmaceutical "bullet" that would target the cancer cells, the other normally dividing cells in the body could be left alone. Building such a truly specific cancer killer requires that we can answer the question, "What makes the cells in a tumor different from all the other normal cells in the body – especially

the actively dividing cells?" To answer that question, we need to understand how tumors arise and how they proliferate so that we can begin designing drugs that can target cancer cells while sparing the normal cells that surround them.

Building Magic Bullets – The Imatinib Story

We have been talking about a model of cancer as a genetic disease, in which mutations in a variety of genes compromise the numerous systems that prevent cell division. Some mutations, such as those in DNA repair genes, act by increasing the mutation rate, thus making mutants in the critical tumor suppressor genes more likely, but other mutations directly inactivate the tumor suppressor genes themselves or activate tumor promoter genes. The question then becomes: What if we could restore the function of the lost tumor suppressor gene or suppress the action of the product of a tumor promoter gene: could we then cure the cancer? Such is the type of molecular biology-based therapeutic approach that most of us describe as the *magic bullet*, that drug that specifically targets a critical defect in the tumor cells themselves without affecting the normal cells.

One magic bullet that already exists destroys a type of cancer cell that we have already mentioned, chronic myelogenous leukemia. As we noted, this cancer is due to the inappropriate activation of a division-promoting gene called ABL. The issue here is that, if one could inactivate the function of the BCR-ABL protein in CML cells, one might be able to stop the tumor growth. Work by numerous investigators had shown that ABL is a *tyrosine kinase*, a protein that chemically modifies other proteins in the cell. Its expression turns on and off at different times depending on signals being received by the cell. The BCR-ABL protein also acts as a tyrosine kinase, but it has lost the ability to respond to the signaling systems that would allow it to turn off when that is what the cell needs. There are many types of tyrosine kinases in our cells, but the BCR-ABL held a special position in the cellular signaling cascade. Suppose one could build a small drug that specifically and uniquely inactivated the ABL protein, leaving the functions of the other tyrosine kinases intact? Could that cure the cancer? A scientist named Brian Druker identified a small compound, initially named STI-571, that can inhibit the BCR-ABL protein.

Working with the pharmaceutical company Novartis, Druker was able to show that this drug works astonishingly well against CML. In the first clinical trials reported in 1998, remissions were observed in almost 100% of the first 31 patients tested! After dramatic subsequent trials, STI-571 (now known as Gleevec™ or Imatinib) sailed through the government's drug approval process in record time. Although there are usually few side effects and most patients get long, substantial remissions with Imatinib, the drug loses effectiveness and eventually fails in patients with end-stage disease. Sadly, even some patients treated early may become resistant to Imatinib as a consequence of new somatic mutations in the tumor cells that survive the treatment. Imatinib also works well against one other type of tumor, known as *gastrointestinal solid tumors (GISTs)*. However, from the specificity of its design, so far it does not appear to have a broad therapeutic range against other types of cancers.

Unfortunately Imatinib can have serious side effects on some people and some people develop resistance to it, presumably as a result of mutations in the BCR-ABL gene fusion. Fortunately there is a second generation of tyrosine kinase inhibitors (TKIs) that also appear to effectively treat CML. The best known of these is Dasatinib, which, like Imatinib, can be taken orally. There are other drugs, including Nilotinib, that can be used if Imatinib fails or is poorly tolerated. Some of these drugs are now being tested as first line treatments for CML,

as are various combinations of these drugs. It seems possible that using such a combination of drugs might deliver a "knock-out punch" to the tumor that it simply cannot mutate around.

Acute Myeloid Leukemia: Wielding a Molecular Lance

Maybe each type of tumor will require its own magic bullet, but the success with Imatinib is informative. At the beginning of this chapter we described the tragic loss of one of Scott's former graduate students, Brenda Knowles, to another type of leukemia, known as *acute myeloid leukemia (AML)*. Unlike CML, AML cannot be attributed to any one single mutational lesion (such as the expression of the BCR-ABL protein). Worse yet, it seems likely that two separate mutations are required to trigger the disease. However, mutations in a gene that encodes another tyrosine kinase called *FLT3* are present in the tumors of about 30% of patients with AML, and most often in those patients with the worst prognosis. Once again, these mutations serve to activate a kinase that should only be expressed in the rapidly dividing *stem cells* of the marrow. In the mutated state, this activated kinase keeps the leukemic cells in a state of rapid cell division. Professor Gary Gilliland and his research team at Harvard University are working on drugs that specifically target the FLT3 kinase. These drugs are now in clinical trials. Others like them will follow. It is simply a real sadness of life that these drugs come a decade too late for Brenda and the others who have died of this disease.

One can imagine other such drugs that would target a molecule that is having specific effects of importance in a particular tumor cell type. In the case of prostate cancer, a drug called flutamide has been developed that can inactivate the ability of the testosterone receptor to respond to testosterone, which offers a tool with which to impair the progress of testosterone-sensitive prostate cancer cells.

Already, women with certain types of breast cancers, those that over-express the *HER2* protein, are treated with a protein called *Herceptin*. Herceptin is an antibody that binds to cells expressing HER2 and slows their growth. Herceptin treatment doesn't work for every tumor, and even when it does work, often it only delays the progress of the disease. Nonetheless, for many patients, it does extend life. These are clearly the first steps in building a diverse molecular pharmacy for the treatment of cancer. Even these limited successes make it clear that if we ever want to treat the vast majority of tumors, we are going to have to be able to understand the genetic lesions that underlie each of those tumors. In the final section of this chapter, we will discuss the development of modern tools for the study of gene expression in tumors that let us assess the activity, inactivity, or hyperactivity of every gene in that tumor.

10.11 CANCER BIOMARKERS

In Chapter 3 we talked about expression profiling and the power of scanning the entire transcriptome – the complete array of genes expressed across the whole genome. We can use expression profiling to compare the mRNA in tumor cells to mRNA from the cells surrounding the tumor and looking to see which genes show a change in level of expression (as manifested by a change in the brightness of the spots corresponding to the gene you are interested in). As chips are being made that have a large fraction of the human genes represented on the chip, it then becomes possible to do what are effectively whole-genome experiments looking for genes whose expression correlates with the presence of a tumor, with a progression from benign to malignant, or with a progression from malignant to highly invasive. Indeed, exactly such strategies have created a new generation of molecular diagnostics for cancer (Figure 10.9).

Expression Profiling

Tumor types can be more specifically identified, and differences in gene expression will be used to detect tumors that already demonstrate gene expression changes associated with increased risk of metastasizing. *Gene expression profiling* will also eventually let a doctor make predictions about optimal treatments. For instance, tumor cell expression of high levels of a gene whose product acts as a pump that pumps chemotherapeutic agents out of the cell is not good news for a patient being treated with chemotherapy. Knowing about this high level of pump activity would be important for a patient whose doctor has alternative treatments to offer. As more research proceeds, this gene expression profiling approach to pharmacogenomics will become an important part of the doctor's clinical testing repertoire, but far more must be learned to understand enough about the clinical implications of such test results.

Polymerase Chain Reaction

Polymerase chain reaction (PCR) also lets us monitor biomarkers that are indicative of recurrence. In the case of tumors such as sarcomas that may shed cells into the bloodstream, we can detect the causative translocation in those shed cells through a PCR reaction, or we can detect the RNA transcribed from the translocated gene through an RT-PCR reaction. This allows new rounds of treatment to begin as soon as possible after recurrence begins. Previously recurrence was not detected until the tumor grew large enough to be detectable by exam or scanning technologies or until the tumor started to cause symptoms. PCR is so sensitive that it can detect even small numbers of cells.

Fluorescent In Situ Hybridization (FISH)

FISH (fluorescent in situ hybridization) allows researchers to assess the expression of

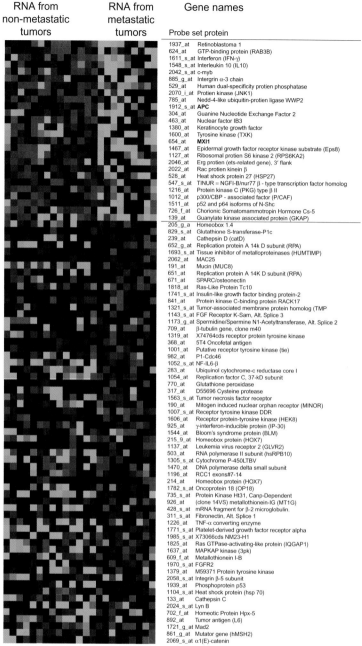

RNA from non-metastatic tumors	RNA from metastatic tumors	Gene names

Probe set protein

1937_at		Retinoblastoma 1
624_s_at		GTP-binding protein (RAB3B)
1611_s_at		Interferon (IFN-γ)
1548_s_at		Interleukin 10 (IL10)
2042_s_at		c-myb
885_g_at		Intergrin α-3 chain
529_at		Human dual-specificity protien phosphatase
2070_i_at		Protien kinase (JNK1)
785_at		Nedd-4-like ubiquitin-protien ligase WWP2
1912_s_at		**APC**
304_at		Guanine Nucleotide Exchange Factor 2
463_at		Nuclear factor IB3
1380_at		Keratinocyte growth factor
1600_at		Tyrosine kinase (TXK)
654_at		**MXI1**
1467_at		Epidermal growth factor receptor kinase substrate (Eps8)
1127_at		Ribosomal protien S6 kinase 2 (RPS6KA2)
2046_at		Erg protien (ets-related gene), 3′ flank
2022_at		Rac protien kinein β
528_at		Heat shock protein 27 (HSP27)
547_s_at		TINUR = NGFI-B/nur77 β - type transcription factor homolog
1216_at		Protein kinase C (PKG) type β II
1012_at		p300/CBP - associated factor (P/CAF)
1511_at		p52 and p64 isoforms of N-Shc
726_f_at		Chorionic Somatomammotropin Hormone Cs-5
139_at		Guanylate kinase associated protein (GKAP)
205_g_a		Homeobox 1.4
829_s_at		Glutathione S-transferase-P1c
239_at		Cathepsin D (catD)
652_g_at		Replication protein A 14k D subunit (RPA)
1693_s_at		Tissue inhibitor of metalloproteinases (HUMTIMP)
2062_at		MAC25
191_at		Mucin (MUC8)
651_at		Replication protein A 14K D subunit (RPA)
671_at		SPARC/osteonectin
1818_at		Ras-Like Protein Tc10
1741_s_at		Insulin-like growth factor binding protein-2
841_at		Protein kinase C-binding protein RACK17
1321_s_at		Tumor-associated membrane protein homolog (TMP
1143_s_at		FGF Receptor K-Sam, Alt. Splice 3
1173_g_at		Spermidine/Spermine N1-Acetyltransferase, Alt. Splice 2
709_at		β-tubulin gene, clone m40
1319_at		X74764cds receptor protein tyrosine kinase
368_at		5T4 Oncofetal antigen
1001_at		Putative receptor tyrosine kinase (tie)
982_at		P1-Cdc46
1052_s_at		NF-IL6-β
283_at		Ubiquinol cytochrome-c reductase core I
1054_at		Replication factor C, 37-kD subunit
770_at		Glutathione peroxidase
317_at		D55696 Cysteine protease
1563_s_at		Tumor necrosis factor receptor
190_at		Mitogen induced nuclear orphan receptor (MINOR)
1007_s_at		Receptor tyrosine kinase DDR
1606_at		Receptor protein-tyrosine kinase (HEK8)
925_at		γ-interferon-inducible protein (IP-30)
1544_at		Bloom's syndrome protein (BLM)
215_9_at		Homeobox protein (HOX7)
1137_at		Leukemia virus receptor 2 (GLVR2)
503_at		RNA polymerase II subunit (hsRPB10)
1305_s_at		Cytochrome P-450LTBV
1470_at		DNA polymerase delta small subunit
1196_at		RCC1 exons#7-14
214_at		Homeobox protein (HOX7)
1782_s_at		Oncoprotein 18 (OP18)
735_s_at		Protein Kinase Ht31, Canp-Dependent
926_at		(clone 14VS) metallothionein-IG (MT1G)
428_s_at		mRNA fragment for β-2 microglobulin.
311_s_at		Fibronectin, Alt. Splice 1
1226_at		TNF-α converting enzyme
1771_s_at		Platelet-derived growth factor receptor alpha
1985_s_at		X73066cds NM23-H1
1825_at		Ras GTPase-activating-like protein (IQGAP1)
1637_at		MAPKAP kinase (3pk)
609_f_at		Metallothionein I-B
1970_s_at		FGFR2
1379_at		M59371 Protein tyrosine kinase
2058_s_at		Integrin β-5 subunit
1939_at		Phosphoprotein p53
1104_s_at		Heat shock protein (hsp 70)
133_at		Cathepsin C
2024_s_at		Lyn B
702_f_at		Homeotic Protein Hpx-5
892_at		Tumor antigen (L6)
1721_g_at		Mad2
861_g_at		Mutator gene (hMSH2)
2069_s_at		α1(E)-catenin

FIGURE 10.9 Use of a Cancer Array gene chip to examine gene expression in cancer. This image shows that certain genes show higher levels of mRNA (in red) in metastatic tumors than in tumors that have not metastasized. Other genes show the opposite pattern. Different kinds of cancers will show a different profile of gene expression levels that can point to where critical biochemical events in the cell are taking place, and eventually assays of changes of this kind will be important in distinguishing different stages of cancer. In this case the RNA samples came from medulloblastoma tumors, and analysis of images like this one allowed the authors to identify genes that play a role in medulloblastoma metastasis. Some gene chips and microarrays can display thousands or tens of thousands of genes, but use of specialized chips that do not contain the full array of human genes lets researchers focus in on where the critical differences are. (Modified from T. J. MacDonald *et al.*, *Nature Genetics*, 2001;29:143–52; with permission from Nature Publishing Group.)

important genes in single cells, thus providing far better tools for diagnosing tumor type in biopsies, detecting metastases, and measuring the response of a tumor to treatment. Also, *chromosome painting* (Figure 10.10) using a large number of colors allows the identification of chromosomes that are damaged or present in altered numbers in tumor cells, which can assist in interpreting how advanced the cancer is and identifying regions of the genome involved in progression of the cancer to later stages. As we can see in Figure 10.10, the story for cells from an advanced stage of cancer can sometimes be very complex and involve translocations, deletions, insertions, and aneuploidy affecting many different genes and chromosomes.

Proteomics

A variety of proteins associated with cancer can be monitored. Use of antibodies that recognize specific proteins lets us tell whether such proteins are present on the surface of tumor cells, or present in bodily fluids such as serum, saliva, tears or urine. The test for PSA (which is associated with prostate cancer) is carried out

FIGURE 10.10 Twenty-four-color chromosome painting identifies multiple translocations and other signs of aneuploidy in a breast cancer tumor cell line. In many cases we can compare the FISH karyotype to that of the surrounding tissue so that we can see just how far the tumor karyotype has migrated away from the normal karyotype by comparing the karyotype in tumor cells to the karyotype of the normal cells immediately surrounding the tumor. (Courtesy of Joanne Davidson, from Davidson *et al.*, "Molecular cytogenetic analysis of breast cancer cell lines", *British Journal of Cancer*, 2000;83:1309–17. Used by permission of the Nature Publishing Group. All rights reserved.)

by using an antibody that recognizes PSA as the basis for detecting the protein in samples from patients. Assays for proteins can be used not only to look at specific proteins known to be produced by tumors, but can also be used to evaluate levels of proteins that the body produces in response to the presence of cancer. Other broad-scale proteomics approaches let us monitor large numbers of different proteins in a single assay.

One especially important class of proteins to monitor is the family of protein pumps, such as p-glycoprotein, that are responsible for pumping specific classes of drugs out of cells. If someone with cancer is being treated with key chemotherapeutic agents, it is important to know whether or not they are producing elevated levels of the protein that pumps such agents out of the cell. Level of p-glycoprotein function can be monitored by measuring the rate at which some specific compounds are being pumped out of the cell. P-glycoprotein levels can affect response to chemotherapy by making it difficult to get enough of the medication into the cancer cells of someone who is producing very high levels of P-glycoprotein.

A Concluding Thought

In our best dreams, 20 years from now your doctor will have available a pharmacy full of drugs that specifically impede the growth of (or, better yet, kill) the tumor cells without impacting their healthy neighbors. This not only has the potential to greatly reduce many of the terrible side effects of current cancer treatments by avoiding attacks on healthy cells but also offers the possibility of getting more complete elimination of cancer cells very early in treatment. Although such drugs may well work only in the early stages of the disease, we anticipate that a better understanding of molecular events in later stages of cancer will lead to breakthroughs in other medicines specific to the properties of later stage cells. Because there are so many ways to turn on cell division and because not all cases are diagnosed early enough, we don't imagine that these drugs will ever cure every cancer or even that every tumor will respond to any one treatment. However, that is not so different from the status of many "curable" things in current medicine, where someone can still die of an infection in a world with plentiful antibiotics. The difference is that we will progress from the current state of affairs, in which a diagnosis of cancer automatically summons up fears of death, to a state of affairs in which survival will be expected and a poor outcome will be the rare event.

We see another plus to very specific pharmaceuticals that target the primary molecular defect in a tumor cell: too many of today's treatments, such as radiation and chemotherapy, affect the immune system along with the tumor cells. If specificity of treatment spared the cells of the immune system, might we also find ourselves able to fine-tune the immune system response to tumor cells to help eliminate any cells missed by the drugs?

We also see great gains to be made in diagnostics. If we can arrive at knowing not just what kind of cancer it is, but also which biochemical pathways have become involved in the tumor's progression, it may streamline targeting exactly the right treatment not just for that cancer but for that stage of that cancer. Of great importance, we anticipate better diagnostics arising from molecular genetics that will reduce the problem of people being diagnosed "too late." Imagine home monitoring systems akin to home pregnancy tests or the blood-sugar monitoring systems used by a diabetic patient, allowing those with especially high risk levels to watch for recurrence of a banished tumor, or appearance of an expected tumor that has not yet manifested itself.

Will the magic bullets yet to come provide a simple outright cure for cancer the way we cure an infection with an antibiotic? Perhaps. For sporadic cases who did not inherit a cancer gene, we expect that cures will be achievable and that curing their cancer once will mostly take care of the problem. Perhaps for individuals who inherited that first hit, a cure will really involve three different stages. First, as for the nonhereditary cases, there will be the problem of how to shoot just the right magic bullets early enough to end the initial cancer. The second stage will involve ongoing diagnostics to detect any new tumors at the earliest possible stage. Third, there will be ongoing medical management to suppress development of subsequent tumors. We can identify key proteins produced by tumor cells and key changes in gene expression events. Using that information, can we concoct a surveillance system to detect the tumor before it gets large enough for anyone to know it is there? We expect that eventually harnessing key features of our own immune systems along with the magic bullets from the pharmacy may allow for many cancers to become a chronic disease, treated over a long lifetime through your local pharmacy. How much of an advance would that be? Of course, we would hope that answers will arise that will offer simple, clean cures, but 20 years from now, if cancer has moved out of the acute life-threatening category into a chronic management category, we will feel as if a dream has come true.

Study Questions

1. What is cancer?
2. What is G0?
3. What are three kinds of cells that keep dividing on a regular basis even though many cell types in the body are resting quietly without dividing?
4. What is a benign tumor?
5. What does the term malignant refer to?
6. What does the term metastatic refer to?
7. What is a tumor suppressor gene?

8. What is a proto-oncogene?

9. What is the two-hit model for the origins of cancer?

10. Sandy and Gary are unrelated individuals with retinoblastoma. Sandy developed one tumor in one eye at the age of 49 , and Gary developed multiple tumors in each eye by the time he was four. What do we expect that their family histories would look like, and why?

11. In one large family, six people in four generations have cancer. Two people have endocrine tumors, one has skin cancer, one has colon cancer, one has lung cancer and a kidney tumor, and one has breast cancer. Why might this be consistent with a defect in a DNA repair genes?

12. A chemical is identified that tends to lead to deletions in the genome. Why might exposure to this chemical represent a greater risk to someone who is heterozygous for a defect in a colon cancer gene than to someone who carries no cancer gene defects?

13. Chronic myelogenous leukemia (CML) is the result of an alteration to the DNA of the individual with CML, and yet CML is not passed from one generation to the next. How can something be gene-based and not be genetic?

14. Two different kinds of leukemia, ALL and CML, can both result from translocation bringing together the BCR gene from chromosome 22 and the ABL gene on chromosome 9. Why can translocation events that bring together these two genes result in two different phenotypes?

15. What are four different ways to end up with loss-of-heterozygosity?

16. What are three environmental risk factors for cancer?

17. What can ETS fusions and PTEN genotypes predict?

18. How can we monitor to detect recurrence of some kinds of cancers as early as possible?

19. What are four things that FISH can show us in cells from an advanced stage of metastatic cancer?

20. What is Imatinib and what is it used for?

Short Essays

1. As you remember from our discussion of retinoblastoma, the heritable form of which allowed Alfred Knudson to develop the two-hit model for tumor-suppressor genes, the RB protein plays a critical role as a tumor suppressor. We can tell when it is not working because of the development of tumor cells. But as researchers started to investigate how loss of RB leads to cancer they found a complex set of mechanisms by which loss of RB affects cell division. What are three different ways that loss of RB can lead to unregulated cell growth? As you consider this question please read "Retinoblastoma, a trip organizer," by Giovanni Bosco in *Nature*, 2010;466:1051–2.

2. Use of whole-genome sequencing to compare the DNA sequence of a tumor and the surrounding tissues can help us identify the underlying cause of a particular cancer. What can we learn by comparing the whole-genome sequence of a primary tumor to the whole genome sequence of tumor cells that have metastasized to another location in the body? As you consider this question please read "Genomics of metastasis" by Joe Gray in *Nature*, 2010;464:989–90.

3. Sometimes finding the genetic defect causing a tumor does not tell us enough. What can we gain from identifying the biochemical pathway involved in a tumor that we cannot gain from identifying the specific gene in that pathway that is defective? As you consider this question please read "Cancer: a tumor gene's fatal flaws" by Julian Downward in *Nature*, 2009;462:44–5.

4. We tend to think of cancer research in terms of a focus on mutations in cancer genes, but interesting new insights are coming out of studies of microRNAs. What kind of role do microRNAs play in cancer? As you consider this question please read "Cancer's little helpers" by Tina Hesman Saev in *Science News*, 2010;178:18.

Resource Project

Once researchers learn about some aspect of cancer or its treatment, clinical trials are needed to test out any proposed use in humans. What are some of the kinds of clinical trials going on for cancer? Go to the National Cancer Institute's website and look at their web page with the title Search for Clinical Trials. Select a form of cancer and look for clinical trials near you. Pick one of the trials that is currently open to new participants. Explain who is participating and what the trial is testing. Who is excluded from participation? What phase of trial is this study and what are they mainly trying to find out?

Suggested Reading

Articles

"Personalized medicine in the era of genomics" by Wylie Burke and Bruce M. Psaty in *Journal of the American Medical Association*, 2007;298:1682–4.

"The molecular pathology of cancer" by T. J. Harris and F. McCormick in *Nature Reviews Clinical Oncology*, 2010;7(5):251–65 [Epub 2010 Mar 30].

"How many cancers are caused by the environment?" by Brett Israel and Environmental Health News in *Scientific American*, May 21, 2010.

"Multiple mutations and cancer" by Lawrence A. Loeb, Keith R. Loeb, and Jon P. Anderson in *Proceedings of the National Academy of Sciences of the U S A*, 2003;100(3):776–81.

"The biology of cancer stem cells" by Neethan A. Lobo, Yohei Shimono, Dalong Qian, and Michael F. Clarke in *Annual Review of Cell and Developmental Biology*, 2007;23: 675–99.

"Biomarkers in cancer staging: prognosis and treatment selection" by Joseph A. Ludwig and John N. Weinstein in *Nature Reviews Cancer*, 2005;5:845–56.

"The phosphatidylinositol 3-kinase–AKT pathway in human cancer" by Igor Vivanco and Charles L. Sawyers in *Nature Reviews Cancer*, 2002;2:489–501.

"Identification of the breast cancer susceptibility gene BRCA2" by R. Wooster *et al.* in *Nature*, 1995;378:789–92.

"Early detection: proteomic applications for the early detection of cancer" by Julia D. Wulfkuhle, Lance A. Liotta, and Emanuel F. Petricoin in *Nature Reviews Cancer*, 3: 267–75.

Books

The Biology of Cancer by Robert A. Weinberg (2006, Garland Science).

Genes and the Biology of Cancer by Harold Varmus and Robert A. Weinberg (1992, W. H. Freeman & Co.).

HOW GENES ARE FOUND

The Gene Hunt: How Genetic Maps
Are Built and Used

The Human Genome. DOI: 10.1016/B978-0-12-333445-9.00011-3

11.1 WHAT IS A GENETIC MAP?

"Ready!" four voices shouted, and Angie opened the door to the room, where she paused and slowly passed her gaze across the contents of the room and the four giggling children clustered in the middle. After a moment's indecision, she moved to the left toward the windows. "Cold," said the other children. Realizing that she must be heading in the wrong direction, she redirected her course towards the fireplace. "Colder," they shouted. She shook her head at this news that she was still heading the wrong way, and continued into the room, moving towards the remaining wall of the room. "Freezing," they chorused, falling over each other in squeals of laughter. She stopped, confused, then turned around to head back towards the door. "You're getting warmer again," called out one of the boys. As she kept heading back towards the door, the volume of their voices rose with each word. "Warmer. Warmer. Warmer. REALLY HOT! BURNING UP!" they shouted as she reached out and touched the doorknob on the inside of the door, the object the children had selected as "it" before Angie came into the room. After laughing over their cleverness in picking an object back at the door as "it," they picked a different child to leave the room and started the game again.

Surprisingly, this game of Hot and Cold is not unlike the process by which human genes are mapped. When searching for a gene, we use a process called a genome scan to test many different positions along the human chromosomes to determine whether that is where our gene of interest is located. However, we usually don't get a lot of information back from each test, since the main question we get to ask is, "Is this particular spot on the chromosome close to the gene we are looking for or not?" It is a yes–no question to which the answer is almost always "no" because our gene is located at only one place and thus almost all of the other positions in the genome are not "it." Once we get close, then we start getting more information about just how close we are, and as we get closer the kind of information we are receiving tells us we are moving towards our target, but if we are not close to the gene, the answer we get back basically just tells us that we are "cold" and looking in the wrong place. Because the genome is so large, we usually have to do many tests for whether we are hot or cold before we find a gene, but usually if we test enough locations, at some point we turn from cold to warm and can begin following the signals from warm to hot to "burning up" at the spot where the gene we want is located.

In this chapter we are going to talk about genetic maps and the ways in which we use data from families to tell us where genes are located and what the genes are that are responsible for the various human traits. To many professors using this book this will look like old history because the ways in which we are doing genetics is changing so rapidly, but this discussion of genetic maps and gene mapping offers important insights into how we think about genetics and how we do genetics. These fundamental concepts still apply whether we are using the technologies from the 1980s, the 1990s, the last ten years, or those yet to emerge in coming years. So this chapter will tell you what genetic maps are, what is on them, how they were produced, and what we can do with them. This chapter will tell you a little bit about what it means when we say that we

have "found" a gene (Box 11.1). And in Chapter 12 we will show you how the Human Genome Project has revolutionized this search for the underlying causes of our traits, the good ones and the deleterious ones alike.

The genomic haystack in which we search for genetic needles is enormous. We are going to start this chapter on gene hunting with a daring assertion: If we can find out where a gene is located, we can both determine its nucleotide sequence and determine the nature of mutations that disrupt it. In fact, we can do that even if we do not know what it encodes, or what the gene product does. We can do this even if our models for disease pathology are incorrect. This approach – called *positional cloning* – led to some of the most inspiring breakthroughs in human genetics in the 1980s and 1990s, including the identification of the genes for cystic fibrosis, neurofibromatosis, and Huntington's disease. Since then, the approach has been modified into a *positional candidate* approach that not only uses position on a genetic map to find a gene but also uses information about gene product function and gene expression that were often not available when the earliest positional cloning experiments were taking place.

We make our prediction that we can find a gene by knowing its position on a given chromosome in part because many genes have been found in exactly this way. Certainly it is not how all genes have been found, but the fact that genes can be found simply by getting a narrow enough fix on the position of a genetic defect tells us that maps of gene locations – genetic maps – are incredibly important things.

What is a map? A geographic map is an ordered array of landmarks on a theoretical representation of a landscape, locations that we can use as anchor points to tell where we are or what we are near. Landmarks can take the form of mountains or bays, parks or statues, famous buildings or street signs. How we detect a landmark depends on what it is; some landmarks like a bay can be seen from orbit and some

---

BOX 11.1

MORE THAN ONE WAY TO "FIND" A GENE

Sometimes, when headlines announce that a gene has been found, the news media mean only that the gene has been roughly positioned on the chromosome. Because the location of a gene is so often a critical piece of information for actually getting our hands on the gene itself, the mapping of its location is important and often newsworthy, especially if the disease is especially common or severe. Other times, when the newspapers announce that a gene has been found, they mean that what has been found is the actual gene itself, complete with the order of As, Cs, Gs, and Ts that make up its sequence and maybe even some insights into protein function and disease pathology. Often, some of the most important findings, such as what the gene normally does or how it causes disease, may take years to arrive at after the gene has been found. Somehow, the same level of front-page fanfare is often missing from these terribly critical later steps. Perhaps it is hard to identify a point in time when we can say, "Before, we did not know what the gene does, and now we know." So even though some of the later steps in the study of a gene may be at least as important, if not more so, most of the fanfare goes with finding the gene, not with finding out how it causes the disease. The functions of genes are often pieced together gradually through a long, slow series of increments of information. When a gene is found, there is an identifiable, discrete moment, a "eureka" moment, when each new person who hears the news experiences that sudden transition from wondering where the gene might be to knowing its exact position in the human genome. When a gene is first found, it has the feel of a breakthrough because it is an initiating event rather than another in a long string of events. But when a disease mechanism is worked out it can be a long slow process lacking in *eureka!* moments that make for great news headlines. So in some ways, the most critical questions we want to answer all come after the finding of the gene, but it is that moment of finding it that seems to warrant a special kind of notice and that provides the tools with which all of those other questions will be asked.

---

landmarks such as a street sign can only be evaluated by someone standing there reading it.

So what is a genetic map? A genetic map consists of an ordered set of genetic landmarkers (called genetic markers) about which we can make queries akin to those we make for a geographic map: am I near my target or not? These genetic markers are each located at a specific position on a chromosome.

Positions on the genetic map that have been genetically defined are referred to by the term *locus*. A locus, or location on a genetic map, can contain a gene that has different alleles that affect a phenotype or it can contain a genetic marker that differs between two individuals without any phenotypic effects. As for the geographic markers, genetic markers take many different forms and how we detect the marker depends on what type of marker it is. This brings us to the question: what is a genetic marker?

11.2 WHAT IS A GENETIC MARKER?

Genetic markers have different alleles that we can distinguish from each other. Although we tend to think of a genetic marker as being some item in the genetic code, some allelic variant present in the DNA that we would detect through a technology such as sequencing, in fact a genetic marker can be anything that varies from one person to the next where the variability in that characteristic has its origins in genetic information. Thus a genetic marker can be a single base difference detectable by sequencing or a large deletion on a chromosome that can be seen through a microscope. But it can also be an ABO blood type detectable through a protein-based assay that evaluates proteins present on blood cells. A genetic marker can take the form of hair or eye color, the ability to taste a particular chemical compound, the size of a variably sized protein detected on a gel, or even a disease phenotype.

A common type of marker used in the development of the first comprehensive human genetic maps was the microsatellite repeat marker. You may recall from Chapter 5 that a microsatellite is a kind of simple repeated sequence that can vary in length from one individual to the next. An advantage of this type of marker is that it can be very easily assayed through quick, easy, cheap PCR reactions. One of the most popular forms of genetic marker in the early twenty-first century is the single-nucleotide polymorphism (which we will discuss further in Chapter 12) because there are such vast numbers of them and they can be readily assayed through high-throughput technologies. But none of the DNA-based markers was available in the early 1950s when the first sections of the human genetic map were being assembled.

The Development of the Human Genetic Map

What kinds of genetic items first appeared on a human genetic map? The earliest human "genetic maps" started with reports on linked pairs of genes, beginning in 1951. Linkage, as you may recall from Chapter 7, is the tendency for things that are close together on the same chromosome to be transmitted together into the next generation. Many of the first reports of linkage involved genes encoding the proteins responsible for human blood types such as the ABO blood type or the Rh blood type, where blood types were originally defined based on incompatability detectable when blood of one blood type is given to an individual with a different blood type. Some of the first reports of the locations of genes responsible for human traits included a 1953 report that an elliptocytosis gene is located near the Rhesus (Rh) blood group locus and a 1955 report that the nail–patella syndrome (NPS) locus is near the ABO blood group locus. By the late 1960s the field saw the emergence of some of the first programs for analysis of human gene mapping data.

Many of the earliest reports of disease gene linkage show linkage to blood group loci and the MHC locus whose genes encode tissue transplantation proteins that play a major role in rejection of transplanted organs. Is this because these blood group genes play a role in human disease? No, it is because those were the genetic markers that were available to be tested.

Thus the earliest days of human gene mapping were rather like the joke about the man looking for his keys under the street light; he wasn't looking there because he thought the keys were there but rather because that was the only place where he could see well enough to look for them.

Accordingly the order in which genes were mapped was not determined by whether or not a particular disease was important or severe, but rather by whether or not the disease gene was located near some known landmark that was being routinely tested by many research groups looking for the locations of the disease genes. Many genes of great importance stayed unmapped for decades simply because they were not "under the street light."

Another set of landmarks that emerged early in the development of the map were large structural aberrations in chromosome structure associated with particular human phenotypes.

Gradually the investigators created a list of chromosomal abnormalities. The presence of extra chromosomes or pieces of chromosomes could be correlated with specific phenotypes; in 1959 Down syndrome was identified as involving an extra copy of chromosome 21. Other phenotypes correlate with deletions of specific points in the genome; in the early 1960s *cri-du-chat* (cry of the cat) syndrome was shown to be associated with deletion of a specific region of chromosome 5. This trait is recognized in babies with feeding problems and an unusual cat-like sound to their cry, who go on to develop severe cognitive and developmental delays, and behavioral problems. Di George syndrome, involving a parathyroid defect, tetany, seizures, and infections, was mapped to chromosome 22 based on multiple observations of both chromosome 22q11.2 translocations and 22q11.2 deletions. There is now a large catalog of phenotypes associated with chromosomal abnormalities, and new syndromes and structural abnormalities continue emerging into the literature.

With the advent of cloning, the scientific community began the long slow process of finding human genes one by one. By the late 1980s the gene mapping workshops of the human genome project were reporting placement of a few hundred polymorphic genes on the map (Figure 11.1). Of perhaps greater importance to overall mapping capability, these scientists identified several thousand restriction fragment length polymorphisms (RFLPs), differences in the human genome sequence that could be detected by whether or not a particular restriction enzyme could cut the DNA at that spot in the sequence. At first, RFLPs represented a big technical breakthrough in our ability to map genes. But these assays were slow and labor-intensive, taking days to carry out and requiring the use of radioactivity to complete the assay.

With the advent of PCR, it became possible to use primers to amplify a piece of DNA and use simple visualization on a gel to tell whether or not the enzyme could cut the amplified fragment. These PCR–RFLP assays could be carried out in hours, not days, but the human genome is large. So many markers had to be tested that it still left the mappers slogging slowly through the genome, looking at one marker at a time

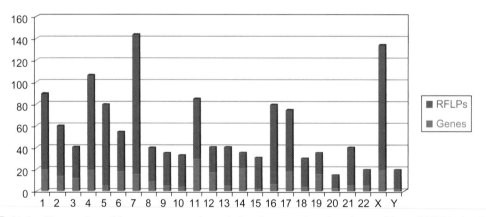

FIGURE 11.1   The number of human genes and restriction fragment length polymorphisms (RFLPs) that had been placed on specific human chromosomes as of April 1988. See Chapter 5 for a review of RFLPs. (After "The dramatic increase in DNA markers" by Kenneth K. Kidd in *Molecular Probes: Technology and Medical Applications*, edited by Alberto Albertini *et al.*, Raven Press, New York, 1989.)

in small numbers of DNA samples in each experiment.

By the late 1980s, as the human genome project was moving ahead with the plan to build a map of the human genome, new markers started to emerge at a fairly rapid pace and the new markers tended to be DNA-based markers. Shortly after the emergence of PCR, a new form of genetic marker came into use – the microsatellite repeat. The earliest microsatellites commonly used in human gene mapping were regions in which the dinucleotide CA was repeated enough times (often more than 18 copies of CA) that its length was slightly unstable, showing length variation in perhaps as many as one in a thousand individuals. This rate is high enough to generate a lot of polymorphic diversity across a population, but the rate is low enough that the lengths could be used fairly reliably as genetic

markers within families. PCR is faster, cheaper, and easier than RFLP assays, and PCR assays can be multiplexed, sometimes allowing for screening of a dozen or more markers in one test tube. Thus these "variable number of tandem repeat" or VNTR markers, led to rapid advances in map development (Figure 11.2).

Once the first draft of the human genome sequence came out in 2001, it became possible to look at variation in sequences between members of populations. This allowed for the identification of points in the sequence that are variable between individuals all over the planet, and to identify other points in the sequence that show private polymorphisms, sequence variants that are commonly found only within a specific population. This led to the identification of huge numbers of single nucleotide polymorphisms (SNPs) throughout the genome. Genome-wide

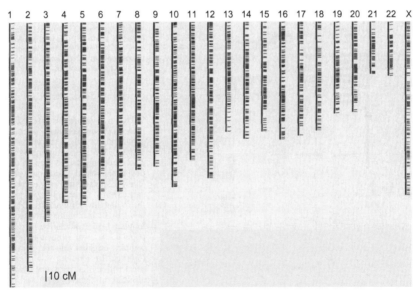

FIGURE 11.2   In 1996 a map was released showing the locations of 5264 markers scattered across the human genome with average spacing between markers of less than a million bases. The markers were all microsatellite repeat markers, short regions in which a very simple sequence is repeated enough times that the sequence length is variable across a population. The density of markers on this map was more than one marker per million base pairs of sequence. See Chapter 5 for a review of microsatellites. (From Dib, C., *et al.*, "A comprehensive genetic map of the human genome based on 5,264 microsatellites," *Nature*, 1996;380:152–4. Used by permission of the Nature Publishing Group. All rights reserved.)

scanning technologies, such as chips containing hundreds of thousands of SNPs, allow for mapping experiments using more than a half-million genetic markers in a single experiment, which we will talk about in Chapter 12.

Informativeness

Informativeness is the property of being informative or providing information. When we look at different kinds of genetic markers we see that some of them almost always provide us useful information in each experiment, while other markers give us useful information less often.

One of the big differences between the microsatellite markers and the SNPs is the number of allelic variants at each marker. SNPs are changes in sequence at one point in the sequence – a position in the sequence with an A in some individuals and a G in other individuals. So any one individual can have one of three genotypes – AA, AG, or GG. But microsatellites, which represent length variations in a simple sequence repeat, can have far more than two different allelic variants; in fact many of them have more than a dozen different lengths in one population. Many of the microsatellite repeats used in building the first comprehensive human genetic maps are heterozygous (have two different lengths) in more than 70% of randomly selected individuals. Thus, depending on allele frequencies, someone might have a 40% chance of being heterozygous at a particular SNP and an 87% chance of being heterozygous at a nearly microsatellite repeat.

Why do we care if someone is heterozygous for a marker? If someone is homozygous for a marker, then that marker is uninformative in that individual. That is, we cannot use that marker to distinguish between the two copies of the marker (and thus the two chromosomes containing those copies) so we can't tell which copy was passed along to each child. So anytime a parent in a family is homozygous for a marker we are

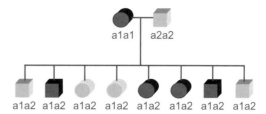

FIGURE 11.3   When the parents are both homozygous, all of the children will have the same genotypes as each other, and we will not be able to tell which of the two parental copies (or the corresponding chromosome) was passed to which child.

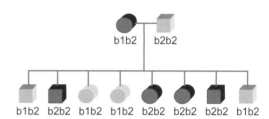

FIGURE 11.4   When the affected parent is heterozygous for a SNP then we can tell for each child which of the two copies they received from the affected parent, so we can ask whether the affected children all received allele b2 and the unaffected children all received the allele b1 as shown here. Does the information shown here indicate that the b2 allele is causing the trait? No, but it does suggest that the b2 allele of this particular marker is located near the causative gene (and in some cases it might even turn out that the b2 allele is causative).

evaluating, we have lost information regarding a landmark on our genetic map (Figure 11.3). If the affected parent is heterozygous we will often be able to distinguish between the two copies (Figure 11.4), but if both parents are heterozygous for the same alleles, some genotypic combinations in their children will be uninformative as to which parental copy they inherited (Figure 11.5).

Distances between genetic markers (measured in centimorgans, also called cM) on such a map are determined by the frequency with which recombination events during meiosis fall between those two items on the map. Items that are very close to each other will rarely have

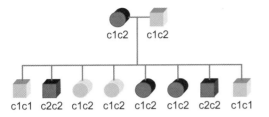

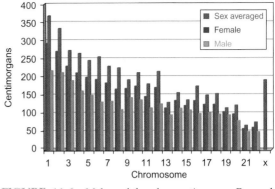

FIGURE 11.5 When both parents are heterozygous for the same alleles of a marker, some of the meioses will be informative (let us tell which allele the parent passed along) and some will be uninformative (leave us unable to tell which allele was inherited from the affected parent). When a child has the c1c1 genotype, we can tell that the child inherited a c1 from the affected parent and a c1 allele from the unaffected parent. When the child has the c2c2 genotype, we know that a c2 came from the affected parent and a c2 came from the unaffected parent. But for the children with the c1c2 genotypic combination, then we cannot tell whether the c1 came from the affected parent and the c2 came from the unaffected parent, or whether it happened the other way around. Sometimes if we use information from other markers located next door to this one we can figure out whether it was the c1 allele or the c2 allele at this marker that came from the affected parent since markers that are very close to each other tend to be transmitted together from one generation to the next.

FIGURE 11.6 Male and female genetic maps. For each chromosome we see that the female map is longer (shows more recombination) than the male map, but the difference between the two maps is not the same. In one study researchers found that the female map of chromosome 8 was twice as long as the male map of that same chromosome, while the female map of chromosome 16 was only 1.2 times as long as the male map. For many purposes we use the sex-averaged map but if we were studying a family in which we had a lot more female meioses than male, we might want to take the sex-specific map into account. (After Broman, .K. W., Murray, J. C., Sheffield, V. C., White, R. L., and Weber, J. L. *American Journal of Human Genetics* 1998;63:861–9. Copyright 1998, with permission from Elsevier.)

recombination events fall between them. Thus they will appear to be linked – transmitting together from one generation to the next at a higher rate than would happen with random assortment. Normally, we end up with a pair of markers, called flanking markers, that frame the area that contains the gene we are seeking. But all of this presumes that we have a map full of genetic landmarks available before we can ask if our gene is near such landmarks, so someone had to build that genetic map in the first place.

Differences between the Male and Female Maps

We can build maps specific to male meioses or female meioses simply by limiting our analysis to the meiotic events that occurred in the germline of the male parent or the female parent. What we find is that the male and female genetic maps show all of their genes in the same order, but we see differences in the distances between markers. The physical lengths of the chromosomes are very similar from one person to the next, but if we compare the genetic distances between landmarks on genetic maps we find that each chromosome is longer (shows more recombination events) on the "female" map than on the "male" map.

On a set of genetic maps assembled by the Marshfield Institute in Wisconsin, the distance between D1S243 and D1S468 is 4.46 cM on the female map and 3.54 cM on the male map, or 4.22 cM on the sex averaged map. If we look at the next interval on the map we find that the distance between D1S468 and D1S2845 is 2.94 cM on the female map and 6.67 cM on the male map, or 4.63 cM on the sex averaged map. Thus even though the overall length of chromosome 1 is longer on the female map, there are local regions within chromosomes where we see more recombination events on the male map than on the female map. So we see that there are very real differences in the rates at which particular intervals recombine during meiosis in males versus females, and the frequency of recombination we will see in one of our studies will depend in part on how many of the meioses in the study happened in males and how many happened in females.

11.3 FINDING GENES BEFORE THERE WERE MAPS

When cloning first became available, and the possibility of isolating individual genes away from the rest of the genome became a reality, researchers had a limited ability to get their hands on specific genes that they especially wanted to study. The easiest things to clone, and thus some of the first things cloned, were those things that were present in high copy number or that were expressed at high levels.

Thus some of the first human genes cloned were ribosomal genes. There are so many copies of ribosomal RNAs present in an actively growing cell that researchers could make a library of cDNA clones (remember that these clones contain DNA copies of the RNA in the cell) and pick out random clones and have a good chance that one of them would contain sequences from a ribosomal RNA gene. In fact, some later techniques for making clone libraries call for a step that gets rid of the ribosomal RNA so that it does not swamp out the other rarer items of interest in the library.

But many of the genes of greatest interest were genes involved in human diseases or other traits, or genes whose proteins had been the object of intense previous study. A lot of these genes were "single copy" genes and many of them are not expressed at such high levels that we can expect to just pluck them randomly out of a library. The early mappers had to have a way to screen for the one clone out of millions or tens of millions of clones representing the genes or expressed transcripts of a cell.

In some cases, we were able to go after a particular gene of interest if we knew of a disease condition in which an abnormal cell type over-expressed that particular gene (and its product). Among the early single-copy human genes that were found were genes that produce known previously studied proteins where a cell line was known that was over-expressing the gene. A cell over-expressing the gene has a large excess of the transcript produced by that gene, and such RNA can serve as a probe that can detect clones containing sequences homologous to the most abundant items in the RNA from the over-expressing cells. When we use hybridization (base pairing between the DNA in a probe and the DNA in a clone) of a probe as the basis for recognizing a target clone in a library of clones, the most abundant sequences in the probe will give the strongest signals so simply picking the clones with the strongest signals gives a good chance of getting what you want. Sequencing can confirm that you found the right thing in cases where the protein and protein sequence are already known. In these cases the genes were not cloned based on a positional cloning or positional candidate cloning approach. Researchers did not start out by finding out where genes were located and then moving to find the gene; instead they started from what they knew about the protein and transcript, cloned the gene, and then later figured out where the gene was located in the genome.

In most cases, over-expression of RNA was not available as a tool for getting at the genes we wanted to find. And redundancy of the code (remember from Chapter 4) made it very hard to find any part of a protein for which the RNA (and thus DNA) sequence could be predicted. Occasional regions of amino acids corresponding to unique codons or that are encoded by very small numbers of codons did allow for design of probes that could recognize the gene encoding that protein. In some cases a pool of related probes would be made representing the different possible combinations of low-redundancy codons that could generate the known protein sequence (see Figure 11.7). Sometimes such pooled probes would succeed in recognizing the desired clone. Once the clone was isolated once, the real sequence could be

RNA PROBE FOR REGION OF UNIQUE CODONS

Protein	Trp	Met	Met	Met	Trp	Met	Trp
Predicted RNA	UGG	AUG	AUG	AUG	UGG	AUG	UGG

POOL OF RNA PROBES FOR REGION OF UNIQUE CODONS

Protein	Met	Trp	His	His	Met	Trp	Cys
RNA 1	AUG	UGG	CAC	CAC	AUG	UGG	UGC
RNA 2	AUG	UGG	CAU	CAU	AUG	UGG	UGC
RNA 3	AUG	UGG	CAC	CAU	AUG	UGG	UGC
RNA 4	AUG	UGG	CAU	CAC	AUG	UGG	UGC
RNA 5	AUG	UGG	CAC	CAC	AUG	UGG	UGU
RNA 6	AUG	UGG	CAU	CAU	AUG	UGG	UGU
RNA 7	AUG	UGG	CAC	CAU	AUG	UGG	UGU
RNA 8	AUG	UGG	CAU	CAC	AUG	UGG	UGU

FIGURE 11.7 For the first probe, all amino acids are encoded by unique codons so there is only one possible RNA sequence that could produce this amino acid sequence. For the second probe, Met and Trp are encoded by unique codons, but His and Cys are each encoded by either of two different codons; with only three such non-unique items in this amino acid sequence, it takes eight different probes to represent all of the different combinations of sequences that the genome might be using to produce this very short region from a protein. Most regions of protein sequence include lots of amino acids that are encoded by four or even six amino acids, so the number of different probes would rapidly become very large, the amount of signal that we could get from any one probe would be diluted, and the chance of cross-reacting with other sequences in the genome would increase. In addition, with each new combination of sequences we risk creating a probe that might cross-react with some gene other than the one we are trying to clone. So development of probes based on the protein sequence was only available if the protein sequence was known and if there was a region within the protein with a cluster of low-redundancy amino acids.

determined and after that experiments could always be based on the correct sequence rather than a pool of theoretical sequences. However, most regions of most proteins make use of amino acids with a lot of codon redundancy, and it is hard to find a protein-coding region that can be found in this way.

After an initial flurry of activity aimed at cloning the genes corresponding to known proteins where tools were available, such as overexpressed RNAs or low-redundancy coding region probes, the field of human genetics was then faced with the problem of how to find the genes responsible for traits for which the protein was not already known. There were two different types of problem to be solved.

One problem was: how do we find all of the human genes, whatever they may be or whatever traits they might be associated with? When this question was first asked it was accompanied by a presumption that, in the long run, it would be important to know every single gene, and not just the gene associated with the particular disease being studied by this or that research group. This turned out to be a huge amount of work and took a huge amount of resources to accomplish. By making cDNA copies of transcripts and cloning them, it was possible to simply pick random clones, one after another, and sequence them. Each cDNA clone represented a copy of a transcript. Finding every single gene turned out to be complicated. By sequencing all of the clones in a cDNA library made from adult heart muscle we can find out genes expressed in heart muscle but there are many other genes not expressed in heart muscle that can only be found by sequencing cDNA clones from each of the other tissues. The problem is that it is not possible to sequence all transcripts from every kind of cell at every point in development under every environmental condition. So this type of mass sequencing project can get to the end and still have not seen transcripts from some cell types under some conditions.

We can also look for theoretical genes – those genes that can be identified by their similarity to other genes but that have not turned up in our experiments. Programs exist that use what we know about existing genes to make predictions of what genes look like, and those programs can search the genome for other sequences that have the characteristics of a gene. And we can compare sequences between species to identify regions of sequence that are highly conserved. We can look for regions of the sequence that look like promoter regions, and we expect that the sequences next to them might be transcribed. One problem was: how do we find all of the human genes, whatever they may be or whatever traits they might be associated with? Once we identify a theoretical gene, we can use reverse transcriptase to make a DNA copy of the RNA, and PCR to go looking for the actual transcript so we can confirm that the gene exists.

This search for the genes was a major focus of the Human Genome Project. A lot of the Human Genome Project was carried out by large genome centers and commercial companies designed to carry out high-throughput processing of clones and information. It aimed to sequence the whole human genome and identify all of the genes and many of the common allelic variants in those genes. Although it is possible, through translation of the sequence into a hypothetical protein sequence, to tell whether any new gene found corresponds to a known protein, there are many non-coding genes and there are many coding genes that make proteins that were not previously known. The story of the Human Genome Project, and how it gave us new tools for the study of population genes, will be told in Chapter 12, but some of our discussions here will presume to use information that came out of those studies.

The other question was: how do we find specific genes whose allelic variants are associated with particular traits such as diseases where we do not already know the defective protein or the defective biochemical pathway? Much of this work has taken place in individual labs run by individual researchers, rather than in big genome centers, but as these projects proceeded, the rate at which they could progress was greatly accelerated by the whole genome approaches and information coming out of the Human Genome Project. These disease-based projects made a lot of use of specific kinds of Human Genome Project information, especially the genetic maps that resulted. And most of these efforts started out with placement of disease genes on maps.

11.4 DEFINING THE THING TO BE MAPPED

Before beginning a Hunt, it is wise to ask someone what you are looking for before you begin looking for it. —**From Pooh's Little Instruction Book, by A. A. Milne**

Ideal Gene Mapping Situation

In an ideal world, we would have good control over a lot of the variables in experiments we design, and we would not have to put up with any flaws like missing data. However, especially when working with human study subjects, we often find that the studies fall short of the ideal. Often with animal model studies we have a lot more control over these variables; for instance, if we are doing a study on obesity or diabetes, we can provide the same diet to all of our study subjects, but when studying human beings there will be a lot of dietary variables that we will have trouble controlling even if we ask people to all eat the same diet. If we are doing a study of a rare trait and we have identified a human family with the trait, we have to confine our study to however many family members happen to exist; we cannot ask people to have more children so we can see whether they develop the trait. If we are studying a family and some family members have passed away, we are limited to working around the missing information. And frequently there are limitations to what kinds of information we can get when studying a human being. If the cell type having a problem is located deep inside the brain, we can only tell limited things about what is happening to that cell during the disease process because (quite reasonably!) we cannot biopsy a human brain for research purposes.

What Actually Gets Mapped?

When we map the location of a trait, have we mapped the trait we think we mapped? How do we know that we have really mapped that trait and not some other trait that is associated for non-genetic reasons? Many situations are complex and lots of people who have one thing in common may have other things in common as well. If we do not have a good grasp of the similarities and differences between our study participants we may miss the fact that the gene we find is involved in one of the other traits that is running through our study. For instance, if we set out to map a gene for vitamin deficiency might we miss the fact that our subjects have a vitamin deficiency because they carry a "taster" mutation that makes them dislike the taste of vegetables and avoid eating vegetables; if this were the case, then we would think we had mapped a gene for vitamin deficiency, expecting to find some protein involved in processing or transport of vitamins, when what we would really have found is something responsible for our sense of taste and responsible for a behavioral trait – eating too few vegetables. If we have done a good job of tracking the environmental variables in our genetic study, and we have done a good job of modeling the factors that could be involved in the disease or the populations we are studying, hopefully that will not happen. But we do not always know what the associated variables are that we should be taking

into account, especially if the trait is very complex and we do not know what is causing it.

We can also run into problems if we are trying to map something too complex. If we want to find the genes involved in heart disease, we can run into problems and fail if we simply pool together people who have heart disease since there are many different kinds of heart disease and many underlying causes. Sometimes we want to go ahead with mapping the complex trait, but sometimes we can make gains in our understanding of the relevant genes if we do one study aimed at mapping genes that control cholesterol levels, and another study to map genes affecting levels of markers of inflammation, and another study mapping genes that affect blood pressure. Once we have identified genes that affect the many different aspects of heart disease, we can then put all of that information into the development of models for what is going on in the complex trait.

Correct Classification of the Phenotype

We also need to be sure that we are correctly classifying everyone with regard to whether or not they have the trait. There are different levels of information we can obtain and some are more likely to be correct than others. The most effective way to assign affected status is for us to examine the study subject ourselves. The next most effective way is to examine their medical records, but often those records may be missing some pieces of information that are very important to our study but that were not important for their medical care. If we simply ask people if they have the trait, they will be correct for some traits that are quite obvious ("Do you have any missing teeth?") but may be wrong about other traits that involve technical information ("Do you have high HDL cholesterol?"). In some cases people who have some of the characteristics of the trait end up confused into thinking that they actually have the whole trait ("I have glaucoma") when the real answer is "I have elevated intraocular pressure and a strong family history of glaucoma so my doctor gave me glaucoma drops."). The least reliable form of information is what we call "family report" where we learn someone's diagnostic status from one of their relatives. A study of glaucoma in Tasmania identified some huge glaucoma families with large numbers of affected individuals. Sometimes they would start studying a new branch of the family and find themselves talking to a new family member they had not met before. They would ask the new study subject "Do you have a family history of glaucoma?" To their surprise, people with dozens of affected relatives would reply "No, I don't have any relatives with glaucoma." What they found was that even in families with a lot of glaucoma, people frequently had no idea that their relatives had this trait. We especially find that very young adults are often mistaken about the health status of their oldest relatives, and people who do not live near each other often are wrong about the health status of their remote relatives.

Some of this results from lack of communication, but some of it results from the "telephone" effect. School children play this telephone game: one whispers to the next who whispers to the next and so on. By the end of a long line of message exchanges, the last person says the message out loud so everyone can laugh at how different the final message is from the original message. Similarly with family health situations, one relative telling another who tells another who talks to us can result in our study receiving the information that Uncle Joe has the glaucoma that is running through the family, but when we examine medical records we discover that Uncle Joe had been struck by a baseball and has a different kind of glaucoma that is caused by compressive injury, and not by a glaucoma gene.

Mapping Modulatory Factors

Consider this scenario. Families for a study of cardiovascular disease have been recruited from a referral clinic at one of the country's most prestigious medical schools, where patients are referred from all over that region of the country by doctors who have decided that they need another opinion on a difficult case. If we use the cases from the referral clinic to find or study a gene for cardiovascular disease, have we really mapped the gene we think we mapped? Stop and consider. The people referred to this clinic might have more than one thing going on in their health. Not only do they have heart disease but they also have something causing their heart disease to be more severe than that experienced by those whose doctors keep them in their local community to care for them. Maybe there is just one factor that causes the heart disease and causes it to be severe, but there might be two different factors, one causing the heart disease and the other causing it to be especially severe or progressive.

11.5 RECOMBINATION AS A MEASURE OF GENETIC DISTANCE

Localizing a gene most often takes place through studies of genetic markers in families or populations. For this discussion we will presume that we now have defined the characteristic we want to map, that we have obtained the phenotypic data we need, and that we have genetic markers and a genetic map available.

The simplest example of mapping occurs when we determine that a given trait shows sex-linked inheritance. This tells us that the gene is on the X chromosome. This gives us our first level of geographic localization: assignment to a chromosome. If we rule out X-linked inheritance then our localization is broader since it could be on the much broader group of autosomes.

Linkage

Things that are farther apart on the same chromosome are more likely to be separated from each other by recombination, and things on different chromosomes will be separated by segregation. Thus, if two mutations are present together on the same chromosome, the closer together they are, the more likely it is that people in subsequent generations will inherit both mutations at once. We recognize linkage between two genes when a particular combination of alleles of those two genes turns up in the offspring more often than expected by chance. Genes that are close to each other will recombine rarely. Genes that are physically far apart will recombine frequently.

If we look at two traits present in the *founder* – the person with both traits from whom the rest of the family is descended – we can ask about co-segregation, that is, how often the two traits stay together or separate from each other as they are passed down through multiple generations of a family. This information can be used as a measure of how close together the two genes are. The earliest genetic experiments looked at traits that are transmitted together from one generation to the next. We rarely find a single family in which two different traits that we want to study are being passed along in the same family.

Let's consider two dominant traits, a hypothetical taster trait (that causes people to detect the bitter taste of a particular substance) and an eyelash trait that involves a double row of eyelashes. If the founder had both traits – the taster trait and the eyelash trait – and married someone who is a nontaster and has normal eyelashes, we could start trying to trace the transmission of these taster and eyelash traits through multiple generations of the family to see whether people with the taster trait tend to have extra eyelashes. If the two genes are right next to each other on the same chromosome and the two dominant alleles are physically located on the same copy of that chromosome, we might see what is shown in Figure 11.8, where almost all of the tasters have an extra row of eyelashes and all of the nontasters have normal eyelashes.

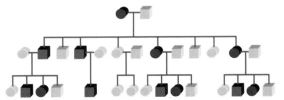

FIGURE 11.8 The hypothetical taster and double-eyelash traits are usually transmitted together in this family. If the left half of the symbol is blue, the individual has the taster trait. If the right half of the symbol is blue the individual has the eyelash trait. In this family every blue symbol has both sides filled. In this case, out of 20 individuals who were at risk of inheriting the taster trait, 11 of them have it. Out of 20 individuals who were at risk of inheriting the double-eyelash trait, 11 of them have it. What is most noticeable is that every one of the tasters also has extra eyelashes and every one of the nontasters has normal eyelashes. This suggests that the two genes are very close to each other on the same chromosome. We also have to remember that the two traits might be caused by one genetic defect. If these two traits are normally each seen without the other, and are only seen together in this family, we would favor the hypothesis that there are two genes located very close together on the same chromosome.

What would we see if the two genes were far apart? Let's reconsider the same fictitious traits and look at the kind of outcome that would have resulted if the two genes were located on different chromosomes (remember how the chromosomes get passed along independently of each other to the next generation) or when recombination events fell between the two genes frequently (Figure 11.9). In Figure 11.8, when the two genes were very close together, most people who inherited one trait also inherited the other. In Figure 11.9, when we hypothesize that the two traits are far apart (on different chromosomes), we see that the chance of any one at-risk individual inheriting one of the traits is the same as what we saw in Figure 11.8, but it is much more rare for one individual to inherit both traits.

What happens when the two dominant alleles are located on the same copy of the same chromosome but there is enough distance between the two genes so that sometimes a recombination

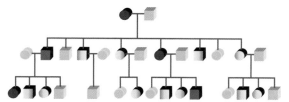

FIGURE 11.9   In this case, in which the two hypothetical genes are actually located on different chromosomes, not
different copies of the same chromosome, we see that out of
20 people at risk for the taster trait, 8 have it, and out of 18
individuals at risk for having double-row eyelashes, 9 individuals have them. In many cases the individual received
one trait or the other but not both. Notice that, occasionally,
both traits get passed to the same individual when simple
chance results in passing a chromosome carrying the taster
allele and a different chromosome carrying the eyelash
allele along to the same person.

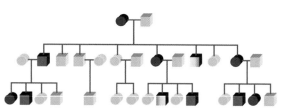

FIGURE 11.10   If the hypothetical taster and eyelash
genes are close together on the same chromosome but far
enough away to allow recombination events to fall between
them sometimes, we will see the effect shown here. If the
left half of the symbol is blue, the individual has the taster
trait. If the right half of the symbol is blue the individual
has the eyelash trait. In this family, where the two genes are
on the same chromosome but far enough apart that recombination can sometimes fall between the two genes, most
affected individuals will be affected with both traits, but
occasionally, someone will inherit only one of the two traits.
Out of 20 individuals at risk of inheriting these defects,
one of them received the taster allele but not the eyelash
allele, and another received the eyelash allele but not the
taster allele. Eight individuals received both the taster allele
and the eyelash allele. This suggests that the two traits are
encoded by different genes that are close to each other on
the same chromosome, but far enough apart to allow for
recombination.

even can fall between the genes? Then we see
what is shown in Figure 11.10, a family in which
the two traits are transmitted together from one
generation to the next in most cases, but every

once in a while someone turns up with only one
or the other of the two traits. The rate at which
children have one trait but not the other gives
some indication of the distance between the two
genes. Since recombination is more frequent in
some regions than in others, the recombination
frequency can only be used as an approximation
of the physical distance.

The Centimorgan Is a Measure of Genetic Distance

By looking at the rate of co-segregation of
two traits (the rate at which they are transmitted together from one generation to the next),
we can estimate the genetic distance between
the genes responsible for the two traits. When
one recombination event (separation of the two
traits from each other as they pass from one
generation to the next) is seen out of one hundred offspring who are at risk for the trait, this
is considered to indicate a genetic distance of 1
centimorgan (Box 11.2), which can also be written as 1cM. Thus two traits that are 10cM apart
are farther apart than two traits separated by a
genetic distance of 1cM. Fifty percent recombination, or 50cM, is the largest genetic distance
that we can measure. We cannot "see" beyond
the 50cM mark on the map. In fact, things that
are 50cM apart on the same chromosome and
things that are 200cM apart along the same
chromosome and things that are on different
chromosomes will all show random assortment
relative to each other, which is to say they will
all show the same result (approximately 50%
recombination because of independent assortment), and we will not be able to distinguish
those cases from each other.

The LOD Score and the Recombination Fraction

Any time we look at co-segregation of two
items in the genome, we end up with two

BOX 11.2

THE CENTIMORGAN

The centimorgan was named in honor of Thomas Hunt Morgan, who received a Nobel Prize in 1933 for his studies of the role of the chromosome in heredity. He made a big breakthrough when he showed that genes are located on chromosomes. And he put forward the idea of the linear arrangement of genes along the chromosome. Although others had provided some of the first reports of data that demonstrate linkage, it was Morgan who put forward the theoretical framework for understanding it. He proposed the concept that linkage in the form of genes transmitting together from one generation to the next could be explained by their being located near each other on the same chromosome. It was in Morgan's lab that Alfred Sturtevant and others used the fruit fly, *Drosophila melanogaster*, to provide experimental proof of linkage.

---

different numbers of importance – the recombination fraction and the LOD score.

The recombination fraction is our best estimate of the distance between the two items. If no recombination occurs we say that we have a recombination fraction of zero. The largest recombination fraction we can measure is 0.5 (which is to say 50% of the possible recombination events), even if the items are on separate chromosomes or very far apart on the same chromosome.

The other number is called the *LOD score*, which is to say the *log of the odds score* (Box 11.3). *The LOD score estimates our confidence in the accuracy of the recombination measurement.* A LOD score of 1.0 indicates that we think the odds are 10 to 1 in favor of the recombination fraction identified being right. A LOD score of 2.0 gives us 100 to 1 odds in favor of being right, and a LOD score of 3.0 gives us 1000 to 1 odds in favor of being right. So if we find a recombination fraction of 1% associated with a LOD score of 5.0, we are highly confident that the two items are close together, but if we find that same recombination fraction of 1% associated with a LOD score of 0.5, we have little confidence in the accuracy of that 1% assessment. In fact a LOD score of 0.5 makes us think that it would be very easy to get the result by simple chance.

For some markers, we will not only fail to see evidence that the two items are next to each other, but in some cases we will actually see evidence that they are probably not anywhere near each other. Evidence that the gene is not in the area we are looking at comes in the form of a negative LOD score. Support for a model of closely adjoining genes takes the form of a positive LOD score. Whenever we see the amount of recombination approaching 50% recombination, we also doubt that the gene is near the tested location.

The standard in the field is that a LOD score ≥3.0 (1000 to 1 or greater odds in favor of linkage) is considered highly significant evidence in favor of linkage – that is, of the two things being close together on the same chromosome. A LOD score ≤−2.0 is considered to be highly significant evidence against linkage – that is to say, about 100 to 1 or greater odds that the gene is not at that location.

BOX 11.3

THE LOD SCORE

To test for how likely it is that the marker we just tested is near the disease gene, we want to find out whether the two items we are mapping are linked (at a distance specified as theta, θ) or unlinked (at a distance of $\theta = 0.5$). After we have tested for recombination and obtained data, we calculate an odds ratio to compare the chance that we would have seen such data if the two items are really linked to the chance that we would have seen such data if they are actually unlinked. After we calculate an odds ratio (the chance given linkage compared to the chance given that they are unlinked), we then take the log of the odds ratio. Thus the LOD score is

$$LOD = Log10 \frac{\text{Chance of these data if the two items are linked}}{\text{Chance of these data if the two items are unlinked}}$$

If we take the odds ratio we are comparing the chance we would get these data if the two items are linked to the chance that they are unlinked. Why do we take the log of this ratio? This lets us express our result as LOD = 8 instead of LOD = 100,000,000 to 1. But in fact one of the most important reasons for converting our odds ratio to a log value is because LOD scores for different families can be added together but the odds ratios cannot. So if we get LOD scores of 0.9, 2.7, −1.3, 1.3, and 0.4 for a group of five families, we can simply add them up to determine that the cumulative LOD score for the whole set of families is 4.0. Because one of the families shows a negative LOD score we would want to go on to ask additional questions about heterogeneity to determine whether there might be more than one locus that can cause the trait, but the LOD score of 4.0 would constitute significant evidence of linkage that would lead us to want to further study that particular point in the genome.

How Many People Do We Have to Include in a Linkage Study?

Each meiosis contributes 0.301 to the LOD score, if we are looking at a family with an ideal structure and complete information. So if we need a LOD score of at least 3.0, we start out using the data from the affected parent to set our hypothesis (for instance that allele 124 is "it") and then we need another 10 phenotyped family members to test that hypothesis and get an LOD score of 3.0 (if allele 124 is completely linked with no recombination). If the marker is far enough away that there is one recombination event then we need at least 17 fully informative family members (people for whom we have complete information) to get a LOD score

of 3.0. As the number of recombination events increases the number of family members we need for a LOD score of 3.0 increases, but it is rare to find families large enough to achieve a LOD score of 3.0 or greater with multiple recombination events. If one member of the family is misclassified with regard to phenotype, then it has the same effect as if there had been a recombination event: we need a larger family to be able to get a LOD score of 3.0 or more. Thus in linkage studies we often try to err on the side of including only those individuals for whom we are quite sure of the phenotype, while not making use of individuals for whom the phenotype classification might be wrong.

It is often hard to come up with families that are large enough to provide a LOD score of 3.0

or greater. The later the onset of the disease, the harder it is to get large families since most family members are too young to have developed the disease (so they cannot be classified for phenotype) and many other family members who had previously developed the disease have passed away (and so they cannot be classified for genotype). Similarly it is hard to get families that are large enough for conditional traits since many family members will never have experienced the eliciting environmental component; in families with malignant hyperthermia most family members cannot be classified because they have not experienced general anesthesia so we do not know what their phenotype would be.

Multipoint Information

We often end up with some uninformative markers, such as markers that are homozygous in everyone in the family. So if the gene is located between two markers, is there any way we can get an idea of where it is without running more markers? Yes. There is a process called multipoint mapping (Figure 11.11) which lets us

extrapolate between markers and make use of combined information from adjoining markers to build a more detailed map. If the maximum LOD scores are seen for the individual markers R and S, we can use information from adjoining markers to tell that the gene is located at a spot about 5 cM in between the two markers. Figure 11.11 shows multipoint analysis across the whole genome for a group of families with open-angle glaucoma; we see that the highest multipoint LOD score is 2.5, a low enough value that we think it could happen by chance when such a large number of markers have been tested. Figure 11.12 shows the data from a genome scan for one large family in which posterior polymorphous corneal dystrophy is transmitted as an autosomal dominant trait; this graph shows one small region of the genome where a peak reaches above the LOD score of 4.0, meaning that we think the odds are greater than 10,000 to 1 in favor of that being the location of the gene. At other points along the chromosome, we see LOD scores reaching as low as −6.0, suggesting odds of more than 1 million to 1 against that being the location of the gene.

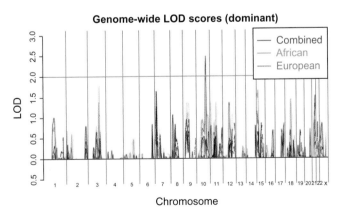

Genome-wide LOD scores (dominant)

— Combined
— African
— European

LOD

Chromosome

FIGURE 11.11   A graph showing multipoint LOD scores across the whole genome for a genome scan that assayed more than 300 markers. The different colors of lines represent data from the whole set of families, the families of African ancestry, and the families of European ancestry. The highest multipoint LOD score was found in families of African ancestry but does not turn up in families of European ancestry. The highest multipoint LOD score of 2.51 is not considered high enough to be significant since a value of 2.51 could occur by chance when so many markers are assayed.

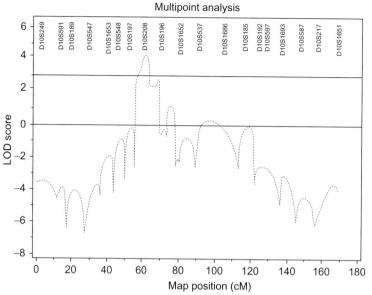

FIGURE 11.12   A graph of data from one chromosomal region from a gene mapping project called a genome scan. Markers from all over the genome are tested for whether or not they segregate independently of the trait. The highest peak on this graph shows the position in the human genome most likely to contain a newly discovered disease locus. The distance along the chromosome is marked across the bottom, and our confidence that the locus is located at that position on the chromosome is compared to the LOD score scale along the side. LOD scores above 3.0 indicate greater than 1000 to 1 odds in favor of the gene being at that location. The peak marking the location of the disease gene on this graph rises above the level of 4.0 on the LOD score scale, so we expect that the chances of the gene being at about the 60 cM mark on chromosome 10 exceed 10,000 to 1 odds in favor of this being the correct location.   (After S. Shimizu *et al.*, "A locus for posterior polymorphous corneal dystrophy (PPCD3) maps to chromosome 10," *American Journal of Medical Genetics*, 2004;130(A):372–7.)

11.6 PHYSICAL MAPS AND PHYSICAL DISTANCES

A Map Based on the Sequence

While genetic distances are all based on statistical estimations of how far apart two items in the genome might be, we actually have a large number of tools that let us look at the actual locations of items in the genome. Some of the earliest maps of the human genome were created by using points in the sequence where restriction enzymes cut as landmarks; size separation on gels allowed us to gain very precise estimations of how far apart these landmarks are. As we gained the ability to sequence DNA we went from knowing that part of an immunoglobulin gene was on a fragment that runs at the 5000 bp position on a gel to knowing that the fragment of that immunoglobulin gene contained in our cloned copy of the gene was actually 5291 bp in length. Once the whole human genome sequence became available we gained the ability to tell exactly how far apart two genes are on the copy of the sequence that is in the database.

Does that tell us how far apart they really are? No, because the length of the human genome sequence is different in each human being.

You will recall our discussions of microsatellite repeats that differ in length from one human being to the next. There are tens of thousands of microsatellite repeats scattered around the human genome, some a few bases longer than the average length for the marker, and some a few bases shorter. By the time we add up all of those small differences we end up with different lengths for each genome. And there are other ways in which the sequence length varies from one person to the next. There are small insertions and deletions at a variety of locations in the genome, and there are copy number variants, places where a copy of a gene has been added or lost. Centromeres are repeats that can vary between individuals. Telomeres vary in length from one individual to the next and in fact differ in length from young to old age within the same individual or even from one cell to the next. Thus we can sequence a region of DNA in one DNA sample from one individual and get an exact distance between two genes contained within that sequence. But if we sequenced the same region in a DNA sample from another individual, we might get a slightly different length between the two genes.

In terms of physical mapping this is not a point of concern. If the distance between two genes is 57,391 bp in one individual and 57,387 bp in another individual, we can presume an average distance of 57,389 bp. An average value serves well enough for our purposes in building a very accurate physical map without always having to know the exact value in every separate person.

Physical Markers Detectable Under the Microscope

Although much of what we know about the role of genes in hereditary diseases starts with genetic mapping, there have also been many insights gained from studies of the physical map of the genome. We can look at the physical map in a variety of ways. We have already talked about the karyotype and the associations of certain phenotypes with abnormal karyotype features. Down syndrome was found to correlate with the presence of an extra copy of chromosome 21. Additional studies of translocations causing Down syndrome due to partial trisomy 21 have allowed for the identification of specific regions of chromosome 21 associated with some of the features of Down syndrome.

Occasionally, genes have been found by looking at the gene and sequences that span a translocation breakpoint. You may recall from Chapter 10 our discussion of the BCR–ABL fusions that can lead to acute lymphocytic leukemia (ALL) and chronic myelogenous leukemia (CML); determination of the sequences spanning the translocation break showed that to one side of the break was a piece of the BRC gene and on the other side was a piece of the ABL gene.

Early studies of Prader–Willi syndrome showed deletions of a small region of chromosome 15 in more than half of the cases, with the deletion originating on the paternal copy of chromosome 15. Studies of Angelman syndrome showed that more than 70% of cases have interstitial deletions spanning the same region of chromosome 15, but showed that the deletion was present on the maternal copy. Detailed studies of the deleted regions in each trait suggested that the molecular mechanisms do not involve the same sequences even though the two deleted regions overlap each other. In one unusual family with a balanced translocation involving the Prader–Willi–Angelman region of chromosome 15, the women with the balanced translocation showed a tendency to produce children with Angelman syndrome. In the same family the men with the balanced translocation showed a tendency to produce children with Prader–Willi syndrome.

Combinations of chromosome banding, chromosome painting, and FISH have allowed detailed studies of such chromosomal abnormalities.

These tools, combined with molecular technologies such as PCR, allow for rapid identification of the boundaries of deletions or the breakpoints of translocations or inversions. PCR amplification of a fragment spanning a translocation breakpoint, followed by sequencing, allows for rapid identification of the genes or sequence regions that have been brought together. Similar treatment can readily identify the boundaries of a deletion and allow us to determine what is missing.

Synteny: Correlating the Human and Animal Model Maps at the Physical Level

While a group of genes are considered linked where genetic evidence of co-transmission between generations occurs, genes that are located in close physical proximity on the same chromosome are considered syntenic whether or not evidence of genetic linkage has been found. As the human genes were being found and the human map and sequence assembled, similar processes were taking place for the important experimental animal model systems. As pieces of the human map started coming together, with the order of genes in small clusters becoming known, researchers became aware of shared syntenic regions – similar clusters of the same genes in the same order occurring in more than one organism. Over the course of evolution, occasional translocation breaks that persisted on into subsequent generations have resulted in separating a single syntenic group into two separate syntenic groups in one organism while lack of such a translocation leaves them in their original order in another organism.

If we look at the sequence of chromosomal regions from different organisms we can see these regions of shared synteny, with the same genes in the same order. We can see a cluster of genes (starting 31.5 cM from the top of mouse chromosome 3) that parallel a cluster of the same genes in the same order on human chromosome 3q25 (Figure 11.13). But sometimes we will see a large syntenic region, such as the one between human chromosome 21 and mouse chromosome 16, in which a long list of genes occurring in the same order is punctuated by occasional genes that are out of order relative to the rest of the list.

On a gross genomic level we can use chromosome painting to look at the large-scale organization of the genomes of different organisms. If we take chromosome paints derived from gorilla sequences, and use them to paint human DNA, we see very large blocks of sequence on each human chromosome that are all painted the same color. Figure 11.14 shows what human chromosomes 1 and 2 look like when painted with gorilla DNA. There seems to be a one-to-one correspondence between gorilla and human chromosomes, with one gorilla chromosome painting one human chromosome in a single color. There are only three human chromosomes that show multiple colors corresponding to multiple different gorilla chromosomes. For example, the long arm of human chromosome 2 ends up painted with colors corresponding to gorilla chromosome 11 and the short arm is the color of gorilla chromosome 12, almost as if two gorilla chromosomes were stuck together to make one human chromosome.

If we look at the DNA sequences of human and gorilla chromosomes we see that the picture is not quite that simple. In the midst of the large blocks of synteny – the same genes occurring in the same order – we see small changes in the sequence scattered throughout that reflect many individual mutations that have happened over the course of evolution since the shared common ancestor to human and gorilla. And in some places within a large block of "color homology" there are local regions of the human chromosome that are inverted relative to the gorilla chromosome.

If we look at human chromosomes painted with DNA from other species we see a wide range of differences in the complexity of the patterns that emerge. In Figure 11.14 we show copies of human chromosomes 1 and 2 painted

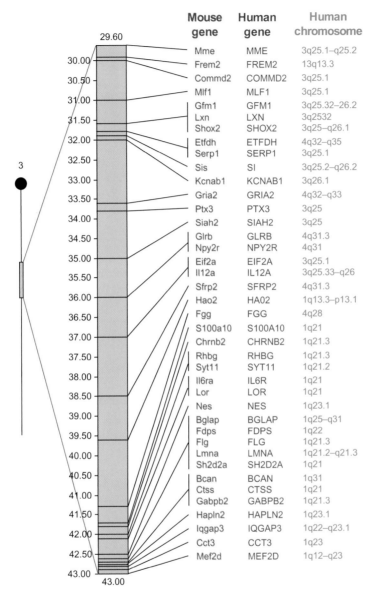

FIGURE 11.13 Mouse chromosome 3 and the regions of various human chromosomes that are syntenic to mouse chromosome 3. Notice that the human DNA that shares synteny with mouse chromosome 3 comes from different human chromosomes, but within any one region of shared synteny the gene order is conserved. These regions of shared synteny are not identical to each other and there are many small sequence differences scattered throughout. (From MGI (Mouse Genome Informatics) at http://www.informatics.jax.org/searches/linkmap.cgi?format=web+map&source=mgd&dsegments=0&syntenics=1&only homols=1&primary_chromosome=3&species=2; accessed May 2010. © The Jackson Laboratory. All rights reserved.)

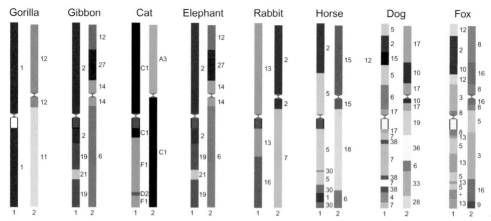

FIGURE 11.14 Human chromosomes 1 and 2. DNAs from chromosomes of eight different species were used for chromosome painting of human chromosomes 1 and 2. Note how simple the pattern is when the chromosome paints come from the gorilla, and how much more complex the pattern looks when we use chromosome paints from dog or fox. (Chromosome painting images created on Chromhome at www.chromhome.org/chromhome; May 2010.)

with DNA from eight different species ranging from the closely related gorilla to more distantly related species such as the dog. Some chromosomes, especially the X chromosome, paint with just one color for chromosome paints from many of the species shown in Figure 11.14. But as we can see from the chromosome painting images, as we move farther away from human on the evolutionary tree we find more gross structural differences between those species and human.

Placement of Clones on the Map

As the Human Genome Project proceeded, huge collections of clones were developed and assigned to positions on human chromosomes. A variety of approaches can achieve such placement but one of the simplest is fluorescent in situ hybridization (FISH). It is also possible to determine which clones overlap each other by comparing the sequences of the cloned inserts. In the course of the human genome project huge numbers of clones, each containing a small bit of the human genome, were ordered relative to each other and relative to their positions along the chromosomes.

The Sequence as the Final Physical Map

Once the sequence of the human genome was completed, much of the guesswork went out of the business of figuring out where anything in the genome is located. The gene responsible for nail–patella syndrome, which had initially been mapped to the long arm of chromosome 9, and then specifically to 9q34, is now known to occupy more than 86,000 bases of sequence beginning at base 129,376,748 and ending at base 129,463,311 (in the May 2010 version of the sequence). In addition to letting us tell exactly where in the genome the gene is located, the sequence lets us tell which strand is the one being transcribed. And if we are looking at more than one gene in the sequence, we can determine the exact relationship of those genes relative to each other without needing to evaluate families or study recombination.

One of the things the human genome sequence shows us is that some genes are transcribed from one strand and some from the other strand. In fact, the very precise placement allowed by the level of detail present in the sequence has allowed us to identify some genes that are contained within introns of other genes. The NF1 gene, which causes neurofibromatosis, is a large gene that harbors three other genes within one of its introns. OMG, EVI2A, and EVI2B are all located within one intron of the NF1 gene. NF1 and these three genes are transcribed from opposite strands, and thus are oriented in opposite directions.

We can only use the sequence to ask about the physical copies of genes. If we want to know the location of the insulin gene, which is involved in diabetes, we can simply look at the sequence to tell where it is located in any species we are interested in. We can tell which genes sit next to it. We can look at the sequence of the gene in different populations to look at how the sequence varies between individuals. We can compare the sequence of the insulin gene in different species to find out which residues in the protein are highly conserved across species.

But none of that tells us what the accompanying phenotype is. If there is another diabetes gene out there that we have not found yet, we cannot find it by just looking at the sequence. The sequence provides us incredibly important tools for the gene hunt, but those tools can help us find the gene only if we use many of the same approaches to the phenotype that we used in the gene mapping problems that used traditional forms of markers. We have to carefully define the phenotype we are looking for. We have to do a careful job of classifying each person in the study. We have to combine those phenotypic details with the sequence information to determine whether a particular sequence variant is the one that correlates with the presence of the target phenotype. So the sequence gives us tools that help us find diabetes genes if we have done a thorough job of characterizing the phenotype of members of a diabetes pedigree or individuals with diabetes in a population of unrelated individuals.

Similarities and Differences between Genetic and Physical Maps

If we examine the order of genes along the genetic map and the physical map, we find that the genes on both maps are in the same order. Things that are near the beginning of chromosome 1 on the genetic map are also near the beginning of chromosome 1 on the physical map, and things that are near the centromere of chromosome 1 on the genetic map are also near the centromere of chromosome 1 on the physical map.

However, if we look at the distances between items on the genetic map and items on the physical map we see differences. In general, if we were to look at an approximate value taken across the whole genome, we conclude that 1 cM of genetic distance corresponds with about a million bases of sequence. However, if we look at this correspondence between genetic and physical distances we find that it varies greatly depending on what point in the genome we are examining. Because there is relatively little recombination around centromeres or near telomeres, the genetic distance in those regions is short compared to the physical distance. And there are areas along the arms of the chromosomes that are hot spots for recombination, showing much higher rates of recombination than the surrounding regions. So we can use the 1 cM to 1 million base approximation but will often find a lot of variation from it depending on which spot in the genome is being examined.

11.7 HOW DID THEY BUILD GENETIC MAPS?

If we look at pair-wise combinations of many different traits, we can arrange things based on the measure of recombination, so that things that rarely recombine are placed close together and things that recombine often are placed far apart. By doing this, we can begin building a map of relative spacing between the genes that cause different traits.

The things we actually want to map are the disease loci, and we want to know which transcribed genes are sitting at that point on the map so that we can test those genes for mutations in individuals who have the trait. So why do we care if we have a framework map of genetic markers that are mostly not part of genes or located in genes?

We often find that two traits we want to map are *not* both present in the same family, so we cannot ask how often recombination events fall between loci that cause the two traits. So if we want to map the disease loci, what we want is some outside point of reference to serve as a landmark so that each trait can be tested for recombination between the trait and the landmarks – the genetic markers.

Thus, if we want to know whether two things are close to each other, we do not have to comb the earth in search of some incredibly rare family that has both of those traits that are usually not found together. Instead, we can pick a genetic marker and ask how far the marker is from the locus responsible for the first trait that is segregating in one family, and then we can ask how far that same marker is from the locus responsible for the second trait segregating in a different family. This lets us place the loci for two different traits relative to the markers of the framework map, and once each disease locus has been placed on the framework map we will be able to tell how far they are from each other even if the two traits are never found in the same family.

This requires that we know where the markers themselves are located. So how did they build the framework maps in the first place? They built them by testing many different markers in pair-wise combinations and each time they asked what fraction of meioses showed a recombination event between the two markers. Using a combination of recombination fractions and LOD scores, researchers were able to

decide which markers were very close together, which were in the same neighborhood, and which were so far apart that they segregated independently.

Once the frameworks maps were available, it became possible to do a genome-wide scan to look for the position of a disease gene. A genome scan evaluates each of many markers from throughout the genome, and for each marker asks the question, how far is this marker from the gene we are trying to map? If we want to screen markers that are separated from each other by a rate of 10% recombination, we would consider this a 10 cM genome scan, and we would expect to have to screen more than 300 markers spread across the genome with a distance of about 10 cM between each pair of markers. DNA from members of a family with a trait would be screened to determine which allele sizes are present for each marker, and for each marker a statistical test would be performed to evaluate the amount of recombination between the marker and the gene encoding the trait.

Putting a Gene Onto a Map

Now that the human genome sequence is known, you might think that genetic maps would be unnecessary and that we would simply look at the sequence to see where a particular gene is located. The human genome sequence lets us tell where things are in the sequence even if they have never been studied in genetic mapping studies. There are tens of thousands of genes that have been identified on the known sequence, but for many of them we do not know what phenotype would result if there were a mutation. And there are many phenotypes known for which we have not yet identified a gene. So the big problem is: how do we get the phenotypes that we see in families matched up with the genes that we see in the sequence? For many genes now placed on the sequence, we will not know what traits they cause until the genetic defects causing the traits are genetically placed onto the map.

There are limits to what we can tell from just looking at the sequence because, although the order of the genetic and physical maps is the same, there are some critical differences. Just looking at the map might give us the correct order of genes and markers along a chromosome, but it does not give us enough information because the rate at which recombination events take place are different for different parts of the genome. On average, we might see that a 1% recombination rate that we call a 1 cM distance on the genetic map would equal about 1 million base pairs of sequence. In some areas of the genome, such as near the centromeres, a 1% recombination rate corresponds with a much larger physical stretch of DNA, and there are other regions of the genome that contain hot spots for recombination where a 1% recombination rate may be seen between things that are separated by distances much, much smaller than a million base pairs. If we look at a particular point on a chromosome, we find that the rate of recombination in that region may be quite different in male meiosis (producing sperm) and female meiosis (producing eggs). But for each combination of a marker and a trait, we can test for whether or not the two always transmit together completely or whether they sometimes recombine.

Knowing the distance between a marker and the gene that determines the trait of interest is only the first step in placing a gene on a map. Let's consider a hypothetical gene that causes the presence of teeth at birth in one large family. When DNA samples from family members at risk of having the baby tooth trait were tested with marker M4, recombination events were seen to fall between the gene and the M4 allele only 4% of the time. If we have a map that shows the location of a set of genetic markers that includes marker M4, we can see that knowing there is a 4% recombination rate is not enough to let us place the gene on the map. The gene could be located 4 map units to the left of M4 or 4 map units to the right of M4 (Figure 11.15).

If we test other markers in the region and find that the baby tooth gene seems to be only 4 map units from M3 but is 18 map units away from M5, we can then place the gene on the map (Figure 11.16). In fact, it often takes testing of multiple markers on the map to get a precise placement, but this theoretical experiment shows how the map placement process works.

So there are many components to our mapping problem: definition of the trait, acquisition of information and enough study subjects, a framework map of markers and DNA samples from the subjects. We now have two different things we want to measure in the course of building a map, or in the course of placing something new onto an existing map: the distance from the trait gene to markers on the map (the *recombination fraction*) and the measure of how likely it is that the data are right (the LOD score mentioned above). If we combine information from multiple different markers, the process is often more complex than what was shown in Figure 11.16 and

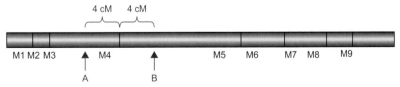

FIGURE 11.15 Finding out the distance from only one marker is ambiguous. In this case, we know that the locus we are studying is located 4 map units from M4, but we still do not know where the gene is because the gene could be located either at position A to the left of M4 or at position B to the right of M4.

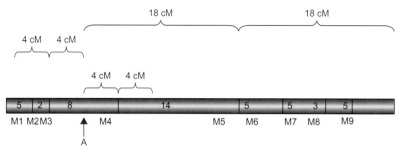

FIGURE 11.16 Adding in information from more than one marker gives us an unambiguous position on the map. Distances along the bar indicate the distance between markers in centimorgans. For each marker, use of information for that marker alone results in two possible locations for the gene, as indicated by the pairs of brackets. Testing marker M3 gives two possible positions. Testing marker M4 gives two possible positions. Testing marker M5 gives two possible positions. But only position A is consistent with the results for each of the separate markers. In a real experiment, the numbers would not be so tidy. The measured distance would be close to the real distance but would not be identical. So we might expect to see 3.8% recombination with M4, 4.2% recombination with M3, and 17% recombination with M5.

may require some sophisticated biostatistical genetic calculations.

11.8 AFTER THE MAP: WHAT CAME NEXT?

Getting from the Map to the Gene

In the early days of positional cloning (Box 11.4), before the completion of the human genome sequence, finding out where a gene was located on the map was the first step in a long search. Once the general location on the map was identified (usually a very large region containing a huge number of genes), many additional experiments in the lab would refine the location to a much smaller region. The genes in the region had to be identified, since at that time most genes were not yet known. Once a gene near the disease locus was identified, sequencing of copies of that gene from individuals with the trait would show whether the gene was mutated in the affected individuals. Once a mutation was found, others with the trait were also tested for mutations. If a gene was found that showed mutations in affected

individuals and it was shown that the mutation co-segregated with the trait in families, it was considered likely to be the cause of the disease and additional studies were initiated to confirm that hypothesis and begin studying how the gene plays a role in the disease. This process of moving from a map position to the identification of the actual gene and mutations usually took years, in the 1990s, but can now move very rapidly since many things that we had to find through lengthy experiments back then can now just be looked up in a database.

After the mapping step, what were the next steps in the gene hunt? Next came:

- Limiting the size of the region we have to search to as small a region as possible. This let us carry out positional cloning, identifying the target gene based entirely on location, but in most cases we ended up needing more information because the positional approach left us with a long list of possible genes that were all located in one genetic neighborhood. Reducing the region usually required that we find additional families and test markers from the disease gene region in family members to look for more recombination events that happen

BOX 11.4

POSITIONAL CANDIDATE CLONING

Although some disease genes were found by positional cloning, many were found by a process called positional candidate cloning. When a retinitis pigmentosa locus was mapped to chromosome 3q23–q24, researchers did not face the usual battle of that era to start finding genes in the region so that mutation screening could be carried out. Instead they looked at the region and found an "old friend," a gene whose product was already known to play a major role in the kind of photoreceptor that shows defects in retinitis pigmentosa. Rhodopsin was already a gene and protein of interest because of its known role in photoreceptor cells, receiving quanta of light and sending the signal to the brain saying "this cell detected light." In retinitis pigmentosa the photoreceptor cell is the primary point of pathology and abnormal photoreceptor functions can be detected. So researchers were able to do a positional candidate experiment, first mapping the disease locus by showing linkage to a marker called D3S47 with a recombination fraction of zero and a LOD score of 14.4! Shortly before that another group had used physical mapping methods to place the rhodopsin gene on the long arm of chromosome 3. So mutation screening was carried out and rhodopsin mutations were indeed found in individuals with retinitis pigmentosa. Researchers all over the world have gone on to find many different rhodopsin mutations in retinitis pigmentosa, and animal models further support the idea that rhodopsin mutations cause the trait.

to fall closer to the gene and thus bound a smaller region. This worked well if the trait was genetically homogeneous and other families all had the disease as a result of this same gene that we just mapped, but often traits are genetically heterogeneous and only some of the families we look at will be linked to our new locus.

- Identifying the known and theoretical genes in the region. This process of finding the genes in a region required years of lab work in the 1980s and requires at most a few hours with a database in the twenty-first century. Often we find ourselves faced with dozens or even hundreds of genes that we need to evaluate as candidates for being the gene we are after, but in the early days of putting disease genes on the map regions frequently lacked information on many of the genes.
- Prioritizing the genes based on functional information about the gene and gene product. A critical piece of information when deciding what gene to screen has often been the function of the gene product, and the issue of whether we can build a model for its involvement in the disease pathology. Another key piece of information has been: where is the gene expressed? When we were looking for genes responsible for retinitis pigmentosa we were more often interested in a gene expressed in adult retina than in a gene expressed only in fetal kidney, but it is actually important to not completely write off a gene that is not expressed in the retina. After all, if a defect affecting one organ causes changes in circulating molecules such as hormones, then there can be an important impact on cells far from where that hormone is being produced.
- Screening for mutations in the most likely genes by sequencing those genes in individuals with the trait. We will often start

by sequencing DNA from members of all of the families that we know are linked to this new locus. We look for co-segregation of an allele with the trait in the family. However, we have to keep in mind that co-segregation of a sequence variant does not prove that the variant is causative. Once we have used recombination events to define boundaries of a genetic inclusion interval that must contain the gene, anything internal to the interval will co-segregate with the trait. So finding a gene that has a likely candidate gene function expressed in the right cell type with mutations that co-segregate make it a likely candidate, but it does not constitute proof that the mutation/gene that we found is causing the trait since there are likely many sequence variants within a bounded genetic inclusion interval and they will all co-segregate.

- Studying the presence of mutations and traits in other families and in populations. One of the next things we do to try to confirm a causative role for the mutation/gene is to screen for mutations in other families. By looking in families we can learn things about penetrance of the trait by determining whether there are family members with the causative mutation who lack the trait.
- Another approach to validation uses screening of populations to ask whether mutations in this gene turn up at a higher rate in a group of cases than in a group of controls. Screening populations also tells us what fraction of the random array of patients walking into clinics with this trait might have their characteristics caused by the gene we have just found. It can also tell us things about prevalence of the mutation(s) in different populations, since particular mutations may vary in frequency from one population to another.
- Showing that an identified mutation is not also present in individuals who lack the trait (or, in the case of recessive disorders,

showing that individuals without the trait have at most one copy of the gene defect). This is important for simple Mendelian traits, although as we will see in Chapter 12, for complex traits we might see a lot of unaffected individuals with the causative mutations without thinking that rules out the gene's involvement in the trait.

- An important step in validation can include construction of transgenic animal models. If we can start with an animal that does not have the trait, and then insert one and only one mutation that we hypothesize is causative, and then see that the animal now has the trait, we consider this strong support for the hypothesis that we have found the gene. With some mutation systems we also have the ability to take the mutation back out again, and show that the trait has now gone away. However, a mouse is different enough from a human that sometimes a gene and mutation that cause a trait in a human do not cause that trait in the mouse. So when we build a transgenic model of the disease, a positive finding provides support for our hypothesis that this gene is "it" but a negative finding does not rule it out as the disease gene.

Gene hunts that used to take as long as a decade can now proceed in a matter of months even using some of these older technologies. For instance, the progeria gene was found less than a year after the international consortium set out to find it.

From Animal Models to Human Disease Genes

In many cases, finding a human disease gene can be complicated by issues of recruiting study subjects. If families are small and rare, if the phenotype is hard to identify, if many individuals with the trait do not realize that they have it, or if the trait is one that people may not admit to having it can be hard to find the gene by simply studying human subjects. In some cultures, doctors say that they have trouble doing genetics in human families because people will not bring their families into the study; they are seeing the doctor because they need help, but they are unwilling to admit that there is anything "running in the family" or that something is wrong with the family genetic background. In addition, when we work with an animal model such as the mouse, we can breed as many animals as we need to do the experiment, but when we study a human family that has only six individuals we are limited to the number of individuals present in that family even though we usually need at least 11 affected family members to complete a linkage study.

Whatever the cause that complicates the recruitment and the conduct of the study as a human genetics study, we sometimes have the option of finding the gene in an animal model, and then using sequence homology to identify the human gene that is the counterpart of the gene we find in the animal model. An especially elegant example of an animal model study that crossed over into human genetics was carried out by Susan Dutcher and colleagues. They set out to identify genes involved in forming cilia. Ciliated structures in humans include the motile cilia (hair-like particles on the surfaces of cells, such as the cilia that help to clear particulates out of the lungs) and non-motile cilia (such as a section of the photoreceptor cell that connects the cell body to the outer segments of the cell). Some organisms have cilia and some do not. By comparing the genomes of organisms that have cilia and organisms that do not, these researchers identified a group of genes that are found in multiple ciliated organisms but are not found in non-ciliated organisms.

When the researchers looked at where the human homologues of these genes were placed on the human genetic map, they found that one of these genes is located in an interval to which others had mapped a gene for Bardet–Biedl

syndrome, a disease that affects a variety of ciliated cell types in the human body. When they tested for mutations in this gene, in DNA from subjects with Bardet–Biedl syndrome, they found mutations and showed that the gene that came out of their ciliated vs. non-ciliated experiment is indeed responsible for some cases of Bardet–Biedl syndrome. As a result they showed that the fundamental defect in Bardet–Biedl syndrome affects ciliated structures in the human body.

If we want to know the phenotype that can result from mutations in the gene we can use homology between species to help us tell things about the phenotype. We can make a transgenic animal that does not have a functional copy of the gene and examine the resulting phenotype to see whether it resembles a known human phenotype. If we find a new rhodopsin mutation and want to know whether it can cause disease, we can make a transgenic mouse with that mutation and evaluate whether or not the retina remains healthy and fully functional. In other cases, we can start with a phenotype of interest, such as a naturally occurring form of hearing loss in the mouse, and map and identify the gene; once we have found the mouse deafness gene, we can screen human DNA for mutations in that gene.

But mice are not humans so the animal phenotype may be quite different from the human phenotype for the same gene or even the same mutation. There are many differences in the genetic background on which we find the mutation in a person and a mouse, and so there are many differences in the sequences of proteins with which our disease gene protein interacts, differences in the physiology of the cells in which the protein carries out its functions, and differences in the developmental steps that take place in each. Mutations that knock out function in the aniridia gene (PAX6) in humans result in incorrect development of the iris of the eye, often with some of the iris missing which results in oddly shaped pupils in the eyes. But the same gene in the mouse causes a different phenotype: mice with defects in PAX6 have a phenotype called "small eye." In both cases the mutation is affecting the eye, but the detailed effect is different.

In some cases, if we want to build an animal model of a human disease, we make a mouse with a known mutation in the gene we are studying, but sometimes we do not get the disease phenotype we were looking for. When we study mutations in genes in the pathway that processes vitamin A, we find that mouse mutations in some of the genes in this pathway show the expected phenotype, but for one of the genes the mouse model shows such a mild phenotype that it is hard to tell whether the mutation is having any effect.

Using the Sequence to Find Human Disease Genes

The production of the human genome sequence has revolutionized how we do genetics and made it possible for researchers studying a particular trait to spend their time and resources on studying the trait instead of spending their energies trying to generate the tools with which to carry out their studies. We are a long way away from eliminating the need for the molecular biology and biochemistry used to identify mutations in disease genes, and some experiments still call for the kind of mapping described here. But the in-between step of finding the genes so we can evaluate whether one of them is "it" has been reduced from years of laborious slogging to a matter of hours fishing for data online. This allows scientists to spend more of their time asking questions of great functional importance, and it allows their operations to cover more territory in less time, with all of the accompanying implications for advances in understanding human health issues of great importance.

In many of the positional cloning experiments there were no good candidate genes in the vicinity to choose from so researchers had to depend entirely on localization to narrow the field until the gene was found, but in other cases if a compelling candidate gene turned out to be in the vicinity it was normally the first one sequenced. In fact, some disease genes have been identified entirely by candidate gene screening, picking a gene based on a known function and testing for mutations without doing the mapping experiments, but experienced gene hunters will tell you that most candidate genes that look like they must be the disease gene based on function turn out not to be "it." Often when a new disease locus has been mapped and we look at the available genes in the region, we can often point to known functions of multiple genes in that region that look like they could play a role in the disease pathology. And most of them are not "it." Nonetheless, the combination of mapping and candidate gene functions has often facilitated disease gene identification.

With the advent of next generation sequencing, as the prices are rapidly dropping and analytical approaches are evolving, the gene hunt is being transformed. So join us in Chapter 12 for a discussion of the Human Genome Project and the next stage in the evolution of human genetics.

Study Questions

1. What is a genetic map?
2. What is a genetic marker?
3. What are three different kinds of genetic markers?
4. What does it mean when we say that we have "found a gene"?
5. What are two advantages of single nucleotide polymorphisms (SNPs) over restriction fragment length polymorphisms (RFLPs)?
6. Why did scientists use blood group markers in so many of the earliest efforts to map human genes?
7. When constructing a genetic map, why is it a problem if the founder is homozygous?

8. Which allele is linked with the trait represented by the blue symbols?

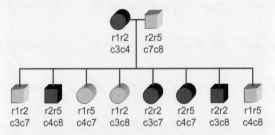

r1r2   r2r5
c3c4   c7c8

r1r2   r2r5   r1r5   r1r2   r2r2   r2r5   r2r2   r1r5
c3c7   c4c8   c4c7   c3c8   c3c7   c4c7   c3c8   c4c8

9. Why can't we tell whether this marker and this trait are linked?

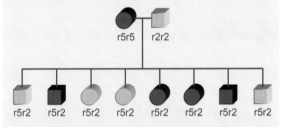

r5r5   r2r2

r5r2   r5r2   r5r2   r5r2   r5r2   r5r2   r5r2   r5r2

10. Draw a map showing where the following markers are located relative to each other:

Marker pairs	Recombination
Q–N	10%
Q–X	15%
N–X	5 cM

11. Where is trait A relative to markers B, F and P?

	Recombination
A–B	9%
A–F	5%
A–P	3%
B–P	12%
B–F	14%
F–P	2%

12. What information do we get from the recombination fraction and what information do we get from the LOD score?

13. Markers D and J show 50% recombination. Where are they located relative to each other?

14. If we use human DNA to paint elephant and gibbon chromosomes, we see that the same pale blue color paints elephant chromosome 27 and gibbon chromosome 21. What can we conclude about the relationship between elephant chromosome 27 and gibbon chromosome 21?

15. In the late 1980s when the Huntington disease gene was mapped, it took years afterwards to find the gene. After the turn of the century, when the progeria gene was mapped, it took less than a year to find the gene. What had changed that made such a big difference in the timelines of these two projects?

16. What information can help evaluate the list of potential candidate genes located in the region to which a gene has been mapped?

17. How can an animal model be used to find a human disease gene?

18. Using genetic markers, you find that the interval containing the gene is flanked by marker G and marker H. You identify the genes located between markers G and H and find a gene with a sequence variant that co-segregates with the disease. Meanwhile another group has tested a different gene and found the same thing for that gene. Why would two different groups find two different genes that co-segregate with the disease?

19. If defects in a particular gene cause the disease we are studying in a human family, why might that same defect not cause the disease in a mouse?

20. Why does the genetic distance of 1 cM not always correspond with the physical distance of 1 million bases?

Short Essays

1. When comparing species as distantly related as mouse and human, we find a lot of similarity, including blocks of shared synteny. How similar are mice and humans and how

far back would we have to look to find a shared ancestor? As you consider this question please read "Mice" in *The Strongest Boy in the World* by Philip R. Reilly (Cold Spring Laboratory Press, 2008).

2. In 1988, before the existence of the big framework genetic maps, geneticists concluded that there is a schizophrenia gene on chromosome 5. How did they first identify this location and what kinds of follow up studies did they do and what did they conclude? As you consider this question please read "Mental illness: How much is genetic?" in *Abraham Lincoln's DNA* (Cold Spring Harbor Laboratory Press, 2002).

3. When scientists mapped the location of a gene for Werner syndrome, a form of premature aging, there were a lot of genes in the region and it was not obvious which gene was causing the problem. How did they figure out which gene it was, and once they had found it, how did they figure out what the gene product does? As you consider this question please read "Human age genes" in *Living with Our Genes: Why They Matter More than You Think* by Dean Hamer and Peter Copeland (Anchor, 1999).

4. When loci are far apart on the same chromosome, it can be hard to tell how far apart the two loci are. How can we figure out how far apart two loci are if they are more than 50 cM apart? As you consider this question please read "The order of the genes" in *The Physical Basis of Heredity* by Thomas Hunt Morgan (Forgotten Books, reproduction of the 1919 publication).

Resource Project

We can look at the relationship of many organisms because the DNA from any organism can be used to create chromosome paints and the chromosomes from any organism can serve as the target to be painted. Go to the Chromhome web application for comparative genomics at the website at Cambridge University (www.chromhome.org/chromhome) to try painting different organisms with DNA from other organisms. When we paint human chromosomes with DNA from other organisms, some chromosomes are painted with only one color and some chromosomes seem to correspond to many other chromosomes from other species. Write a report of up to one page in length about what you see when the human X chromosome is painted with DNA from different species, and about how much we can conclude from the painting results regarding the relationship between X chromosomes in different organisms.

Suggested Reading

Articles and Chapters

"Mapping the cancer genome. Pinpointing the genes involved in cancer will help chart a new course across the complex landscape of human malignancies" by F. S. Barker and A. D. Collins in *Scientific American*, 2007;296(3):50–7.

"Forming the fly lab: Contributions of A. H. Sturtevant and C. B. Bridges" in *Mendel's Legacy: The Origin of Classical Genetics* by Elof Axel Carlson (Cold Spring Harbor Laboratory Press, 2004).

"Limb and kidney defects in Lmx1b mutant mice suggest an involvement of LMX1B in human nail patella syndrome" by H. Chen *et al.* in *Nature Genetics*, 1998;19:51–5.

"Positional cloning moves from perditional to traditional" by Francis Collins in *Nature Genetics*, 1995;9:347–50.

"The search for a gay gene" in *The Science of Desire: The Search for the Gay Gene and the Biology of Behavior* by Dean Hamer (1991, Touchstone).

"Progress toward personalized medicine for glaucoma" by Sayoko Moroi and colleagues in *Expert Review of Ophthalmology*, 2009;4:145–61.

"Dogs" in *The Strongest Boy in the World* by Philip R. Reilly (Cold Spring Laboratory Press, 2008).

"Chromosome mapping with DNA markers" by R. White and J. M. Lalouel, in *Scientific American*, 1988;258:40–8.

"Mechanistic insights into Bardet–Biedl syndrome, a model ciliopathy" by Norann A. Zaghloul and Nicholas Katsanis in *Journal of Clinical Investigation*, 2009;119(3):428–37.

Books

The Science of Desire: The Search for the Gay Gene and the Biology of Behavior by Dean Hamer (1991, Touchstone).

The Physical Basis of Heredity by Thomas Hunt Morgan (Forgotten Books, reproduction of the 1919 publication).

The Strongest Boy in the World by Philip R. Reilly (2008, Cold Spring Laboratory Press).

Mapping Fate by Alice Wexler (1996, University of California Press).

The Human Genome: How the Sequence Enables Genome-wide Studies

OUTLINE

The Human Genome. DOI: 10.1016/B978-0-12-333445-9.00012-5

THE READER'S COMPANION: AS YOU READ, YOU SHOULD CONSIDER

- The historical steps that led to the Human Genome Project.

- Why some of the most important steps can't be easily listed on a timeline.

- The different objectives of the Human Genome Project.

- The difference between *in vivo, in vitro,* and *in silico.*

- The rate at which genetic information production has increased.

- What the human reference sequence is.

- Why we can't tell you how long the human genome sequence is.

- Which is really the smallest chromosome.

- How many human genes there are.

- What is important about the other genome projects.

- How the genes are arranged in the genome.

- What is in the genome besides the genes.

- How the sequence varies between individuals and across populations.

- What factors can lead to allele frequency differences in different populations.

- Why populations in Africa have more haplotypes than European populations.

- What types of technologies have emerged from the Human Genome Project.

- How some things can be associated even though they are not causative.

- What a GWAS is.

- When we use association and when we use linkage.

- How we deal with multi-testing issues in a GWAS.

- Why a power calculation is important.

- What are "-omics"?

- What role copy number variation can play in a trait.

- What is paralogy?

- What has been found by re-sequencing of whole genomes.

12.1 THE HUMAN GENOME PROJECT

On July 20, 1969, many of us watched, spellbound, as Neil Armstrong stepped off the bottom rung of the ladder of the lunar lander and made the first human footprints in the powdery dust on the surface of the moon (Figure 12.1). "That's one small step for man, one giant leap for mankind," resounded around the world and down through the subsequent years. All of us who witnessed it felt the magic of that astonishing moment that transformed mankind from an earthbound species to one that

FIGURE 12.1 Neil Armstrong holds a singular place in history as the man who first stepped onto the surface of the moon. Many of us old enough to have watched the moonwalk live remember the moment clearly and can compare stories about that day. The event was focused on Armstrong but that historic step was the product of work on the part of a huge team working for many years to put him there. Similarly the leaders of the human genome project who announced the completion of the human genome sequence were backed by the efforts of large teams of scientists at human genome centers around the world. (Photo courtesy of NASA.)

had walked on the surface of another celestial body. We had watched for years as launch after launch headed us inexorably towards that moment when we all held our breath while Armstrong stepped off. Armstrong's famous declaration of a leap for mankind told us that humanity had just crossed over a historic divide, separating all of prior earthbound history from all of subsequent history. Often, the real watershed moments in history can be harder to pinpoint, as we have seen in the course of the Human Genome Project. Map after map after map of the human genome emerged as we moved through the 1990s, each a bit more complete, but there was never a discrete moment when we went from not having a human gene map to having one. Similarly for

the sequence – it gradually became more and more and more complete. There was a day on which the sequence was announced around the world, but it did not spring into existence, it grew base by base over a period of 13 years. Most of science proceeds by baby steps, punctuated by occasional leaps forward, but the moments are rare when we can point to a large discrete advance akin to the moment when Armstrong stepped onto the lunar surface.

What Is the Human Genome Project?

This chapter on population-based approaches to finding human genes begins with a discussion of the Human Genome Project, which has provided us with so many of the tools, technologies, research strategies, and information that we use in our current approaches to finding and studying human genes. While there are still many studies taking place that involve classical linkage studies of human families, such as we discussed in Chapter 11, there has been a big move in the direction of genome-wide studies in populations. Before we start telling you about how we do such studies, or why, let's first consider where the tools and information for such studies came from.

What is the Human Genome Project? A government description of the project says it is "a worldwide research effort that has the goal of analyzing the structure of human DNA and determining the location of all human genes." It was the brainchild of a group of scientists who saw the possibility that thinking on a large scale and using high-throughput methodologies could help the field to leap-frog past the slow march of individual labs studying one gene at a time. Goals of this project included:

- Determination of the complete sequence of the human genome.
- Identification of all of the genes.
- Development of databases to acquire and disseminate information.
- Development of new technologies and forms of analysis.

- Transfer of technologies to the private sector.
- Exploration of the *ethical, legal, and social issues* (known as ELSI) surrounding the use of the information being generated.
- Similar analysis of the genomes of a selected set of experimental animal model systems.

Over the past 30 years, genetics research has been building towards the unveiling of the human genome sequence. Like the new and successive launches that carried the space program towards the moonwalk, major break-throughs in molecular biology have paved the way step by step towards this genetic "leap for mankind." From the 1950s, when the locations of the first three human genes were mapped, to the 1970s when the first human genes were cloned and sequenced, we have moved forward to 2010 when databases at multiple locations in the world house copies of the sequence of the whole human genome. In the spring of 2003 a paper announcing a high-quality version of the *human genome sequence* was published but the sequence did not spring full grown, like Athena emerging from the head of Zeus; rather it was built continuously over a period of years so that there was no dramatic moment when we moved from a time when the sequence did not exist to a time when it did.

By 2010 we have arrived at a time when we have a reference sequence of a little more than three billion bases of DNA, along with a lot of accompanying annotation and informa-tion about the sequence. There are more than a thousand genetic tests available. More than 1800 disease genes have been identified. Hundreds of clinical trials are in process that make use of information or products that have arisen from the Human Genome Project.

These advances have earned a lot of interest, but they have not captivated the imagination of the whole world in quite the same way the moonwalk did. Those of us who spent child-hoods playing at being astronauts and walking on the moon do not usually find our children putting on white lab coats, picking up toy pipettes, and dreaming of finding a gene. This is in part because a single publicized moment that shows someone stepping out of a spacecraft is fundamentally more glamorous and easier to identify with than putting in long hours pipet-ting. But the real impact of the human genome sequence is vast, and those of us who see our day-to-day lives in science revolutionized by the existence of the completed sequence look back at the scientists who proposed the Human Genome Project and wonder if they realized what aston-ishing things would come of their daring dream.

The Human Genome Timeline

You might think that the source of the human genome project originated when the US National Institutes of Health and the US Department of Energy began formally funding the Human Genome Project. Or that it started during the decade before that when the concept of mapping and cloning the whole human genome was first proposed. But the foundations of the project can be found much farther back in time.

Our timeline presents some of the landmark events along the way (Box 12.1). In some senses we might see the human genome project as beginning in 1683, with van Leeuwenhoek's dis-covery of bacteria, since bacteria form the basis of the cloning systems used in the human genome project. Or we might consider the project to have started in the 1800s with the discovery of DNA, evolution, and rules of inheritance. How did we get from that far time, when no one knew how inheritance works, what evolution is, or even that DNA is the repository of genetic information, to the twenty-first century when "The Sequence" is joining other resources, such as the periodic table of the elements, as bodies of knowledge that we can access without even worrying about how anyone came to know such things?

At the beginning of the twentieth century, as the lost works of Mendel were being redis-covered and introduced to the scientific world,

BOX 12.1

100 EVENTS ON A HUMAN GENOME TIMELINE

- 1665   Hooke discovers cells and coins the name "cell".
- 1683   van Leeuwenhoek discovers bacteria.
- 1839   Schwann publishes cell theory of animals.
- 1859   Darwin describes natural selection in his book *On the Origin of Species*.
- 1865   Mendel publishes his work on inheritance in his paper "Experiments in Plant Hybridization".
- 1869   Miescher discovers the chemical substance DNA.
- 1879   Flemming first observes mitosis and chromatin.
- 1900   DeVries, Correns, and Tschermak rediscover Mendel.
- 1905   Bateson and Punnett discover linkage.
- 1908   Garrod attributes alkaptonuria to an inborn error of metabolism.
- 1910   Morgan discovers sex linkage.
- 1913   Sturtevant provides the first linkage map.
- 1916   Calvin Bridges proves the chromosome theory of heredity.
- 1927   Herman Muller uses X-rays to induce mutations artificially.
- 1931   Creighton and McClintock show recombination is exchange between chromosomes.
- 1941   Beadle and Tatum propose the one gene–one protein model.
- 1944   Avery, McCarty, McCleod show that DNA is the "transforming principle" in pneumococcus.
- 1944   Barbara McClintock discovers transposable elements.
- 1950   Chargaff shows amount of A = amount of T and amount of G = amount of C.
- 1952   Hershey and Chase show DNA in a virus carries the genetic information.
- 1953   Watson, Crick, Franklin, Wilkins elucidate the double helical structure of DNA.
- 1955   Tijo finds that there are 46 human chromosomes.
- 1958   Meselson and Stahl show semi-conservative replication of DNA.
- 1958   Kornberg carries out *in vitro* synthesis of DNA.
- 1959   Crick proposes the Central Dogma of Molecular Biology.
- 1959   Lejeune shows extra copy of chromosome 21 in Down syndrome.
- 1961   Robert Guthrie creates first test for phenylketonuria.
- 1961   Jacob and Monod lac operon and concept of regulatory proteins binding to sequences in DNA.
- 1961   Brenner, Jacob, and Meselson discover messenger RNA.
- 1966   Nirenberg, Khorana, and Holley elucidate the genetic code.
- 1970   Caspersson and Zech invent chromosome banding.
- 1970   Smith, Arbor, and Nathans isolate the first restriction enzyme.
- 1970   Temin and Baltimore discover reverse transcriptase.
- 1972   Berg and Mertz produce the first recombinant DNA.
- 1972   Human and gorilla DNA found to be 99% similar.
- 1972   Khorana creates the first synthetic gene – an alanine tRNA gene.
- 1974   Belmont report on human subjects in research.

(Continued)

BOX 12.1   (cont'd)

- 1974   NIH forms Recombinant DNA Advisory Committee (RAC) to oversee cloning.
- 1975   Holley completes RNA sequence of MS2 bacterial virus.
- 1977   Sanger invents enzymatic sequencing.
- 1977   Maxam and Gilbert invent chemical sequencing.
- 1977   Roberts and Sharp discover introns.
- 1977   Boyer uses expression cloning to produce synthetic insulin.
- 1978   First single copy human gene cloned – insulin.
- 1978   Birth of Louise Brown, the first baby conceived *in vitro*.
- 1980   Botstein, White, Skolnick, Davis propose use of RFLPs to build a genetic map.
- 1982   First genetically engineered drug released – insulin.
- 1983   Gusella maps the HD gene using DNA markers.
- 1984   Cantor and Schwartz sequence Epstein–Barr virus.
- 1985   Kary Mullis invents polymerase chain reaction.
- 1986   Gene for chronic granulomatous disease is first gene found by positional cloning.
- 1986   Smithies, Capecchi, and Martin develop site-directed mutagenesis.
- 1986   Cloned mouse is first cloned mammal.
- 1986   Lee Hood produces the first automated DNA sequencer.
- 1987   Leicester, England, first DNA criminal conviction, first DNA criminal exoneration.
- 1986   DMD gene found by positional cloning.
- 1988   National Center for Biotechnology Information created.
- 1988   Office of Human Genome Research established.

- 1988   Yeast artificial chromosomes developed as cloning vectors.
- 1989   Jeffreys DNA fingerprinting and use of DNA in forensics and paternity.
- 1989   Collins and Tsui clone cystic fibrosis gene by positional cloning.
- 1989   Concept of the sequence tagged site (STS).
- 1989   Fluorescent labeling of DNA.
- 1990   Anderson treats SCIDS with adenosine deaminase gene in first gene therapy trial.
- 1990   BLAST algorithm for comparing sequence information published.
- 1991   First human genome centers established.
- 1991   Venter proposes expressed sequence tags (ESTs).
- 1991   Microarray chip technology invented.
- 1992   Second generation human genetic map released.
- 1993   Ambros, Lee, and Feinbaum discover microRNAs.
- 1994   FlavR SavR© tomato, first genetically modified food on market.
- 1995   First bacterial genome sequenced (*Haemophilus influenza*).
- 1996   Yeast genome completed.
- 1997   Roslin Institute clones Dolly the sheep.
- 1997   *E. coli* genome sequenced.
- 1998   *Caenorhabditis elegans* (round worm) sequence completed.
- 1998   Thompson cultures human embryonic stem cells.
- 1999   First human chromosome completely sequenced (chromosome 22).
- 1999   Baulcombe discovers RNA silencing.
- 1999   *Drosophila melanogaster* genome sequence completed.
- 2001   Successful RPE65 gene therapy in the Briard dog.
- 2001   Official announcement of the completed human genome sequence.

(Continued)

BOX 12.1   *(cont'd)*

- 2002   Mouse and rice genome sequences completed.
- 2002   Sequence of malaria genome and genome of the mosquito that carries malaria.
- 2003   High quality human genome sequence completed.
- 2003   GloFish©, the first genetically engineered pet.
- 2004   "Next generation" sequencing is added to repertoire of genomic tools.
- 2004   First RNA interference clinical trial treats macular degeneration.
- 2000   ChIP-on-chip uses chromatin immunoprecipitation to identify protein binding sites genome-wide.
- 2005   Chimpanzee and dog genome sequences are completed.
- 2005   GWAS finds CFH variant associated with age related macular degeneration.

- 2005   First version of HapMap catalog of human haplotypes and sequence variants.
- 2007   First progeria clinical trial.
- 2007   First individual human genome sequence – Craig Venter's – by traditional sequencing.
- 2007   Completed sequence of James Watson's genome by next generation sequencing.
- 2008   Genetic Information and Non-descrimination Act is passed.
- 2009   First RPE65 clinical trial in humans starts.
- 2009   First clinical trial using human embryonic stem cells to treat human spinal chord injury.
- 2010   First synthetic genome – *Mycoplasma genitalium.*
- 2010   Lupski and colleagues use whole genome sequencing to find a CMT neuropathy gene.

---

no one knew what Mendel's proposed genes were actually made of. Because the presence of only four letters in the alphabet seemed to lack the complexity needed to encode the amount of information in the genome, the scientific world originally dismissed DNA as the possible source of genetic information. There was rather universal surprise when experiments in the 1940s showed that DNA is in fact the repository of genetic information. By the 1950s, discovery of the secrets of DNA replication and transcription via base pairing answered some of the questions about how genetic information could be copied and transmitted from cell to cell and from one generation to the next. By 1961 the unveiling of the triplet code solved the problem of how to spell out 20 amino acids using only four letters.

As Armstrong stood on the moon in 1969, no one had yet deciphered the order of As, Cs, Gs, and Ts in even one human gene. That awaited the invention of *DNA sequencing*, the technical process by which we read the order of genetic letters in a piece of DNA. Two different methods for determining DNA sequence emerged into the scientific world during the 1970s, along with *gene cloning*, the process by which an individual gene can be separated away from the rest of the genome and replicated in many identical copies. The first pieces of sequence from human genes and chromosomes started coming out in the 1970s. At that point, during the very first years of the human gene hunt, deciphering of a few hundreds or thousands of letters spelling out a tiny segment of the genome was regarded as a huge achievement. Julia learned early in her graduate school research project that having even a piece of a human gene in the late 1970s was enough to get her research published in one of the top scientific journals in the world.

In the 1980s, polymerase chain reaction (PCR) gave us the ability to study DNA without first cloning it, and artificial chromosomes gave us the ability to clone enormous pieces of DNA thousands of times the size of the first things cloned in the 1970s. Advances were punctuated by the emergence of concepts such as *sequence tagged sites* (*STSs*) and expression sequence tags (ESTs), pieces of sequence that could be entered into a publicly accessible database as a mechanism of transferring genetic information between labs without having to physically transport any actual samples of DNA.

The human genome project grew out of discussions during the 1980s and officially came into existence in 1990 when the US National Institutes of Health and the US Department of Energy began funding a biological science program of unprecedented scope. The 1990s were swept along with the appearance of genome centers using robotic and high speed computers to achieve high-throughput progress towards the completion of the sequence. The new millennium saw the emergence of *microarrays* and *next generation sequencing* as tools for being able to survey the sequence of the chromosomes and the expression of tens of thousands of genes in one experiment. In the course of 50 years, we had gone from not even knowing which chemical structure houses the genetic information to a point at which we know so much about our genetic blueprint that we can now do meaningful virtual genetics experiments inside of computers without ever touching a pipette, experiments involving millions or even billions of pieces of information. In the course of reaching this point, we saw not only the invention of new scientific techniques and equipment, we also saw some fundamental changes in how we approach science.

To look at the rate at which the sequence was unveiled, we can examine the amount of sequence present in the publicly accessible database called GenBank that is curated by the National Center for Biotechnology Information. This electronic information repository contains the human genome sequence in addition to sequence information on all of the other organisms being studied. The information contained in GenBank comes from not only our close relatives, such as primates and mice, but also from much more distantly related organisms, such as fruit flies, zebra fish, yeast, bacteria, and even viruses. In 1982, the first year for which information is available, there were 606 files of sequence data containing 680,338 base pairs of sequence. By 1992 there were more than 200 million base pairs of sequence. By 1998 there were more than 2 billion base pairs. By the year 2000 the number of files in GenBank passed 10 million, and the amount of sequence available from all organisms passed 11 billion base pairs! By 2002, this had grown to more than 22 billion base pairs of sequence, according to the National Center for Biotechnology Information. By 2009 the original section of GenBank, with its individual sequences of individual genes, had 106,533,156,756 bases of sequence in 108,431,692 sequence files, and additional sections of the database housed 148,165,117,763 bases of sequence in 48,443,067 sequence files. Over the course of 25 years, we had progressed from a stage at which massive efforts by a team of people working together for years produced a small piece of one gene a few hundred bases in length, to the point in the twenty-first century when the sequence was being read so rapidly that new computer technologies and mathematical algorithms had to be developed to handle the massive rate at which the sequence was expanding (Figure 12.2).

Thus our timeline of the Human Genome Project (Box 12.1) starts long before the formal declaration of the project, and continues on beyond the completion of the sequence itself.

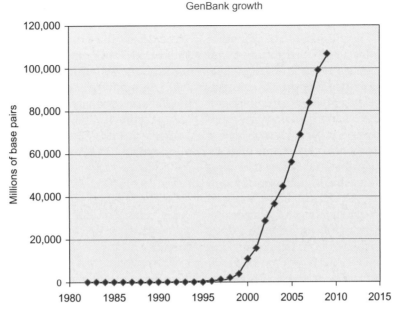

FIGURE 12.2   The rate of growth of sequence data in GenBank from 1988 to 2009 increased to the point that new technologies were needed to handle the massive influx of information. (After GenBank Statistics http://www.ncbi.nlm.nih.gov/genbank/genbankstats.html; accessed October 2010.)

We have listed many landmark events, but this timeline does not include all of the crucial advances towards the completion of the human genome project. Consider again the discrete nature of the step onto the moon and compare it to the gradual accumulation of information that became the final version of the sequence; history faces a problem when faced with how to list the many parts of the project, such as the development of the overlapping clones of the human genome, that were done by many people over the course of many years and do not lend themselves well to placing on a timeline even though these steps were just as important as some of the discrete steps that we do list.

Who Carried Out the Human Genome Project?

Most of the sequence from the public project was produced by the six countries participating in the formal international human genome consortium: China, Germany, France, Japan, the UK, and the US, but over the course of the Human Genome Project many advances in many different aspects of human genetics and genomics were ongoing in countries all over the world. It should be noted that one of the most important things about the project was that the information was not only freely traded between the participants but was made available to the rest of the scientific world, going into publicly accessible databases almost as fast as it was produced. This allowed scientists all over the world to make use of the information in their individual projects as rapidly as the sequence became available.

As government-funded projects flourished, industry joined in with an interesting mix of collaboration and competition with the publicly funded projects. As the initial groundwork of getting the clones and getting them mapped was finished, the real work of sequencing took off and turned out to be a run to the finish line between the International Human Genome Project effort on the public side of the race and Celera Genomics, Inc., a biotechnology company. Of interest to our understanding of the origins of the human genome sequence is the fact that the public and private efforts to generate the sequence used different strategies. The public enterprise built overlapping sets of clones, mapping them relative to each other, sequencing them, and placing pieces of sequence in place based on the locations of the clones from which they were derived. As overlapping pieces of sequence were generated, software for pairing and aligning sequences allowed the fusion of multiple separate sequence runs into one single region of sequence. The Celera Genomics project dispensed with the process of placing the clones relative to each other; they simply sequenced the clones at random and allowed the sequence alignment software to place the regions of sequence in place relative to each other based on where pieces of sequence overlapped with each other.

The result of the push to complete the sequence was an announcement by Bill Clinton, the 42nd President of the United States, that the human genome had been sequenced. He was joined by Francis Collins, who was then the Director of the National Human Genome Research Institute and now heads up the National Institutes of Health, and by Craig Venter, the founder of Celera Genomics (Figure 12.3).

Many in the human genetics community laugh about this being the only government project ever to come in under budget and ahead of schedule. But, oops, we have to stop for a disclaimer here; we have not evaluated those other government projects, so we can't say that none of the others set such a record, but we do know that the Human Genome Project actually did get there sooner, and cheaper, than anyone had expected. This is a marvelous testimonial to the amazing levels of energy and dedication

FIGURE 12.3 Since their press conference announcing the existence a draft version of the human genome sequence, Craig Venter (left) and Francis Collins (right) have continued to play major leadership roles in the field of human genomics. (Photo by Alex Wong/Newsmakers, February 12, 2001. © Getty Images, reproduced with permission.)

and brilliance that the researchers brought to the project, but it is also the result of technical advances that let the project move more rapidly as it progressed.

Who Owns Your Genome Sequence?

The human genome sequence actually exists in a set of databases around the world, located in the countries that produced the sequence. The sequence is curated by places such as GenBank at the National Center for Biotechnology Information (NCBI), the University of California at Santa Cruz Genome Browser, and the Sanger Center in the European Union. A variety of important sites worldwide contain programs for genomic analysis and databases of information derived from the sequence, including the sites that house the sequence and other locations such as GeneCards at the Weizmann Center in Israel and GenomeNet in Japan.

While these genome centers own the computers and the software, they do not legally own the sequence of the human genome in the sort of sense we normally consider for inventions or something that someone has authored. In the course of this race, major questions were raised about just who does own the sequence of the genes locked up inside every cell in our bodies.

To consider the question of who owns the sequence, or holds the patent on the sequence and its use, we have to consider what makes something patentable? In the United States we can patent something if it is useful, if it is something new and not previously known, if it is more than an obvious advance on the previous state of knowledge, and if it is described in adequate detail that someone with appropriate expertise could use it for the intended purpose. Traditionally, naturally occurring versions of things are not patentable, so we cannot patent maple syrup or the nitrogen fixing bacteria that live in the root nodules of some kinds of plant. But if we come up with an invention that improves either of those items in some non-obvious way – inventing a fertilizer that causes maple trees to produce more syrup or that lets nitrogen-fixing bacteria live in the roots of plants that don't normally fix nitrogen – then we could patent the use of that fertilizer for the purpose of bringing about those improvements.

Questions of whether genes can be patented have been around for decades. The very idea of patenting human genes horrified many researchers, even as others were rushing out to file patents on things they had cloned and sequenced. Why should we care whether someone obtains a patent for something? In the US, the existence of a patent has benefits – it can induce pharmaceutical companies to invest money in development of a new drug – but also can be detrimental – limiting access to treatments or diagnostics and in some cases even preventing further study of the gene by academic groups seeking to understand the disease process or develop new treatments or diagnostic tests.

After the US Patent Office has issued more than three million genome-based patent applications in the US, and issued patents on as many as 20% of the human genes, a Federal Court has now overthrown patents held by Myriad Genetics on the BRCA1 and BRCA2 breast cancer genes. In 2009, the American Civil Liberties Union (ACLU), representing a group of patients plus organizations representing about 150,000 scientists and health professionals, brought this legal challenge to the BRCA1 and BRCA2 patents held by Myriad Genetics. The suit claimed that the 15 Myriad patents on these genes were not valid because these genes are products of nature (non-patentable under US patent law 35 U.S.C. 101), the patents violate the First Amendment because of the prevention of the free transfer of information; and the patents grant a monopoly. The American Civil Liberties Union claimed that Myriad's patents are illegal, but in the past the US patent office has issued many comparable gene patents. When the Federal Court struck down the Myriad patents on BRCA1 and BRCA2, the judge said that the gene sequences used in the patents do not differ noticeably from the naturally occurring forms of the genes. Another recent legal ruling in favor of gene patents suggests that patents with an emphasis on the test, rather than the gene, may still find a home within US patent law. Meanwhile, there are still two additional rounds of judicial review above the level of the Federal Court that ruled on this case, so we expect that there will be several more rounds of legal battle on this issue after this edition of our book comes out.

The Myriad BRCA1 and BRCA2 case is likely to set an important legal precedent and impact thousands of patents that have been issued by the US patent office. Clearly the system currently walks an uncomfortable line between protections that allow product development and restrictions that block access to and use of products and information. As long as further exploration of patented genes within the non-profit context of academic research is allowed, we can hope that the feared stifling of ideas and developments will not arise from letting a company patent the use of part of your genetic information.

12.2 THE HUMAN GENOME SEQUENCE

It's a history book – a narrative of the journey of our species through time. It's a shop manual, with an incredibly detailed blueprint for building every human cell. And it's a transformative textbook of medicine, with insights that will give health care providers immense new powers to treat, prevent, and cure disease. —**Francis S. Collins, Director, National Institutes of Health**

What is the human genome sequence? Simply put, it is the order in which the letters of the genetic alphabet, As, Cs, Gs, and Ts, are arranged along the chromosomal DNA strands that are tens, even hundreds of millions of letters in length. The whole length of the sequence, from one end to the other, measures more than three billion As, Cs, Gs, and Ts arranged one after another to spell out what we are made of and how our cells operate. It starts at the top of chromosome 1 and runs down each chromosomal arm in the order in which the chromosomes were originally numbered according to their apparent sizes.

If we look at the Entrez Gene website we find a listed human genome sequence length of 3,101,788,170 bases, but we are going to start by telling you that is not how long your genome is. Or Julia's. Or Scott's. There is in fact no single number that qualifies as *the* length of the human genome. With the tens of thousands of simple sequence repeats and the many copy number variants, the length of the genome is different in every human being. In fact, with small differences in length between any two copies of a chromosome located in an individual, and with variability of the lengths of telomeres over the course of aging, we cannot even specify one exact length of the genome for one individual.

What is the reference sequence? It is not the sequence of any one human being, although some individual sequences have been done, starting with the sequences of Craig Venter and James Watson. Rather, the reference sequence is pooled data on multiple different individuals. In trying to talk about particular points in the human genome sequence, we usually do not need to agree that a particular base is exactly 25,849,578 base pairs from the tip of chromosome 3 in every person; we just need to all agree that we are going to say that that is the coordinate of that particular base pair so that we know we are talking about the same base when we discuss it. We need one more piece of information to be sure that we are communicating correctly: we need to know which version (or Build as it is called) of the sequence we are looking at.

Is the sequence truly complete? No, actually it is not. There are still some gaps but these gaps contain highly repetitive sequences that current technology is having a hard time getting across. For instance, current technology does not do a good job of cloning and sequencing across the highly repetitive material of the centromeres. Do we think anything of interest is lost in those gaps, or do we think the gaps just contain "junk"? In fact, there is evidence on the X chromosome that part of one of the genes may be lost inside of one of these gaps, but we are able to tell it is there by looking at comparable regions of the X chromosome from closely related organisms. A lot of functionally important information has been found, but might there still be a thing or two to be learned once those gaps are closed? Quite likely, but the current estimate is that the current sequence gives us a large fraction of the information we need to get out of the sequence.

How Big Are the Chromosomes?

One of the revelations that came out of the sequence is that those who named the chromosomes based on their size did not get the order entirely right (Figure 12.4). Yes, chromosome 1, with more than 240 million base pairs of sequence, is still the largest chromosome.

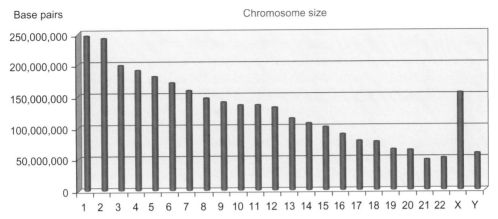

FIGURE 12.4   Sizes of the human chromosomes based on the sequence.

No, chromosome 22, with almost 50 million base pairs of sequence, is not the smallest chromosome. Instead, chromosome 21, at almost 47 million base pairs of sequence, wins that honor. We find ourselves wondering whether the very small size of this chromosome, and the small number of genes located on it, helps to explain why this is the one autosome for which trisomy is truly viable.

Comparison of different kinds of information reveals a variety of interesting things about the human genome. For instance, the shortest chromosomal arms undergo recombination at about twice the rate (relative to their length) found for the long chromosomal arms, something that probably assists in ensuring that at least some recombination takes place on these short chromosomal arms to help hold chromosomes together during meiosis. On the other hand, recombination takes place at a much lower rate around the centromeres. Although these ideas have been around and did not emerge with the completed sequence, having the whole sequence has allowed a much more complete overview of where the major differences in recombination rates fall.

The completion of the sequence has allowed development of technologies for screening the genome for copy number variants, such as regional deletions. This type of screening is facilitating studies of the involvement of local and large regional differences from the reference sequence, such as regional deletions found in some cancers. It is also allowing detailed studies of the changes in genome architecture that take place in some tumor cells that develop increasingly unstable genome structures as they become metastatic and the disease progresses. This approach has been helpful in identifying genes that play a role in sarcomas, meningiomas, pheochromoctyomas, and 1p36 deletion syndrome. This ability to screen for missing sequences also facilitates studies of forms of mental retardation that result from large regional deletions.

The X and Y Chromosomes

The Y chromosome is one of the smallest chromosomes. With a sequence of almost 58 million base pairs in length it is less than half the size of the 155 million base X chromosome with which it pairs. The X chromosome has almost 3000 genes, and the Y chromosome has a little more than 700 genes. By comparison of the X and Y chromosome sequences to the sequences of those same chromosomes in a variety of other organisms researchers have been able to paint an evolutionary history that shows that many sequences present on the ancestral chromosome were gradually lost from the Y along the way to result in a chromosome that is very different from the X, but that still contains some genes that are present on both the X and the Y (see Chapter 7).

According to the paper that reported the completed sequence of the X chromosome, the X chromosome contains about 4% of the genes in the genome, but almost 10% of Mendelian diseases for which the gene location is known turn out to result from defects in genes on the X chromosome. According to this same paper, there are 113 different genes on the X for which mutations have been shown to cause hereditary diseases, with there being 168 different phenotypes associated with those genes.

While it might be tempting to ask why there is an excess of disease causing genes on the X, the real question is: is there in fact an excess of such genes? Consider this: it is much easier to recognize X-linked inheritance, which can often be recognized simply by looking at the mode of inheritance in the family, than it is to recognize the locations of autosomal genes, each of which can be mapped only through detailed experimental studies. So we have to consider an alternative hypothesis: that a higher fraction of the real disease genes have been recognized on the X, while many diseases caused by autosomal genes have yet to be localized and identified.

The Mitochondrial Chromosome

The sequence of the mitochondrial chromosome was actually completed in 1981, long before the rest of the genome, with some revisions added in 1999. So determination of the mitochondrial chromosome sequence was not part of the human genome project, and yet this chromosome is part of the human genome and the 16,569 bases of sequence is part of the human genome sequence. As you will recall from Chapter 2, thousands of mitochondria,

organelles of roughly the size and shape of a bacterial cell, exist inside of the human cell and serve to produce energy for the cell's use. Each of these organelles has its own mitochondrial chromosome that is tiny compared to the human genome or even the smallest of the human chromosomes. The 37 genes of the mitochondrial chromosome produce some of the proteins used by the mitochondrial energy factory, but there are other genes in the nucleus of the cell that also encode mitochondrial proteins that make up the rest of the energy production machinery of the mitochondria. The proteins produced by the nuclear mitochondrial genes and the proteins produced by the mitochondrial genes work together to create energy in the course of turning complex molecules such as sugars into simple substances such as carbon dioxide and water. Because of the mitochondria's critical role in energy production, it is important that the mitochondria not be forgotten in any broad-scale view of human genetics and genomics.

A growing number of inherited diseases have shown mitochondrial inheritance and mutations in one or another of the genes on the mitochondrial chromosome. But we have to keep in mind that as the mitochondria are replicated with each cell division during the growth and healing processes of the organism, that mitochondria can accumulate mutations that were not inherited. The result is *heteroplasmy* – presence of more than one mitochondrial sequence simultaneously in the same cell. Mitochondrial heteroplasmy has been found to be associated with a variety of late onset diseases such as Alzheimer disease and Parkinson disease.

12.3 THE OTHER GENOME PROJECTS

The human organism is not the only organism for which we know the genome sequence. The National Center for Biotechnology Information houses data for more than 3000 different genomes from eukaryotic organisms (remember them – the ones with nuclei?), more than 3000 different bacteria, 100 archaea, and 3000 viruses. By 2010 NCBI was listing more than 800 completed bacterial and archaeal genome sequences out of almost 2000 genome projects in progress; these projects include the genomes of important pathogens, such as ones that cause cholera and meningitis. As more bacterial genomes are being completed, additional work is going into sequencing of different strains with important phenotypes, with significant findings on the "pathosphere" emerging from comparisons of strains that cause different diseases or symptoms. While work progressed on bacteria and the even smaller genomes of some important viruses, the list of more complex organisms being sequenced has grown.

Out of 796 eukaryotic genome sequencing projects completed by the year 2010, we find that 38 are complete, 333 are being assembled, and 425 are still being sequenced. Out of 81 mammalian projects, two have been completed – chimpanzee and pig – but many others, including human, are in final assembly stages, letting us draw some conclusions when we try to draw comparisons across species but leaving us with some unanswered questions. Some of these projects target animals that are important for agriculture, such as the cow and sheep, while others such as the dog and cat are important to breeders' interests in developing and maintaining purebred animals and to veterinarians interested in curing diseases affecting our pets.

A variety of other organisms such as fungi are also targets for genome sequencing. Not only are some of these fungi important model systems for the investigation of fundamental biological questions, but some of them are also commercially important, such as yeast used in brewing and bread-making. Another 74 projects targeting parasites include organisms that cause sleeping sickness, diarrhea, and leishmaniasis.

Out of more than 100 plant genome projects, six have been completed, including important agricultural organisms such as rice. Plant genome projects have been driven by agricultural needs to improve the nutritional composition of grains and the ability of plants to survive pests and disease. At the same time, advances in our ability to manipulate genomes have led to debates over genetically modified foods.

Of special interest is a cross-over between two of these projects. One project is aimed at sequencing the genome of the malaria parasite, and the other is aimed at sequencing the genome of the mosquito that houses the parasite during part of its life-cycle. Because malaria is endemic in many large regions of the globe and causes so much morbidity and mortality, important advances are expected to come out of this combined host–parasite project.

How Do Animal Genome Projects Inform Human Studies?

Several important things arise from the availability of so many genome sequences. Sequence similarities and clustering of genetic groupings in related organisms allow scientists to rapidly use findings from one species as a basis for studies in other species. The result is that a gene found in a fruit fly, studied cheaply in large numbers of lightning-fast experiments, can lead to an understanding of something important in the mouse that begins the development of pharmaceutical products destined eventually for human use. Those same homologies have helped with identification of some human genes not found by other methods. We can compare findings for different organisms to study evolution in a lot more detail than was possible previously. And animal models with fully characterized genomes serve as the first stage of testing of therapeutics destined for clinical trials in humans.

Thus, although many of the ongoing plant and animal genome projects are justified on some levels by direct applications of the information to that particular organism, in many cases the most important findings for any one organism will be those that impact our understanding of things going on in other organisms, too. This is especially true for the animal model systems used in research, including the mouse, the fruit fly, the worm, the slime mold, and the zebra fish, which allow for very powerful genetic and biochemical studies to develop materials and information that can then be applied to studies of other critters, including we humans.

But these model systems get us something else incredibly important, beyond the "applied" situations such as therapeutic testing. They gain us the ability to ask questions about the most fundamental biological processes, questions that we often cannot ask with human as our "experimental animal." Look at the Human Genome Timeline in Box 12.1 and you will see breakthrough after breakthrough that was done using a model organism, not a human. We did not first find that DNA was the genetic material by looking at people; those studies came from studies of bacteria and bacterial viruses. We did not first observe linkage in humans; it came out of work on fruit flies. Telomeres were discovered in a unicellular organism called *Tetrahymena*. Cloning was invented using bacterial systems. RNA silencing was discovered in worms. The ability to knock out genes or introduce mutations into the genome was developed in mice and non-human organisms. The use of green fluorescent protein to label specific cell types in studies of development arose from studies of jelly fish. The discovery that viruses play a role in causing cancer arose from studies in chickens. Important advances in our understanding of programmed cell death have come from studies of creatures such as slime molds. Cell division cycle genes responsible for critical aspects of cell division were first identified in yeast. The list of fundamental discoveries that have come out of non-human studies is impressive in its scope and the results of such studies impact all areas of biological research and

human medicine. Once these, and many other, advances were made using model organisms, it was possible to translate those findings into human studies and applications. As researchers continue working with such model organisms, the existence of genome sequences for those organisms provides them with critical tools for their work.

Comparative Genomics

When we do comparative genomics, comparing genome-wide information between different species, we start to get an overview of the relationships between different species. Some of the questions we want to ask are complicated questions about the level of sequence similarity between two organisms, a question that gets harder to ask for organisms that are less like each other. But other questions are turning out to be pretty straightforward – which organisms have the smallest genomes? Some of the viruses. Which organisms have the largest genomes? The jury is still out on that one but it may not turn out to be the human genome (see Figure 12.5).

A long string of genetic letters sounds pretty boring, so we have to ask: did they find anything interesting along with this boring pot of alphabet soup? In fact, the result of the genome project has been a delicious series of surprises, and we expect new revelations to continue for years as people continue mining information from the sequence.

12.4 THE GENES IN THE HUMAN GENOME

How Many Genes Are There?

At the turn of the millennium, as the human genome sequence was pouring out of the automated sequencers at an amazing pace, one of the first questions on everyone's mind was,

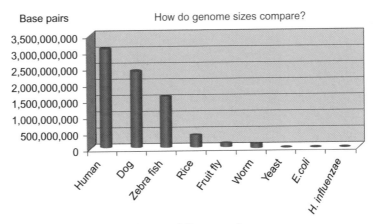

FIGURE 12.5 Comparison of the genome size for nine different species.

"Just how many genes are there?" The scientific community was expecting to find something like 50,000 genes, 100,000 genes, or even more, since this was the best estimate available. So everyone was surprised at the answer that came out of the analysis of the sequence.

The current estimate for the number of genes in the human genome, which you might have heard on the news, is often listed as being a little over 20,000. But hold on a minute. Twenty thousand genes? What are we tallying when we count these genes? In earlier chapters we have told you about a lot of different kinds of RNAs: messenger RNA, tRNA, rRNA, microRNAs, silencing RNAs, and a variety of other categories of non-coding RNAs. So it is important to be careful when someone names the number of genes in an organism to ask, "the number of what kind of genes?"

It turns out that the number "a little over 20,000" refers to messenger RNAs that encode proteins and that have been validated as real genes. If we consider all the categories of genes, and not just the messenger RNAs, the number of human transcribed genes is over 30,000. It is likely that more non-coding genes will be identified, and that some more of the potential messenger RNAs will be validated but we do not expect the number of genes to change enough

to get back into the range of those earlier estimates of over 100,000 genes or more.

The findings brought up other questions: can 30,000 genes really produce all of the elegant complexity of a human being? For those thinking in more concrete terms, can 30,000 genes even produce as many different proteins as there appear to be in a human cell? If you think back to Chapter 4, you may recall that one gene can make more than one final mRNA (and thus more than one final protein product) through alternative splicing, and that the titin gene alone makes thousands of different splice variants. So one of the revelations arising out of the human genome project is that some of the complexity turns out to be due to large amounts of alternative splicing. Thus this small number of genes can account for a much larger number of proteins produced by many cell types and developmental stages of a human being.

But the low number of human genes in some ways seemed a bit of a blow to the egotistical humans of this world, since it is not even twice the number of genes found in a fruit fly (Figure 12.6)! Was the expectation that we would have more genes really just a matter of hubris? Consider the relative sizes of the genomes (Figure 12.5) and the relative number of genes

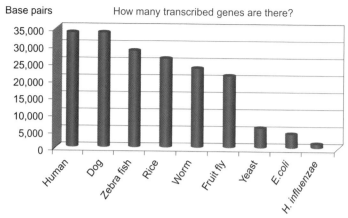

FIGURE 12.6   The number of transcribed genes does not show as big a difference between higher and lower organisms as we might expect based on our human-centric view of the world. From this chart it would appear that the big gap might be between single-celled and multi-cellular organisms, and not between smart, tool-using humans and our "lesser" brethren on this planet.

(Figure 12.4) in human and fruit fly. The fly has a gene about once in every 5528 base pairs of sequence, but the human genome has a gene about once in every 91,506 base pairs. In fact, if genes were spaced as densely in the human genome as they are in the fly there would be hundreds of thousands of genes, and if they were spaced as densely as they are in yeast there would be more than a million human genes!

Identifying the Genes

So how were these 20,000 (or 30,000) genes identified? It's not as if we can just look at a string of As, Cs, Gs, and Ts and say, "Look at this patch of sequence: it is a gene." However, there are a number of ways to tell where genes are located within a whole genome-worth of sequence. It is of course easy to recognize the genes that had been studied and were well known before the whole genome sequence was done. For many genes that had not previously been studied we find a new gene by discovering that a piece of sequence from a randomly sequenced cDNA copy of a transcript matches

a place in the whole genome sequence. We can compare the known sequence of our known gene or our randomly selected cDNA to the whole genome sequence, using sequence comparison software such as the BLAST program, which is used to search for sequences in GenBank Sequence Repository at the National Center for Biotechnology Information (NCBI).

Computer programs have been developed that can screen genomic DNA sequence for regions of sequence with properties similar to the properties of known transcripts, properties such as having an open reading frame flanked by probable splice sites, or having codon usage that is similar to the pattern of codon usage for that organism (using the commonly used codons more often than the rarely used codons). However, keep in mind that the programs that have been developed to search for genes are limited in their ability to detect genes and can't find everything that really is a gene, in part because the programs are developed based on genes we already know about. Thus they might miss new classes of genes that we do not yet know about. Also, it is much harder to predict non-coding genes, which we are discovering are much more common than we thought possible 10 years ago. Predicted genes all have to be tested experimentally to be validated before we can end up saying that they are genes.

Programs can also recognize sequences that are conserved across species, showing similar sequences between human and other primates, or in some cases even between human and fly or yeast. We commonly find that many genes show regions of conservation, at a higher level of conservation across species than exists in the surrounding non-gene sequences. Some genes like histones are highly conserved even if we look across widely unrelated species, while other genes are different enough from one species to the next that the comparison programs do not detect them based on sequence homology.

What Are We Missing?

Why do we think that we do not have all of the genes if we have sequenced the cDNA copies of the transcripts? Consider this: what if a gene is only expressed for a few hours at a critical point in development of the fetal kidney? Do we expect that transcript to be reflected in the cDNAs we have sequenced from adult heart, brain, skin, lung, etc.? How many different cell types would we have to look at, and in how many different cell types, to be sure we have seen all of the genes? What if a gene is only expressed under some rare circumstances, perhaps in response to some unusual environmental condition? Then it would be unlikely to be in our collection of sequenced transcripts. And what if a gene is only transcribed at very low levels because the cell does not need very much of the gene product? How many cDNA clones would we have to sequence from each tissue to be sure that we had not missed anything rare? And finally, what if there are classes of genes that do not resemble the known genes and cannot be identified by the programs that predict genes, and what if some of those genes

are not well-conserved across species? In all of these cases we expect that there would be some things still missing from our collection of genes.

Gene Deserts

We would offer one last caution. One of the curiosities that came out of the human genome sequence was the discovery of *gene deserts*, regions lacking transcribed sequences or anything that looks like a gene. Does the genome really contain large regions devoid of genes? Or are there genes with properties that differ enough from the known genes that the computer programs don't know how to find them? A gene desert in the fruit fly genome offers us a lesson relative to our concept that we know all of the genes or can tell how to recognize them. There is a mutation in this fruit fly gene desert that results in a fascinating phenotype that can be seen in one generation after another in company with the genotype. In the strictest early sense of a Mendelian gene, a region in the middle of this gene desert acts like a gene. So let us once more remind you of Scott's favorite rule: if it's a gene, it can be mapped, and if it can be mapped, it's a gene. We don't know quite what this region is doing or why the mutation causes a problem, but clearly there are things out there in the genome that are probably genes, and we apparently do not yet know how to recognize them.

Does this mean that we think we are still missing a lot of genes? No, at this point we think there remain some novel genes to be found, but we also think that we have a good general idea of the number of genes in the human genome. We suspect that there are indeed genes in the human genome that the computer programs don't yet know how to identify, and genes in the human genome not yet represented in the collection of sequences taken from transcripts because they have not looked at transcripts from every possible cell type, stage of development, or environmental condition. Maybe the number of genes we can't detect yet is small, but we can't really know just what we will eventually learn about the gene deserts. We also can't yet know how large the number of genes will end up being by the time we look at every different cell type at each stage of development and in response to a variety of conditions that alter gene expression. Certainly, we do not expect the number to be vastly different from the current estimate, but we also expect that the genes have not all been found yet. The number of human genes is currently a bit like the ocean with the tide lapping at the shore, with a core of information that is now a solid constant, surrounded by a fraction of changing information that comes and goes as additional experiments tell us more about things that we thought we already knew.

How Are the Genes Distributed?

If we look at the distribution of genes along the chromosomes, we find that the ratio of gene per chromosomal size is not equal. We also find that the proportion of mRNAs to non-coding RNAs is not the same from one chromosome to the next. If we look at the distribution of genes along chromosome 1 we see that genes are also

not equally distributed within a chromosome. For instance, chromosome 11, which is the eleventh largest chromosome, has the sixth largest number of genes in the human genome (Figure 12.7). Moreover for a given chromosome, we see that some areas are rich in genes and others are relatively gene-poor (Figure 12.8).

We can also see that both strands of each chromosome end up having many different genes transcribed. For one gene we select, one strand gets used as the template strand for making RNA, but for a neighboring gene, the other strand might be used as the template. Because of the polarity of the two strands (they point in opposite directions), genes that get read off of one strand read in the opposite direction from genes read off of the other strand. Thus a chromosomal region containing six genes might show a pattern of transcription similar to that shown in Figure 12.9.

Sometimes the situation can be more complicated. Take the case of the very large *NF1* gene that is responsible for a disease called *neurofibromatosis*. The NF1 gene covers about 350,000 base pairs on chromosome 17. It has 59 exons that become part of the 13,000-base mRNA produced by the gene. In one of the

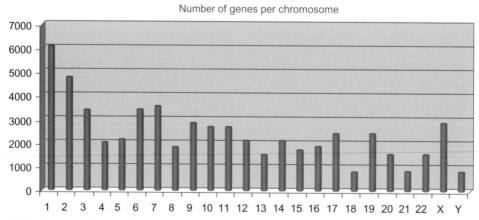

FIGURE 12.7   Number of transcribed genes per human chromosome, including both coding and non-coding genes.

Density of genes on human chromosome 1

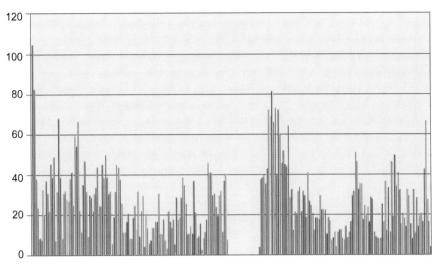

FIGURE 12.8   Genes are not equally distributed along the length of a chromosome.

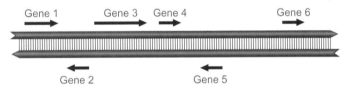

FIGURE 12.9   Transcription of several genes from one chromosomal region. Arrows mark genes, with genes 1, 3, 4, and 6 all being copied from one strand of the DNA going in one direction, and genes 2 and 5 being copied from the other strand of DNA going in the other direction. Notice that some genes are longer than others and that the amount of space between the genes is not always the same.

introns, if we look at the opposite strand, we find three small genes, OMGP, EVI2B, and EVI2A, that produce proteins that are not involved in causing neurofibromatosis. In other words, some of the genes are located within the introns of other genes!

Gene Families and Paralogy

We see that some families of genes (genes with related sequences and/or functions) are located in clusters in the same region of a chromosome. The long arm of chromosome 2 contains a cluster of at least a half-dozen developmentally important genes called HOX genes, transcription factors that regulate fundamental developmental steps such as formation of limbs. Additional clusters of HOX genes are found on other chromosomes.

Some members of gene families seem to be located near each other to allow for co-ordinated regulation of gene expression (Figure 12.10). A good example would be the two clusters of globin genes, each of which goes through an orderly developmental progression from expression of fetal versions of the protein to expression of the adult alpha- and beta-globin forms.

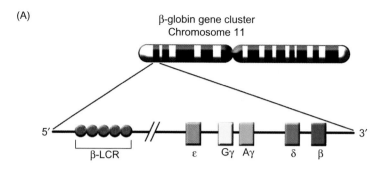

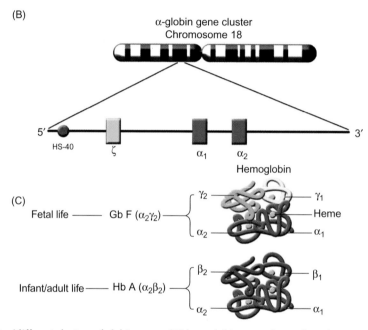

FIGURE 12.10   Two different clusters of globin genes: (A) beta-globin genes, located on chromosomes 11, and (B) alpha-globin genes, located on chromosome 16, each undergo co-ordinated developmental regulation of expression of genes within a cluster, changing which genes are expressed over the course of the lifespan. (C) The change in expression pattern results in a different form of hemoglobin in the adult than what is present in the fetus. Because the fetal genes are still present in the adult, a strategy for treatment of sickle cell anemia turns up the expression of a fetal globin gene in adults with two copies of the HbS allele. (Reproduced from P.S. Frenette and G.F. Atweh, "Sickle cell disease: old discoveries, new concepts, and future promise", *Journal of Clinical Investigation*, 2007;117(4):850-8, Fig. 2. © The American Society for Clinical Investigation.)

Other families of genes are scattered all over the genome and nowhere near each other. For instance, the red and green color opsin genes are located in one cluster on the X chromosome, but the blue opsin gene is on chromosome 7, and the rhodopsin gene is on chromosome 3. For many other categories of genes, such as channel genes, whose products transport ions, and transcription factor genes, there are many genes in the category and they are scattered all over the genome.

In some cases, we can see *paralogous* regions of the genome, which are regions that appear to contain members of the same gene family, the gene families arrange in the same order on each of the separate chromosomal regions. For instance, there is a region of chromosome 1 that has a cluster of genes that include LMX1A, PBX1, and RXRG. On chromosome 9 there is another region of related sequence that includes related genes that include LMX1B, PBX3, and RXRA. Another related region on chromosome 6 includes PBX2 and RXRB among the genes shared with one or the other of the first two regions. And if we look in the mouse we see syntenic regions that mirror this structure, showing some of these genes clustered in the same order.

The model for how these genes are related to each other says that all three regions are derived from one ancestral sequence lost far back in the mists of time, and that three different copies of that ancestral sequence ended up scattered to three locations within the genome where the different copies of the ancestral genes diverged as a result of mutation. In fact, if we look in the mouse sequence we see evidence of these same three diverged copies of the ancestral sequence, which suggests that the duplication and divergence started in some long-lost ancestor to both mouse and man!

For some gene families there are only a few copies dispersed around the genome, but some gene families include a large number of members; there are hundreds of copies of tRNA genes and pseudogenes derived from them. Altogether, more than half of human genes can be considered paralogues because they are members of gene families consisting of multiple members whose related sequences are hypothesized to be derived by duplication and divergence of an ancestral sequence.

What Is Out There Besides the Genes?

If we add up all of the genes and chromosomal structures we can account for, the amount of DNA they encompass falls far short of the total. Less than 2% per cent of the DNA is used up on the sequences that make it into the final spliced transcripts. Only about a quarter of the DNA is taken up with sequences that get spliced out of transcripts in the course of making mRNA. What might the other approximately 75% of the DNA be for? Some people talk about "junk" DNA doing nothing or at best serving as filler or spacer sequences. In some cases, they might be right, since it appears that sometimes all that is needed is to keep two points on a chromosome a precise distance apart from each other without it much mattering what the sequence is that fills up that space.

As mentioned in previous chapters, there are many other functionally important items out there besides the genes, including structures such as centromeres and telomeres. There are also evolutionary relics such as *pseudogenes* (nonfunctional copies of genes) and copies of viral genes left behind after a virus infection. Some of the non-gene sequence takes the form of repeated sequences covering more than 50% of the sequence (Figure 12.11). Some families of repeated sequences occur in huge numbers, with ~1.5 million short interspersed nuclear elements (SINES), 850,000 long interspersed nuclear elements (LINES), hundreds of thousands of microsatellite repeats, and more than 700,000 places where a transposition event has introduced a copy of a sequence into a new position in the genome.

While we might dismiss many of these sequences as "junk" in fact questions are being raised about whether some of these sequences might play functional roles in the cell. There are many other functions we have not accounted for in our discussions – regional control of gene clusters, replicating DNA, interaction with scaffolding proteins, propagation of X-inactivation along the X chromosome and more. For some of these functions, the DNA sequences and chromosomal locations of those sequences are known; for others, they are still being sought. It's not such an outrageous bet that some of the supposed "junk" will eventually account for some of the functions that are not yet accounted

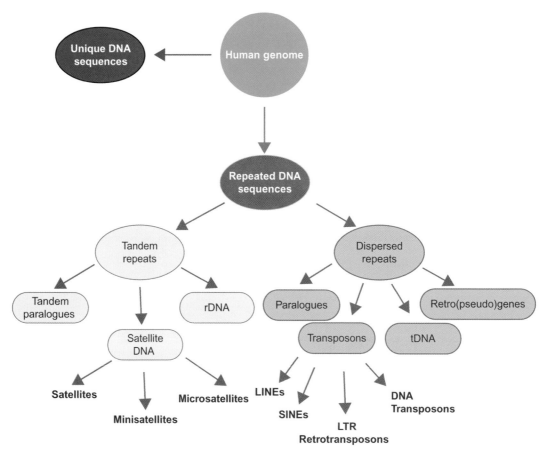

FIGURE 12.11   The human genome sequence consists of a combination of unique sequences and repeated sequences. Repeats include tandem repeats (the same sequence more than one time in a row adjoining each other) and dispersed repeats (the same or related sequences located at different locations in the genome). The number and arrangement of repeats in the current sequence is derived from ancestral sequences through a wide array of different mechanisms. (After Richard *et al.*, *Microbiology and Molecular Biology Reviews*, 2008;72:686–727, with permission from the American Society for Microbiology.)

for, but we bet that some of them will turn out to play functional roles in ways that we don't yet know enough to ask about.

12.5 HUMAN GENOME VARIATION

We Are All Amazingly Similar

The first complete genome sequence of an individual was that of Craig Venter (Table 12.1).

About 3.7 million places in his genome were identified as varying between the two copies of the genome housed in Dr Venter's cells. We also find that 99.5% of the sequence is identical between the two copies, meaning that 0.5% of the sequence varies between the two copies of his chromosomes. This is a higher level of variation than was expected from early human genomic studies. In both the Venter sequence and the James Watson sequence, the next completed whole genome sequence, we see hundreds of thousands of small

TABLE 12.1   Features of three versions of the human genome sequence

Genome sequence	Reference sequence	Craig Venter	James Watson
Year finished	2003	2007	2008
Cost	$2.7 billion	$100 million	<$1.5 million
Coverage	8–10 X	7.5 X	7.4 X
No. countries	6	3	1
No. SNPs		3.7 million	3.2 million
Time to finish	13 years	4 years	4.5 months
Sequencing approach	Traditional	Traditional	Next generation sequencing

insertions and deletions, many a few bases in length, in addition to the huge number of SNPs.

HapMap is a project aimed at identifying variability in the human population. Single nucleotide polymorphisms were determined from a set of 270 individuals from four populations, the Yoruba population in Nigeria, the Chinese population in Beijing, the Japanese population, and a population of European ancestry from the United States. As of 2007 the HapMap project has identified more than 3 million places in the human genome where the sequence is different in different members of the population, although not everyone is heterozygous at every one of those positions.

Although human blueprints are amazingly similar, there is more than enough diversity to allow SNPs to serve as very powerful tools for the investigation of many different questions. Panels of SNPs can be used to screen individuals with complex diseases to look for regions of the genome where small differences in the sequence turn out to be associated with the disease. As described in Chapter 10, SNPs can be used to evaluate loss of heterozygosity in different stages of tumors. SNPs are used in anthropology and archeology to look at the genetic relationships between ancient samples and current populations. SNPs are also used in tracing relationships of current populations to shared ancestral populations. SNPS are

gaining importance in forensics in the identification of remains. SNPs will also become increasingly prominent in diagnostic situations and in the arena of pharmacogenomics, where the dream of every doctor is a test that will identify which drugs will work best for any particular patient.

Variation Between Populations

For some alleles we see a lot of variation in allele frequency between different populations. Consider the frequency of lactose intolerance, which is prevalent in Western Europe and rare in Southeast Asia. Similarly, for some traits, such as blonde hair and blue eyes in northern Europeans, we see substantially different frequencies in different populations. When we consider differences in disease we see a broad array of examples. If we consider just the different types of glaucoma we find that primary open-angle glaucoma is most common in individuals of African Ancestry (at four times the rate of individuals of European or Asian ancestry), while angle-closure glaucoma is more common among Chinese and normal tension glaucoma is most common among Japanese.

How would such differences come about for alleles and traits that are present in all of these populations? Some of the mechanisms are *active*, such as natural and sexual selection processes that favor some alleles over others for making it into the next generation. Remember the Arctic foxes from Chapter 1, which underwent a substantial change in phenotype over the course of a few tens of generations when artificial selection was carried out, with the parents of each new generation coming from the animals closest to the final desired phenotype. For the sickle cell allele we can see how active selective forces influence the frequency of an allele that confers improved survival capability in areas where malaria is a huge problem. But for other traits such as hair color, how active could the selection processes be? Consider this: sun exposure assists the body in making vitamin D, and this process works more effectively in those individuals with the palest skin. So how much selective advantage would there be for pale-skinned populations in Scandinavia where low sunlight levels for part of the year could lead to vitamin D deficiency?

Other mechanisms are *neutral,* such as random genetic drift, a process akin to drawing beads from a barrel. If we start with eight different genes that each have exactly equal numbers of the two different alleles, and look at the slight variation in moving from one generation to the next we see small fluctuations. Some of these result from the fact that some individuals in a generation do not have kids, so they do not pass alleles to the next generation. Some individuals from that generation have more children than the others and contribute a larger fraction of the gene pool for the next generation. And the alleles they contribute are randomly selected, with a fresh draw of alleles for each meiosis. Some of those fluctuations move the frequency of an allele up slightly and some of those fluctuations move the allele frequency down slightly. If we do this many times over ten generations we will see that for many genes the random walk between values varies only a little, going up and down and keeping the allele frequency near the original allele frequency. But for some genes, such as gene 2 in Figure 12.12, the random walk between increased allele frequencies and decreased allele frequencies includes more decreases than increases. And for some genes, such as gene 8 in Figure 12.12, the random walk between closely adjoining values includes more increases than decreases. The fact that an

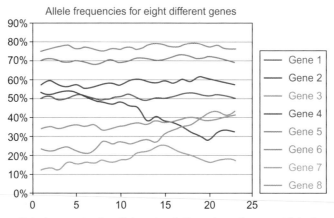

FIGURE 12.12   Drift in allele frequencies for alleles of eight hypothetical genes. Allele frequencies shift by very tiny amounts, some up, some down, to keep the allele frequency near what it was in the first generation. Random drift edging up and down keeps many alleles at about the same frequency for many generations in a row, while an occasional gene will end up with multiple small shifts all moving either up (Gene 8) or down (Gene 2).

allele frequency starts a trend in one direction does not mean that it has to keep trending in that direction (gene 7).

A third mechanism for ending up with a population with different allele or disease frequencies from the original population is migration. If a large population has 20 different alleles of a microsatellite repeat marker, and a small fraction of that population leaves to establish a separate population in a new location, the migrating group carries only a fraction of those 20 alleles along. So the new population might start out with only 12 of the alleles. Looking at microsatellites turns out to be a bad way to try to monitor this kind of thing because microsatellites have a high mutation rate and so over time the number of alleles in the new population could potentially change fairly rapidly. If we look at SNPs instead we are faced with the fact that each SNP tends to just have two alleles so it is hard to tell very much when we try to compare two populations.

However, information about SNPs that sit near each other on the same chromosome can be combined to create a kind of local genetic fingerprint, called a haplotype, that is characteristic of a particular section of chromosome in any one human being. By comparing haplotypes we can look at where the individual differences are, and we can figure out which of the allelic variants are on the same chromosome together (Figure 12.13). So even though any one SNP might be fairly uninformative, the overall fingerprint of the haplotype for a region is highly informative, and there are enough different haplotypes in a population to let us tell a lot about different populations.

A limited number of different haplotypes trace back through long tracks of history to shared ancestry in the far past. However, because recombination events have exchanged pieces of DNA between chromosomes during meiosis, we see that Person A and Person B may share the same haplotype for a spot at the end of chromosome 1 but have different haplotypes for a spot 30 million base pairs further down the chromosome. Meanwhile, at the 30 million base mark Person B shares the same haplotype with Person C. By studying these blocks of SNPs that have traveled together through time, scientists are developing tools to assist in mapping of genes involved in complex disease, while also finding ways to look back through time at genetic events that may have happened thousands of years ago.

One of the things we see from such studies is that African populations tend to have more different haplotypes in any given region than other populations. This is expected. All of our ancestries trace back into Africa, and subpopulations that left there moved out into Europe, the Middle East, East Asia, and then the Americas. When one population migrated away from a larger original population, the migrating group carried along only a fraction of the haplotypes present in the original population. So the population that migrated away from the founder population started out with a reduced number of haplotypes and, while an occasional new haplotype might be created through mutation, there is no expectation that new haplotypes will be created in the population the left at a higher rate than the rate at which new haplotypes are being made in the original population. We expect to see fewer

Haplotype 1 **T T C A G C C T C A T T T A C G**
Haplotype 2 **T T C A C C G T C A C T T A C G**
Haplotype 3 C **T C A G C G T C A C T T A C G**
Haplotype 4 C **T C A G C G T C A C T T A C G**
Haplotype 5 C **T C A C C G T C A T T T A C** A

FIGURE 12.13 Five different haplotypes from the same region of chromosome. Some bases are conserved between the different haplotypes (black) and some bases are different as a result of a mutation in an ancestor. Notice that haplotype 2 is derived from haplotype 1 through two changes that introduce a G at base 7 and a C at base 11. Haplotype 3 seems to be derived from haplotype 2 through a single change of the first base to a C. Haplotypes 4 and 5 have each been derived from haplotype 3, each arising because of a single base mutation.

haplotypes in "younger" populations in Europe where the population is derived from a small founder population that carried with it only a subset of the total available haplotypes present in the population from which they originated (Figure 12.14).

12.6 GENOME-WIDE TECHNOLOGIES

During the early years of molecular biology, we could only look at one gene at a time, and even looking at that one gene took a lot of time and resources. So as we came to know about larger and larger numbers of genes, the available technologies limited our ability to look at all of the genes in a single experiment. As the Human Genome Project has proceeded, a variety of genome-wide technologies have emerged. Some have come directly out of the project itself. Others have been developed by those who saw ways to make use of the information coming out of the Human Genome Project. As the Human Genome Project provided increasing numbers of markers, there was

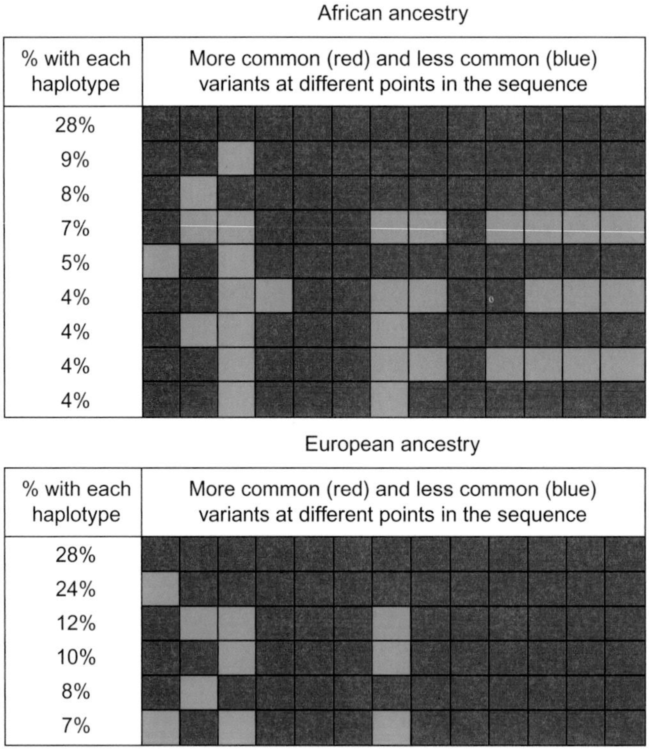

FIGURE 12.14 Haplotypes in two different populations. Each line represents a different haplotype showing a genetic fingerprint in the region of one gene, the LMX1B gene, with the same haplotype being the most common in both populations. Percentages show what fraction of the population has that haplotype. A larger number of haplotypes is thought to indicate a population that was founded farther back in time than the population with fewer haplotypes. These are not the only haplotypes present in these populations but these are the common haplotypes. Blue squares mark one base of difference in the sequence, with the whole haplotype spanning thousands of base pairs of sequence. (Courtesy of Goncalo Abecasis, University of Michigan, and Abigail Woodroffe.)

increased need for technologies allowing high throughput such as:

- Design of PCR to multiplex – simultaneously analyze a dozen or more markers in one reaction.
- Multi-well plates containing rows of 96 wells, 384 wells or 1536 wells arranged in rows.
- Multi-channel pipettors for handling whole rows or plates at once instead of one sample at a time.
- Programmable robotics that allowed rapid handling of the large numbers of samples in such plates.
- Programmable instruments that can continue loading samples and analyzing them all night.
- Instruments that could read sequence and other test results directly into a computer.
- Chip-based technologies for assaying expression of a whole genome worth of genes.
- Automated sequencing systems and Next Generation Sequencing.
- Bioinformatics and improved database technologies.
- Comparative genomics making use of information and resources from multiple genomes.

Now that we have all of these technological advances and genome-wide arrays of information, what are some of the critical questions we can ask that we could not ask before the Human Genome Project?

12.7 GENOME-WIDE ASSOCIATION

In the last chapter we talked about linkage mapping using genetic markers and human families with the trait of interest. Genome scans often screened hundreds of markers in the course of looking for a region of the genome linked to the trait in family studies. But other genome-wide strategies are available that make use of populations of unrelated individuals instead of families.

Association

What is a test for association? In the simplest sense it is a test for whether two things occur together more frequently than we would expect by chance alone. In the case of genetics we go about testing for association by asking whether a population with the trait of interest has a particular genotype more frequently than a population that lacks that trait. When we hypothesize that a particular allele of an SNP called SNP1 is associated with trait A, we can look at how frequently SNP1 is found in a population with trait A as compared to a normal population that lacks trait A. In the example in Figure 12.15 we find that 60% of the population with trait A has SNP1 but only 20% of the normal population has SNP1. If we had randomly selected 100 individuals from the normal population (with its 20% SNP1 frequency), we think that there is less than a 1 in 10,000 chance that we would have gotten 60 individuals with SNP1. We express this by saying that we find a p value less than 0.0001. Usually we want to do a more complicated test that takes into account the three different genotypic combinations: individuals who have one copy of SNP1,

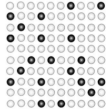

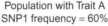
Population with Trait A   Normal population
SNP1 frequency = 60%   SNP1 frequency = 20%

FIGURE 12.15   When we look at the frequency of SNP1 in different populations we find that it is associated with Trait A, that is that it occurs much more frequently (60%) in that population than it does in a normal population that lacks Trait A (20%). Notice that there are people in the normal population who have the associated genotype but do not have the trait. That is because this genotype contributes to the trait but is not by itself capable of causing the trait without other contributing factors.

individuals who have two copies of SNP1, and individuals in whom both copies are the normal allele. We also want to be able to take into account things like the penetrance of the trait – the extent to which the genotype results in the trait. This is how we look at things if we have one candidate gene and we want to evaluate its association with a particular trait.

Genome-wide Association

For many complex traits, however, we do not yet know which gene(s) to test as candidate genes. So we end up wanting to search the whole genome to determine which genes are associated with the trait. A common genome-wide association study might use a half million different SNPs (or more!). We can analyze the data two different ways. We can do an association test for every separate SNP, testing a half million SNPs or more. Or we can arrange the SNPs into haplotypes and then test regional haplotypes for association. In many such studies of common complex traits, the result is not the identification of a causative gene but rather the identification of multiple genes or regions of the genome that are present in the individuals with the trait more often than in individuals without the trait. Additional experiments are then needed to determine what kind of role each of those genes plays in the trait, and it is quite common to find that only some of the genes identified in an initial association finding turn out to be reproducibly associated.

We cannot offer you some absolute statistical indicator that says, "yes, this gene is involved." For linkage we have our LOD score of 3.0 as a standard for whole genome linkage, and for many simple experiments that test for association between two simple variables we can point to p values less than 0.05 or 0.01 as expected cutoffs for something to be considered significant.

Why do we not have a simple indicator of which items in a GWAS are associated and which are not? Let's start by considering one of the biggest issues, the problem of how to interpret the results of many different tests in one experiment (multi-testing), and the problem of how to cope with variables that are not completely independent of each other.

Multi-testing

One of the biggest problems we face with a genome-wide association study is the multi-testing problem. Consider this. If you flip a coin and it turns out to show heads we don't think there is anything particularly remarkable about it. If you flip a coin five times and they all come up heads you might think you were pretty lucky. If you flip that coin 50 times and it always comes up heads you are probably going to think it did not happen by chance and that there must be some other explanation, such as the coin having heads on both sides. But if you flip the coin 50 times and somewhere in the set of flips you get five heads in a row you are unlikely to find that remarkable. You would likely think it due to chance and you would not be likely to find yourself looking for evidence of something that influences which way the coin lands.

Similarly with testing genetic markers, we expect that if we test too many markers some of them will show data that make it look like they are associated when they are not. That is to say the numbers will have occurred by random chance, because we did a lot of tests, and not because the two items are actually associated. What we are looking for is an event that happens often enough that we think it could not have happened by chance.

There are standard statistical methods for compensating for multi-testing and telling whether an event is likely to be significant amidst a large number of tests. Such methods do not work well for data of the kind produced in a GWAS. Why? Because those methods assume that we are looking at many separate events where the outcome of one event does not depend on the outcome of any of the other events. When we flip a coin we do not expect that the outcome of that coin flip will be affected by whether the previous coin flip was a head or a tail.

But when we test whether a particular SNP allele is present in an individual, the likelihood of a particular allele being present is not independent of the likelihood of a particular allele being present at the SNP next to it. Why? Because they are physically located on the same piece of DNA and the chance that those two alleles would separate from each other during meiosis is proportional to the distance between them (and the probability that a recombination event would fall between them). Because there is a limited number of haplotypes in the population, there are limited numbers of haplotype combinations for the alleles in any given region. So we know that if we get a particular allele at one SNP, we can predict the probabilities of the adjoining alleles based on the fraction of haplotypes in which those SNPs are found together. Thus the chances for particular alleles at two adjoining SNPs are not unrelated and we consider them to be dependent variables, not independent variables. However, we cannot treat our data set as if it contained only fully dependent variables because some of the markers are far enough apart that they do segregate independently, and some are just far enough apart that they separate sometimes but not randomly.

The multi-testing issue is huge when we do a half-million tests, so we need some way to deal with it. One approach is to do a simulation of the experiment in a computer to generate an empirical p value. We take the frequency of genotypes, or haplotypes, in the population and sample from that population by pulling individuals at random and then asking for their genotype. And then we do it again. And again. If we repeat that test by running this simulation over and over within a computer, we can get a good idea of how often we could start with a population with the same characteristics as

our starting population, and yet somehow end up getting the data we actually observed. The empirical p value simply tells us how many times we expect to have to draw those 100 individuals from a population with 20% allele frequency before we would expect to get a batch of 100 individuals in which 60 of them have the genotype that we hypothesize might be associated. If we run 10,000 simulations of the experiment, and we get 10 individuals with SNP1 once in every hundred runs, then we would say that there is a 1 in 1000 chance that the observed result might have happened by chance and not because there is a real association. But if we do 10,000 runs and never see a value as high as 60% then we think that there is less than a 1 in 10,000 chance that the data we observed happened by chance alone.

Two-stage GWAS

One of the ways we can test the validity of a GWAS finding is by doing a two-step experiment. If we identify an association between an allele called SNP1 and trait A in an initial GWAS study, we can take samples from a second population of people with trait A and test them for the frequency of SNP1. If the original finding was valid, we can usually expect that we will also see an association between SNP1 and trait A in that second population. Even if the allele and the trait are associated in the first place there are some situations in which we might fail to see the confirmation in the second population. For instance, SNP1 in the first population was the result of a mutation that happened after migration separated the two populations.

Another situation in which we might see an association in one population but not another would happen when one of the two populations lives in an environment that lends a selective advantage that would lead to an increase in the frequency of SNP1. An example of an association that we would see in a population from equatorial Africa but not in a population from Finland would be the sickle cell allele that confers malaria resistance. However, if we recruit a study population from Ghana, and then recruit a second population for confirmation from the same region of Ghana, we would have a much higher expectation that our observation from the first population would be confirmed in the second population.

Association and Causality

Another issue with association studies is the issue of causality. The fact that two things are associated does not mean that one of them caused the other. Both may be the result of some third item that we have not discovered. So if we observe that individuals with a particular kind of rash tend to have elevated levels of a particular cytokine, we might find that the elevated cytokine level and the rash are associated but we do not know that the elevated level of cytokines causes the rash; if we knew the full set of contributing factors to the trait we might discover that a virus causes the elevated cytokine level and also causes the rash with there being no direct causal relationship between the cytokine and the rash. In other cases, an associated item may be one of many factors that have to be present before the trait occurs, so that it contributes to the existence of the trait without being a simple direct cause of the trait.

Linkage Disequilibrium

In some cases a trait may be found along with a particular genotype because that genotype is close to the causative allele but is not itself causing the trait. We say that such a neighboring allele is in *linkage disequilibrium* with the causative genotype. What do we mean by linkage disequilibrium? It is the occurrence of the two alleles together at a higher rate than we would expect by chance. Although linkage disequilibrium does not always happen because of linkage, a very common scenario by which linkage disequilibrium can happen occurs when two alleles are located on the same chromosome, so close together that recombination events rarely or never fall between the two. Imagine this scenario. A chromosome exists that contains allele A. A second mutation occurs very close to allele A, a mutation that we will call allele B (Figure 12.16). This creates an individual who has two different alleles involved in two different traits that are right next to each other on the same chromosome. When this individual has children and grandchildren, the ones who receive allele B will almost always receive allele A. If we were to do a GWAS that includes allele A in the screening panel of SNPs but does not include allele B, we will find allele A present in the individuals with trait B at a frequency higher than expected by chance. This does not mean that allele A is causing trait B, but we have to consider two very different models for what might be happening:

> Model 1: allele A is associated with trait B because allele A causes trait B.
> Model 2: allele A is located right next to the causative allele and is in linkage disequilibrium with it.

In other cases a high-risk allele may contribute to the presence of the trait but only if other factors are also present. Thus if someone has several high-risk alleles for age-related macular degeneration, then adding in smoking and a diet low in lutein and xanthein could help tip the balance towards age-related macular degeneration; someone with the same high-risk alleles who has never smoked and who consumes a diet high in lutein and xanthein might be found in the normal control population even though both individuals have the same genetic risk factors. These might not be the only environmental risk factors, so someone who takes all the appropriate precautions regarding known risk factors may still end up among the cases eventually.

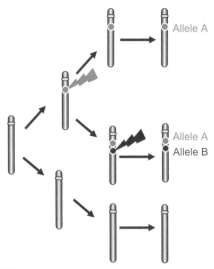

Allele A

Allele A
Allele B

FIGURE 12.16 If we do a GWAS looking for genes associated with trait B, we do not look at all of the SNPs in the genome. If allele A is in our screening panel, but allele B is not, the two alleles are so close to each other that we will find an association between trait B and allele A, but we will not find a simple one-to-one correlation of allele A with trait B. Some individuals with allele A will have trait B and some will not, but individuals with trait B will all have allele A. The situation in real life is more complicated because there will usually be multiple independent mutation events leading to trait B, with each of these mutation events happening at different points in history, with each occurring on a different genetic background, some with allele A and some without allele A.

Power to Detect an Association

How large a population is needed to tell whether a genotype and a trait are associated? Consider the populations in Figure 12.15. If we drew five people from the case population and three of them had SNP1, and we drew five people from the control population and one of them had SNP1, would we be able to use this information to conclude that the real frequencies in the two populations are different? No, our sample size is too small and we cannot conclude that a draw of 3/5 had to have come from a population with a different composition than

the population from which the 1/5 draw came. Then how many people would we need to sample before we could draw a conclusion regarding whether two allele frequencies are really different? It turns out that we can do a *power calculation*, a statistical test that will let us know how large a population we must look at to be able to detect a particular level of difference between the two populations.

To do a power calculation we need to know the frequency of the risk allele in the population from which our study population is drawn, and we need to consider what level of risk elevation we think is associated with the risk allele. Often we do not know what level of risk will turn out to be associated with the allele (called the genotype relative risk), so we test a range of genotype-relative risk values. Usually we want to identify a study population size that will allow us to detect association even if the level of risk increase is modest. If we are only interested in identifying association of alleles that render a high level of genotype-relative risk then we need a much smaller study population than we would need to identify association of an allele of modest genotype-relative risk. Clearly, the smaller the difference in frequency we are trying to detect, the larger the population that will be needed to be able to tell that the populations are actually different.

Genes and Environment

The statistical methods used in a GWAS let us take into account multiple variables relative to each study subject. This includes environmental factors such as smoking or alcohol use, and also the various medications that each subject is taking. In 2007 NIH started a $40 million dollar per year program to investigate a combination of genetic and environmental factors impacting complex human traits such as glaucoma. The NIH Genes, Environment and Health Initiative (GEI) supports GWAS with environmental information included. This program not only supports the primary GWAS but also supports other studies such as replication studies to confirm results and fine-mapping studies to better localize a gene after an initial GWAS finding. Genetics Association Information Network (GAIN) is a public–private partnership involving NIH, Pfizer Pharmaceuticals, and others that set out to use GWAS approaches to study six common traits of major public health importance. Traits that were investigated include schizophrenia, attention deficit hyperactivity, kidney disease in juvenile diabetes, psoriases, bipolar disorder, and major depression.

Completed GWAS

A large number of genome-wide association studies have now been done. The traits investigated in this way are often complex traits of great public health interest that were not solved by linkage studies. The results from such studies have been widely variable.

In the case of age-related macular degeneration, a relatively small study with 96 cases and 50 controls screened a little over a 100,000 SNPs to find association between an AMD allele and the CFH gene. They found that this CFH allele accounts for about 40–50% of AMD.

GWAS studies of diabetes have given results that are more like what is being found in many studies. By doing a meta-analysis (a form of analysis that can handle data derived from multiple different studies), the authors of this study were able to combine information on more than 40,000 cases and almost 100,000 controls. The result was the identification of a dozen new genes that increase risk of type II diabetes. Some of the genes that they found are involved in functions relevant to diabetes, including functions of the beta-cells of the pancreas, and affect insulin function. What is so notable here is that after finding so many genes using so many study subjects, less than 10% of diabetes is accounted for!

Strategies

As genetic studies shifted from linkage towards GWAS, we hear some people talking as if GWAS is a new technology replacing linkage in the same way that Next Generation Sequencing replaced traditional sequencing. We want to emphasize here that this is NOT the case. GWAS is a different strategy, and we do not use it because it is a new technical trick. We use it because it answers some kinds of questions that linkage cannot answer, just as linkage answers some kinds of questions that would not be effectively handled by GWAS (Table 12.2).

"Fourth Quadrant" Strategies

Other approaches are needed when we get to the fourth quadrant in Table 12.2 – the situation involving rare alleles where the trait is very rare and/or shows low penetrance. Some of these approaches will not work for all of the rare allele, low penetrance situations but they do allow us to get at some of the situations that are hard to get at with either linkage or a standard GWAS. One of these situations is seen when the deleterious genotype is not a common allele in a gene but rather is an array of different

TABLE 12.2   Choosing a mapping strategy

		TRAIT PENETRANCE	
		High	Low
ALLELE PREVALENCE	Common	Linkage or GWAS	GWAS
	Rare	Linkage	Other strategies

mutations located within the same gene. A standard GWAS will not work if no one allele is common enough to be represented in the panel of markers used for GWAS screening and most of the causative alleles are so rare that they do not occur in the panels of SNPs used in GWAS screening. We can try an allele-sharing approach or a whole genome sequencing approach.

12.8 ALLELE SHARING AND SIB PAIR ANALYSIS

The concept behind allele sharing is that we know the rate at which we expect siblings who share the same trait to share alleles. If we find that siblings share alleles in a particular gene at a higher rate than predicted by chance then we wonder if this gene might play a role in the trait. If the first sibling inherited allele 1 from mom, then we know that the next sibling has a 50% chance of also inheriting allele 1 from mom (Table 12.3). A similar chance of allele sharing happens for the alleles that come from dad. If

we see excess allele sharing we would think that this shared allele is near the gene of interest but we would not expect that this is telling us whether this allele is causative. It might just be telling us that this allele is in linkage disequilibrium with the causative allele.

If we look at a lot of pairs of siblings (sib pairs) who have the trait of interest and find that sib pairs share alleles near the myocilin locus more frequently than would be predicted by chance, there are two possible explanations for why they would share alleles so often. One possibility is that the allele is associated with the trait (because it is in linkage disequilibrium with the causative allele), and the other possibility is that there might be a lot of homozygous parents whose children would all receive this allele. This latter possibility happens a lot with SNPs where there are only two alleles available in the population. The Human Genome Project, and its studies of allele frequencies in human populations, helps us evaluate the possibility that there are a lot of homozygous parents in our study. In this case, the experiment

TABLE 12.3   Expected frequency of allele sharing between affected siblings

	Alleles sib 1	Alleles sib 2	No. alleles shared IBD[*]	Frequency
0 alleles shared	A1A3	A2A4	0	1/4
	A1A4	A2A3		
	A2A3	A1A4		
	A2A4	A1A3		
1 allele shared	A1A3	A2A4	1	1/2
	A1A4	A2A4		
	A2A3	A1A3		
	A1A4	A2A3		
2 alleles shared	A1A3	A2A4	2	1/4
	A1A4	A1A4		
	A2A3	A2A3		
	A1A3	A2A4		

[*]= identical by descent.

works quite well if we use microsatellite repeat markers that we know have a high probability of being heterozygous in the parents of the siblings. The Human Genome Project also offers us information on haplotypes so that we can test for association with a particular haplotype rather than testing for association with an individual allele.

Why Not Linkage or GWAS?

So if sib pair analysis will detect association, why was a GWAS not a good way to look at this problem for myocilin with its many different causative mutations? Because a GWAS looks at unrelated individuals, but the allele shared by sibling pair number 1 might not be the same allele shared by sibling pair number 2. Note that in Table 12.3 we can see a sibling pair sharing alleles A1A4, because the myocilin allele in that family arose on a genetic background containing allele 4. But in another sibling pair the shared alleles might be A2A3 because the myocilin allele in that family had arisen on a chromosome that contains allele 3. So if we looked in a population, we would see each of the four alleles represented at about the normal frequencies in the population because each new myocilin mutation arises on a chromosome without regard to what allele is present at the point in the sequence being used to test for association. But if we look in affected siblings, we might see that allele 1 is shared more often than expected based on the population frequency of allele 1, and allele 2 is shared more often than predicted based on the population frequency of allele 2. And so on.

Why would we do linkage in some situations, and look at allele sharing between siblings in another? Linkage analysis is very powerful under the right circumstances. But for linkage analysis we have to have families that are large enough to allow us to achieve a statistically significant result. But when a disease is complex or very late onset, although we may find some family history of the trait we do not expect to find large families and a simple Mendelian mode of inheritance. However, for many traits it is easy to find families in which at least two siblings have the trait.

Parametric vs. Non-parametric Analysis

So here is a key thought: when family structure allows for parametric linkage analysis (analysis that requires that we be able to specify parameters like mode of inheritance) such as traditional linkage analysis, it is more powerful than non-parametric analysis (which does not specify the parameters needed for a linkage study). But if we are wrong in the parameters we specify, if we are wrong about the mode of inheritance or rate of penetrance or other key variables, then we can fail badly in our attempt at gene mapping. The charm of non-parametric analysis is that it works even when our assumptions about mode of inheritance are wrong. It works even if we are unsure of the other parameters needed for linkage analysis. But – it has less power to detect the gene(s) we are after. And many of the problems we want to solve are so complex that even sib-pair analysis has limited ability to find our answers.

12.9 COPY NUMBER VARIATION AND GENE DOSAGE

Although much mutation screening has traditionally focused on the classical mutation categories such as missense and nonsense mutations present in coding sequence, a category of major importance that has emerged is copy number variation (CNV). The main mechanism by which CNVs impact phenotype is through alteration of gene dosage, which as we have discussed before should be two copies of each autosomal gene and two copies on the X chromosome in females and one copy on the X or Y chromosome in males. A CNV that adds or removes a copy of a gene is altering the normal dosage and is expected to affect the level of gene product. For some genes, such as those encoding enzymes, copy number variation may have little effect, while other genes such as transcription factors will more often be highly sensitive to copy number variation.

A simple obvious example of CNV is deletion of one copy of the gene, leaving the individual with one copy instead of two. We also see simple duplications of a small local region of the genome leaving the individual with an extra copy of one or more genes in this local region. In some cases the local rearrangements are more complex and may involve combinations of deletion, insertion and/or inversion.

More than 38,000 copy number variants longer than 100 base pairs in length have been observed using genome-wide scanning technologies (Figure 12.17). While the millions of single nucleotide polymorphisms (SNPs) outnumber the thousands of CNVs, in fact the CNVs actually cover a larger total expanse of genomic sequence than the SNPs.

The rate at which new CNVs arise appears to be orders of magnitude higher than the rate

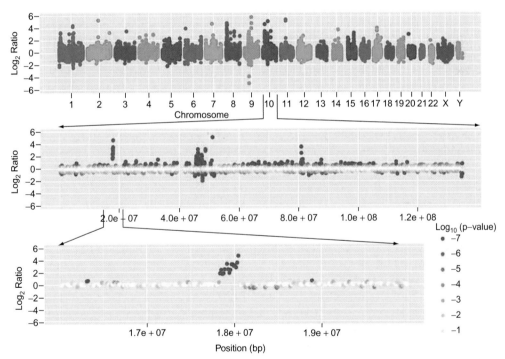

FIGURE 12.17 Diagram showing a comparison of the numbers of copies of sequences across the genome between the Venter whole genome sequence and the Watson whole genome sequence. The bottom panel shows a small region of chromosome 10 that includes a region in which one of the sequences has an extra copy as compared to the other sequence, as shown by the cluster of red dots above the line that represents the normal 2-copy level for the other genes in the surrounding region. In the middle panel, showing all of chromosome 10, we can see that there are at least three places along chromosome 10 that show this kind of CNV. The top panel shows all of the human chromosomes, with the information so compacted that it is hard to tell any details but it is clear that many but not all of the chromosomes have some regions of copy number variation between the two sequences. (Reproduced from Chao Xie and M.T. Tammi, *BMC Bioinformatics*; 2009;10:80, Fig. 5. © 2009 BioMed Central Ltd, by permission of Springer Science+Business Media.)

at which "traditional" mutations occur. While the rate of point mutations is distributed fairly evenly across the genome, some regions of the genome experience new CNVs at a higher rate than other regions. For example, the rate of new CNVs affecting the PMP22 gene on chromosome 17p11.2 is more frequent than the rate of new CNVs affecting the DMD gene on Xp21.2, and new CNVs affecting the DMD gene happen more often than new CNVs encompassing the LMX1B gene on 9q34.

For some genes and traits, the trait can result from any copy number error – increase or decrease. Rieger syndrome is a complex and variable trait that involves abnormalities of the iris and cornea, missing or malformed teeth, umbilical abnormalities that may include umbilical hernia, and glaucoma. Multiple different Rieger syndrome loci have been found, but one of them offers us an important example of what can happen with copy number variation. For the gene FOXC1 on chromosome 6p25, Rieger syndrome results whether one of the chromosomes carries a copy of FOXC1 that is deleted or duplicated!

In other cases, duplication and deletion of the same region result in different traits. James Lupski and colleagues have shown that rare CNVs involving 17p11.2, involving PMP22, have been associated with multiple different traits involving peripheral neuropathy, some through duplication, some through deletion, and some through more complicated rearrangements of the region. While many of these CNVs involve the transcribed gene, there are some additional CNVs that appear to affect only an adjoining region with implications that regulatory sequences may be involved.

When PMP22 is part of a tandem duplication where two adjacent copies are found, one right after the other, the phenotype that results is CMT1A, one of the forms of Charcot–Marie–Tooth neuropathy (CMT). There are at least 40 different loci in the human genome that can

cause CMT, and inheritance can be autosomal dominant, autosomal recessive, or X-linked. More than 100,000 people in the United States have some form of CMT. The form called CMT1A is autosomal dominant. It starts in children and teenagers and involves slow nerve conduction velocity. It affects both motor function and sensation in the hands and/or feet, but does not affect the central nervous system. Duplications of PMP22 can cause CMT1A, and missense mutations in the same gene have been reported to cause CMT1A of greater severity than those resulting from the segmental trisomy that contains one copy of PMP22 on one copy of chromosome 17 and the duplicated copy of PMP22 on the other copy. For the cases in which the segmental trisomy occurs, RNAi to reduce PMP22 expression levels is among therapeutic approaches being explored.

If PMP22 is deleted, the resulting phenotype is another form of neuropathy called hereditary neuropathy with pressure palsies (HNPP). This trait is characterized by the development of neuropathy in response to pressure. Effects may include weakness, atrophy or sensory loss. What is affected will depend on what part of the body is under pressure. A deletion of a specific 1.5 million base region of chromosome 17 (spanning PMP22) is found in about 80% of those who have HNPP while a missense mutation is found in the rest of the cases.

CNVs have been implicated in many different traits, many of them traits that can result from either a mutation in a particular gene or a dosage change in that same gene that was deleted or duplicated. Of great interest are observations involving CNVs and some of the major public health problems. In the case of Alzheimer disease, it can result from point mutations in the APP gene but researchers have also found duplications of APP that co-segregate with Alzheimer disease. It has been suggested that duplication of SNCA may be among a complex array of causes of Parkinson disease.

In studies of mental retardation, many different CNVs have been seen around the genome, and some patterns are starting to emerge such as multiple different CNVs that localize to one specific region of the X chromosome. In the case of autism, newly arisen CNVs have been implicated because they occur at a frequency higher than expected by chance, with some of them involving genes that had already previously been implicated in autism. CNVs have also been implicated in a broad array of other phenotypes including lupus, schizophrenia, and even susceptibility to HIV.

The CNVs in the human population are not yet all known, so the search for additional CNVs (and disease association of those CNVs) continues. Evaluation of copy number variants now makes use of the whole human genome sequence to scale the technology up to the genomic level. Thus it is now possible to screen the whole human genome in one experiment and end up with a fairly detailed picture of where the copy number differences occur (Figure 12.17). If we use one color of dye to label genomic DNA from a "reference" sequence and a different color of dye to label genomic DNA from the individual being tested, we can hybridize both to the same chip containing sequences from throughout the genome. Where our test sample has the same number of copies as the reference sample, we will see the same strength of signal from both dyes. If our test sample has a greater or lesser number of copies at one point in the genome, we will see more or less dye-labeled DNA hybridized to sequences derived from the duplicated or deleted region of the genome.

12.10 WHOLE GENOME SEQUENCING

We have arrived at what some call the post-GWAS era, where whole genome sequencing has become a tool for the study of individuals, families, and populations. In Table 12.1 we told you about the first whole genome sequences. The original reference sequence cost 2.7 billion dollars and took 13 years to complete. The Venter sequence cost 100 million dollars and took 4 years. The Watson sequence cost <1.5 million dollars and took 4.5 months to complete. As this book goes to press, the cost of a human whole genome sequence has dropped below $10,000, and the price is rapidly dropping towards that magical, long-awaited moment when we have available the Thousand Dollar Genome.

You might think that the Thousand Dollar Genome would mean that we would see a huge explosion of whole genome sequencing of large numbers of people. However, consider this – at a time when the price of the sequence is rapidly plummeting from 10,000 dollars down towards 1000 dollars, analyzing that sequence is still very expensive. So whole genome sequencing is now here but the rate at which it is used will increase gradually, as the cost of analysis falls and ease of use increases, rather than exploding onto the scientific scene overnight. Such vast quantities of information are generated that hardware needs more speed and capacity than previous forms of genetic analysis had required. Huge amounts of time on the part of statisticians are needed to assemble the many short runs of sequence into an assembled complete sequence, identify all of the places in the genome that are heterozygous, and identify all of the known polymorphisms present in the sequence, before we can even start to ask any of the real biological and genetic questions we have about that sequence. For example, once we have eliminated the huge number of polymorphisms that are common in the population, we can start to evaluate the variants that are held in common by individuals with the same trait but that are not present in the general population.

What kinds of questions are being asked as we move into the era of whole genome sequencing? Let's consider some of the first experiments of this kind that are being done.

Exome Sequencing

The exome is the collection of all the exon sequences in the genome. Re-sequencing an entire genome produces massive amounts of data and comparably massive strategic and analytical complexities. In some cases the desired question can be addressed much more cheaply by doing a genome-wide analysis of SNP data. In some cases, more detailed information can be obtained while limiting the size of the analytical problem by doing whole exome re-sequencing. Screening of the exome allows identification of sequence variants affecting the protein coding sequence, but would be expected to miss regulatory mutations such as promoter mutations.

An important success of exome sequencing can be seen in a large 2010 study of 208 X-linked mental retardation families carried out by Tarpey and collaborators from more than a dozen institutions around the world. Mental retardation is a major public health problem with serious personal, societal, and economic consequences. Mental retardation is a broad category that actually encompasses almost a thousand different traits, with about 80 mental retardation loci known on the X chromosome alone. Many genes for X-linked mental retardation remain to be identified. In this study traditional sequencing was carried out spanning about a million bases of coding sequence from the coding exons on the X chromosome, with about 75% of the sequence successfully read in each sample. The study found 1858 coding sequence variants including 983 missense mutations, 22 nonsense mutations, 15 frame shift mutations, 13 splice mutations, and an array of silent changes. Many of the differences between any two samples turn out to be recurrent – occurring in more than one person,

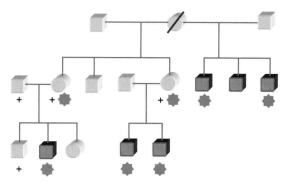

FIGURE 12.18 Mental retardation gene ZNF11 identified by sequencing exons on the X chromosome from probands from 208 families. The blue symbol marks a mutation in ZNF711 in one of the mental retardation families and + represents the copy of the sequence that lacks the mutation. This mutation co-segregates with the trait, and two different families in the study had truncating mutations in this gene. This mutation is not found in the normal population. (After P. S. Tarpey *et al.*, *Nature Genetics*, 2009;41:535–43, with permission from Nature Publishing Group.)

but there are also many rare non-recurring variants not found in other study participants. The inclusion of data on co-segregation in the relatives turned out to be a very important step in ruling out some of these variants as causes of the trait. Evaluation of a set of about a thousand normal controls helped to rule out some of the other variants as causative. At the end of this study, only about 25% of the samples showed causative mutations, possibly because there was not complete sequence coverage in all of the samples. This study identified nine new X-linked mental retardation genes, including ZNF711 (Figure 12.18). They also reported that "loss of function of 1% or more of X chromosome genes is compatible with apparently normal existence".

Finding a Causative Mutation

In the case of CMT, Lupski and colleagues have used whole genome sequencing to identify mutations in SH3TC2 as the cause of an autosomal recessive form of CMT in one large

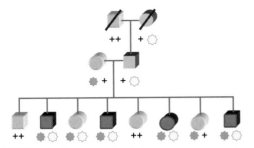

FIGURE 12.19 Use of whole genome sequencing to identify the gene causing Charcot–Marie–Tooth (CMT) in a family showing autosomal recessive inheritance. Four of the siblings who received one SH3TC2 allele from their mother (blue symbol) and a different SH3TC2 allele (yellow symbol) from their father are affected with CMT. + represents a copy of the sequence that lacks either of these two mutations. The father and paternal grandmother who are heterozygous for one of these alleles show axonal neuropathy but do not have CMT. (After Lupski *et al.*, "Whole genome sequencing in a patient with Charcot Marie Tooth neuropathy", *New England Journal of Medicine*, 2010;362:1181–91.)

sibship (Figure 12.19). First they did whole genome sequencing on DNA from the proband. As expected, millions of SNPs were observed in the proband's DNA and more than a half million SNPs turned out to be unique SNPs that had not been seen previously. Screening for copy number variants in the proband identified 234 CNVs, none of them affecting genes known to be involved in CMT; 220 of these CNVs overlap with known CNVs that are considered part of the normal variation in the human genome, but none of the CNVs affected any of the known CMT genes or genes known to be involved in other forms of neuropathy.

After dismissing most of the SNPs based on informatics approaches, such as those SNPs known to be normal human variants, the researchers focused on 54 SNPs in coding sequence of 40 genes known to play a role in neuropathy. Two of these SNPs turned out to be in the same gene, SH3TC2, which was already known be a cause of CMT. This finding of mutations in both copies of SH3TC2 was

expected for this autosomal recessive trait, and when the rest of the family was tested both of these alleles were present in each affected family member. Some family members with only one mutated copy showed lesser neurological effects. One of the mutations causes an axonal neuropathy when only one copy of the gene is defective, and the other mutation was found along with carpal tunnel syndrome, a trait that is also found in some cases with PMP22 CNV.

Finding a New Disease Gene

In a similar project, Sobreira and colleagues at Johns Hopkins University identified the cause of metachondromatosis, a disease whose cause was not previously known. Metachondromatosis is a rare simple Mendelian autosomal dominant trait involving characteristic bony "bumps" on places such as the hands and feet. A whole genome sequence of an affected family individual from a five-generation family revealed the expected large number of sequence variants including many common SNPs present in the normal population. Linkage data from other family members let the research team focus on the region containing an 11 base pair deletion in PTPN11. You will recall from our discussion in Chapter 5, an 11 bp mutation, which is not a multiple of three, is expected to throw off the reading frame and lead to an abnormal protein sequence and an incorrect stop point beyond the point of the mutation. The researchers confirmed their finding by identifying a nonsense mutation in the same exon of the same gene in an affected member of another family. In both families they showed that the mutation co-segregated and they showed that these mutations were not present in a large population of normal controls. This approach combines the power of whole genome sequencing with the power of linkage to rapidly address the cause of a rare trait, and offers the possibility that a lot of other rare diseases may rapidly join the ranks of those traits that have been "solved."

Chip-based Technologies Meet Next Generation Sequencing

Just as knowledge of the human genome reference sequence lets us readily re-sequence genomes in search of causative mutations, knowledge of the sequence of the genes allows us to re-sequence the transcribed sequences to determine which genes are being expressed at high levels and which are being expressed at low levels. In Chapter 3 we talked about the use of microarrays to look at the levels of expression of many genes. As a result of the Human Genome Project, microarrays now give coverage of most of the coding sequence genes. Exon arrays, which cover all of the exons instead of all of the genes, even let us separately monitor levels of different exons within the same gene so that we can monitor alternative splicing events in addition to overall expression levels of the gene. And now, with the advent of Next Generation Sequencing, we can sequence the transcripts from a cell sample in a process sometimes called deep sequencing. We can add up the number of times we get a "hit" on each gene and draw conclusions about which genes are expressed at higher levels and which genes are not expressed at all.

Another important chip-based technology is called ChIP-on-Chip. ChIP uses chromatin immunoprecipitation, a process that lets us retrieve pieces of DNA bound to a protein of interest, such as a particular transcription factor. Because it is common for a transcription factor to bind to many locations in the genome, such as the regulatory sequences in promoters, we end up wanting to be able to take a genome-wide view of where the binding sites are. By using ChIP to pull the bound DNA fragments away from the rest of the genome, we can then hybridize to a microarray chip to figure out which pieces of DNA we retrieved. And more recently, with the advent of Next Generation Sequencing, we can now skip the microarray

step and directly sequence the DNA fragments that were retrieved. This lets us do experiments in which we isolate RNA from a tissue to look at sequence-based expression profiling and then we do ChIP followed by sequencing to identify where the transcription factors are bound. This lets us start to evaluate how binding of a particular transcription factor relates to global patterns of gene expression.

Whole Genome Diagnostics

Another area in which whole genome processes are becoming important is diagnostics. Obvious uses for this approach would apply to traits for which a large number of different genes are known, where knowing which gene is involved could make a difference in medical decisions being made. There are more than a hundred different genes known to play a role in retinal degeneration. Whole genome sequencing and evaluation of sequence variants in all of those genes might pinpoint the gene in question, but in many cases we would be able to use the approach taken with metachondromatosis – a combination of whole genome sequencing and some limited amount of linkage involving other relatives. Alternatively, as the price of sequencing is coming down, we may reach a point of simply sequencing multiple family members and looking for sequence variants that hit one of the retinal degeneration genes *and* that are present in each of the affected family members. We expect that as we reach the point of doing whole genome sequencing on multiple family members we will start turning up complex situations in which more than one retinal degeneration gene may be involved. You will recall from Chapter 9 the story of digenic inheritance involving both ROM1 and peripherin, or the story of the various genes involved in triallelic inheritance of Bardet–Biedl syndrome. We expect that there are more such situations waiting to be discovered as whole genome sequencing

Cancer Diagnostics

becomes one of the routine tools in the geneticist's toolkit.

One of the most important near future uses for whole genome technologies will be cancer diagnostics. Analysis of this kind is already showing that there are usually multiple cancer gene defects in a tumor by the time a doctor starts asking diagnostic questions about the type of cancer. Comparison of the whole genome sequencing of DNA from the patient and the patient's tumor can identify the sequence changes that have taken place during tumorigenesis, while genome expression profiling looks at differences in gene expression between the tumor and the cell type from which it is derived. However, the information that comes out of such screening is not simple. In any one tumor we expect to see mutations in many genes, in fact in multiple cancer genes, because mutation continues taking place as the cancer cells divide, so the mutations present by the time we detect the tumor go beyond the mutation that led to the cancer initially. We do not always expect to find a simple situation involving nothing but a mutation in the primary cancer gene that started the process. And we expect to see changes in expression of many genes. So what use will this information be? Let's take a look at what some recent studies have been able to find out about one form of cancer – acute myeloid leukemia (AML).

In the whole genome sequence of DNA from cancer cells and normal tissues in a 50-year-old woman with AML, ~750 point mutations were found, 12 in coding sequence and another 54 mutations in other conserved sequences. Of these 64 mutations, four were found among DNA samples from another 188 AML samples. In another AML study, 201 copy number variants were found to be present in the tumors but not the somatic tissues among samples from 86 patients with AML. Some of the regions affected by copy number variation turned up recurrently among the samples that were screened, and about half the patients had no CNV located in the tumor but not in the accompanying somatic tissue. Other research groups have found two characteristic profiles of microRNAs typical of different classes of AML, and some expression profiling patterns have been identified that seem to be characteristics of AML. All such genomic screening seems to find many changes in any one person, with some of the changes in any one person or tumor seeming to be held in common with tumors from another individual with the same type of cancer.

So whole genome sequencing, expression profiling, microRNA profiling, and copy number variant analysis will each yield a complex of excess information from screening of samples from any one person. But complex analytical methods are starting to pull out patterns that will be usable for classifying particular tumor types as we accumulate more information on large numbers of samples and individuals.

One of the important problems that might be solved by this kind of analysis is correct assignment of tumor type and stage. While many tumors are correctly assigned by classical methods such as cytology and knowledge of the site at which the tumor originates, tumors are not always correctly classified. In the case of metastatic cancer it can sometimes be difficult to correctly assign the original tissue of origin of the tumor. And correct classification is incredibly important in a world where the optimal treatment for one tumor type may be quite ineffective against another tumor type. While the answers coming out of these genome-wide approaches are complex and still being sorted out, researchers are moving towards a time when such information may be combined with classical methods for tumor classification to improve decisions about treatment.

Pharmacogenomics

Pharmacogenomics, the use of genomic approaches to study pharmacologic responses, is one of the many "omics" that have exploded onto the scientific scene as our knowledge of the human genome has expanded (Box 12.2). First let's consider a traditional pharmacogenomics study and then look at how we can approach such things now that we have The Sequence.

Many early pharmacogenomics projects were hypothesis-driven projects. The researchers started out knowing things about the drug and its mechanism of action, and they looked at genes that they knew or suspected would interact with the drug to affect its function. At least 25 different genes have been implicated in causing asthma, which affects 7% of the US population, and is responsible for thousands of deaths in the United States every year. Asthma causes chest tightness and difficulty breathing, and acute attacks may be relieved by use of an inhaler containing a class of drug called a beta-adrenergic agonist, but these inhalers do not work for everyone. This can be a terrible problem if it is first discovered by someone in the middle of an acute, severe attack of asthma. Researchers looked at frequencies of alleles in the gene encoding the beta-adrenergic receptor, the protein with which the drug interacts.

With the advent of genome-wide technologies, a much more global view of the problem became possible. Let's consider recent studies of tipifarnib, a drug that acts well on some AML cases but not others. Expression profiling compared cells before and after treatment with the drug, in individuals who responded to the drug and in individuals who did not. Eight genes were identified whose expression correlates with responsiveness to the drug. Expression of one of these genes, AKAP13, is by itself a major predictor of tipifarnib efficacy. While the studies so far have been small and more exploration of the subject is needed, identification of

---

BOX 12.2

THE MANY "OMICS" OF THE GENOMIC ERA

The suffixes -ome and -omics have been added to many words to indicate global and complete sets of information on a topic. Genomics is the complete set of information on a genome. Proteomics is the complete set of information on the proteome, the collection of all of the proteins in an organism. Transcriptomics is the complete set of information on the transcriptome of an organism. Exomics deals with the exons. Metabolomics deals with the metabolites. Phenomics deals with the complete body of information on phenotypes in an organism. Almost any category of information that can be conceived on the scale of the whole organism can be turned into an "omic." Some are obvious: kinomics for all of the kinases, lipidomics for all of the lipids, oncogenomics for cancer. But then we come across less obvious "omics" such as connectomics – information on structural and functional connectivity in the brain – and metagenomics, also sometimes called ecogenomics – the study of genetic samples in environmental samples such as soil. As the concept gets carried to the extreme we even find things like bibliomics – the collection of information in a bibliography. We expect the list of "omics" to continue expanding as other researchers appropriate the concept to their particular field or body of data.

these genes, and the pathways and functions in which they are involved, has provided new insights into the mechanisms of drug action. Such work, on many tumor types and drugs, moves forward in the hope that it might lead to improvements in treatment targeting that would improve life expectancy.

What's Next?

People sometimes ask what we geneticists are going to do with ourselves now that the Human Genome Project is done. After all, when we look back over our human genome timeline we see ourselves standing at the end of a long historical cascade of events. We find ourselves standing here, on the shoulders of giants who came before us, those who laid down all of the separate cobblestones in the road that led us here. After all, if the sequence is known and the genes have been found, aren't we at an end?

Now that the human genome sequence is done, now that we can read the genetic information inside the human cell all the way from one end to the other, we really stand at the beginning, not the end (Box 12.3). Consider all of the effort that went into the creation of the Periodic Table of the Elements that is used by chemists. At the time it came into being, many of those who contributed the information that

made it possible probably felt like they had slogged long and hard and deserved to sit back for a rest and say, "It's done." But most students of science these days did not experience the creation of this amazing tool. We are moving towards an era in the near future when the human genome sequence, and the catalog of the genes and phenotypes and sequence variants, will simply be another one of those tools used by people who will blithely look up the needed detail without giving any more thought to how we know a particular genomic fact than we now give to how we know the properties of the different elements in the Periodic Table.

So we race at an amazing pace towards the post-Genome Project era. Experiments that once took weeks, months, or even years can now be done quickly, sometimes being completed in a matter of hours or days and taxing our ability to handle the massive amounts of information produced. Sometimes we can bypass the lab bench altogether and ask our question entirely within the computer. Thus has genetics progressed from *in vivo* (in the living organism) to *in vitro* (literally "in glass") to *in silico* (perfomed on a computer). With the sequence in hand, we can finally begin asking things we could not have dreamed of asking back in 1969, when Armstrong walked on the moon and no one knew the sequence of even one human gene.

BOX 12.3

THE HUMAN GENOME SEQUENCE: A STARTING POINT FOR GENE DISCOVERY

"This is the beginning of genomics, not the end. Critical understanding of gene expression, the connection between sequence variations and phenotype, large-scale protein–protein interactions, and a host of other global analyses of human biology can now get seriously underway." *Francis S. Collins, Director, National Institutes of Health*

Study Questions

1. How many genes are predicted in the human genome? Why is this number lower than previously expected.
2. Are human genome sequences patentable?
3. What is the reference sequence used in genome studies?
4. What are the other regions in the genome, or the so-called "desert regions?" How do you know if sequences should be considered a gene?
5. What is mitochondrial heteroplasmy and how does this contribute to disease?
6. How are so many genes expressed in a tissue-specific manner leading to control of individual genes important for cell function?
7. What is HapMap and what useful sequence information can be obtained from this project?
8. How has genomic technology allowed us to answer new and previously not easily explored questions? What are examples of genomic technology used in this manner?
9. What can whole genome analysis provide and how does this information affect diagnostics?
10. What is pharmacogenomics and how is this revolutionizing drug and disease treatment regimens?
11. How is exome sequencing different from whole genome sequencing?
12. The PMP22 gene can be responsible for CMT1A neuropathy even if there are not mutations present in the sequence of the gene. How is this possible?
13. PMP22 can also cause hereditary neuropathy with pressure palsies. How can the same gene be responsible for these two different traits?
14. If a trait is quite common but not highly penetrant, how would we best go about searching for the gene that is responsible?
15. Allele A is associated with Trait X but Allele A is not causing Trait X. How can this be?
16. What is a test for association?
17. Why is it especially important to take multi-testing into account in a genome-wide association study?
18. What are three of the technologies that allow for high throughput studies?
19. What are three ways for an allele to shift frequency in a population?
20. How many places in Craig Venter's whole genome sequence are heterozygous?

Resource Project

Many of the biggest discoveries providing insights into the most fundamental biological processes have come from studies of animal model systems. This becomes especially obvious when we look at breakthroughs that have been awarded the Nobel Prize. Go to the Nobel Prize website and prepare a half-page report on a Nobel Prize awarded for work in an animal model system and explain why this result is important to humans.

Short Essays

1. Genome-wide associations have been hailed for providing breakthroughs in our understanding of the underlying basis of complex genetic traits, but they can be a real challenge to carry out. What are some of the factors that can make a difference in how successful such studies are? As you consider this question please read "Human genome wide association studies" by Tim Keith in *Genetic Engineering and Biotechnology News*, January 15, 2007.
2. One of the main objectives of the Human Genome Project was the completion of the human genome sequence, but as the sequence has become available for use in research many more advances have emerged or can be predicted in the near

future. What are five advances that came out of the Human Genome Project and what are five advances that are predicted to emerge from the project in the near future? As you consider this question please read "Human genome ten: Five breakthroughs and five predictions" by Ker Than in *National Geographic Daily News*, March 21, 2010.

3. Evidence from caves in the Middle East shows that about 80,000 years ago modern humans and Neanderthals lived in the same region. What does whole genome sequencing tell us about the relationship between these two groups and how does such a study help point to recently evolved genes? As you consider this question please read "Close encounters of the prehistoric kind" by Ann Gibbons in *Science*, 2010;328:680–4.

4. The ENCODE project has been carrying on a detailed analysis of the human genome sequence to identify all of the functional elements of the sequence within focused regions of the human genome, and has raised interesting questions about what a gene is and what junk DNA is. As many new transcripts are being discovered some think these are all signs of new genes. What other explanation do some researchers offer to explain what else some of these transcripts might be? As you consider this question please read "DNA study forces rethink of what it means to be a gene" by Elizabeth Pennisi in *Science*, 2007;316;1556–7.

Suggested Reading

DVD

NOVA: *Cracking the Code of Life* (WGBH Boston, 2001)

Cracking the Ocean Code (starring J. Craig Venter) (Discovery Communications, Inc.)

Articles and Chapters

"It's in our genes. So what? DNA takes you only so far" by Sharon Begley, in *Newsweek*, December 7, 2009, p. 36.

"Current-generation high-throughput sequencing: deepening insights into mammalian transcriptomes" by B. J. Blencowe, S. Ahmad, and L. J. Lee in *Genes & Development*, 2009;23(12):1379–86.

"Beyond genome-wide association studies: Genetic heterogeneity and individual predisposition to cancer" by A. Galvan, J. P. Ioannidis, and T. A. Dragani in *Trends in Genetics*, 2010;26(3):132–41. Epub 2010 Jan 26.

"The Great Human Migration: Why humans left their African homeland 80,000 years ago to colonize the world" by Guy Gugliotta in *Smithsonian Magazine*, July 2008.

"Whole-genome sequencing in a patient with Charcot–Marie–Tooth neuropathy" by J. R. Lupski *et al.* in *New England Journal of Medicine*, 2010;362:1181–91.

"RNA world – the dark matter of evolutionary genomics" by P. Michalak in *Journal of Evolutionary Biology*, 2006;19(6):1768–74.

"HapMap and mapping genes for cardiovascular disease" by Kiran Musunuru and Sekar Kathiresan in *Circulation: Cardiovascular Genetics*, 2008;1;66–71.

"How to make a Dodo: Biologist Beth Shapiro has figured out a recipe for success in the field of ancient DNA research" by Andrew Curry in *Smithsonian Magazine*, October 2007.

"Darwin's surprise" by Michael Specter in *The New Yorker*, December, 2007, pp. 65–73.

"A Planck walk" by Stephen Pincock in *The Scientist*, 2008;22:46.

"Lessons learnt from large-scale exon re-sequencing of the X chromosome" by F. L. Raymond, A. Whibley, M. R. Stratton, and J. Gecz in *Human Molecular Genetics*, 2009;18(R1):R60–4.

"The 2% difference: Now that scientists have decoded the chimpanzee genome, we know that 98 percent of our DNA is the same. So how can we be so different?" by Robert Sapolsky in *Discover Magazine*, April 2006.

"Mice and humans with same anxiety-related gene abnormality behave similarly" in *ScienceDaily* (January 19, 2010; www.sciencedaily.com/releases/2010/01/100114153002.htm.

"Genes from Ebola virus family found in human genome" in *Scientific American*, July 30, 2010.

"The human genome: Big advances, many questions" by Steve Sternberg in *USA Today*, July 7, 2010.

"A decade later, genetic map yields few new cures" by Nicholas Wade in *New York Times*, June 12, 2010.

Books

Genomic Disorders: The Genomic Basis of Disease by James R. Lupski and Pawel Stankiewicz (2006, Humana Press).

HOW GENES PLAY A ROLE IN TESTING AND TREATMENT

Genetic Testing and Screening: How Genotyping Can Offer Important Insights

THE READER'S COMPANION: AS YOU READ, YOU SHOULD CONSIDER

- The difference between screening and testing.
- What a medical geneticist is.
- What a genetic counselor is.
- Other forms of gene testing besides prenatal testing.
- What kinds of traits are included in neonatal screening programs.
- Why a test would be designed to be a biochemical test.
- Why a test would be designed to be a genetic test.
- What we can tell from alpha fetoprotein levels.
- Why the Quad test needs to include information on how far along the pregnancy is.
- What might lead someone to consult a medical geneticist.
- How far into a pregnancy amniocentesis can be done.
- How far into a pregnancy CVS can be done.
- What a karyotype can tell us about a sample from amniocentesis and CVS.
- Where you can turn for information about genetic testing and genetic diseases.
- What direct-to-consumer genetic testing can and cannot tell you.

When he came into the world, he was greeted with all of the hope, love, and eager nervousness that greet so many newborns who go on to fill their parents' lives with chaos and delight. However,

immediately following his birth, they knew there was a problem. In place of lusty cries at birth, instead of kicking feet or wiggling arms, he presented a picture of complete stillness, laying there silent and unmoving as a crisis erupted around him. They were not surprised that he was small, since prenatal ultrasounds had indicated short femur length that they had waved away at the time as something consistent with the modest height of some of his relatives. They had not expected the low-set ears, the high palate, the muscle weakness, the heart defect, and the unresponsive stillness. Amidst a flurry of medical tests, anxious discussions, and fearful waiting, his parents were told that he needed heart surgery if he was to survive. Without a diagnosis, the doctors could not tell his parents whether the surgery would be enough to save his life or whether he would die anyway of other problems besides his heart defect. A decision had to be made, and it turned out that the rush of medical events forced his parents to make that decision before they could obtain the karyotype results that might have better informed their choice. They needed to know: did their son have an extra copy of a chromosome or piece of chromosome, and if so, would that information tell them that he had greatly reduced hopes of survival even if he had the operation? In agony, his parents agreed to surgery in the hope that there was hope to be had beyond the surgery. While he recovered, while his pain was reflected in his parents' anguish, the genetic testing results came in too late to prevent the surgery. Yes, he had an extra copy of part of chromosome 18, and while he would likely not live to see his first birthday, no one could say for sure how long he would be with them. With a sense of guilt over having subjected him to a painful surgery that his parents would not have agreed to if they had realized it could not save him, his parents took him home to watch him around the clock. They hovered over him, willing him to breathe each next breath, watching as formula moved through a tiny tube down his nose to his stomach because he could not coordinate the movements needed to drink. So he continued for five months, until his body could do no more and he stopped breathing for the last time.

The kind of genetic testing used to diagnose tri-somy 18 takes time, and in the first hours and days of his life, when the heart defect became a crisis that could not wait, decisions had to be made at a faster rate than the genetic testing could be completed. His par-ents have never gotten over some of their regrets. They wished they had followed up on the report of short leg bones that resulted from one of the ultrasound tests. They wish they had had genetic testing done before he was born so that they would have been fully informed when they had to decide about the surgery. That he was born so ill, they could not prevent. That they were uninformed when they had to make decisions, that is something they wish they could have done differ-ently. Although this is not the usual course of events, it shows us that many of the cases in which we end up wishing we had more information available are ones we don't anticipate. It also shows us that information gained from prenatal genetic testing can have impor-tant uses other than making decisions about whether to continue a pregnancy.

13.1 WHAT IS MEDICAL GENETICS?

The field of medical genetics is practiced by doctors with subspecialty training in medical genetics and genetic counselors trained in coun-seling people about genetics. Medical genetics offers help for every stage of genetics in our lives. Prenatal testing and medical diagnosis assist in cases in which a trait seems to run in a family or resembles a known genetic trait. The medi-cal geneticist can also end up playing Sherlock Holmes, sifting through clues to arrive at an answer to a medical mystery, or even providing information about a trait to someone who is unaf-fected but concerned because of family history or for other reasons. Although many problems in this field come to the doctor's attention at birth or during childhood, medical genetics also plays an important role in diagnosis of traits in adults with later-onset conditions or whose correct diagnosis was missed during childhood.

Medical genetics specialists have an MD degree and specialty training in an area such as pediatrics or internal medicine, *plus* fellowship subspecialty training in the diagnosis and treat-ment of genetic disorders, birth defects, and other types of malformations. As more and more is learned about the genetic causes of human health problems and effects of teratogenic agents (things that can cause abnormal embryonic development), the role of the medical geneticist is becoming increasingly important and speci-alized. And as we move into the era of direct-to-consumer testing, the increasing array of available information will require sophisticated interpretation of results; the real impact on phe-notype will often go far beyond the predictions arising from any one test and encompass a broad array of complex interactions between genes and environment. Your family physician can give you some information about genetics, but as we move towards an era of whole-genome sequenc-ing, the role of medical geneticists will take on increasing importance relative to adult onset traits. You can get some information from your pediatrician, or an internal medicine specialist, but there are many traits that are rare enough that your family doctor may have never actu-ally seen an example of this genetic disorder in his or her medical practice. You may find your-self wanting to see a medical geneticist if you have questions about a complex, severe, or rare medical condition in yourself, your child, or some other family member. Or you might need to make major decisions about having additional children after a child with a birth defect and/or genetic disease has been born or decisions about continuing a pregnancy. You might need a genetic test done, or to undertake surgery or other major intervention because of a genetic disease. A medical geneticist's extra training includes not only information on genetics and birth defects but also training in techniques and resources for sorting out some very complex health puzzles with underlying genetic and environmental causes.

BOX 13.1

WHY SEE A MEDICAL GENETICIST?

There are a broad array of different things that might mean you would benefit from a trip to a medical geneticist. Although some of these traits are seen in infants, others may not turn up until much later in life. Examples of some of the most common items include:

- Advanced maternal age.
- Previously affected child.
- Presence of a chromosomal anomaly in a parent.
- Parents are possible carriers of a genetic trait.
- Family history of a neural tube defect.
- Some kinds of environmental exposures.

You might go to a medical geneticist because you fall in a category that is at elevated risk for a high-risk pregnancy, such as being an older mother (Box 13.1). You might be concerned about whether you have been exposed to a teratogen that can cause a birth defect, even if that kind of birth defect doesn't seem to run in your family. You might be seeking help because you have had many miscarriages and want to find out why. The problem that leads you to a medical geneticist need not be life-threatening nor does it need to concern a baby or a pregnancy. Sometimes medical genetics specialists solve medical mysteries that don't get brought to them until the patient is an adult. As you can see, medical geneticists deal with prenatal, pediatric, and adult situations, and they deal not only with genetic disorders but also with birth defects and other situations in which the genetic origins may not be obvious to you. You might not have to have a referral from another doctor, but you should find out whether the specialist's clinic or your health plan requires a referral. (We might add that medical geneticists are quite rare!)

The other half of the medical genetics team is the genetic counselor. These health care specialists have a master's degree in genetic counseling. Their education has trained them to be able to assist you with the medical, genetic, psychological, and social repercussions of whatever it is that took you to the medical geneticist in the first place. The genetic counselor works with a medical geneticist or other physician to help determine the origins of a trait that runs in your family, the causes of a birth defect, or the probability that something present in your relatives could turn up in your children. Part of the genetic counselor's job is to help assess risks to individuals who are not yet born or not yet known to be affected. They will also educate you to be sure that you understand any tests that are offered. They play a later role in helping interpret the outcomes that result from genetic testing and screening. Once results are obtained, the genetic counselor explains the results and deals with questions and concerns you may have about what you have found out.

The purpose of the medical geneticist and the genetic counselor is not to tell you whether you should have a genetic test done or whether you should have a child or continue a pregnancy. Their role is to be sure that you are armed with all of the information that you need so that you can make the best, most informed decision possible for you and your particular circumstances. As the amount of genetic information available increases and the complexity of the choices for testing or dealing with test results increases, the role of the medical genetics team becomes increasingly important.

13.2 SCREENING VS. TESTING

Evaluation of genotypes in individuals and populations can be carried out in a variety of different ways. The first choice, because it seems like the obvious thing to do if you want to know about someone's genotype, is of course to simply determine the DNA sequence, and then to infer the effects a sequence change will have on the structure of a protein produced by the gene. But of course, as we have discussed in Chapter 5, we cannot always predict whether or not a particular sequence change will have an impact on function. Sometimes we cannot tell because we do not know whether a particular amino acid substitution will be functionally relevant, but sometimes we cannot tell because we are dealing with other classes of mutations, such as changes in the sequence of non-coding genes, changes in the sequence of regulatory RNAs such as micro RNAs, or changes in regulatory regions such as promoter regions where only some of the bases are actually functionally involved in regulation of gene expression.

We have a second choice in screening, which is to look directly at the protein itself. We can ask whether the protein is present in normal levels. We can ask whether it localizes to the right place in the cell or tissue. Or we can directly assay some function of the protein such as enzymatic activity.

And a third choice is to assay for some biochemical product of the pathway in which the protein (or non-coding RNA) is active. Frequently when evaluating samples from humans we find ourselves using the first choice – looking at the DNA sequence – or the last choice – looking at some intermediate metabolite produced by the biochemical pathway.

Screening vs. Testing

In this chapter, we will use two different terms for this process of assaying something about patients' samples we want to evaluate: screening and testing. What is the difference between a screen and a test?

A screen is an assay that is carried out without indicators such as symptoms or markers of disease. The common types of screening are population-based and administered to most members of a particular group, such as newborn babies or pregnant women. Usually a screen applies to everyone in that group and is not used selectively with regard to the risk factors of individuals within that group. Thus we talk in terms of newborn screening, which looks at all newborn infants for genetic defects causing a metabolic defect of genetic origin, with the sample for screening coming from blood from the baby's heel. Although gene sequencing and mutation screening can detect some of the things we want to identify, in fact many of the assays used in newborn screening are biochemical assays of enzymatic activity or assays for the levels of intermediate metabolites produced in the pathways in which those enzymes participate. As we will discuss below, such screening takes place in the United States on a routine basis and is not focused on a subset of babies with symptoms or characteristics they observe in that one specific child.

A test, on the other hand, is an assay requested by a doctor on an individual basis, usually in response to some medical information or risk factors indicating the need for the test. A test is ordered for an individual on the basis of that individual's particular health history, symptoms, and risk factors. It may take the form of a biochemical test for enzyme activity. It might measure the level of an amino acid or sugar or lipid that is the product of a biochemical pathway.

Biochemical vs. Genetic Assays

Why do a biochemical assay instead of a genetic assay? Time, cost, and test effectiveness are all factors. Multiple steps in a biochemical pathway lead to the final end product, so if we measure products of a pathway we are potentially evaluating the function of multiple enzymes that act in that pathway. Thus by measuring levels of the final product in a pathway we may use a single assay for that product instead of separately assaying each separate enzyme in the pathway.

Even if there is only one enzyme involved, a biochemical assay may still offer advantages since it takes only one rapid biochemical assay to detect a functional deficit that might be caused by 15 different mutations in 15 different affected individuals. Defects in nitrogen metabolism that disrupt the urea cycle, such as ornithine transcarbamylase deficiency (OTC), can hit any of more than a half-dozen different enzymes. We could assay each separate enzyme, but it is so much easier and cheaper and faster to simply monitor ammonia levels in the blood to identify accumulation of the typical byproduct of a defect in this pathway than to screen each gene separately.

So why not always just do biochemical tests? In some cases, there may not be a simple biochemical test that is adequately specific and sensitive to tell us what we need to know. Also, we often want to identify the problem before the development of functional deficits so that we can intervene before damage occurs, which may mean that we want to do the testing before the biochemical imbalance becomes large enough to measure through biochemical testing. Sometimes we don't do a biochemical test because the locations of the cells expressing the genetic defect could only be assayed through something invasive, difficult, and expensive, such as a liver biopsy. If we suspect a defect in one of the opsin molecules of the retina, a biopsy of the retina would be invasive and damaging. But if we simply sequence the gene, we may take the sample from any cells in the body. So if we take our cell sample from blood or from swabbing the inside of the cheek, sequencing of the gene may give us the information we need even if we are sampling cells that do not express the gene.

Screening Populations vs. Testing Individuals

What types of items warrant screening on a population basis instead of testing in individuals? Screening tends to include things of moderate to great severity for which a feasible and cost-effective assay is available and for which intervention such as diet or medication could help to prevent damage. Many of these traits are recessive diseases that can easily turn up in individuals with no known risk factors, and only rarely is there family history that would suggest testing of the individual. In many cases, an indication that testing is needed becomes evident only after byproducts of the defective pathway have started to accumulate. Testing on an individual basis can take place once the infant has developed symptoms of the disease, but for many of the kinds of things being assayed in the newborn screening programs it is very important to find out that the infant is affected and begin intervention before the symptoms develop in order to minimize risk of permanent damage or death.

One dramatic success is newborn screening assays for phenylketonuria (PKU), which occurs in 1 in 20,000 infants and can cause profound neurological and cognitive damage. The successes result from a combination of factors. The test can be done rapidly and in a cost-effective way using the blood spot from the newborn screening program described below. Once a PKU infant is identified, dietary intervention can make a major difference in the child's health prospects, and subsequent monitoring can be done from blood samples or even from urine. The PKU diet limits intake of phenylalanine, in part by limiting intake of protein-rich foods that contain phenylalanine and avoiding aspartame and the phenylalanine to which it can be converted. However, each of our bodies does require some phenylalanine, so it cannot be completely eliminated from the diet. Although PKU *screening* is a standard item in newborn screening panels, it is recommended that siblings of PKU children should have blood levels of phenylalanine *tested* quite soon after birth to be sure of catching PKU at the earliest possible time in these children who are elevated risk.

There are two main sets of tools available for prenatal diagnosis:

- Noninvasive or minimally invasive screening routinely done on pregnant women receiving good prenatal care
  - Ultrasound
  - Maternal blood screening
- Invasive tests done in response to risk factors, symptoms, or screening results that suggest elevated risk of the baby having a problem that could be detected through such a test
  - Amniocentesis
  - Chorionic villus sampling (CVS).

13.3 PREIMPLANTATION GENETIC SCREENING

After nine years of being unable to conceive because of blocked fallopian tubes, Lesley Brown and her husband, John, turned for help to two British gynecologists who were working on the development of human in vitro fertilization (Figure 13.1). In late 1977, they took an egg from Leslie, fertilized it with John's sperm, and let the resulting embryo begin growing in the lab. Two and a half days later they implanted the fertilized egg into Lesley's uterus and then began waiting to see whether Lesley would become pregnant. For the first time in about 80 tries, this process of in vitro fertilization worked for a human couple. In July of 1978 the world saw the birth of Louise Brown, the first human child conceived by in vitro fertilization. And four years later, the Browns did the same procedure again, giving birth to Louise's sister, Natalie.

It is impressive that we can carry out genetic screening of an individual shortly after the fertilization event that creates the embryo. This is only possible for situations in which *in vitro* fertilization creates the embryo outside the body before it is implanted into the mother's womb (Figure 13.1). The other thing that made such screening possible was the development of polymerase chain reaction (PCR), which allows for being able to assay a genotype from one or very few cells.

As *in vitro* fertilization was being developed, one of the surprising discoveries was that it is possible to remove cells from a very early embryo without causing any harm to the embryo. Sampling without causing damage is possible because the fetus forms from cells inside the embryo but the cells on the outside

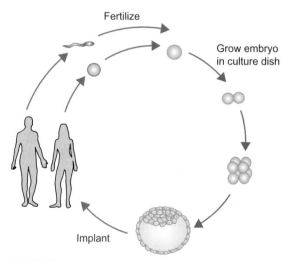

FIGURE 13.1 *In vitro* fertilization. In the lab, an egg removed from the mother is combined with sperm from the father, and the resulting embryo is grown in an incubator for three to five days. Often this process produces multiple embryos. Because the embryos do not all result in a baby, more than one embryo is often implanted at the same time. Sometimes even implanting multiple embryos does not result in pregnancy; in other cases all of the embryos make it and this occasionally results in the high profile cases we hear about in the news where a family doing *in vitro* fertilization ends up with four or more babies in one pregnancy. Extra embryos can be frozen and saved to be implanted in the future.

do not become part of the baby. In 1966 researchers showed that they could remove a small bit of material from the outside of a sheep embryo, from a region on the outside of the embryo that does not end up becoming part of the fetus. This allowed for the earliest form of embryo genotyping: examination of chromosomes to assign the sex of the embryo.

After the first birth resulting from *in vitro* fertilization, similar births occurred in country after country around the world until this process finally moved out of the research world and into fertility clinics world wide. This made having children possible for couples who previously would have had no hope of children, such as women with damaged fallopian tubes that could not deliver the egg to the uterus.

In 1983, with the announcement of PCR, *in vitro* fertilization also made preimplantation genetic testing possible. With this genotyping advance, it became possible to select a few cells from the trophoblast (Figure 13.2), the outer region of the embryo, for genotyping of sequence variants in the DNA. To do this, we have to already know what gene we want to assay and we have to have done genotyping in other family members to determine which mutation is causing the trait in the family requesting testing. At the end of this assay, we can tell whether or not the embryo has the mutation that is present in the affected relatives, but we cannot yet do a broad assay of many different genes.

So if Duchenne muscular dystrophy runs in the family, and we know the specific mutation, we can test this embryo for the presence of that mutation. Only a small amount of DNA is available so we cannot test for a lot of different things. Because multiple embryos are created,

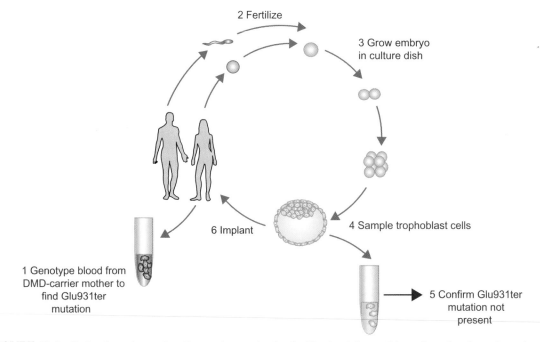

FIGURE 13.2   Preimplantation testing. By carrying out *in vitro* fertilization it is possible to show that the embryo does or does not possess the mutation that causes the trait of concern in the family. In this case, where the mother is a known carrier for Duchenne muscular dystrophy, she is the one who is genotyped. In other cases, the initial genotyping may involve testing of the father or both parents.

we expect that some embryos will turn out to carry the causative genotype and others will not. Embryos that are free of the causative genotype can be placed where they will implant into the uterine lining and begin forming a fetus.

In some cases preimplantation testing can solve an additional problem. Sometimes a couple wants genetic testing in a case where it is not known whether the parent is affected or not. The parent wants to ensure that the child is not affected but does not want to know whether the at-risk parent will eventually develop the disease. This situation can arise for something like Huntington disease where onset often takes place after any reproductive decisions have been made. When there is not yet a cure and the trait is lethal, there are many individuals who do not want to know that they will eventually develop the trait. And yet they want their child to be protected. It is possible for a clinic to do *in vitro* fertilization, test each embryo, select those that lack the causative allele, and implant them in the mother without telling the couple whether or not they had to leave out some embryos because they carried the causative mutation. This lets the couple protect their children against Huntington disease without risking getting bad news that they are not ready to receive.

Timing is critical for this genotyping. We have to be able to genotype the embryo, and arrive at a conclusion, within a very short period of time, after cell division has provided spare cells for testing, and before the embryo has to be implanted. If we continue trying to grow embryos in the lab for more than about five days, the growth slows and stops. So we are very limited with regard to which types of testing we can do. We can carry out an assay to ask a question, where we already know exactly what genotype we are looking for. We cannot ask a broad array of open-ended questions to address a vague list of possible risks.

Many pregnancies do not go through *in vitro* fertilization. Many pregnancies were not planned, and even when a pregnancy was planned, many people come to the idea of testing only after the pregnancy is already going on. In other cases the cost can be beyond the reach of families with limited resources. And another factor is this – *in vitro* fertilization, not surprisingly, does not always work. So there end up being a lot of different kinds of situations in which testing does not take place before implantation.

Even in those cases where preimplantation screening has ruled out some feared mutation that runs in the family, there are still a variety of other things about a pregnancy that have to be monitored. We can do preimplantation testing to rule out a Duchenne muscular dystrophy mutation and have that offer no protection against other things that could go wrong that can be monitored by other kinds of testing once the pregnancy is under way. This includes some kinds of developmental problems like spina bifida, as well as genetic problems such as aneuploidy.

To help monitor these other situations that were not covered by the preimplantation testing, there are standard types of prenatal screening that go on during the first trimester (first three months) and second trimester (second three months) of a pregnancy.

13.4 PRENATAL DIAGNOSIS DURING THE FIRST TRIMESTER

A variety of standard screening processes go on during a pregnancy. Some of these involve standard kinds of tests that we encounter in a doctor's office even if we are not pregnant, such as assays of blood pressure. Others involve tests of the mother's serum for proteins that are normally of concern only during pregnancy. And special attention is paid to older mothers because age is a risk factor for a variety of problems, including giving birth to a child with Down syndrome or other aneuploidy.

Minimally Invasive Screening

Routine screening takes place in the office of the obstetrician or family practice doctor who is following the course of the pregnancy. You do not have to consult a medical geneticist to end up with routine prenatal blood tests or ultrasound monitoring of the progress going on in the pregnancy.

Most children are, in fact, born healthy and without major genetic anomalies such as extra copies of chromosomes. Results of blood tests can provide reassurance throughout the pregnancy, and assist in being sure that the mother's health is protected along with that of the baby. Blood tests can help provide peace of mind to couples who fall into one or another high-risk group, such as older mothers or parents who have already had a child with a birth defect.

However, since couples who elect such testing often fall into known risk groups, there obviously will be cases in which the blood tests come out to be outside of the normal range of values. In such cases, this information is crucial in indicating the need for the additional testing that we will discuss in the next section.

Ultrasound

When Shannon found out that she was pregnant, her doctor told her that part of the regular process of following the health and progress of her baby would involve ultrasound images to get a look at the baby. He told her that once the baby was old enough they would be able to see the heart beating and would be able to monitor things about the baby's health. She was quite surprised when an early ultrasound at about six weeks showed that she was carrying twins! After having grown accustomed to the idea that a child was going to come into their lives, she and her husband happily readjusted their world view to include two children. But when they did an ultrasound six weeks later there was only one baby. She was devastated! As the pregnancy had progressed she had been mentally focusing love and attention on two developing babies, and now one was gone! What had happened to the second baby? The doctor explained that this "vanishing twin" phenomenon is quite common. Only a small fraction of pregnancies that

start out with twins end up resulting in the birth of twins. A variety of factors including aneuploidy can be responsible for those losses. As is commonly the case, the other baby was not hurt by the loss of the twin and Shannon's daughter was perfectly healthy when she was born.

Ultrasound uses sound waves with a frequency higher than the upper range of human hearing to produce an image of structures inside of the body (Figure 13.3). This kind or imaging lets the doctor tell important things about the pregnancy. In some cases an ultrasound indicates that a delivery by caesarean section will be needed, for instance if the mother is carrying conjoined twins. Some ultrasound findings, such as herniation that leaves substantial amounts of intestine outside of the baby's body, may indicate that the delivery needs to take place in a hospital with a neonatal intensive care unit. Some results tell doctors that they need to have special levels of support available for the baby at the delivery, such as when breathing assistance is needed at birth because of lung abnormalities. In rare cases ultrasound may identify a heart defect, such

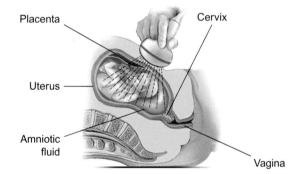

FIGURE 13.3 Ultrasound can show critical features about a baby's development including whether organs are all intact and developing normally. It can also show the sex of the baby, and it can provide information such as head measurements and developmental landmarks that help a doctor tell how far the pregnancy has progressed. (Picture from Illustration © 2000 by Nucleus Communications, Inc. All rights reserved. http://www.nucleusinc.com.).

as hypoplastic left heart with restrictive septum, which can be repaired through prenatal surgery on the fetus. In addition to evaluating measurements such as femur length and head size, ultrasound can evaluate the beating heart. Nuchal translucency predictive of Down syndrome shows up as a region of increased fluid at the back of the neck of the fetus on an ultrasound image.

The Combined Blood and Ultrasound Screen

The risk of having a baby with Down syndrome can be predicted based on screening during the first trimester for a combination of information from ultrasound (the nuchal translucency test) plus a blood test that measures levels of two proteins in the mother's blood: PPAP-A and β-HCG. PPAP-A and β-HCG are proteins present in the mother's serum that vary from normal levels when certain kinds of fetal anomalies are present, such as some kinds of aneuploidy.

When the ultrasound result is combined with the blood test result, about 82–87% of Down syndrome cases are correctly predicted, but there is a false positive rate of about 5%. A false positive rate is the rate at which babies without Down syndrome are predicted to have Down syndrome. Thus we do not simply rely on the results of the blood test to make any decisions about the pregnancy. Instead, we take the results of this test as an indicator of whether additional testing is needed.

13.5 PRENATAL DIAGNOSIS DURING THE SECOND TRIMESTER

When there is additional testing during the second trimester, 88–95% of Down syndrome cases are correctly predicted, with a 5% false positive rate. The Quad test measures maternal serum levels of α-fetoprotein (AFP), estriol, β-HCG, and dimeric inhibin (DIA).

Levels of these serum proteins change the course of a pregnancy and our interpretation of what a particular level means can be incorrect if we have incorrectly estimated how far along the pregnancy is. As additional ultrasound tests are done it is possible to estimate how far the pregnancy has progressed, which is important to the interpretation of the serum levels of the blood tests since the levels of the tested serum proteins vary over the course of the pregnancy. Thus whether a particular value on one of the blood tests would be considered as predicting increased risk depends on how far the pregnancy has progressed. As you can see in Figure 13.4, there is overlap in values between the normal fetuses and those with Down syndrome or other problems that can be detected with prenatal testing. In the case of AFP, which is produced by the fetus, a low concentration

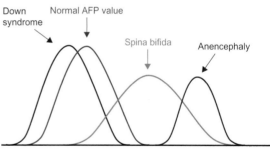

FIGURE 13.4 Distribution of levels of AFP in the mother's blood can suggest the need for additional testing. There is considerable overlap between normal values and the lower values observed on average for Down syndrome. There is some overlap between normal values and the elevated values seen in cases that later result in the birth of a baby with spina bifida. As you can see here, there is a region of overlap representing values observed in a small fraction of all three categories of infants – Down syndrome, normal, and spina bifida. Since these values change over the course of a pregnancy, interpretation of the values can be thrown off if there is an error in the estimation of how advanced the pregnancy is. (After Wald *et al.*, *American Journal of Medical Genetics*, 1988;31:197–209, with permission from Wiley; and Brock, *et al.*, *Prenatal Diagnosis and Screening*, p. 161 (1992, Churchill Livingstone) with permission from Elsevier.)

of this protein in the mother's blood is a possible indicator for Down syndrome, whereas a very high value might be indicative of a neural tube defect such as spina bifida or anencephaly. Each of the individual blood tests has similar overlaps between normal and abnormal levels, but when values for each of the different serum proteins are combined they start to have substantial predictive power.

The combination of ultrasound and blood tests is not enough to give perfect prediction of which fetuses have chromosomal anomalies or neural tube defects. Too many cases of Down syndrome are missed and there are too many "false positives," but values outside of the normal range justify additional testing, so we do not have to rely on these tests alone.

13.6 AMNIOCENTESIS AND CHORIONIC VILLUS SAMPLING

Sandra had married later in life than she had originally planned. As an older mother she found herself at increased risk of carrying an aneuploidy fetus. When the doctor recommended that she have amniocentesis she was unsure of what to do about it. After all, if she found that she carried a baby with Down syndrome, would she actually do anything other than continue with the pregnancy? Sandra and her husband had heard a lot about Down syndrome and thought that they could probably be good parents to such a child. They were a bit shaken up to learn that some of the things the test could detect involve things far more severe than Down syndrome. Neither one of them had ever heard of trisomy 13 or trisomy 18, with their terrible health consequences. Sandra finally decided that she could not even tell what she would do in such a case as long as it was all hypothetical. She was reassured to learn that having the test was not committing them to any particular course of action if the test results showed aneuploidy, and that they could spend time talking to a genetic counselor to help sort through their options. In Sandra's case, she never did find
out what decision she might have made since the test results came back normal, and she gave birth to a healthy son, but she still is not sure what she would have done if they had gotten a different answer.

If ultrasound and blood testing indicate increased risk of aneuploidy, then additional more invasive testing in the form of amniocentesis or chorionic villus sampling (CVS) will be recommended. These tests are invasive and risk harm to the fetus. Although it is a low level of risk, any risk has to be weighed very carefully in the balance of things. So we want to use the blood tests to identify as many women as possible who do not have any indications that they should be having the invasive tests done.

Many of the things predicted by the combination of blood and ultrasound tests involve chromosomal anomalies such as trisomy, so karyotyping is the next form of analysis that is needed where elevated risk of trisomy has been identified based on results of blood tests. In addition to using information from the blood tests, doctors will use additional information in making decisions about doing karyotyping. They use information on the mother's age, presence of an anomaly such as a balanced translocation in one of the parents, or other indicators that there is increased risk that the baby might have an anomaly that could be detected by karyotyping.

Some parents may request such testing even if their doctor does not feel that they are at elevated risk, but it is important in such cases to weigh the risks of the procedure as well as the risks of a baby with a problem. Remember that the oldest mothers are at the highest risk of an aneuploidy pregnancy, but most aneuploidy babies are born to young mothers because the young mothers are the ones who give birth to most of the babies.

Consider this situation: Mom is 37 years old and the AFP level is on the low end of normal. The combination of information from first and second trimester ultrasound and blood tests suggests that this woman might be carrying a fetus with an autosomal trisomy. At this point we want to do a test – an assay carried out on this one specific woman in response to the results that came out of the initial screening. This is an assay that will not be done on all pregnant women, only on those for whom we have individual, specific findings suggesting elevated risk and a need for further testing. To do this test, we need access to a reasonable number of fetal cells.

Currently, there are two standard methods for getting these cells: amniocentesis and CVS. Both techniques can be done after the 10th to 12th week of pregnancy, with CVS being available a few weeks earlier than amniocentesis. Both techniques provide the necessary cells for a variety of genetic tests, most notably karyotyping.

Amniocentesis

Amniocentesis is a process by which a doctor removes a sample of amniotic fluid that surrounds and cushions the fetus so that cells in the amniotic fluid can be tested (Figure 13.5A). Ultrasound is used as a guide while the doctor inserts a needle through the abdomen into the uterus and removes a small amount of the fluid. Amniocentesis is commonly done between the 15th and 20th week of the pregnancy and often right around week 16. At some medical centers earlier amniocentesis is possible. With the use of ultrasound, the doctor can see the tip of the needle: it looks like a very bright star on the ultrasound image. Because the doctor is also able to see where the fetus is on the ultrasound screen, it is possible to aim so as to miss the fetus, and the chances of damage to the fetus are minimal.

For these reasons, this procedure is now considered quite safe; the risk of miscarriage has been estimated by one study to be about 1 in 400 and has been estimated by another study to be about 2–4 in 400; knowledge of miscarriage rates in individuals who do not have this test tells us that some of those miscarriages might have happened anyway even if the test was not done. In fact, about 1 in 4 pregnancies end in

a miscarriage, many of them so early that the woman may not even realize that she was pregnant. The risk of miscarriage drops after the 12th week.

A physician may encourage a pregnant woman to undergo amniocentesis or CVS if she is over age 35, if she has already had a baby with a genetic defect, if she or her husband is known to carry a translocation, or if family history suggests elevated risk of a genetic disease such as Duchenne muscular dystrophy or Tay–Sachs disease. Under circumstances such as these, her risk of an autosomal trisomy or specific genetic defect begins to exceed the risk of miscarriage from the procedure alone. The disparity between the two risks increases as the mother gets older.

The amniotic fluid withdrawn by this method provides a rich source of both fetal cells and fetal proteins. The cells can be cultured using newer "microdrop" methods, and metaphase spreads needed for karyotyping can be obtained in a few days to a week. Cells can be even more quickly analyzed by FISH assays, although such techniques provide information only about the specific chromosomes for which probes were available or used. Sufficient cells are available for both DNA analysis and biochemical testing. Discussion of these sorts of analyses will be deferred while we consider another means for getting fetal cells, chorionic villus sampling.

Chorionic Villus Sampling

CVS, which can be performed between the 10th and 12th week of pregnancy, is diagrammed in Figure 13.5B. The doctor inserts a flexible needle, known as a catheter, through the center of the cervix and into the uterus. The doctor uses a needle through the abdomen or a cannula through the cervix to sample the chorionic villi. This tissue divides mitotically very actively, and thus metaphase cells for karyotyping can be obtained quickly. The tissue can also be subjected to other tests, such as mutation screening.

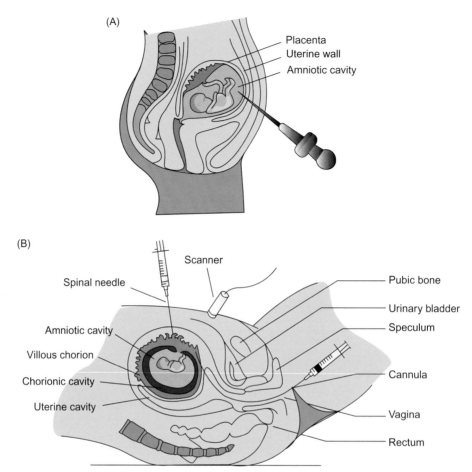

FIGURE 13.5   (A) Amniocentesis, usually performed at 15 to 20 weeks of pregnancy. (B) Chorionic villus sampling, performed at 10 to 12 weeks of pregnancy, may be carried out either through a catheter via the vagina or though a transabdominal needle. In both cases, the noninvasive ultrasound imaging allows visualization of the fetus throughout the procedure. In addition, monitoring of both maternal and fetal status afterwards can help provide peace of mind that the procedure was completed safely. CVS samples extraembryonic tissues, and since extraembryonic tissues can suffer from elevated levels of chromosomal anomalies compared to the fetus, CVS samples can sometimes carry chromosomal anomalies not present in the fetus. Amniocentesis samples skin cells shed by the fetus into the amniotic fluid and thus directly assays the fetus's genotype. The subsequent testing on the sample that is retrieved can take time, but the term limit for terminating a pregnancy is 24 weeks, which is after test results are available.

There appear to be regional differences in use of CVS vs. amniocentesis, with some areas preferring one procedure and other areas preferring the other. CVS may be preferred because it is often done earlier in the pregnancy than amniocentesis. Amniocentesis may be preferred because some workers report a lower success rate of correct karyotyping with CVS, apparently because it samples extraembryonic tissues that have an elevated rate of chromosomal anomalies. Although there have been troubling reports that CVS might induce an elevated frequency of limb anomalies, this appears to have been disproven. Both methods provide the same three things: metaphase cells for karyotyping, fetal DNA for DNA analysis, and cells and enzymes for biochemical studies.

13.7 ANALYSIS OF FETAL CELLS

Karyotyping

We have already told you about karyotypes and shown you a picture of a normal human karyotype in Chapter 6. Such figures are obtained by taking dividing cells from the fetal sample and lysing them (breaking them open) on glass slides so that the individual mitotic chromosomes spread out in a loose field. After the slides are stained with dyes, the resulting clusters of chromosomes from each cell are photographed and examined. Skilled cytogenetic technicians begin with large photographs of each metaphase spread. Then they carefully cut the picture of each chromosome out of the photograph and match each pair of chromosomes side by side on a piece of mounting paper. The folks who do karyotyping in the cytogenetics labs are truly gifted at pattern recognition.

Some disorders, such as autosomal trisomies, and sex chromosome anomalies, such as Turner (X0) and Klinefelter (XXY) syndromes, are easily picked up by this method. However, a good karyotype can also identify more subtle aberrations, such as deficiencies or duplications for small regions of the genome, translocations of material between chromosomes, inversions of the material on a given chromosome, and the fusion of two ends of a given chromosome to form a ring.

Although such anomalies are not common in humans, they do occur and they often have phenotypic consequences. For example, being heterozygous for a deletion of material on the short arm of chromosome 5, which is to say possessing only one copy of that material, results in a disorder known as *cri-du-chat* (or "cry of the cat") syndrome. This syndrome, recognizable because such infants mew like kittens when they cry, results in severe mental retardation and craniofacial anomalies. We have also noted previously the role of deletions in causing Prader–Willi and Angelman syndromes. Note that if the right culture conditions are used, it is possible to recognize the presence of "fragile sites," such as those seen in regions at which chromosomal breakage is observed in *fragile X syndrome* patients.

One might also note that karyotyping sometimes reveals abnormalities, such as an XYY karyotype, whose effects are not clearly understood or defined. Other abnormalities may be found, such as balanced or reciprocal translocations that may not affect the health of the child but might well affect his or her ability to produce children. Some care is required in explaining such outcomes to the prospective parents. This point also applies when karyotyping anyone. A student who accidentally discovered that she carried a balanced translocation as the result of using her own cells for a routine lesson in how to do karyotypes found that even though she was phenotypically normal, the translocation offered a potential hazard to her fertility and the health of any children that she would have. Although it was possible to tell her that there are potential health hazards for some of her offspring, simply looking at the rearranged chromosome structures under the microscope cannot indicate what form those health problems might take. Great care must be taken in considering what to do with such information when we know that there is a genetic anomaly but we really do not understand what the consequences might be.

Karyotyping will also, by default, tell you the sex of the fetus. But then, this information is also usually available as the result of routine ultrasound screening before the parents ever arrive at having a karyotype done. Prospective parents differ in whether they want to be told what sex was revealed by screening or testing. Some parents say that they prefer to be surprised in the delivery room regarding the sex of the baby and thus ask that the report of the results of a normal karyotype be limited to "everything looks fine." Other parents want advance information on this point and are happy to be able to know things like what kinds of clothing to buy or what color to paint the nursery. One even finds the occasional case of a split decision between parents.

Once the fetus is identified as having a complex chromosomal rearrangement, an important question is: Exactly which chromosomal bits are involved in the rearrangement? Much is known about the consequences of extra copies of even small regions of a particular chromosome, so if we can go beyond telling that there is a rearrangement to telling exactly which chromosomal regions are going to have too many or too few copies, we may be able to make some predictions about what the consequences will be. In some cases a simple karyotype is enough to identify the additional piece of DNA, but FISH (the use of a specific probe from one specific chromosomal region) or chromosome painting can also help us to get a specific answer to our questions about the baby's karyotype, especially if the anomaly is very small. There are some specific regions of the genome associated with well-characterized syndromes, so in some cases, important questions can be asked by FISH-ing with probes specific to the region in which deletions are known to cause traits such as velocardiofacial syndrome or diGeorge syndrome.

DNA Analysis

In a real sense, this section is simply a summary of the last 12 chapters. There are many other situations in which testing of an individual will take place in response to specific known situations such as when there is already someone in the family with the trait of concern.

We can make some predictions based on the modes of inheritance that we learned about in the first several chapters. Where the couple has already had a child with a recessive trait, then we know that both parents are carriers and the next child has a 1 in 4 chance of having the trait. If the couple has a child with an X-linked recessive trait, then the mother is a carrier and we expect that approximately half of her sons will be affected. If it is an autosomal dominant trait and one of the parents has the trait, then we predict that about half of their children will have the trait.

There are thousands of traits for which the gene is known, and much is known about the pathogenic mechanisms of many of these diseases. When a pregnancy is planned in a family where the causative gene is known, testing can take place ahead of time to identify the specific mutation that runs in the family. If the mutation is known then this allows for several kinds of testing options.

Suppose a couple's first child had Duchenne muscular dystrophy (DMD) and the couple came to you asking about the DMD status of their second, unborn child. Karyotyping might provide some reassurance because a female fetus is very unlikely to be affected. If the fetus is male, the answer is less clear-cut because the odds of the child being affected are 50% if the mother is a known carrier.

However, if you already know the nature of the DMD mutation borne by the first child and have a sample of fetal cells available from the second child, it is straightforward to determine whether this child carries the mutation. Now you can provide truly useful information to the parents. Very similar things can be done in the cases of quite a number of other diseases, such as cystic fibrosis, when the mutation in the family is already known. Tests are being developed rapidly for a host of other disorders for which tests were not previously available. Even if the gene is not yet cloned, a closely linked DNA marker can sometimes be used to diagnose the genetic state of the fetus if DNA from other family members is also tested. There is nothing special about fetal DNA, in terms of its chemical properties; any of the tests described so far can be used to assay the genetic state of a fetus.

Our ability to do genetic testing on the fetal cell samples will be affected by whether the primary genetic defect has already been determined for affected family members and carriers before tackling prenatal diagnosis. For many, the test for a known mutation can be fast. However, as stated earlier, many genes are quite large. An open-ended search for an unknown mutation somewhere within one of these genes can take some time, precious time that you do not want to spend during the period in which prenatal testing takes place.

For some genes, such as the CFTR gene that can cause cystis fibrosis when mutated, tests have been developed that can identify a large number of known mutations but cannot detect rare mutations or new mutations not previously observed. For other genes, however, the development of testing has not advanced as far, and it could take too much time to determine the primary mutation on the time scale of the prenatal test.

Thus, if you are concerned about a genetic defect and want to include mutation screening as part of your family planning, it will work out much better if you start asking questions before you are pregnant. You may be told that a standard test is in place that can do everything you need done at the time of the prenatal test, or you may be told that you qualify for some type of "preimplantation" testing. However, depending on what the gene is and what the genetic defect is, beginning your inquiries ahead of time might give you important choices that might not be available if you wait until week 12 of the pregnancy. Of course, there are many cases (new mutations or recessive diseases that you don't realize are lurking in the genomes of both you and your spouse) that you don't even know you should ask about until the first child with a problem is born into a family. Even then, asking your relatives questions about the family medical history can sometimes offer a warning. If you have a cousin with cystic fibrosis and your spouse had a great uncle with cystic fibro-

sis, you should be asking yourself whether you want to talk to a member of a medical genetics team before the first pregnancy. Sometimes the test result will be the happiest one of all – that you are not carriers. However, if that is not the answer, being informed can save you from later saying, "If only I had asked."

This ability to do genetic testing is not a panacea. Indeed, the facility of such tests becomes most worrisome in terms of the material we have presented on the inheritance of complex traits. One worries about people in the near future attempting to test fetuses for DNA markers associated with traits such as mental illness, obesity, beauty, strength, intelligence, or sexual orientation. It is perhaps a rather personal set of prejudices, but we draw a distinction here between testing for traits an individual will quite certainly express and which will greatly diminish their quality and length of life, such as DMD or fragile X, and those traits that they *might* express and whose effects on their quality of life are hard to assess or fall within the range of normal human variability.

Aren't there diseases for which the responsible genes are neither mapped nor cloned? Yes, there are. Some things we cannot yet test for. But for other things there may be biochemical or enzymatic tests available even if not all associated genes are known.

Be Sure Your Information Is Correct

Our ability to assess the genetic health of a fetus is impressive, and our capabilities expand daily. To the extent that truth is good, knowledge is power, and informed choice is better than uninformed choice, advances in the technologies of prenatal diagnosis can greatly improve our quality of life by providing us with better information from which to make better choices. As more and more information emerges from the Human Genome Project and a large number of research projects involving specific traits, it becomes more and more important to understand enough

---

BOX 13.2

ORGANIZATIONS

There are many other places that you can turn to for information. An Internet search finds organizations such as the Alliance of Genetic Support Groups, the March of Dimes, and the National Organization for Rare Disorders. Some individuals who are trying to take an active role in communication about disorders in their families have established Web pages that present information or reach out to others with similar problems. There are several ways to locate an organization that provides information or support relative to a particular disease. For many different diseases, organizations raise funds for research, provide support groups, and provide information about the disease. One example is the Foundation Fighting Blindness, which supports research, carries on educational programs, has local chapters throughout the country, and holds national meetings attended by patients, family members, caregivers, and educators who want to understand more about forms of retinal degeneration, such as retinitis pigmentosa, macular degeneration, or retinoschisis. Often your doctor's office will have information about such organizations. If they don't, try checking with a medical geneticist or other specialist who sees many cases of the trait in question. Even just looking under Social Service Organizations in the yellow pages of the phone book for a large city can connect you with a variety of organizations that can help you get information or support. Looking in the phone book for a town with a population of 100,000 yielded organizations that deal with cancer, lung diseases, multiple sclerosis, kidney disease, sickle cell anemia, epilepsy, and birth defects (the March of Dimes).

---

about genetics to be able to interpret the information being provided by the press. It also becomes important to be good at evaluating the information and the sources providing it. There are, after all, many things on the World Wide Web that are completely false, in addition to many important, valid sources of good solid information. Although your doctor, medical geneticist, or genetic counselor will be the most reliable source of information when trying to make decisions about genetic issues in your own family, you will often find yourself getting information from other sources. There are a variety of information resources out there that can give a good overview of many traits or in some cases individual traits that are the focus of that organization (Box 13.2). The GeneClinics site offers a set of GeneReviews that provide rigorous, detailed information on specific genetic traits, although they cover only a small fraction of the traits we know about. Online Mendelian Inheritance in Man (OMIM) provides a surface overview of the state of research on a large number of genetic traits. Other good sources of information can be found at websites for organizations such as the National Institutes of Health, the Department of Energy, the Centers for Disease Control and Prevention, and the Food and Drug Administration, all of which include some information for the public.

Where to Turn for Genetic Testing

Many of the tests requested by your family practitioner or the medical geneticist will be done by clinical laboratories in your own hospital or nearby (Box 13.3). These tests are done under clinically certified conditions specially designed for use in the context of a medical

BOX 13.3

CLINICAL LABORATORIES

Once a gene has been identified, researchers may spend years developing tests for that gene before it becomes available through a commercial clinical lab. Once optimal screening processes are worked out in the research labs, once it has been determined just exactly what can and cannot be predicted based on the test, and once the patent lawyers are done squabbling (and perhaps even before), the test will become available commercially. However, even at that point there are no guarantees that you will find out what you want to know, since finding out that you have the mutation can still leave you with many unanswered questions. For some genes, mutation screening results can provide clean, simple answers, but for many genes, sequence changes fall not into the two expected categories, sequence variants that cause disease and sequence variants that don't, but rather into three categories, those that cause disease, those that don't cause disease, and those for which we cannot tell whether they will cause disease. Even when we know that a mutation can cause disease, we may not know whether having the mutation assures you of ending up affected or whether it just puts you at higher risk. Hopefully, long-term studies of

mutation screening results in many people will increase our ability to make predictions based on the results of mutation screening. Currently, however, for many genes and mutations, having the results explained by someone with clinical genetics training will greatly assist in being sure that you know as much as possible about what your test results mean. The critical things to figure out in trying to arrange for a test include determining whether the test is available from a clinically certified laboratory and determining if there is only one source of the test or if you will need to make a choice between several labs. Other issues will, of course, include things such as cost and whether the test is covered by insurance, and what kind of sample you will need to provide. There are other, more complex issues, such as what level of predictions can be made from the test that might influence your decision about whether to be tested or not. A clinically certified lab will carry out the test according to a rigorously defined set of standards and protocols designed for use in this type of clinical screening. Normally, the doctor involved in carrying out the sampling procedure will also determine which lab will do the testing.

practice. However, some very specialized testing, such as screening for mutations in a particular gene, may be done at only a few places in the country. If information on the gene responsible for your genetic disease is recent, the testing might not be available from any of the clinically certified labs and might only be available on a research basis, conducted by the labs that are doing research on that gene. The best source of information about where and how to obtain genetic testing will be your local genetic health professional and not the Internet.

13.8 SEX SELECTION

In some cases where the trait is sex-linked, genetic testing can use information on a *haplotype* (a kind of genetic marker fingerprint for a region of a chromosome) or even a known mutation to distinguish between affected and unaffected fetuses, allowing parents to make decisions on a fully informed basis for each pregnancy. Sometimes, if the gene has not yet been identified or information on an affected haplotype cannot be obtained from other family members, the sex of the fetus may be the only information available to parents trying to decide what to do. Testing for sex is sometimes done in cases of severe sex-linked disorders for which no specific test is yet available. The concept is that, in cases in which the mother is a known or obligate carrier of an X-linked disorder, the parents may decide that if they only have daughters, they will not have to face the 50% chance of having a child with the trait in question.

Can you imagine a situation in which a family might be faced with deciding whether to simply have only girls, even though half of the boys would have been unaffected? Duchenne muscular dystrophy (DMD), with its terrible consequences for the child and the relatives who love him, can be a problem to diagnose even now that the gene has been cloned. The gene is enormous, and available genetic tests only detect some kinds of mutations. If a couple has a son with DMD and the mutation is not known by the time a decision has to be made about starting or continuing a next pregnancy, the couple has several different choices: they might decide to have no more children and focus their attention on the needs of the child they already have. They might decide to have more children and hope that the next child is healthy, but take their chances. Or they might decide to have only daughters because they know that their daughters are not usually expected to have DMD. As more and more disease genes on the X chromosome are identified and testing methodologies improve, we will move gradually away from having anyone face a decision to select for the sex of a child simply because they cannot test for the mutation they actually want to know about.

A much more problematic situation arises in some cases in which couples seek prenatal testing for the sole purposes of ensuring that the child they bear will be of the "right" sex. To some, it might seem absurd that anyone would have an abortion simply because the child's sex is not the sex they were hoping for. To others,

the necessity of producing a child of the "right" sex may be a matter of grave importance. In some cases of cultural pressures to have sons while having few children, doctors find themselves faced with parents who want the doctor to carry out testing for the sex of the fetus so that the couple can use this information in deciding whether or not to continue the pregnancy. In other cases, a family that has had five sons may be found putting pressure on the doctor to help them assure that the next child will be a girl.

Although ultrasound can determine the sex of the child without invasive genetic testing, thus limiting some risks, there are major ethical problems with choosing to terminate a healthy pregnancy should the fetus be of the "unwanted" sex. *Currently, sex selection simply because a couple wants a child of a particular sex is not supported in the medical genetics community.* Moreover the American Medical Association has advised physicians that sex selection of this kind, in the absence of any accompanying health problems, is not something that physicians should do. The issue of sex selection raises the larger overall issue of what constitutes an allowable basis for anyone to elect discontinuation of a pregnancy. Clearly, different doctors and different prospective patients may hold different views on these subjects. There are certain levels at which the right of individual autonomy in health decisions leaves each individual to decide where they draw the line. But when the decision not to continue a pregnancy is based on sex alone or on things regarded as cosmetic or falling within the normal range of human variability, little if any support will be found for such choices among health professionals in the United States.

13.9 NEWBORN SCREENING

Current newborn screening programs do not screen for all of the known disorders that are detectable and "fixable" through medical or dietary intervention. However, some technological advances allow for screening for multiple different biochemical defects in one test, so we are hopeful that the number of disorders covered in these newborn screening programs will improve. The National Newborn Screening and Genetics Resource Center lists more than 50 different newborn screening tests that include hearing tests and assays for PKU, cystic fibrosis, congenital adrenal hyperplasia, sickle cell, plus a broad array of disorders in metabolism of fatty acids, organic acids, amino acids, and more. The March of Dimes has a list of 29 items that it recommends be included in newborn screening programs.

In the United States, each state independently decides which tests are part of the newborn screening panel. Other countries also have substantial neonatal screening programs. In some parts of the world, programs to screen babies at birth for things such as PKU have made a dramatic difference for babies lucky enough to be born where such programs are in place. However, there are many places in the world where none of these tests are done, and in many places where some kinds of newborn screening programs exist, there are still known tests that are not being done for many such metabolic deficiencies that can kill, cripple, or lead to mental disabilities, such as OTC deficiency. For some disorders, we do not screen because there is no way to remediate the defect even if we can identify it. For some problems that can create a crisis in the life of an infant, newborn screening programs are all that is needed to dramatically alter that child's future for the better.

By consulting sites such as the March of Dimes it is possible to find out about the tests they recommend and which states offer what. It is possible for parents to request that additional testing be done over and above the routine newborn screening, although the additional testing would have to be paid for by the family. One of the virtues of newborn screening programs is

that they are done on almost everyone and do not usually involve any cost to the family. But it is a problem that this screening slides right past the notice of many new parents who do not realize what is being tested or why. If you live in a state where some tests are not part of the newborn screening program, asking your doctor about newborn screening programs can help you understand your options. These include options for ordering tests that are not included in your local screening programs.

Newborn screening is carried out using blood spots, a tiny bit of blood stored on a piece of filter paper. Much of the testing in newborn screening programs takes place through screening of levels of intermediate metabolites – the products produced by the various steps in the biochemical pathways that keep our cells running. The blood is collected two to five days after birth, so results of screening may not be available until after the baby has already been discharged and gone home. This timing can be especially critical for a child born with a trait such as ornithine transcarbamylase (OTC) deficiency or PKU, since ideal handling of such a child would place him on a heavily modified diet starting immediately at birth. However, there are realistic limitations to how rapidly any program can expect to screen for a lot of different disorders in every single child born in a hospital across the whole country. As it is, running the program this way results in substantial health benefits for many children who are identified by the newborn screening program.

13.10 ADULT GENETIC SCREENING AND TESTING

Adult genetic testing covers a huge territory. While some kinds of testing deal with the kinds of medical situations we have talked about in this book so far, there are a variety of other testing situations that have nothing to do with disease risk (Box 13.4).

BOX 13.4

OTHER USES OF GENETIC TESTING

Non-medical uses of DNA testing seem to be proliferating. DNA-based testing has freed some men who had been wrongfully convicted of rape, and the surprisingly high frequency of cases shown by DNA evidence to be wrongful convictions has led some states to institute major changes in their criminal justice system. In other cases DNA evidence has helped secure convictions, and it is now possible to lift DNA evidence from amazingly small sources of evidence – from a lip print on a soda can, from sweat in a hat band, or from grass on which a body has been resting. In Russia, DNA screening allowed the determination that bodies found in a grave were those of the last Russian tsar and his family, including the finding that the Princess Anastasia, who had been rumored to have survived, was indeed among those who died. Immigration programs in some countries have started using DNA testing to evaluate whether people being brought into the country on the basis of being a relative are, in fact, related as purported. Paternity testing has sealed the connection of some fathers to their children and sent others on their way out of the child's life. In one recent story, DNA-based testing was used to show that a rescued child was not related to the parents who were sure this must be their missing child. In anthropology, screening of mitochondrial DNA sequences from around the world have created a view of ancient migration patterns that support the idea that humanity traces back to a small number of women in southern Africa, with much talk of tracing us all back to Eve. It is even possible now to have genotyping done that will tell you about your ancestry with surprising accuracy!

Adult genetic testing situations normally take place in response to medical situations in the individual or in response to information on diagnoses that have taken place in some member of the individual's family. When someone in a family is diagnosed with Huntington disease, other adults in the family who are too young to show symptoms are faced with deciding whether or not to be tested for a repeat expansion in the HD gene. When someone develops early memory problems, mood changes and problems with co-ordination, testing for a repeat expansion in the HD gene can help to solidify the diagnosis without waiting for the neurological symptoms to progress to full-blown tremors.

Some genotyping situations in adults are really just a reprise of the kinds of situations faced in children. In some cases, an individual may have a trait that normally would have been detected in childhood, but perhaps have such a mild, slowly progressive version of a trait that it might not have been diagnosed in childhood. When symptoms then manifest in the adult, genetic testing can help to confirm the diagnosis and in some cases can help inform the situation if it turns out that the individual has a mutation in the known disease gene but has a different mutation than the ones seen in the early onset cases. Because phenotypes can sometimes vary considerably, individuals with early onset of a trait may sometimes arrive at adulthood without having been correctly diagnosed. As more and more genes become available for testing it is becoming possible to use genetic testing to help sort out these cases with unusual presentations.

Whole Genome Screening

A genotype assay that is now readily available is a genome-wide assay for single nucleotide polymorphisms (SNPs). In research of this kind SNP assay is being used for gene mapping and discovery via the kind of genome-wide association study (GWAS) that we talked about in Chapter 12. Large numbers of studies are currently doing association studies to try to determine which SNPs are playing a role in a large number of different phenotypes. As the results of these studies are emerging, identified SNPs start being available for use in risk estimation. While this is adding a large amount of information to the array of genotyping tools available, in many cases we find that the identified SNPs have only moderate predictive capability and that there is still a large component of many traits that remain unaccounted for. One possible explanation would be that there are large environmental components still to be identified, but another possibility is that the unidentified components are the kinds of rare genotypes that are not readily detected by this kind of association study. Additional experiments are emerging onto the genetic scene involving sequencing of whole genomes that should be able to find these rare genotypes. For some diseases like diabetes, we expect that by the time it is all sorted out it will all turn out to be very complicated, involving many different loci and many different alleles at some of those loci, all complicated by a variety of environmental factors. This is going to really complicate efforts at risk prediction. But even if we do not have all components of a disease identified, as we increase the number of identified components, we improve our ability to predict and we improve our ability to diagnose.

Direct-to-Consumer Testing

One of the things that has come out of the growing supply of genetic information is commercially available testing that can provide risk assessment for some traits. While a lot of important genetic testing options are available through a medical geneticist, some kinds of tests are available on a direct-to-consumer basis.

One of the first forms of direct-to-consumer testing that became available was paternity testing. Small amounts of DNA from a child and a father can be compared to determine whether half of the genetic material in the child matches genetic material in the father. Testing of this kind is used by courts, families, and sometimes even immigration services for some countries when they try to decide whether someone is related to the family that is trying to claim them as a relative. This kind of testing makes use of a small amount of genetic information since confirming relationships does not require a lot of information.

A variety of direct-to-consumer companies are emerging. They allow customers to sign up on line and mail in a sample. Some of these companies will do testing and offer risk assessment for a variety of common complex traits including some for which we can change diet or behavior if we know that we are at risk of the trait, such as diabetes or heart disease, and some kinds of cancers. Some of these companies test for genotypes that predict how you will respond to particular classes of medications. Other companies look at genotypes that can be used to estimate where your ancestors came from.

For less than a thousand dollars it is possible to get testing results for a variety of common complex traits, and some companies also will provide information on what your genotype says about your ancestry. One company offers information relative to 28 traits such as cancer and diabetes as well as 12 different classes of medications. For each gene tested, the companies, indicate whether the customer is at low, moderate or high risk of a particular trait.

There are major limitations to what can be interpreted based on direct-to-consumer testing. For many of the listed traits covered by the direct-to-consumer genotyping companies, only some of the genetic components are known.

In some cases they are doing tests for a handful of alleles for a disease where all identified genotypes so far may account for less than 10% of the disease risk. Another issue with such services is that they tend to do their testing in the absence of a variety of other testing results that a doctor would have available to modulate interpretation of test results. So they are evaluating risk of cardiovascular disease without having information on your weight, smoking history, or current cholesterol levels. One of the companies explains that while they are offering risk estimates for breast cancer their testing does not include the BRCA1 and BRCA2 genes since they are not the highly prevalent players in the game of breast cancer roulette.

One of the companies indicates that they have genetic counselors available to answer questions about test results, but such services are not provided by all of the companies. While the particular limitations vary from one company to the next, it is important to consider whether the company provides genetic counseling before and after testing, whether they are testing for the traits of greatest concern in your family, and whether you are more of less interested in some of the specialized services such as ancestry information and feedback on genetic effects on responses to medications. Even if you decide to do such testing on a direct-to-consumer basis it is a good idea to talk to your doctor about such testing and any recommendations the company makes relative to your health care and recommended alterations to your lifestyle and diet.

The Thousand Dollar Genome

As the price on whole genome sequencing is dropping – plummeting really – we will soon face a time when it will be possible to simply sequence the whole genome as part of clinical care. People in the field joke about the idea that the "thousand dollar genome" will be available shortly, more than 3 billion bases of your genetic sequence for a thousand dollars or less. Given the vast numbers of sequence differences from one person to the next, the genetics community faces the problem of how to sort out what any of it means. We expect that there will be a period of time when we will be able to determine large amounts of genetic information about someone without knowing what a lot of it means. The problem of sorting out additive risk for traits influenced by combinations of alleles at dozens, hundreds or even thousands of genes will be massive in its scope. Already important insights are arising from the first whole genome sequences that have been done, but there are huge numbers of alleles in each of those genomes for which risk associations are not known even though many of them may well have an effect on risk.

Research-based Testing

In the early stages of development of a genetic test, the test may not be available from a clinically certified lab. There can be a lot of reasons for this. If some aspect of the technology still needs to be worked out, the test may be feasible but not yet meet clinical standards. In some cases, the test protocol meets clinical standards, but not enough is yet known about what can be predicted based on the kinds of results that are produced.

For instance, if a mutation is associated with the trait but does not seem to be an outright cause of the trait, researchers may elect to continue more investigation of the gene before making testing available for general clinical use so that they do not end up putting uninterpretable results into the hands of doctors and patients who will then not know what to do with information that is ambiguous. For instance, there is a sequence variant in the optineurin gene that is found more frequently in people with glaucoma than in people who do not have glaucoma. However, there are many people with glaucoma who do not have it, and many people who have it who do not have the

BOX 13.5

RESEARCH LABORATORIES

Genetic tests that become available to patients through clinically certified laboratories are usually first developed by researchers working in laboratories, first to identify the gene and then to learn about how much can be predicted about the phenotype that will be caused by any particular gene or mutation. The research lab may be at a university, in a hospital, at a pharmaceutical company, or at a research institute. Such labs are headed by MDs or PhDs, who design and direct the research conducted by a team of researchers. Medical genetics research includes a broad array of topics, such as searches for genes, studies of how mutations are caused, investigation of how chromosomes are duplicated and distributed to the correct daughter cells, research on animal models of human diseases, and testing of improved methods for diagnosing genetic anomalies. Because the testing techniques first emerge out of research done in these labs, they are the only source of such testing available before commercial development of a test. And because they are research operations

and not clinical or commercial labs, you would not normally expect to get results back from the test. The result is limited availability of testing. People who participate in medical research help contribute to the body of knowledge that has to accumulate before a test can move from the research environment into certified clinical testing. Research labs will normally not be found doing prenatal testing unless the subject of the research is about some aspect of prenatal testing. If you want to receive results that will offer risk assessment and affect clinical decisions, hospital and commercial labs that are certified for doing medical testing can provide test results in a medical setting. If you have a genetic test done through your clinic, it will likely be done in a certified hospital lab, but if your test is done as you participate in a research program, you might or might not end up getting any results back from the test. The informed consent paperwork for a research study should tell you whether you can expect to receive an answer from the testing that will be done.

disease. For now, it is considered a risk factor – something that tells you that you have a higher chance of the disease than the general population without indicating that you are certain to develop the disease. Until more is understood about what role this sequence variant is playing, further research is needed and results are not clinically very helpful.

In other cases, a test may involve a trait that is rare for which there has been no incentive to work out the test under the more involved conditions required for clinically certified testing. In such cases, a clinical test would actually be possible and informative but would require that

someone make an investment in converting the test from research to clinical status. Each of the tests that become available in a clinic started out in the research lab that found the gene and sorted out the genotype/phenotype relationships (Box 13.5).

Participation in research-based testing will start with an informed consent process to tell you about the study objectives as well as the risks and benefits. If the project only involves mutation screening, you would only have to provide a blood sample or saliva sample from which testing could be done. When testing is still at the research stage, your participation in the research can help advance understanding of the underlying cause of the trait. But there may be limits to how much a test result can tell you about your own situation. For instance, if there are only two known genes causing your trait of interest and it is clear that most genes for your trait have not been found yet, there will be little insight to be gained from learning that you do not carry mutations in either of the two known genes.

13.11 ETHICAL, LEGAL, AND SOCIAL ISSUES

The Conditional Pregnancy

Years ago, women attempting to bear children waited to seek a pregnancy test until their second missed period. They then went in for the so-called "rabbit test," and a positive result was often sufficient to warrant announcing the happy news. These days, the home tests available at most supermarkets or drug stores can be used on the first day of the first missed period. However, earlier knowledge, coupled with increases in prenatal diagnostic techniques, has not always resulted in earlier announcements of impending births. Rather, women are increasingly aware that, on average, more than one-sixth of human pregnancies will result in miscarriage before the end of the twelfth week. Perhaps not surprisingly, some couples are then waiting to announce the pregnancy either until after the end of the first trimester or until they have seen a healthy fetus developing on a sonogram.

Couples are also becoming increasingly more guarded and concerned about genetic disorders, as well. Some of these couples who are concerned that a negative result will lead them to terminate the pregnancy prefer to keep the news of a pregnancy private until they are sure that they will go through with the pregnancy. Some women say that they also consciously try to avoid accepting or "bonding" with the pregnancy for fear of becoming attached only to have to lose the pregnancy through miscarriage or following the adverse result of one or another test. Clearly, as our technology gets better, the number of disorders that can be analyzed will increase dramatically. One cannot help but wonder just how "conditional" pregnancy can become, and what kinds of psychological effects this concept of conditional pregnancy will have on those who wait for the moment when they can decide that they can finally believe that they are going to have a baby.

Genetic Matchmaking and Other Strategies

What can we do with genetic testing information to inform our reproductive decisions? Certainly there is the obvious: decide whether or not to continue the pregnancy, a choice that we hope will start to gain some alternatives in the future, such as perhaps curing the condition. We hope that there will come a day when the prenatal test will primarily serve to find out whether stem cell development of a replacement organ needs to be started before birth, or treatment in utero needs to be started to prevent a developmental mistake from causing damage, or something about the embryo's environment needs to be manipulated through changing the mother's diet. In some cases, testing can even be done before implantation, although this approach is costly and does not always result in successful establishment of a pregnancy. However, there are other uses, such as in the case of the parents of the little boy at the beginning of this chapter who just wanted to be able to act from full and correct information when making decisions about their child's life and quality of life. For others, testing offers no information that the parents would use, and it is important that people who do not want testing have the option to decline it (Box 13.6).

BOX 13.6

THE OPTION TO NOT TEST

Even a brief conversation with Jill shows her to be expressive and intelligent, with her striking prettiness often lit up by warmth and sympathy and humor that have to have been assets in her nursing career. However, when Jill was growing up, people who did not know her well sometimes decided that she was a bit stuck-up or standoffish. If you were to meet her, you would wonder how anyone could think her manner anything but friendly. That is, you might wonder, unless you looked down and saw her constant canine companion, a beautiful yellow lab wearing a sign saying, "Please do not pet me, I am working." Jill was born deaf and as a teenager began losing her sight. She was "mainstreamed" in the public school system, learning to speak words she could not hear and to "hear" spoken language through a combination of lip-reading and other cues. She was adept at carrying on a conversation with someone sitting across from her at the lunch table or standing and talking to her in the hall. However, if someone who did not realize she was deaf asked, "Hi, how are you doing?" as they passed her in the hall, they would get no reaction because she had no idea she had been addressed. As her visual field gradually shrank, she might not see if someone who was not right in front of her waved and then looked confused when she did not turn to wave back. Jill has Usher syndrome, a combination of traits that starts out with deafness and later adds progressive vision loss from retinitis pigmentosa. She and her brother, who also has Usher syndrome, participated in a study

of Usher syndrome that identified the genetic defect causing their hearing and vision loss. Researchers identified the gene and mutation responsible for their combination of visual and aural deficits. Jill has never gone back to ask the details of what they found out. She says that the information that came out of that study is of little use to her. It will not tell her anything about herself that she does not already know, and if she were ever to have a child, she would not have prenatal testing done because she would go ahead and have the child whether or not there were hearing or vision problems. This option to not carry on genetic testing is an important aspect of the rights of the individual to make their own reproductive decisions, so it is important that the system for educating and counseling people about genetic information should include the ability to meet the needs of those who would continue the pregnancy no matter what the outcome of the test. In such cases, the main question becomes whether there is anything that such testing can do that can meet some other aspect of the family's needs, such as helping them prepare ahead of time to be able to meet some special needs present right at birth. There will be many cases like Jill's, where testing is not going to tell her anything that she needs to know sooner than she would be able to find out the old-fashioned way. The system needs to continue accommodating the desires of people like Jill to not do prenatal testing right along with the needs of those who want such testing done.

Let's back up a step and consider this: what impact might there be on courtship and marriage if you could look into a genetic crystal ball before picking a mate? What if you could look at your latest heartthrob and see that they are also your perfect genetic complement – having a different set of genetic flaws than your own, or at least genetic flaws that you would not mind passing along to your children? Sounds great? What if you looked at that same "love of your life" and saw instead that the two of you share some terrible recessive genetic flaws right along with your passions and your values and your hobbies?

Use of genotyping information in mate selection has led to substantial reductions in the rate with which children are born with *Tay–Sachs disease* in some communities, not through abortion or even people refraining from having children, but rather through selective arrangement of who will have children with whom among those at high risk of bearing such children. However, if we are going to put such genetic information into the equation when picking a mate, is the implication necessarily that we might have to walk away from our soul mate based on a printout from a genetic testing company? Perhaps not. As genetic and reproductive technologies improve, giving up on having children, walking away from the love of your life, or having an abortion will be increasingly pushed aside by alternatives that let us fix the problem instead. Other technologies will allow for testing in the context of *in vitro* fertilization to pre-select embryos free of the defect in question to be implanted in the mother's womb. Parental genotyping before reproduction will improve the odds that the children who are born will be healthy, or at least free of the identifiable defects for which their parents are carriers, and will improve the likelihood that neonatal health management will be improved in situations involving serious genetic illness that impacts the first days of life.

The "Maybe" Result

As we learn more about complex diseases, issues in genetic testing become more complex. Clearly, testing for a mutation that causes some terrible disorder that involves pain and death for an infant falls at some opposite end of a spectrum from testing for cosmetic traits, such as eye color or a cute nose. However, even in trying to talk about the relative ethical dilemmas involved in considering such a spectrum of trait severity, the way we talk about it implies something else very important. It implies that the test we do will provide us with some absolute answer: if the child has the mutation, they will have the trait, and if they do not have the mutation, they will not have the trait. Test results don't always give such simple answers.

As we begin looking at complex traits, we find that, in addition to the ethical dimension to the problem, there is a practical dimension. What do we do with "maybe" results? What do we do with a test result that says, "This child will have a 50% increase in risk of heart disease over the risk to the general population"? What do we do with a test result that says, "There is an 80% chance that this child's IQ will fall on the low side of normal"? What do we do with a test result that says, "This child will be at increased risk of a life-threatening illness, but only if exposed to identifiable environmental items that will be difficult but not impossible to avoid"? What do we do with a test result that says, "This child absolutely will develop the disease in question, but the disease severity can range from lethal to barely even annoying, and we cannot tell you where on the severity range this child's clinical course will fall"? What do we do with a test result that says, "This child has a 90% chance of developing Alzheimer disease 70 years from now at a time when the unknown future of biomedical advances might (or might not) turn this frightening, uncurable, fear-inducing fate into something treatable over the counter"? The fields of medical genetics and bioethics struggle with issues of whether there are identifiable places where clear lines can be drawn, but we expect that many of these issues will remain very fuzzy for a long time to come as we struggle with the implications of information that tells us, "maybe, but maybe not," and, "eventually, in a future so far away that we cannot know what this really implies."

The Line that Can't Be Crossed

The medical establishment is unlikely to develop any eagerness to assist in terminating pregnancies based on complex or cosmetic traits, especially those that are not even sure to happen. However, physicians are more likely to be responsive to wishes of parents who want to terminate a pregnancy involving a condition that will lead to a very early, painful death for an infant. For many of the other areas out in the middle ground, it is the job of the medical genetics community to educate the parents, to offer insights into possible consequences of different choices, to provide them with long-term in addition to short-term views of the situation they face, and to inform them about the best-case and worst-case scenarios. In the end, though, the decision falls to the parents, not to the medical genetics professionals.

Every case is as unique as the individuals caught up in the situation. The combination of the parents, their stages in life, their individual personal and medical histories, and the context in which they live, can result in very different decisions being made relative to exactly the same test result. Something people have to be able to take into account in making these decisions is that the person they are today, and the issues that drive their decisions today, will likely change over the course of a lifetime. Thus they may need assistance in seeing just how their current circumstances, as well as their possible future situations, color their decisions.

People have substantial differences in what they can tolerate, different limits beyond which they cannot go. Some hit their limit when faced with terminating a pregnancy. Others hit their limit when faced with watching their baby die a slow painful death. Some hit a limit when faced with raising a child whose needs they know they cannot meet in a society that too often fails to live up to the ideal of offering a loving alternative home for such a child. So many factors contribute, including financial status, presence or absence of a social safety net to help a family cope, mental illness, alcoholism, and the presence of other family members who are desperately ill. A diagnosis of Duchenne muscular dystrophy in a child might well mean one thing to a young, happy, healthy, financially secure couple going through their first pregnancy in the context of a very supportive extended family but mean something quite different to a couple struggling with unrelated heath problems in the parents, major financial problems, lack of support from family or friends, and another child in the family suffering from more advanced stages of the disease.

Even trisomy 18, which might seem like a simple case because it represents the severe end of the phenotype spectrum, can elicit a broad range of responses. In the story at the beginning of the chapter, the parents wished they had had the information they needed to save their baby from unnecessary pain. Other parents, when receiving the news of trisomy 18 as a result of prenatal testing, terminate the pregnancy because they feel that it would be tantamount to child abuse to put an infant through what they know is coming. However, we know a woman who bore her trisomy 18 daughter and went home to sit and rock with her until she passed away, a font of maternal love and calm, and a picture of surprising acceptance of a situation that would have left others raging in anger and frustration.

Some see a decision to terminate a pregnancy as a selfish decision that does not consider what the child's perspective might be. Some, though, who see themselves wishing they could die quietly in their sleep instead of lingering on in lengthy pain, have that same perspective of their child's life. Several years ago, Julia met a young man who considered his parents to be guilty of criminal negligence for failing to terminate the pregnancy that produced him, and a whole host of medical problems, when they knew that any child of theirs would be at 50% risk of inheriting the trait that ran in the family. Yet his father must have had a very different view of the situation, suffering the consequences of that same medical condition himself and yet deciding that the positive aspects of his life so outweighed the problems that he would be willing to pass both life and this trait along to his child.

For each of us, there is an ethical brick wall that would stop us dead in our tracks if we ran into it. Fortunately, most of us are never faced with our own brick wall, and thus we do not even know where our real limit would be if we were put to the test. For those who get the news that there is a problem, there are amazing differences in what that brick wall turns out to be for different people.

For some, the act of terminating a pregnancy would be the thing too terrible to contemplate, to take a life or to fail to at least give the child a chance. For others, to continue the pregnancy is the thing they would never be able to forgive themselves for, to bring a child into too brief a life, one filled with pain in which their needs are desperate but cannot be met. Either decision represents a heartbreaking way for the parents to say, "I love you."

None of us will live with the consequences that fall upon someone else's family when they make reproductive decisions of this kind, invariably profound and painful in their far-reaching implications. So the role of modern medical genetics is to allow people to make decisions that are as fully informed as possible. However, the role of modern medical genetics is not to make the decisions for the family. Those who will live day in and day out, with the child or with the absence of the child, are the only ones who can know just where they will reach that boundary beyond which they cannot go.

Study Questions

1. What is a karyotype and how is the karyotype determined?
2. Describe pre-implantation genetic diagnosis and what it is used for.
3. How is *in vitro* fertilization (IVF) performed?
4. What is alpha fetoprotein and how is it used in genetic diagnosis?
5. What are SNPs and what information can be gained from large-scale SNP analysis?
6. What are the standard forms of prenatal screening and prenatal testing?
7. What approaches are used to detect small genetic abnormalities and why is this information valuable?
8. Why would you need to see a medical geneticist?
9. What maternal serum proteins can be used to evaluate risk of neural tube defects or aneuploidy in a fetus?
10. If we have a genetic testing result for a particular gene does that necessarily tell us what phenotype will result?
11. What are three different things that direct-to-consumer testing companies test for?
12. What is the difference between a test and a screen?
13. Neonatal screening asks about traits like phenylketonuria but does not ask about other traits like Huntington disease. Why would we do newborn screening for the one but not the other?
14. Once a gene has been found and a test is known, how can someone find out whether this gene is involved in their trait if there is no test available from the clinical testing labs?
15. What are two non-medical uses of genetic testing?
16. When evaluating risk associated by assaying the AFP level, why does it matter that we have a correct estimation of how advanced the pregnancy is?
17. When doing pre-implantation genetic testing, why is it possible to sample cells from an embryo without causing damage?
18. Why is it a good idea to get your medical information from a doctor or an official source such as the GeneReviews rather

than from Internet sources found through keyword searches?

19. What risk factors besides tremors might indicate a need to test for Huntington disease repeat expansion?

20. What is genetic matchmaking and what can it accomplish?

Short Essays

1. We tend to think of genetic testing as something that we do in a medical situation to look for disease risk, but genetic testing also has important uses in agriculture. Since the bovine genome sequence was completed, what sort of role does genetic testing play in artificial selection in the bovine population? As you consider your answer please read "What the cow genome tells us" by Greg Boustead in *Seed Magazine*, June 8, 2009.

2. We often think of infections as being something caused entirely by the bacteria, viruses, or parasites that cause a disease. In some cases, other factors can also influence risk of the disease. What can genetic testing tell us about the risk of contracting malaria? As you consider this question please read "A new NOS2 promoter polymorphism associated with increased nitric oxide production and protection from severe malaria in Tanzanian and Kenyan children" by M. R. Hobbs and colleagues in *The Lancet*, 2002;360:1468–75.

3. Direct to consumer testing (DTC) is the latest genetic fad, but California has now made it illegal for a company to offer genetic testing without an order from a doctor. Why would the department of Public Health in California require that a doctor be involved in any genetic testing that takes place? As you consider your answer please read "DTC genetic testing: 23andme, DNA Direct and Genelex" by Leslie Pray on the website for *Nature Education* 1(1), 2008.

4. Sometimes genetic testing results give information that can lead to treatment, but sometimes the most important result from such a test can be a change in how we behave in response to our test result. Does mutation screening have any real effect on those who are told that they carry an allele that confers higher risk of the deadly cancer melanoma, and what factors determine whether a test result has a positive effect on self-screening behavior? As you consider this question please read "Genetic testing has mixed impact on skin self-exams" by Susan London on the *Internal Medicine News* website July 1, 2010.

Resource Project

Once DNA evidence became available it turned out that a surprising number of individuals serving prison time for major offenses were innocent and wrongly convicted. In 2010 The Innocence Project website reported an 18% rate of wrongful convictions subsequently overturned, and up to 2004 more then 6% of capital cases had been found to be wrongful convictions according to the Center on Wrongful Convictions at Northwestern University. Visit the website for The Innocence Project to find out whether your state allows a defendant to appeal for DNA testing after a conviction. Write a brief essay on the conditions under which your state allows DNA testing. Discuss the number of states that do not allow post-conviction DNA testing and the number of states that provide any compensation for individuals who were wrongly convicted.

Suggested Reading

DVD

60 Minutes: A Not So Perfect Match (April 1, 2007)
60 Minutes: Roots (October 7, 2007)

Articles and Chapters

"Courts consider who owns the human genome: Myriad Genetics owns the patent over certain breast cancer genes, effectively giving them ownership over any test involving the genes" by Jane Bosveld in *Discover Magazine*, January–February 2010.

"Family ties: African Americans use scientific advances to trace their roots" by Whitney Dangerfield, Smithsonian.com, February 1, 2007; http://www.smithsonianmag.com/people-places/DNA.html.

"Genetic testing: Practical, ethical and genetic counseling considerations" by Regina E. Ensenauer, Virginia V. Michels, and Shanda S. Reineke in *Mayo Clinic Proceedings*, 2005;80:63–73.

"Personal Genome Project: These holidays, give the people who have everything the one thing they don't have: a map of their own DNA," by Eric Jaffe, Smithsonian.com, December 12, 2007; http://www.smithsonianmag.com/science-nature/wishful-genome.html.

"Beware genetic snake oil: A new report calls for increased FDA scrutiny of the genetic testing industry" by Claudia Kalb in *Newsweek*, April 3, 2008.

"Mapping the god of sperm" by Rachel Lehmann-Haupt in *Newsweek*, December 16, 2009; http://www.newsweek.com/id/227104/page/1.

"AP impact: Testing curbs some genetic disease" by Marilyn Marchione on February 18, 2010; http://news.yahoo.com/s/ap/20100218/ap_on_sc/us_med_gene_testing.

"Neanderthal Man: Svante Paabo has probed the DNA of Egyptian mummies and extinct animals. Now he hopes to learn more about what makes us tick by decoding the DNA of our evolutionary cousins," by Steve Olson, *Smithsonian Magazine*, October 2006; http://www.smithsonianmag.com/science-nature/neanderthal.html.

"Infancy" by Philip R. Reilly in *Is It in Your Genes? The Influence of Genes on Common Disorders and Diseases that Affect You and Your Family* by Philip R. Reilly (2004, Cold Spring Harbor Laboratory Press).

"Pregnancy" in *Is It in Your Genes? The Influence of Genes on Common Disorders and Diseases that Affect You and Your Family* by Philip R. Reilly (2004, Cold Spring Harbor Laboratory Press).

"Medical sleuth: To prosecutors, it was child abuse – an Amish baby covered in bruises, but Dr D. Holmes Morton had other ideas," by Tom Shachtman in *Smithsonian Magazine*, February 2006; http://www.smithsonianmag.com/science-nature/people-feb06.html?c=y&page=1.

Magic Bullets: How Gene-based Therapies Personalize Medicine

THE READER'S COMPANION:
AS YOU READ, YOU SHOULD
CONSIDER

- The difference between gene therapy and gene-based therapy.

- How we introduce genes into the human body.

- The advantages and disadvantages of viral vectors.

- Why we test gene therapy in animal models.

- Why we use different approaches to gain-of-function and loss-of-function traits.

- How animal models can advance gene therapy.

- Where you can go to get information about ongoing clinical trials.

- Why some genes are the first chosen for gene therapy trials.

- What is special about gene therapy in the eye and brain.

- Why ADA deficiency was one of the first traits treated with gene therapy.

- What a phase I clinical trial is designed to test.

- The difference between gene therapy and gene-based therapy.

- Why some gene therapy targets downstream pathology instead of primary cause.

- How small RNAs can help suppress RNA levels from gain-of-function alleles.

- Why it may be harder to treat traits that are very sensitive to gene dosage.

- How gene therapy can be used to treat situations that do not involve a mutation.

- How gene therapy can be used to treat cancer.

- How gene therapy can be used to help patients tolerate traditional cancer treatments.

- What role nanotechnology plays in gene therapy.

- How a bacterial recombinase can determine where a gene integrates in the human genome.

- How recombinant proteins can be used to treat genetic traits.

- Why ADA gene therapy stopped for a while, and why it is in progress again.

- What determines which traits are treated.

- What determines who participates in gene therapy clinical trials.

14.1 REPLACING A LOST GENE OR FUNCTION – THE RPE65 STORY

If I have seen further it is by standing upon the shoulders of giants. —**Sir Isaac Newton**

The scene on the screen in the conference room looked just like a home video, a movie showing a beautiful Briard dog named Lancelot walking into a dimly lit room. The speaker presenting the video explained that the last time Lancelot tried to navigate such a room, he could not do it without bumping into things constantly. The room before Lancelot seemed crowded, with disarranged furniture crammed into the space and scattered about. The audience in the conference room watched, spellbound, almost holding their breaths, as Lancelot made his way through the room, carefully avoiding objects as he swung his head around in an odd manner to scan the area ahead of him with one eye. He daintily picked his way through the obstacle course, never touching so much as a table leg. The film stopped. As the lights came up a few quiet spontaneous cheers could be heard over the applause that broke out around the room.

One young man pounded his fist emphatically down onto his knee in time with his head which nodded up and down. Several of the rational, objective researchers in the room had lumps in their throats and tears in their eyes as they listened to the conclusion to the presentation. Gene therapy treatment of Lancelot's right eye when he was four months old had effectively cured a canine model of Leber congenital amaurosis (LCA), a severe form of early childhood blindness that is incurable and may be diagnosed in humans in the first year of life. Those attending the talk had just witnessed a medical miracle: a "blind" dog that could walk through a crowded, unfamiliar room and successfully avoid contact with objects. Lancelot could see with his treated eye!

Lancelot and some of his relatives develop vision problems because of a naturally occurring defect in a gene called RPE65. Since both copies of the gene are defective, the obvious approach to gene therapy was to put a good copy of the RPE65 gene into the cells of Lancelot's eye. The strategy proved valid when the three blind puppies who were treated turned out to be

FIGURE 14.1 Lancelot, the Briard dog, visits the US Congress to highlight gene therapy research. (Photo credit: Foundation Fighting Blindness.)

cured, and they stayed cured! Since then many more of Lancelot's relatives have been similarly treated and cured. The movie starring Lancelot has played to audiences of scientists from around the world, and Lancelot has even visited Capitol Hill to attend a congressional briefing on gene therapy. To the scientists in the conference room, the concept of using this approach to cure blind children was emotionally compelling in addition to being scientifically attractive. Since then, the idea of applying such cures to humans has moved beyond the theoretical as the first human RPE65 gene therapy trials have led to improved vision in study participants with LCA. The general approach looked as if it might even be usable for some other recessive forms of inherited retinal degenerations, too.

However, many gene therapy projects have not been so successful. Why can't all of the other diseases in need of gene therapy simply be treated in the same way as the Briard dogs were treated? Not all diseases can be treated this way because there are a broad array of technical and strategic issues to be sorted out that differ from one disease to the next and from one gene to the next. In this chapter, we introduce you to how gene therapy works and to some of the issues that keep gene therapy researchers in their labs burning the midnight oil in search of answers.

After great expense of time and resources on the part of many really, really smart people, we finally know the sequence of the human genome (and many other genomes, as well). The genes have been found (well, many of them, anyway). We are starting to find out what some of the gene products do. Biochemical pathways are coming together that provide us broad conceptual insights into a variety of pathogenic processes. Those of us who consider this a beginning, not an end, now face the critical question: What do we do with all of this knowledge? How do we convert all of these advances into help for people who are not adequately helped by the current state of medical knowledge?

The hope that comes from successful gene hunts points in the direction of gene therapy, the therapeutic use of the discovered genes themselves, and not just the knowledge gained from finding those genes. Explorations of the idea of human gene therapy began even as the first human genes were being cloned. One of the first genes proposed as a serious target for gene therapy was adenosine deaminase, but the gene therapy field has expanded to the point that the NIH Clinical Trials website lists more than a thousand ongoing gene therapy trials out of more than ninety thousand clinical trials of all kinds. These gene therapy trials target many different diseases. In this chapter we will talk about a variety of gene therapy strategies that are currently being used, others that are being developed, and gene-based therapies that target something about a gene without actually putting a copy of that gene into the body.

For the purposes of this chapter, we start our discussion of gene therapy after the genes in question have already been found. We start with traits for which we know for sure exactly what needs to be added or replaced, information that came out of mapping in families, association studies in populations, and animal model studies including knock-out or knock-in animals that have either lost the gene or gained a specific mutation in the gene. By the time we start working on gene therapy, we know what the gene is, what it does, where it is expressed, and how it fails in the disease situation, and we have an animal model in which to carry out the first rounds of gene therapy. This lets us do a very detailed study of both safety and efficacy before we first try it out on a human being.

In the case of Lancelot, we are dealing with the RPE65 gene. This gene produces a protein that carries out a critical step in the visual cycle pathway, which you may recall from Chapter 9 as the pathway that processes vitamin A, essential for photoreceptor function. The defect in Lancelot's eye is a simple monogenic trait caused when both copies of the RPE65 gene have been lost or become defective. Lancelot the Briard dog provided the proof of principle: that it is possible to simply put in a new copy of the RPE65 gene and get restored function

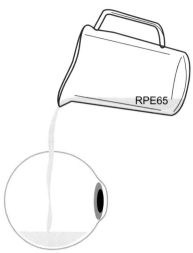

FIGURE 14.2 Gene replacement therapy adds back a functional copy of a gene in cases in which the disease results because defects in both copies of the gene cause loss of the cell's ability to carry on the functions normally handled by the product of that gene. In the case of the Briard dog Lancelot, many good copies of the REP65 AAV gene therapy construct were added into his eye in the vicinity of the retinal pigment epithelium cells that lacked the RPE65 protein activity that normally takes place there. Those copies of the RPE65 gene construct were carried into the cells by an adeno-associated virus (AAV) gene therapy vector.

(Figure 14.2). By using recombinant DNA technologies, the researchers were able to insert the RPE65 gene into a gene therapy vector made from an adeno-associated virus to create the construct.

Why is an understanding of Lancelot's blindness and cure of such great importance to the human population? Defects in RPE65 cause a form of severe, early vision loss in children who have a trait called Leber's congenital amaurosis (LCA). According to the Foundation Fighting Blindness, children with LCA often show substantial visual deficits while they are infants and electrophysiology tests show little or no detectable function of the retina. As one of these children grows up, she might be able to see well enough to count the fingers on a hand held in front of her face. Another child might be limited to detecting bright lights and the motion of a hand moving in front of his eyes. But the importance of gene therapy for an RPE65 defect reaches far beyond the impact on these individual children. Successful gene therapy for RPE65 provides a strong proof of principle – that gene therapy can work when we get the details right. However, this does not mean that the details that make gene therapy work for RPE65 will work for Huntington disease or cancer.

This study of Lancelot and his relatives is typical of how gene therapy development proceeds. First a gene was identified and shown to be the cause of the disease through studies of animal models and human subjects with the disease. Then gene therapy was tested in an animal model. Finally, once enough was known about the mechanisms of disease pathology, the cell types involved, and the events taking place in gene therapy of the animal model, human gene therapy trials began.

The Human RPE65 Gene Therapy Clinical Trial

Because gene therapy of Lancelot and his relatives was so successful, testing of RPE65 gene therapy in humans has begun. As we write this book, the National Institutes of Health lists seven different clinical trials, six of them in phase I testing of safety and side effects of the treatment, and one of them moving into phase II where testing of larger numbers of subjects will allow further evaluation of vision improvement. To start such a study, very small numbers of subjects receive the gene therapy construct in a test of the safety of the construct. Secondarily, researchers also evaluate visual function. Inclusion criteria for the study call for subjects to have substantial visual impairment but to not be completely blind so that they can carry out visual testing to determine what effect the treatment is having on their visual abilities. So far, not only has the treatment proven to be safe, but the success goes far beyond that.

Among the first 12 subjects treated, the gene therapy treatment was found to be safe down to as young an age as eight. *Even though the first phase was just a safety phase, the initial studies now show improved vision in 20 subjects, with the greatest improvements showing up in the youngest subjects!* Additional projects will move the studies into phase II to further evaluate safety and efficacy, and additional phase I trials will test safety in other groups such as even younger subjects. So far the vision research community is very excited at evidence that gene therapy for RPE65-defective LCA appears to work, and the outcome appears to persist.

When we succeed in doing gene therapy for a particular trait, or targeted at a particular tissue or pathway, we learn things that let us improve our ability to treat other traits with similar features such as affected tissue or mode of inheritance. There are many other genes that can cause LCA when defective, including other recessive forms of LCA. In developing RPE65 gene therapy researchers learned to work with a gene therapy vector (something we will discuss in more detail later in this chapter), and once they have the vectors and delivery systems worked out then it will potentially be much faster to develop replacement of the next LCA gene. There are several kinds of recessive retinal degeneration caused by defects in both copies of a single gene that could likely respond to almost exactly the same therapeutic protocol, with almost the only change from Lancelot's treatment being the choice of which gene to put into the eye. We expect that as more details of RPE65 gene therapy are worked out, the information gained will apply not only to treatment for disease caused by RPE65 defects but will begin helping with design of gene therapy for other forms of LCA and for many other single-gene loss-of-function vision defects (Box 14.1). One of the biggest next challenges will be moving beyond LCA to treat other more complex situations such as Usher syndrome – which affects both the eyes and the ears.

BOX 14.1

RESTORING FULL COLOR VISION

Many of the traits we talk about treating are rare single-gene traits, but some of the problems looming on the gene therapy horizon are not so rare. A common form of colorblindness is found in about 8% of men. There are a variety of different mechanisms that can cause such colorblindness but in many cases the man with the color vision defect is simply failing to make either the red opsin (the photoreceptor protein that sees red light) or the green opsin (the photoreceptor protein that sees green light). Researchers at the University of Washington recently did gene therapy in an animal model of colorblindness, male spider monkeys who do not make a red opsin and cannot see a full spectrum of colors. By adding a human red opsin gene, they were able to give the monkeys the ability to detect colors they have never before been able to see. It took about five months after the treatment before the monkeys started showing the ability to detect red, and two years later they were still able to see red. This exciting result suggests that it may become possible to use a similar approach to treat color vision deficits in humans. This is especially happy news for Scott and millions of men world wide who wonder what the world looks like to those who make three different kinds of color opsins.

14.2 REPLACING A LOST GENE – ADA DEFICIENCY

In 1972 David Vetter was born with severe combined immune deficiency (SCID). Up until this point in time, children with SCID routinely succumbed to fatal infections, as had David's older brother, who also had SCID. When David was born he was transferred into a plastic containment bubble, and was raised there in sterile isolation until the age of 12. In an effort to free him from his terrible, isolated existence, a bone marrow transplant was performed. Although at first the transplant seemed to be working, he succumbed to one of the complications that sometimes accompanies bone marrow transplants, cancer caused by a virus that had been undetected in the donated bone marrow. David had grown up with friends who had to interact with him across the divide of his sterile barrier. He grew up with family who loved him, but who never got to touch him until he was dying. David's life, lived just out of reach of the people who loved him, has been the subject of books,

documentaries, and movies. Science has now taken steps beyond that primitive use of simple mechanical barriers to save the life of the child often referred to as the "bubble boy." One of the first treatment steps was the administration of the actual ADA enzyme as a medication, and work then progressed to use of gene therapy for SCID due to adenosine deaminase (ADA) deficiency. The story of ADA gene therapy reflects the kinds of advances and setbacks that have kept gene therapy moving forward while keeping it from moving out of the research arena into the offices of all of our local family doctors.

ADA deficiency is a life-threatening trait that results from a defect in a single gene. Bone marrow transplant continues to offer a potentially permanent cure, but it can be difficult to find a donor who is an adequately close match, and a transplant risks outcomes such as the one that took David Vetter's life. Since the time when David Vetter lived in his bubble, there has been progress with ADA deficiency on several fronts. Children with ADA have been treated with

PEG-ADA, a version of the ADA enzyme that helps to clear out some of the toxic metabolic intermediates and improves immune system function but does not restore complete health. The first case of human gene therapy, carried out in 1990, was treatment of ADA deficiency. ADA deficiency was an attractive target for development of one of the first gene therapy projects for several reasons:

- The severity of the trait cried out for a new treatment approach.
- The therapeutic strategy was simple – replace a single missing gene and gene product.
- The gene was known and had been cloned.
- The gene was small enough to be put into a gene therapy vector.
- Treatment was expected to work even if they did not fix 100% of the cells.
- They had good assays for whether or not the treatment was succeeding.
- Treatment could target a very accessible set of white blood cells, not something complex and inaccessible buried in the brain.

The First ADA Gene Therapy Treatment in 1990

The toddler Ashanti DeSilva had ADA deficiency and the treatments with the PEG-ADA version of the enzyme were gradually having less and less effect. Introduction of the ADA gene into circulating blood cells resulted in cells that could produce the enzyme, but circulating cells in the blood are a transient population of cells that has to be renewed from the bone marrow. Over the next few years Ashanti stayed healthy but needed the gene therapy treatment repeated periodically along with supplemental treatment with PEG-ADA.

The next big breakthrough in ADA gene therapy came when treatment was carried out on cells from the bone marrow. This approach seemed to be succeeding, producing a self-renewing population of cells that had been corrected for the genetic defect. Then, two of the patients who had had their ADA deficiency repaired developed leukemia! Investigation showed that the leukemia was the result of something specific to the particular viral vector that had been used. Newer vectors look like they have eliminated this problem and there is optimism that gene therapy for ADA deficiency will be able to keep moving forward.

As each new study of this kind stumbles and picks itself back up to keep going, the goal is to save the lives of those who desperately need help. These studies can only proceed through the participation of the study subjects who help to test whether the treatments are safe and then whether the treatments work. There are now thousands of gene therapy clinical trials ongoing and many more that have been completed. Out of all of these we can tally four deaths, but have you ever thought of how many people die on a regular basis in response to a standard often-used medication? How many have died because they had an allergic reaction to a drug, or because they lost control of body temperature in response to general anesthesia? We lack the data that would let us compare the rate of deaths from gene therapy to the rate of deaths from other kinds of studies, or the rate of death of individuals with similar health histories who are not participating in clinical trials. Each study participant is a hero, someone brave and determined, engaged in a fight for the lives of those who need the treatments that are being developed (Box 14.2).

14.3 TARGETING DOWNSTREAM DISEASE PATHOLOGY

In some cases, we may be trying to compensate for a problem that is too genetically complex to tackle at the point of the disease gene itself; in other cases, the trait may not even be genetic in its origins. In such cases, we may need to simply

BOX 14.2

A HERO AMONG US

In 1999, when Jesse Gelsinger was 17, he had a goal that was amazingly different from that of his high school classmates in Tucson, Arizona. Across North America, seniors in the spring of 1999 were talking about what colleges they would attend, applying for jobs, planning weddings, and deciding whether to enlist in the service. While they planned educations and careers, Jesse was waiting to turn 18 because that was the magical age that would let him become a human subject in a gene therapy research project. How did this young man come to such an extraordinary, selfless view at a time when many his age were focused on themselves and the complex transitions going on in their lives? Some of the answer comes from Jesse's own medical history. Jesse suffered from a mild form of the same recessive disease, ornithine transcarbamylase (OTC) deficiency, that kills the severely affected babies he wanted to help. Jesse could identify with the danger to these infants, even though he had never met them, because Jesse himself could not make enough of the OTC protein, which is part of the urea cycle that is used to remove excess nitrogen that enters our bodies when we consume proteins. If the urea cycle doesn't work, the nitrogen from the proteins accumulates in the form of ammonia that can cause brain damage. Ammonia production can be limited by a low-protein diet and medications, but the one baby in 25,000 who is born with OTC deficiency can usually be expected to go into a coma within days of birth. Even with medical help, many OTC children suffer permanent brain damage; many die before they are one month old, and almost half die before the age of five years. However, in some individuals like Jesse the disease is less severe because only some of their cells carry the genetic defect. Seventeen other people before Jesse had walked safely away from participating in the phase I test of safety of the OTC gene therapy protocol. Jesse responded to the treatment by going into multiple organ failure. There remain questions about why he died, whether or not the particular viral vector choice might have played a role, and whether partial OTC-deficiency played a role in Jesse's death. The result was that the OTC gene therapy trial was discontinued. Jesse's act of heroism ended up improving the whole field of gene therapy by causing re-evaluation and changes throughout the field. But OTC gene therapy in humans has not resumed. We wonder whether Jesse would see the OTC gene therapy program itself as the second casualty in this terrible circumstance. Jesse was a true hero among the many heroes who have made gene therapy possible, and we look forward to the day when his dream finally comes true, when babies with OTC get to live healthy, normal lives.

bypass the whole issue of which gene (or what else) is causing the disease, or even how many genes are involved, and target some other aspect of the disease pathology, things that take place downstream of the initiating events of the disease (see Figure 14.3). Sometimes what is needed is to add a different gene that can supply a function that improves the body's ability to put up with the damage being caused, or that provides a mechanism to assist the body in recovering from damage that has been caused.

An exciting example of this kind of "end-run" gene therapy are some of the approaches being developed to treat cardiovascular disease. Going after the downstream pathology becomes especially important for some phenotypes such as

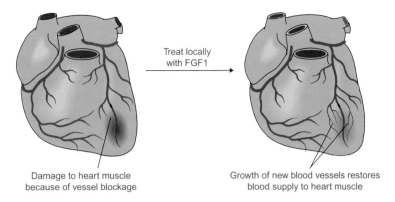

Treat locally
with FGF1

Damage to heart muscle
because of vessel blockage

Growth of new blood vessels restores
blood supply to heart muscle

FIGURE 14.3 End-run gene therapy. In some cases, gene therapy can be used to treat a disease without going after the primary causes of the disease. This artist's conception shows how uses of growth factor FGF1 can cause new blood vessel growth in a local area of the heart to restore blood supply to a region previously supplied by a blocked vessel. Successes of this kind have been seen in animal models, and some early human studies in gene therapy of cardiovascular diseases are ongoing.

cardiovascular disease because the patient frequently becomes available for treatment of any kind only after damage has taken place – often a heart attack that has damaged the heart muscle. A complex array of genetic and dietary features lead to a heart attack, and the medical community is already working on that end of the problem. But we will continue needing a way to intervene in those cases where damage has occurred. We can accomplish a lot if we target some of the downstream problems by looking for ways to clear plaque from arteries or improve the health of the damaged heart muscle.

Researchers have shown that a growth factor called FGF1 can be used to stimulate local growth of new blood vessels to supply heart muscle in cases in which blockage is reducing the blood supply to the heart (Figure 14.3). In patients with damaged heart muscle, the combination of genetic and environmental factors that could have caused this is likely complex and different for different individuals. Yet a single treatment approach that goes after the secondary problem of getting a blood supply to the heart could completely ignore the difference in underlying causes among the patients yet still

successfully restore oxygenation of heart muscle. Introduction of genes encoding other growth factors such as vascular endothelial growth factor (VEGF-1) are also being developed for this purpose.

Other gene therapy projects target many different aspects of cardiovascular disease. One study used gene therapy to provide an APOE gene that produces an APOE protein that helps reduce "bad" cholesterol, resulting in disappearance of plaque attached to blood vessel walls. Some studies are working on ways to deliver gene therapy into the walls of blood vessels to help clear plaque and improve the health of the vessels. Other studies are looking at how gene expression is changed by the use of medications commonly used to treat heart disease.

14.4 SUPPRESSING THE UNWANTED GENOTYPE – USE OF siRNAs AND miRNAs

In the case of gain-of-function mutations (remember the concept of the monkey wrench) such as Huntington disease, we can't use the

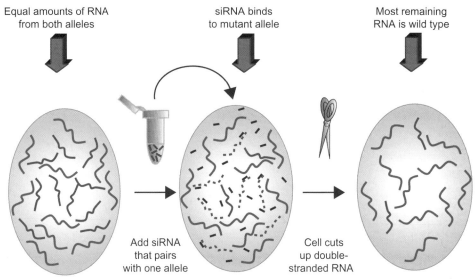

Equal amounts of RNA
from both alleles

siRNA binds
to mutant allele

Most remaining
RNA is wild type

Add siRNA
that pairs
with one allele

Cell cuts
up double-
stranded RNA

FIGURE 14.4   Gene suppression therapy. If the problem can best be solved by reducing the amount of a gene product (or its activity levels) a variety of technical approaches can be used. RNA interference is one of them. Small interfering RNAs (siRNAs) can trick the cell into digesting and getting rid of RNA to which it binds and thus reduce the amount of the gene product in the cell. If the siRNA is homologous to the causative mutation and binds to it, but not to the normal sequence, then the cell will selectively chew up the RNA to which the siRNA is bound (the RNA carrying the mutation) while leaving the wild type RNA relatively untouched. The outcome will be a cell that still has a normal amount of the wild type RNA for that gene, but that has reduced amounts of the RNA from the gene copy with the mutation. This works well where the cell can tolerate some reduction in total RNA from that gene and where improvement can result from reducing the amount of the RNA with the mutation without completely eliminating it.

approach we used for RPE65. There is already one good copy of the gene present in the cell and putting in more good copies of that gene may not help the situation. However, the situation can be helped by therapeutic approaches aimed at getting rid of the unwanted monkey wrench or the by-products of its misbehavior. So if the problem involves a toxic by-product, the use of gene therapy techniques to reduce the amount of a specific RNA can lead to reducing the amount of gene product being made. As you may recall from Chapter 3, small interfering RNA technology can reduce the amount of transcript coming from the offending gene; the treatment adds many copies of an RNA that is so small that it is readily taken up by the target cells. The sequence of this small RNA is complementary to the sequence of the mRNA produced by the disease gene allele. Because of the sequence complementarity, the *small interfering RNA (siRNA)* can bind to the mutant transcript and get the cell to destroy the RNA coming from the disease gene (Figure 14.4). In some cases, it is conceivable that the siRNA can be designed so that the transcript from the disease allele will be destroyed at a higher rate than is the transcript from the normal allele, allowing for the possibility of reducing the amount of a toxic byproduct while still allowing for some normal protein to carry out the normal function. Other strategies work at the level of the gene product, by adding in a gene whose product will chemically activate or inactivate the problem gene product.

In some cases the problem is more complex than the simple presence of the defective allele. For some genes we not only need to get rid of the monkey wrench, but we also need to keep a normal level of the transcript. Some projects working on this strategy are using two approaches together – siRNA to reduce the level of the defective alleles plus gene therapy to restore the overall level of transcript from the normal allele to its normal level. This is only needed for genes that cannot tolerate a reduction in overall RNA levels.

MicroRNAs (*miRNAs*) have been shown to play a role in a large number of different biological processes. In the case of heart disease, miRNAs have been shown to be associated with the development of cardiovascular problems such as cardiac arrhythmias, heart failure, and atherosclerosis. Among approaches to cardiac disease being explored we find the use of anti-miRNA oligonucleotides (AMOs). These AMOs are short nucleotides homologous to the miRNA shown to cause a problem. When the AMO binds to its complementary miRNA, it effectively reduces the final amount of protein product resulting from the mRNA. One of the problems with the development of AMOs is that they are small enough that their short runs of sequence can often match up to sequences from many different mRNAs in the body, not just the target we are after. Because therapeutics makes use of artificially synthesized oligonucleotides, and because we have the whole human genome sequence available, we can select regions of sequence that are unique to the target mRNA when we design our synthetic AMO; we are not limited to using the miRNA sequence that is available in a human cell.

14.5 GENE SUPPLEMENT THERAPY – MORE OF THE SAME

In some cases, tissues in the body simply need to be making more of something they already make. The item to be supplemented is not missing and the gene is not mutated. One of the situations in which this approach is

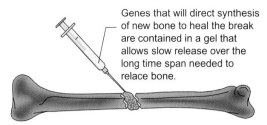

Genes that will direct synthesis of new bone to heal the break are contained in a gel that allows slow release over the long time span needed to relace bone.

FIGURE 14.5   Gene supplementation therapy. An example of this strategy is the use of gene therapy agents that can induce cells in the bone to manufacture new bone. This is especially important in cases of severe fractures and fractures that do not heal well. By embedding the gene therapy agents in a gel at the site of the break, it is possible to have slow release of the DNA and gradual expression of the relevant genes over the extended time period needed for bone healing. By using a bone morphogenic protein at the site, bone growth is stimulated locally without such activity going on at unwanted locations elsewhere in the body.

being used is to get cells to make the proteins necessary for the formation of new bone material (Figure 14.5). In these cases, the patient does not have a defect in bone formation but rather has an injury of some kind that is more than his own body can heal easily. Gene therapy treatment of skin cells with bone morphogenic protein before placement of the cells into a region of bone erosion in periodontal disease can lead to formation of new bone in the region. Another approach places the gene therapy agents and cells into a gel placed at the point of a break in a bone, with gradual release over time resulting in sustained expression of the genes being used in the treatment.

There are many other phenotypes that call for this kind of supplementation. Any defect caused by hemizygosity – having only one functional copy of the gene instead of two – could benefit from this kind of strategy. This includes a variety of transcription factors, where diseases such as aniridia or Rieger syndrome result from having only one good copy of the gene. While the idea of restoring copy number seems like a simple fix for this problem, it is important not to overshoot the amount, not to end up with multiple additional copies. In the case of Rieger syndrome we see that the disease not only results from lacking a copy of the gene but can also result from having an extra copy of the gene. When we give someone a new copy of a gene, it does not simply replace the existing copy, so the new gene goes into the cell in a new context where the regulation of expression may be different. So as we do gene supplementation therapy we face the added problem that for some genes we add back the dosage has to be exactly right, not just in terms of having enough but also in terms of not having too much.

14.6 STRATEGIES FOR CANCER THERAPY

Gene Therapy to Target Tumor Cells

In some cases, especially with cancer, what we really want is to be able to destroy specific cells while leaving the surrounding cells intact. There are a lot of different ways to kill cells, and we can see some of the diversity of possibilities when we look at available cancer therapies. Similarly, gene therapy approaches to cancer use many different strategies. While some focus on killing cancer cells, other strategies may simply aim to give the cell back the ability to control the cell cycle and regulate its growth.

One of the approaches that has made it as far as phase III clinical trials in the United States, and is in clinical use in China, is the delivery of p53 into tumor cells. As you will recall from Chapter 10, loss of p53 is a primary cause of some cancers, and p53 is also lost in secondary steps of some cancers as tumor stages progress. Adding p53 back through gene therapy has been reported to be beneficial in treatment of cancers of the head, neck, and lungs. There remain a variety of problems with such delivery since we normally cannot get effective introduction of the construct into every cell in a tumor.

An especially ingenious idea was developed by researchers who want to use a "suicide vector" approach to destroy malignant brain tumor cells while leaving the surrounding brain cells untouched. Brain cells are not usually thought of as growing or dividing. Some viruses will infect either dividing or non-dividing cells, but there are types of viruses that infect only actively dividing cells. By selecting the kind of virus that infects only actively dividing cells, we can target the gene therapy into any actively dividing cells in the treated region while not treating the non-dividing cells that surround them. Use of such a virus lets us target an aggressively growing tumor while leaving the brain cells surrounding it untouched. Administration of an antiviral drug called gancyclovir will expose many of the brain cells to gancylcovir, but the drug will be harmless to most of the cells in the brain. It will specifically kill only those cells that have taken up the virus, so the tumor cells will die but surrounding tissues will remain intact (Figure 14.6).

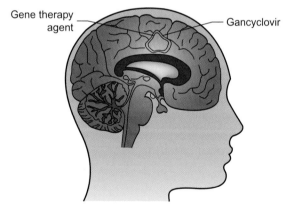

FIGURE 14.6 Magic bullet therapy. Many different strategies are being developed for being able to target therapy in such a way that only the tumor cells die while the normal cells remain healthy. One strategy is to use two different therapeutic agents that are each benign alone and kill cells only where both agents are present. Use of a gene therapy virus that can only infect dividing cells will tag tumor cells while sparing surrounding nondividing cells. A secondary treatment kills only tagged cells. This strategy would not work in many tissues of the body.

This concept, that cell death will occur only where two separate events coincide, resembles a process in current use in cancer treatment. In this process, low-level radiation administered from multiple different directions spares the surrounding tissues while killing only those cells present at the point where multiple radiation beams come together at the same place to result in a dose high enough to kill the cells. Other research groups are trying a variety such approaches that call for cell death to occur only where two different events come together, thus sparing any cells that are exposed to only one or the other of the two items. And one of the ways in which gene therapy approaches can best help limit delivery is through use of viruses that in some way selectively infect the tumor cells as compared to cells of the surrounding tissues.

Gene Therapy to Improve Effectiveness of Traditional Cancer Therapies

An intriguing concept in cancer therapy is to increase the effectiveness of chemotherapy by doing gene therapy that lets the patient tolerate a higher level of chemotherapy. If we put a gene encoding a protein that pumps specific chemicals out of the cell (a pump protein) into bone marrow cells to increase their resistance to the effects of anticancer drugs, while not putting that same gene into the tumor cells, we can increase the therapeutic dose without increasing the damage to the bone marrow that is usually one of the worst complications of treatment (Figure 14.7). Clearly this approach will not work for any blood-based cancers such as leukemia, but could be a real boon to anyone with a solid-tissue tumor who has to undergo chemotherapy.

When we look at what gene could be added to provide such protection we are struck by the idea that we might even be able to get this effect without having to add in a gene! How could that be? One of the genes that we would most like to be able to add to the bone marrow in someone who has to undergo chemotherapy for cancer elsewhere in the body is a gene encoding a pump protein. But we know that the body has natural mechanisms for up-regulating expression of such pump proteins. In fact, this up-regulation in tumor cells can sometimes be a problem that can interfere with treatment. So if we can learn how to regulate expression of this gene, then the optimal approach would be to down-regulate the pump protein in the tumor cells and up-regulate it in the bone marrow and lining of the gut. This then becomes a gene-based therapy that is not actually gene therapy in a classical sense.

A similar strategy, whether through gene therapy or through regulation of expression of genes such as those encoding pump proteins, might reduce the hazard of living or working in a contaminated environment, including environments

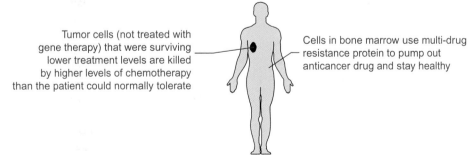

Tumor cells (not treated with gene therapy) that were surviving lower treatment levels are killed by higher levels of chemotherapy than the patient could normally tolerate

Cells in bone marrow use multi-drug resistance protein to pump out anticancer drug and stay healthy

FIGURE 14.7 Supplemental gene therapy. Another use of supplemental gene therapy is to boost the ability of the patient to survive higher levels of chemotherapeutic agents being used to attack the tumor cells.

that increase our risk of cancer. During our lives, we suffer a variety of exposures that can be directly harmful or can increase our risk of things such as cancer. As we learn more about the normal mechanisms used by the body to eliminate toxic substances, more about biochemical pathways that can convert toxic substances into safe (or safer) substances, and more about ways to get compounds pumped out of cells or excreted from the body, we gain the potential to use gene therapy to protect us from exposure or to clean up our internal environments once we are exposed. The same pump proteins that can affect whether therapeutic levels of chemotherapy are getting into target cells or other cells in the body can also serve to help pump out some toxic compounds coming in from the environment.

We have presented only a couple of examples of things being tried, but we want you to understand that for cancer, as for many of the other traits we discuss in this chapter, a wide variety of approaches are being tried. We are struck by the intelligence and creativity that seems to go into the design of many of the new strategies being developed.

14.7 GENE-BASED THERAPY INSTEAD OF GENE THERAPY

Use of Recombinant Proteins as Therapeutic Agents

Sometimes the answer is gene therapy, and sometimes the answer is gene-based therapy where we are using genes to produce the therapeutic agent without actually putting the gene itself into the patient. This concept of gene-based therapy rather than gene therapy is not even terribly new. One form of gene-based therapy is the use of a recombinant protein, that is to say a protein that has been produced through use of an expression vector that contains a copy of the gene encoding that protein. For many years, recombinant tissue plasminogen activator (TPA)

has been produced outside of the body and then injected so that it can help reduce clotting in individuals such as those having heart attacks or strokes. Recombinant human insulin, made from a cloned version of the human insulin gene, allows diabetics to use human insulin instead of pig or cow insulin. A variety of other growth factors, blood clotting agents, and other products that our own bodies normally make can now be synthesized outside of the body and then added back in. In forms of hemophilia, as different individuals have been found to be lacking a specific blood clotting factor it has become possible to purify or produce a recombinant form of that blood clotting factor to be used therapeutically. Knowing that insulin is what is missing in children with type I diabetes lets doctors supply them with insulin, the protein product of the insulin gene. In each case, what we are talking about is not gene therapy, because we are not putting any genes into any of these patients, and yet it is gene-based therapy because the therapy is designed directly from the understanding gained when it was determined what gene or gene product was missing.

Use of Drugs That Can Affect Transcription or Translation

One of the approaches being used to alter gene expression for therapeutic purposes involves inducing the cell to turn on an alternative gene to compensate for the gene that is functioning properly. As you may recall from Chapter 5, sickle cell anemia results when both hemoglobin copies are the HbS allele, resulting in red blood cells that take on a sickled shape and get stuck in capillaries. Studies in a mouse model of sickle cell anemia showed that gene therapy could add back a normal globin allele and convert sickle cell anemia to sickle cell trait. But for many years it has been known that sickle cell anemia could be helped through the use of a gene-based therapy that is not itself gene therapy; by using a chemical called hydroxyurea doctors are able to turn on expression of a fetal form of hemoglobin to help compensate for the defective form of adult hemoglobin that is the hallmark of the disease. Thus at this point the development of the pharmacologic approach using hydroxyurea is being used even as gene therapy for sickle cell anemia is being worked on.

In the case of Duchenne muscular dystrophy some treatment approaches being tried may generalize to use for other traits. Clearly these drugs are not yet optimal but they are both promising enough that work on them is continuing. The demonstration of effective treatment with these drugs would lay the groundwork for similar approaches to other traits. One drug being tried is an anti-sense oligonucleotide that binds to the transcript and leads to efficient skipping of exon 51, an exon in which many Duchenne's mutations are found. Boys with exon 51 causative mutations constitute 13% of the Duchenne muscular dystrophy population, and skipping of this exon restores an open reading frame to the protein. Targeting of exon 51 is expected to be especially effective since studies of large deletions spanning this region show that onset of symptoms is delayed until very late in life. Studies in animal models suggest that this type of treatment can be tolerated (and work) for a long period of time. A lot of work will be needed to develop a similar exon-skipping drug and protocol for each of the 76 exons in the gene that have mutations that cause Duchenne muscular dystrophy. In phase I clinical trials muscle biopsies show the presence of dystrophin protein, something these boys were not previously making and studies have moved onto a stage I/II clinical trial to continue testing safety while looking at efficacy!

Other researchers are looking for ways to get the muscle cells to read through nonsense mutations that cause truncation of the dystrophin protein. The first drug found, an antibiotic called gentamicin, results in production of dystrophin but it is too toxic to take long term. Screening of large numbers of compounds has led to the identification of other possible drugs

that seem to be able to get this read-through effect for dystrophin without interfering with use of the normal stop codons present in other genes. It remains to be seen whether they can be tolerated for continuous treatment lifelong. Variations on this approach are also being tried for cystic fibrosis.

Drugs That Target the Biochemical Defect

Recently researchers set what may well be a new record for the elapsed time from the gene hunt to the treatment. In 2003 Francis Collins and his colleagues reported that mutations in lamin A (LMNA) cause progeria, a severely premature aging syndrome. Progeria is so rare that there are usually fewer than twenty individuals in the whole United States who have this trait at any one time. The average life span of these children is 13 years, but some of these children may occasionally live as long as 20 years. Those affected are very short and often under-weight for their height. They have large, bald heads, prominent eyes, an unusual gait, and many other distinctive features. Although many of these features of appearance are how we so readily recognize these children, the key issue in progeria is the cardiovascular disease that ends up taking their lives. By 2005 researchers knew that a key to the disease process was a modification of the truncated lamin A protein produced in the bodies of these children; if a farnesyl chemical group is present on the protein, the protein remains anchored in the membrane. The result is messed up nuclear membrane architecture that has a typical appearance called blebbing. Researchers found that if they used a drug called a farnesyl transferase inhibitor (FTI) to keep the protein free of this chemical modification, they could help restore the nuclear architecture. As a sign of how successful this was, the nuclear blebbing disappeared. Researchers went on to show that treatment with FTIs could help prevent the cardiovascular disease in a mouse model of progeria, and could even undo some of the damage in progeria animals who had already developed cardiovascular disease. In 2007, just four years after the cause of the disease was first reported, the first clinical trial began that uses this gene-based therapeutic approach – treatment with FTIs. There are now three different clinical trials under way to test a combination of three drugs together, a statin to take a traditional approach to cardiovascular problems such as cholesterol, a farnesyl transferase inhibitor to directly address the underlying cause of the disease, and a bisphosphonate to deal with other problems of aging such as osteoporosis. Although there are some aspects of the phenotype that may not yet be solved by this treatment approach, the mouse model suggests that this may help give these children more years of life, and years with higher quality of life. For three of the children in this study who are older than 13, this treatment breakthrough has come just in time for kids feeling like they were living on borrowed time.

14.8 DELIVERING GENE THERAPY

One of the tricky parts of gene therapy turns out to be getting the gene into the cell. We end up balancing a trade-off between the efficiency with which we get the DNA into the cell and the side effects that take place as a result of the delivery method. Some of the methods with the lowest side effects are so much less efficient than viruses that they cannot achieve a therapeutic level of gene transfer. Does this mean that we give up on them? No, it means that we keep working on the technical development of these approaches in search of breakthroughs that could offer new options.

Use of Viral Vectors

One of the earliest gene delivery strategies involves the use of *viral vectors*. Such vectors are engineered to remove their disease-causing properties and to give them the ability to carry human genes along with their own DNA. There are several advantages to the use of such viruses. They greatly enhance the ability to get DNA into the cells, and such efficiency turns out to be critical to the success of such projects. Another advantage is that viruses can sometimes help restrict which cells get targeted for treatment. This lets us put the treatment into some cell types while keeping it away from others.

Viral vectors also allow for mass production. No, we are not seeking mass production for some industrial economic reasons (although the economic issues are worth considering, too!). Rather, being able to do mass production allows for kinds of quality control that we cannot do if we manufacture one dose of something each time we do a treatment. If you can make one gene therapy construct, prepare huge amounts of it, and do extensive testing on it, you can still have enough at the end to use in treatments. This lets you know that the batch you are using to treat people with is the same batch that passed all of the safety testing.

The biggest disadvantage to viral vectors is the tendency of the body to mount an immune reaction. In some cases, if what is desired is the destruction of a particular cell type, the use of vectors that invite an immune reaction may actually enhance the therapy, but it would not be useful in situations that call for repeated treatments because the virus would not be able to get to the target cell once the body develops an immune response to the virus.

There are differences between the different viruses used in vector construction. Adeno-associated virus integrates into the genome to give stable, long-term expression, and it can infect both dividing and non-dividing cell types. It can only be used for some gene therapy projects since it cannot accommodate really large genes, and the processes for producing it are less efficient than for some of the other vectors. Adenoviruses can take larger inserts but tend to produce a much stronger immune reaction than the adeno-associated viruses. Other viruses offer differences in sizes of genes that can be accommodated, how immunogenic they are, whether they can infect particular cells types like nerves, and how stable or transient the resulting gene therapy will be. This gives gene therapy designers a lot of choices, but it is not possible to do a simple mix-and-match selection of an exact profile of features the researcher wants. There are a limited set of viruses currently under development and in most cases the advantages end up having to be weighed against disadvantages.

Non-viral Delivery Systems

There are other ways to get genes into cells that don't use viral vectors. In some cases, DNA copies of the gene can be packaged into *liposomes*, lipid packets that surround the DNA and help carry it into the cell. In other cases, direct injection of DNA can be carried out but would only get the DNA into a very limited set of cells. There are some promising approaches being used that call for removing cells from the body, carrying out the delivery of DNA in a cell culture dish, and returning the cells to the body.

One very important alternative delivery system uses a plasmid instead of a virus as the delivery system. By including a nuclear localization signal normally present on a virus, the researchers are able to get the DNA to go on into the nucleus. To get around the lack of an efficient viral system for getting the DNA into the cell and then into the nucleus, DNA bound to a protein called phi31C is delivered into the cell through use of hydraulic pressure to push the DNA into the cell efficiently. The phi31C bacterial recombinase is a protein that specifically inserts DNA into one of a very small number of locations in the human genome. Once inside the cell, the gene being introduced is stable and can't come back out or move to a new location. The small number of ph31C sites in the genome are not located in any of the oncogenes or other coding sequences of concern, so this system helps to overcome one of the problems we face with gene therapy: that up until now we have had little control over where the gene ended up once it got into the human genome. Thus each new gene therapy event that involves stable insertion of the gene into a chromosome risks having that insertion event interrupt a crucial gene such as a cancer gene. Through the use of phi31C Michelle Calos and her collaborators are putting genes into a variety of cell types including skin, muscles, and white blood cells. The phi31C recombinase system is especially important because it lets researchers target genes to locations that are known to be safe, and to keep the gene there once it is inserted. This strategy is offering a new approach to gene therapy for hemophilia A.

One of the most intriguing new areas of gene therapy design is coming out of the field of nanotechnology. Dendrimers are large, branching, tree-like macromolecules that are water-soluble and have the ability to surround and encapsulate other molecules. Some kinds of dendrimers have been developed for use in drug delivery, and can serve a similar purpose for getting a copy of a gene into a cell. When biodegradable polymers containing DNA dendrimer complexes are used, it is possible to apply these polymers to the target tissue and limit the transfer of DNA on a very tightly localized basis.

14.9 DO WE HAVE TO TREAT THE WHOLE BODY?

Controlling Delivery

In some cases, if something is missing that is used by every cell in the body, we would like to be able to carry out a treatment that will restore the gene throughout the body. But in some cases, even if most or all cells make a gene product, we may be able to get away with putting the gene back selectively into some place like circulating blood cells or the liver and find ourselves solving the problem without having to treat every cell.

In many cases, we would actually prefer to avoid treating the whole body if we can, partly to help limit the immune reactions going on, partly because treating fewer cells means less risk of rare side effects, and partly because in some cases there will be cell types in the body that actually need to not be expressing the gene we are trying to get into one specific organ or cell type. Even for genes that are expressed throughout the body, disease resulting from a defect is often specific to a few organs or even one specific cell type. So we would prefer to limit the gene therapy agent very specifically to just the cells we want to treat.

One way to limit which cells end up with the gene therapy agent involves the selection of the type of gene therapy vector. Some vectors will treat only actively growing cells, whereas others will treat cells in any state of growth. Some vectors are derived from viruses that already have some specificity in terms of which cells in the human body they prefer, such as viruses that preferentially infect cells of the central nervous system. If we were wanting to treat the eye, we would want to ask whether we could build a vector from a virus known to infect the eye. If we wanted to treat cystic fibrosis, we would want to build our vector from a virus that infects lung cells. Now, in most cases, we do not have the luxury of starting with viruses that show absolute specificity for just the cell we want to target, but we can again do a least a bit of limiting where our treatment goes, depending on the vector we select.

Controlling Expression

If we include a promoter region in our construct, we can further limit the localization of expression beyond what was accomplished by the gene delivery process. So far, in studies of transgenic animals, all too often a promoter region placed artificially into a cell does not grant a pattern of gene expression identical to the natural pattern usually directed by that promoter. The promoter will give very specific expression in just one cell type when present in its natural location on the chromosome but the transgenic version of the promoter will not give expression in all cells of that type, and it may also give some expression in other cells when present as part of an external construct added to the cells. This may be happening in part because the *endogenous promoter* (the one that was there in the first place) is affected by other regional things, such as the structure of the chromosome in the local region, existing methylation pattern, or other sequences present at some distance from the promoter such as enhancer sequences. Thus, although use of a promoter specific to a rod cell may allow us to get something expressed in some of the rod cells, we do not yet have a way to exactly mimic the natural pattern of expression for that gene.

One strategy for treating only the cells you want to treat is to remove the target cells from the body, treat them in culture, and then return them to the body once they are fixed. This can be done with blood cells if you just need to end up with some treated cells and do not need to fix every single cell. However, if you need to treat every cell in the liver, this approach will not work.

Controlling Immune Reactions

Another strategy for limiting delivery is to deliver into a localized region. In treating the liver, some efforts to limit delivery involve injecting into vessels that feed directly into the liver, but this still results in some of the gene therapy agent ending up in other parts of the body. In treating bone, the clever use of a gel to hold the gene therapy agent in a localized position seems to help.

The eye and brain are expected to be good targets for gene therapy because there are some ways in which they are isolated from the rest of the body. The normal immune surveillance experienced by most of the body does not extend to the eye and brain. Thus some kinds of immune reactions that eliminate the gene therapy agent or kill the treated cells elsewhere in the body can potentially be avoided for eye and brain. On the other hand, it is a well-known phenomenon that the eye can end up being attacked by the immune system if it attracts too much attention from the immune system, so testing of gene therapy approaches to the eye and the brain have to be explored very carefully.

14.10 WHAT ARE THE BIGGEST PROBLEMS WITH GENE THERAPY?

The RPE65 gene therapy story so far looks like a big success, and the things that have helped it to succeed make it look like similar strategies for treatment of other forms of retinal degeneration might also expect such happy outcomes.

Immune Surveillance

One of the distressing early findings in many gene therapy efforts was that, in many cases, positive results from treatment ended up being transient. In cases in which a repeat effort at treatment was tried, often the result the next time was much reduced or even nonexistent. This turns out to be the result of the action of the immune system that normally protects us from infection by bacteria and viruses. The immune system is very good at rapidly mounting a defense against such an infection. In many cases, the mechanism for getting the gene into the cell is an altered virus that can carry the gene into the cell. The use of cloning technology has allowed researchers to create gene therapy vectors that are derived from viruses that normally infect human cells but that have had the genes removed that make the virus able to cause disease. In place of the removed genes, the researchers place the gene that is due to be introduced into the cells. However, the protein coat that protects the viral DNA as it moves through the bloodstream and into the cell is the same protein coat that normally stimulates your *immune system* to attack the virus and keep you from becoming ill. And our bodies have evolved to be very efficient at recognizing and attacking infectious agents such as viruses.

Researchers have worked to change those viral coat proteins to make them less visible to the immune system, but the ability of the body to eliminate viruses is rather amazing. The first time the gene therapy agent is administered, the viral particles avoid being eliminated but stimulate the beginnings of an immune response. If expression of the introduced gene drops off over the course of six months, the body is effectively well immunized against that virus by the time another attempt at treatment is made. The next time the same gene therapy construct is injected, the ability of the immune system to remove the virus may be so effective that none of the constructs will ever reach the cells that need to be treated.

The immune system may also recognize treated cells as foreign, which would result in the body trying to destroy the treated cells. This has turned out to be a problem in the treatment of Duchenne muscular dystrophy. During the 1990s, as frantic parents were asking how they could get their children into gene therapy trials, saying that they would be willing to do even very risky things rather than just sit and watch their children die, it was not possible to move ahead with trying gene therapy on the children because of concerns that the treatment for this particular trait not only would not cure them but actually might make them worse if the immune system were to attack the treated cells. Researchers have been working at changing the gene therapy vectors to make them less likely to induce an immune response. There are now vectors that have fewer problems with creating an immune reaction, and progress has resumed now that researchers have developed a mini-version of the DMD gene that could fit into the vectors that have serious size limitations.

A very different issue is the problem of what happens to the human genome when a new gene is put in from the outside in such a way that the transgene integrates into chromosomal DNA. If the gene integrates into a region of *junk DNA* between genes, it might have little effect other than curing the cell's metabolic defect. However, if the transgene integrates into a gene in the chromosome and disrupts it, the consequences will depend on which gene gets disrupted. If it disrupts one copy of a gene encoding an enzyme involved in metabolism, it may have little effect or perhaps at the worst it will kill that one single cell.

However, if the gene therapy agent is delivered *in vivo* into a large number of existing cells in a human organ, each cell becomes a separate integration event. So even if a large number of cells receive transgenes that integrate safely between genes, it would take only one transgene integration into certain kinds of cancer genes to cause a problem. In the long run, optimal design of gene therapy will need to gain the ability to control where the transgene integrates, or at least to prevent certain kinds of integration events. After the initial tragic events when a seemingly successful ADA gene therapy trial turned up with several cases of leukemia, researchers have figured out why this event happened and have developed an alternative vector that does not cause this problem. Thus in spite of some early set-backs, the field is continuing to advance in overcoming the worst barriers to progress.

Amidst the problems with immune responses and transient expression, the RPE65 treatment that used a viral vector has produced effective results that required only one round of treatment. There were no problems with an immune response. No problems with transient expression. Why did the RPE65 trial not fail like so many of the others? Why did it not elicit an immune reaction against the delivery system?

Many gene therapy programs are succeeding in pushing past these issues. Some groups are working to improve the vectors. Some groups are switching to different vectors. And some groups are developing and using delivery systems that completely bypass the problem of using a virus at all. Use of liposomes and nanotechnology dendrimers offers promising alternatives that may some day render the use of viruses unnecessary.

And yet the RPE65 trials have been succeeding while using viral vectors. So we have to ask, is there something different about gene therapy in the eye? The key to the success of the RPE65 gene therapy efforts may lie in the unique properties of the eye and the central nervous system. The eye, like the brain, is not subject to immune surveillance, the process by which the body monitors and protects most of the other organ systems of the body. Each is protected against some of the problems that have plagued other gene therapy projects because the immune system does not handle the brain and eye in the same way it handles the rest of the body.

The fact that the RPE65 trials seem to be working to provide functional correction of the problem and persistent expression of the genes actually suggests to us that a lot of different ocular traits might be amenable to gene therapy that has not yet paid off systemically.

14.11 SO, WHOM DO WE TREAT?

Which Traits Do We Treat?

Many may well wonder why their particular trait does not have a gene therapy clinical trial taking place. They may well wonder who is getting treated if they are not. There are a lot of very serious problems out there that all need to be solved, and many of them do not yet have gene therapy efforts going on. So if we are not developing therapies for some of those problems, then whom do we treat? The answer is a rather pragmatic one. You would think it would be simple – solve the most severe things first, solve the things that affect the most people first. But the real answer is not that simple.

The real answer is that a variety of factors go into determining which of the problems are tackled first, factors that can be determined by figuring out how many of the following questions can be answered with the word "yes":

- Has the gene been found?
- Is the trait genetically simple?
- Is the trait free of environmental complications?
- Is the trait severe?
- Is the underlying pathology understood?
- Do we have an available animal model for preliminary studies?
- Do we have a cell culture model and biochemical assays?
- Do we know which cell types we want to treat?
- Can we limit treatment to avoid tissues that need to be left untouched by therapy?
- Is the gene small enough to fit into existing delivery systems?
- Do we have some leeway on dosage if expression levels vary?
- Do we have a clever new idea that seems like it would apply to this particular situation?
- Have we learned things from other past studies that would make this one easier?
- Are the cells we need to treat still alive in the target population?
- Do we expect the treatment to be free of unsafe complications?

Clearly there is no simple formula for putting together all of these factors. One of the important factors is that other intangible thing – where do new ideas arise? In some cases a trait may be highly meritorious as a target for gene therapy according to the above list, but if the people working on it have not been able to come up with a good set of strategies for how to go about it, then the research is not feasible, no matter how important the problem. In other cases,

something may not be the most severe trait, but it may be the ideal trait for trying out some brave new idea in how to get gene therapy to work, to develop new methods that would apply to a lot of other traits in need of treatment.

So what it amounts to is this: is the available combination of genes and strategies one that can make gene therapy development feasible? Some of the most desperate cases may not have gene therapy development going on because something about the needed therapy is not yet feasible. In some cases, traits that seem less terrible in their consequences may have ongoing gene therapy development because they seem as if they would be much easier situations to treat or would teach us something important that would move the overall field of gene therapy forward. By working on these more feasible cases, advances in gene therapy take place, teaching us important things that we need to know to be able to tackle some of the more difficult problems. And no matter how terrible a trait, and no matter how desperately everyone would like to see it cured, if it is too complex in its origins, if we do not know enough about the causes, we may not yet be able to develop gene therapy for it.

Who Are the Study Subjects?

One of the other key issues deals with the selection of specific individuals to participate in gene therapy trials. In some cases, the inclusion and exclusion criteria for a study may include only some stages of a disease. This can result in people who are excluded and don't understand why they can't join the study. In some cases, they may be excluded because the therapy that is currently feasible is not expected to work on their stage of the disease. In other cases, their stage of the disease may be considered to be at much higher risk of potential hazards of the study. Once again, the determining factors are often quite pragmatic. In the treatment of Huntington disease, one might imagine that the most advanced cases might offer the most compelling arguments for treatment, as well as the greatest opportunity to demonstrate gains from the therapy. However, when we look at the disease pathology we see that cells in the brain are dying, and quite frankly if the basis of the treatment is to put a neuroprotective gene into brain cells to help keep them alive, it is simply not going to work in cases where those cells are no longer there to be protected. Other strategies aimed at getting cells to grow and regenerate might work well in that same case, but that is irrelevant if the gene therapy trial you are wanting to join requires that you still have cells that you no longer have. So often simple issues of what can and cannot be made to work will over-ride the seemingly dominant issue of who most needs the treatment.

In the first round of a clinical trial, when a small number of individuals are tested to determine whether the treatment is safe and perhaps to pin down the appropriate dosage, there are questions about who is most appropriate to treat. To many of us, it seems obvious that those with the most to lose without treatment and the most to gain from treatment would logically be the ones to take the risks in these early tests of safety. In a gene therapy study aimed at treating OTC deficiency, a bioethicist ruled that the most appropriate participants would not be infants at high risk of dying of OTC deficiency. Some might think it appropriate that those with the most to gain (or lose) would be the ones to take the largest risks. Instead it was decided that the pressures that the child's desperate health status place on the parents to put the child into the study, combined with the inability of the child to decide for himself if he is willing to be a study subject, seemed to make it ethically unacceptable to include these children in the first gene therapy tests. Why? Because the consent to participate in the study would be considered to have been given under undue pressure. To some on the outside of the study, this seems surprising. Anyone participating in such studies

is under great pressure to participate because of their health status, and anyone watching from the outside would wonder at how this supposed ethical dilemma is balanced against the ethical dilemma of expending a potentially lifesaving treatment on some unaffected individual who cannot benefit instead of offering it to an incredibly ill child who could potentially be saved if the therapy turned out to work.

Clearly, the complex situation in which a patient dying of cancer agrees to a treatment becomes incredibly more complex when the decision is being made by parents if the child cannot decide for herself. However, on some levels the issue is the same and the reasons in favor of participating are the same. A whole field of bioethics has grown to include very active consideration of very complicated situations such as these, and each new trait and treatment protocol seems to raise new questions about how to walk the fine line between treatment risk and disease risk, between informed consent and undue pressure to participate, or between death from nonintervention and the risk of death if there are unforeseen consequences of the intervention.

Where To From Here?

There are those who sometimes discuss the idea of treating the germline, going beyond the individual treatments that are now being tried; they propose making changes that could pass along to the next generation so that we do not have to keep re-treating each new family member. However, treating the germline is not currently on the horizon. Any manipulation of the germline has the potential to reach far beyond the health of any one individual to impact the human population, and the technical problems of messing with the germline substantially exceed any problems we have presented here. Any delving into artificially directed "evolution" of the human genome calls for vast wisdom and ethical insights that are still being developed. So for now, the field is focused on the most immediate issue, finding somatic cures for individuals without touching the germline in hopes of moving beyond research to use in real medical settings.

Research is the first key to solving the problem. For gene therapy to arrive in your local doctors' offices, much work remains on the part of people with many different kinds of expertise. Some of the smartest people in the world are working on the development of these technologies. Geneticists go after the right genes to use. Biochemists characterize the gene products and sort out the pathways. Molecular biologists design constructs that bring together human and viral DNA. Nanotechnology researchers are developing coated delivery systems. Stem cell researchers work to develop the ideal cells for use in bioengineering. Cell culture workers and animal model researchers test out preliminary ideas to pioneer new approaches and identify where improvements are needed. Virologists work to develop the vector systems for delivery of the genes. Immunologists study immune responses against the vectors. Biostatisticians evaluate the outcomes to help us tell whether something has actually worked, and help tell us how many subjects are needed in a study to be able to get a meaningful answer. Doctors work to improve systems for delivery of treatments and for monitoring the health status of treated individuals. And gradually, like a building being erected, the many pieces of the treatment puzzle are coming together towards a finished product.

But the other key to the whole process of developing gene therapy is the patients themselves, an often-unmentioned group who seem to us to be the real heroes in this story. The gene therapy story is about them, and the answers we seek are for the benefit of the many who cannot currently be helped by traditional medicine. Through a partnership of the patients who need the cures and the researchers developing the cures, eventually we will arrive at that seemingly magical moment when babies born with a terminal illness can be treated and sent home to grow up along with the other children who were born healthy, just as Jesse Gelsinger wished.

> Will you make me some magic with your own two hands?
> Could you build an emerald city with these grains of sand? —*Jim Steinman*

Study Questions

1. What is gene therapy?
2. What is adenosine deaminase (ADA) deficiency?
3. What are the pros and cons of AAV versus AV gene therapy vectors?
4. What is nanotechnology and how may this be useful in gene therapy?
5. How can miRNA or siRNAs be useful for alternative gene therapy approaches?
6. What major problem with gene therapy does not happen when gene therapy is delivered to the eye, and why is gene therapy in the eye different?
7. Why is integration in gene therapy a concern?
8. Who should be involved in a gene therapy trial?
9. What are two different strategies for using gene therapy to treat cancer?
10. What are two gene therapy strategies other than simple replacement of a missing gene?
11. Why was RPE65 a good candidate as a target for gene therapy?
12. Why was the treatment of the Briard dog Lancelot of importance to humans?
13. Which trait was the first human trait on which gene therapy was tried, and why was it the trait selected?
14. What went wrong with the ADA gene therapy trial and why have efforts to treat ADA continued?
15. What causes damage in OTC deficiency?
16. What kind of "end-run" approach to gene therapy can be used to assist healing following a heart attack?

17. What good does it do for gene therapy to add in more of something that is already present in the body?

18. How can gene therapy help heal a bad break in a bone?

19. Why are the strategies for gene therapy of cancer so different from the strategies for treatment of metabolic disorders?

20. How else can a cloned gene help us treat a disorder if we do not use it for gene therapy?

Short Essays

1. In humans, some aspects of visual function in the brain develop through use of the visual system. As scientists consider how to treat color blindness there has been concern that putting the missing gene back into the cells of color blind men would not be enough because their brains would not have the capability to use the information. How has our understanding of this situation been informed by the recent gene therapy treatment of color blind monkeys? As you consider this question please read "Monkey see monkey juice" by Evan Lerner in *Seed Magazine*, September 18, 2009.

2. Researchers carrying out gene therapy for severe combined immunodeficiency (SCID) found that the gene therapy construct that they used caused an acute T cell leukemia in some of the study subjects. They thought the leukemia was the result of a very specific combination of genes involved in the vector and the chromosomal integration event. However, another effort at gene therapy for chronic granulomatous disease (CGD) using a different, but related, vector led researchers to think that they had accidentally generated a model for human T cell leukemia. What did the researchers learn and what did it tell them about gene therapy vector design? As you consider this question please read "Gene therapy activates EVI1, destabilizes chromosomes" by Cynthia E. Dunbar and Andre Larochelle in *Nature Medicine*, 2010;16:163–5.

3. Hematopoietic stem cells are the precursors to the various cell types found in human blood. The HIV virus infects one of the differentiated blood cell types, the CD4+ T cells, but CD4+ T cells that carry a mutant form of the CCR5 receptor protein are resistant to HIV infection. How can gene therapy be used to fight HIV in infected individuals, and why is it not enough to treat the CD4+ T cells? Why will it be difficult to turn this approach into something that can be easily applied to large populations of infected individuals? As you consider these questions please read "Can HIV be cured with stem cell therapy?" by Steven G. Deeks and Joseph M. McCune in *Nature Biotechnology*, 2010;28:807–10.

4. Since RNA interference was first discovered in the worm *Caenorhabditis elegans* researchers have tried applying it to many different problems in human biology. Why does RNA interference turn out to be especially appropriate for some of the dominant neurodegenerative diseases that result from simple sequence repeat expansions? As you consider this question please read "Allele-specific RNA interference for neurological disease" by Edgardo Rodriguez-Lebron and Henry L. Paulson in *Gene Therapy*, 2006;13:576–81.

Resource Project

There are a variety of resources that let us check on what is happening with clinical trials in the US and elsewhere in the world. Go to the Gene Therapy Net website and look at the Clinical Trials Databases section. Check on what is happening with gene therapy for cystic fibrosis in the US and two other countries and write a brief essay comparing what is happening with gene therapy for this trait in these three countries.

Suggested Reading

Articles

"Making a play at regrowing hearts" by Kenneth Chien in *The Scientist*, 2006;20(8):34.

"Whither gene therapy? Success has been mingled with failure; a few technical modifications could make the method safer" by Alain Fischer and Marian Cavazzana-Calvo in *The Scientist*, 2007;20:36.

"Kenyon's ageless quest: A San Francisco scientist's genetic research renews the ancient hope for a way to slow aging" by Stephen S. Hall in *Smithsonian Magazine*, March 2004 (Available at www.smithsonianmag.com/science-nature/quest.html).

"Egg sharing in return for subsidized fertility treatment – ethical challenges and pitfalls" by Boon Chin Heng in *Journal of Assisted Reproductive Genetics*, 2008;25:159–61.

"Gene therapy may switch off Huntington's" by Bob Holmes at NewScientist.com, March 13, 2003 (www.newscientist.com/article/dn3493-gene-therapy-may-switch-off-huntingtons.html).

"To build a killing machine: David Kirn can't turn his back on a century-old quest to pit oncolytic viruses against tumors" by Andrew Holtz in *The Scientist*, 2007;21(5):48.

"Gene therapy in a new light: A husband-and-wife team's experimental genetic treatment for blindness is renewing hopes for a controversial field of medicine" by Jocelyn Kaiser in *Smithsonian Magazine*, January 2009 (www.smithsonianmag.com/science-nature/Gene-Therapy-in-a-New-Light.html).

"Latest developments in viral vectors for gene therapy" by Kenneth Lundstrom in *Trends in Biotechnology*, 2003;21:117–22.

"Interspecies SCNT-derived human embryos – a new way forward for genetic medicine" by Stephen Minger in *Regenerative Medicine*, 2007;2:103–6.

"Gene therapy to improve the body's ability to eliminate cancer" by R. A. Morgan, M. E. Dudley, J. R. Wunderlich *et al.* in *Science* 2008;314(5796):126–9.

"Re-engineering humans" by S. Jay Olshansky, Robert N. Butler, and Bruce A. Carnes in *The Scientist*, 2007; 21(3):28.

"Subtle gene therapy tackles blood disorder" by Danny Penman at NewScientist.com, October 11, 2002 (www.newscientist.com/article/dn2915-subtle-gene-therapy-tackles-blood-disorder.html).

"The return of the phage: As deadly bacteria increasingly resist antibiotics, researchers try to improve a World War I era weapon" by Julie Wakefield in *Smithsonian Magazine*, October 2000 (available at www.smithsonianmag.com/science-nature/phenom_oct00.html).

"DNA nanoballs boost gene therapy" by Sylvia Pagán Westphal at NewScientist.com, May 12, 2002 (www.newscientist.com/article/dn2257-dna-nanoballs-boost-gene-therapy.html).

Fears, Faith, and Fantasies: How the Past and Present Shape the Future of Genomic Medicine

Once, human beings were as children, needing simple tales and naïve visions of pure truth. But in recent generations the Great Creator has been letting us pick up his tools and unroll blueprints, like apprentices preparing to work on our own. For some reason He's permitted us to learn the fundamental rules of nature and start tinkering with His craft. That's a fact as potent as any revelation. Oh, it's a heady thing, this apprenticeship and the powers that go with it. Perhaps in the long run, it will turn out to be a good thing. But that doesn't make us all-knowing. —**David Brin in Kiln People**

As we come to the end of this book, we offer you three different perspectives on human genetics, with a special concern for some of the scientific and ethical pitfalls that are ready to waylay some of the best intentions. Our backdrop for this discussion will be three hypothetical tales that effectively put the same fictitious young man into a time machine and transport him to three different eras – the beginning of the twentieth century, the beginning of the twenty-first century, and the beginning of the twenty-second century – to examine how each of those three societies might react to him. While we have limited patience for some of what we see in the past and limited ability to project into the future, we hope that this discussion will let you follow some ethical issues and technical capabilities along the road from the past to the future, from an era in which we understood little about what determined our biological fates to an era when we may perhaps reach a point of being able to

513

intervene in such fates. We stand at a crossroads where public policy decisions in the next few decades will decide the fates of many people who urgently need the help that genetic technologies could provide them. It will be critical that our society as a whole find ways to proceed that avoid the major errors of the past while optimizing our ability to do the most good wherever possible.

The title of this chapter derives from our very mixed reactions to what we see along this road of genetic progress. The term "fears" reflects our concerns that, even as our society proceeds with the best of intentions, the course along the future road could still be influenced by attitudes similar to those that caused grievous errors and injustices in the no-so-distant past. The term "faith" reflects our belief that most of those involved in trying to sort out what can be done and what should be done have genuine interests in keeping the road running in an ethical direction, and that the current trends in scientific culture demand high standards of scientific ethics that should help guard against some of the errors of the past. The term "fantasies" applies to the many truly wonderful possibilities that loom in such a near future that we expect some of them to become real within the next few decades. This combination – fears, faith, and fantasies – carries us forward, mindful of the historical mistakes to be avoided, earnest in wishes to find ways to make genetic technologies work for good, and excited at the possibilities unfolding as some of the fantasies become realities.

15.1 FEARS – A TALE OF EUGENICS

Those who cannot remember the past are condemned to repeat it. —**George Santayana in Reason in Common Sense**

It is 1931, and Allen's life is in an upheaval that he can barely understand. He had spent much of his early childhood at home, receiving little education because the local school could not cope with someone

the teachers had declared an imbecile. His uneducated parents had no idea how to help teach someone who could not talk and rarely seemed to understand anything complex that was discussed. With little knowledge of the world outside of his parents' small farm, Allen has now found himself in trouble with the law. He was arrested for assault after he took a swing at the leader of a group of older boys who were taunting him (as they often did). Now, he finds himself standing in fearful confusion before a judge, and he cannot understand that this judge is making pronouncements of profound and far-reaching implications for the rest of the course of Allen's life. The judge feels a need to protect society from this subnormal and obviously violent individual. Allen offers no response as the judge informs him that he is to be placed in an institution (for his own good) and sterilized (for the good of humanity). When he is led away to face this terrible dual fate, he finally realizes that something is not right, but it is too late. His efforts to squirm free of the bailiffs simply confirm to the onlookers that the judge was right. In 1991, looking back on this tale from a perspective he only achieved many years after the event, Allen can only shake his head in disgust and residual anger and a very intelligent understanding of what had happened so long ago. Allen is not an "imbecile." Allen is deaf. As a mixed blessing, Allen's deafness was diagnosed by a doctor that he met in the hospital where he was forcibly sterilized. Unfortunately, Allen spent far too long in an institution before his parents figured out how to get him released and into an education program for the deaf. With a grace born of long practice and passionate conviction, his hands dance through the explanation that his form of deafness is recessive and so there was little expectation that he would have passed it along to his children. He pauses before adding that if he had fathered any children, he would have preferred that they were deaf children growing up in the nurturing center of the deaf community so that they would not be part of the cruel culture that could do the terrible things to people that had been done to him.

When Scott was about 16, he came across a section on marriage laws in an almanac and found that about seven states still had laws proscribing epileptics from marrying. This left a profound impression on Scott because he has epilepsy. These laws were fossils of the *American eugenics movement*. They were based on the now-discredited ideas that epilepsy is associated with insanity and imbecility and that epileptics are dangerous. These rather odd views are themselves probably remnants of medieval beliefs suggesting that seizures were an exposition of demonic possession. The science was bad, but the laws were made anyway, something that should serve as a caution to all lawmakers who proceed to make laws on subjects they do not really understand.

We all stand on the threshold of a revolution that will change the lives of our species forever. We have learned how to assess at least some of the information contained in our genes, and we will continue getting better at this particular trick, much better. We are also rapidly developing the skills necessary to modify our genomes, surely in our somatic cells, possibly in our germlines as well. There are many potential benefits of this technology, but there are also some major potential pitfalls. As an example, we want to

digress into a bit of a history lesson. Specifically, we want to talk about a subject called the *science of eugenics* (Box 15.1). *Eugenics* is a term for the selective breeding of the human population for purposes of "improving the quality" of the human race.

All of us will watch in the near future as decisions are made about a variety of potential uses of information arising from genomic science, and will have to make decisions about how involved we will become in these decision-making processes. So we give some historical context for some of the decisions that will be made by talking about government-sponsored eugenics programs in the not-too-distant past, programs that made laws that deprived people of their ability to be parents. We're going to talk about governments that made laws about sterilization, incarceration, and even about people's right to be alive. And the programs we are talking about happened here in the United States; they were not created and carried out by obvious monsters, by evil people vilified by the culture for their crimes, but by people who were seen by their contemporaries as good people, as pillars of

---

BOX 15.1

THE AMERICAN EUGENICS MOVEMENT

A great deal of historical information exists to offer details of the eugenics movement's scientific investigations (and their flaws), the legal cases (and their impact), and the sociological arguments (and their underlying prejudices). Eventually, the theories and advisements of this movement made their way to Nazi Germany, where they were picked up by a growing eugenics movement of even greater and more horrifying scale than what happened in the United States. A group at the Cold Spring Harbor Laboratories in New York has established an archive project to assemble information on the history of the American eugenics movement and its inherent flaws. For some people in the United States it can be quite an eye-opening experience to realize that a movement based on such a foul combination of error, fraud, and prejudice could have been allowed to develop so much power over the lives of the people of this country. For some people the eye-opener is that this did not happen in some remote past or distant land but rather right here in our own sociological backyard.

their community. They supposedly did these things in the name of the public good, and that makes it very scary.

In 1903 an organization called the American Breeders Association was formed. The association set out to bring Mendelian ideas to the United States. Much of what they did dealt with horses and other animals, but they also began to follow up some rather theoretical work on human breeding begun by a man named Galton in England. Shortly after the formation of the American Breeders Association, the American eugenics movement began. This movement evolved under the guidance of a federally funded agency located at Cold Spring Harbor, New York, called the Eugenics Record Office. It was run by a Harvard professor named Charles Davenport, a man considered to be one of the great liberal minds of his day.

The agency was interested in collecting data about the human population. They trained people to go out into the country and find pedigrees so they could gather evidence about how certain human traits or "diseases" were transmitted. The government was paying a lot of money to record-keepers to find appropriate families, so the record-keepers were motivated to find evidence of inheritance, a form of conflict of interest that current good scientific practices try to prevent.

One particular bit of data they collected that stands out was two pedigrees that showed that "seafaringness" is an X-linked trait. But consider this: what is the chance that such a pedigree would not look X-linked at a time when virtually all sailors were men? There were pedigrees for idiocy, silliness, nomadism (the love of wandering), vagrancy, criminality, and more. Remember, people were paid to go out and create these pedigrees. They went to places such as prisons and mental institutions. There was purportedly a great deal of fraud and serious error involved.

Fueled by such information, the American eugenics movement quickly built up real steam. Eugenics booths and education programs were set up in county fairs and schools all over the country. Some brochures for the movement urged people to "wipe out idiocy, insanity, imbecility, epilepsy, and create a race of human thoroughbreds such as the world has never seen." These views on heredity seemed to fit with general common sense. People knew that some of these traits or behaviors did tend to run in families and that certain traits tended to occur in some families more so than in others. In this sense, the American Eugenics Office was providing the so-called "evidence" to buttress well-established prejudices.

Unfortunately, a lot of these flawed data were used as justification for new laws. Several states passed laws prohibiting people with certain traits from marrying. It seemed reasonable to many that one way to make a better society was to simply prevent marriages that were predicted to produce certain categories of "defective" progeny; state governments passed laws that idiots, criminals, and epileptics couldn't get married. In fact, related laws started being applied to other groups, including girls who had gotten pregnant out of wedlock. Eugenics-based marriage laws quickly became the norm in our growing country.

Buttressed by the eugenics movement, and fueled by prevailing racial prejudices, 34 states also passed laws making marriage illegal between people of different races, the so-called *anti-miscegenation laws*. People worried and talked openly about the so-called dangers of "racial degeneracy." These things were not just happening in Nazi Germany; they were happening in the United States.

Soon the laws would go beyond regulating marriage. In 1907, Indiana passed the first law requiring involuntary sterilization. It mandated that people with certain traits, including epilepsy, be sterilized; soon other states would follow suit with similar laws. By the 1930s, more than 30 states had passed laws of mandatory sterilization for an incredibly large number of traits. Between the 1920s and the 1940s, it is estimated that 30,000 to 35,000 people were sterilized involuntarily.

This number is very likely to be a gross *underestimate* because not all cases were reported. The people in the eugenics movement were deadly serious and were backing up their politics with the surgeon's scalpel.

Things got even worse around the 1920s and 1930s. Life got hard for people, and prosperity's infinite view was changing. Immigration was on the rise from all parts of the world. People in the US worried that some of these new arrivals were genetically inferior and that these genetically inferior people were bringing undesirable genetic traits into the country. Much of the testimony that helped the passage of the Immigration Restriction Act of 1934 was centered on arguments that high fractions of immigrants coming from certain countries were "feebleminded." Indeed, a progenitor of the IQ test was administered to newly arriving immigrants and suggested an enormous frequency of feeblemindedness among people from certain countries. People felt that such individuals should be denied entrance into the United States because they were genetically inferior. However, these IQ tests were being administered in *English* to people who had just arrived in the US and spoke not a word of English. We have to wonder how anyone could arrive at the conclusion that such a test would tell us anything, yet these data served as the basis for one of the most restrictive immigration acts in history, which stayed on the books until 1960.

As painfully crazy as all of this must seem to you, it is important to realize that these laws had wide backing throughout American society, even at the highest levels. A landmark case on involuntary sterilization went to the United States Supreme Court in 1924. The case was decided by none other than Oliver Wendell Holmes, known then as an independent vital force for social reform. Holmes was known as a kind and intelligent man, but let us quote from the decision in which Holmes and his court upheld the rights of states to sterilize supposedly genetically inferior individuals against their will:

We have seen more than once that the public welfare may call upon the best citizens for their lives. It would be strange then, if it could not call upon those of us who already sap the strength of the state for these lesser sacrifices, often not to be felt as such by those concerned, in order to prevent our being swamped with incompetence. It is better for all the world if instead of waiting to execute the degenerate offspring for crime or to let them starve from imbecility, societies can prevent the genetically unfit from continuing their own kind. The principles that sustain the compulsory vaccination are broad enough to cover the cutting of fallopian tubes. If we are willing to ask the best of us that they lay down their lives in the defense of their country, in the defense of liberty, and in the defense of freedom, why can not we ask from the weakest of us to voluntarily deprive themselves of the right to reproduce children?

Again, these laws were passed and supported not by obvious monsters, but by people considered by those around them to be very good people. They did what they did in the name of right. That's what scares us: that those who were considered to be very good people, acting for what they considered the good of others, with the full support of the church and of the state, were able to do so much evil to so many with so few voices being raised. It is easy to recognize and prevent evil when it comes with a gruesome countenance, a threatening manner, or overt ill will, but it slips by all too easily when concealed beneath civilized manners, a quiet demeanor, and gracious speech.

In the end, the wave of immigration changed this society. People finally began to realize, after twenty or thirty years, that people from all parts of the world are more or less the same everywhere and that everyone had an enormous amount to contribute here. Common experience belied the messages of the eugenics movement.

More importantly, real genetics was blooming as a science. People were getting an idea of what genetics could and could not do. Good scientists were trying to do serious human genetics. They were discovering that nothing was as simple as the eugenics people said, and they were also discovering that the eugenics movement pedigree data could not be replicated. So by the end of World War II, most of the activities of the eugenics movement went away. What is left are reports in the history books, some fossilized laws about marriage in a handful of states, and far too many people who can remember what was done to them against their will.

15.2 FAITH – A TALE OF ETHICAL, LEGAL, AND SOCIAL ADVANCES

And though the nightmares should be over some of the terrors are still intact. —**Jim Steinman** in **Objects in the Rear View Mirror May Appear Closer Than They Are**

It is 2010. Alan is six months old, and his recent diagnosis of deafness has led to some very heated debates among his relatives, none of whom is deaf. Their debates about how to raise him are finally resolved with the decision to hire tutors and teach him to be bilingual, communicating via sign language and also through lip-reading and speaking. Alan's mother has found a nearby deaf family whose children include a deaf toddler that she hopes might become a playmate for her son as he grows up. She has started taking other members of the family along to her sign language classes and is fast becoming friends with the deaf mother of the children her son will play with. She plans to find additional deaf families for him to interact with, not only so that he will have playmates but also so that his adult role models will include deaf adults in addition to his hearing relatives. Since the genetic change causing Alan's deafness has been identified, relatives want to engage in the debate about whether there should be genetic testing during the next pregnancy, but they are gently edged out of a decision that Alan's parents feel falls to them alone. Some relatives pressure them to start inquiring about cochlear implant surgery, but Alan's parents insist that Alan will have to participate in such a decision once he is old enough. One grandfather has been angry ever since Alan was born. His calls and visits focus on his demands that the family enroll their child in an upcoming gene therapy

trial that he has heard about. Alan's mother insists that such treatments still carry risks and benefits that Alan should get to weigh for himself when he is older. The grandfather tries begging, bribing, and even threats as he makes it clear that he himself would gladly take such risks to avoid a marginalized existence in a society that still stumbles too often in its interactions with people not cast in a standard mold. Alan's mother counters that her son's existence is in no way marginal, and that he has every bit as much potential to live a good life and become a productive member of society as the rest of the babies in their town.

This whole scene actually epitomizes medical genetics at the beginning of the twenty-first century. We can diagnose things that we cannot change. There is a wealth of different ways we can help even if we cannot undo the event that happened in the first place. And people have very different views of what constitutes the "right" answer. If we compare this story to Allen's tale a century ago, we see major advances on ethical and technical levels. Even if we can't do everything we would like to, we can treat Alan with respect and offer him options that will let him fully develop his abilities in ways that he will get to choose. Unfortunately, similar children will not always be treated with so much respect or gentleness at this point in history, but the possibility of fair and reasonable treatment is increasing in the early twenty-first century. We see the trend, not always practiced, towards thinking that the individual should be the final source of decisions about her own welfare, but we still see signs of residual paternalism. Sometimes decisions have to be made before the child is old enough to decide for herself – we can't outwait some kinds of childhood cancers to let the child decide later in life. But we see some who want to practice paternalistic approaches that take those decisions away from the person who needs to make the decisions, even in cases where the decision could wait. So things have improved substantially in the past hundred years, but we also see that not everyone has dispensed with the problematic views of the past.

The eugenics movement grew out of a combination of bad science and prejudices that led people to buy into and act on that bad science because it was telling them what they wanted to hear. What protects us against more of the same? One of the best current protections we have against the problems of the past is the system of peer review that evaluates every step of a research project involving human subjects research. This not only protects against the kind of bad science with which the eugenics movement rationalized their activities, but it also works to ensure the highest standards of ethics in the conduct of science.

One of the vanguards protecting the quality of the science is the peer review system. To get funding to do the work, a project must be reviewed by a panel of other scientists who determine whether the question to be asked is valid, the methodology appropriate and feasible, and the objectives important. And while a piece of science is going on, informal systems of communicating about projects include each scientist presenting their work to other scientists who can raise questions about the methods and results. And once the work has been done, any piece of science that gets published in a reputable journal is subjected to detailed (and sometimes, to the authors, painful) scrutiny by others in the field. If major flaws are found, the paper is likely to be rejected. Whether reviewing a proposed project or reviewing a manuscript that presents completed work, the reviewers are charged with finding any flaws they can in a piece of science, whether in the data, the analysis, or the logic being used in model building.

Science is also pushed to higher standards by practices calling for replication of published work. If someone claims to have found a particular result, others using the same methods should be able to get very similar answers. Many journals require that authors agree to provide their cell lines or other materials to others who want to try to reproduce their work, so anyone who publishes knows that someone else could come

along shortly and publish a paper indicating that their result cannot be replicated. In addition, funding agencies require a rigorous review of proposed work, including a detailed analysis of the feasibility of the work and the validity of the models and approaches. The result of all of this is that almost no one carries on research in a vacuum, and everyone's work is subject to a great deal of scrutiny by others.

The *National Institutes of Health* (*NIH*) is the largest funding agency for biomedical science in the United States. The NIH plays a major role in maintaining the quality of science by demanding that the scientists they fund meet the highest standards, and then backing their demands with a large financial stick. If someone is suspected of fraudulent scientific practices or of science that does not meet NIH standards, NIH conducts an investigation that can result in barring a scientist from receiving NIH funding, sometimes for a few years and sometimes for life. In some cases, investigations have lasted for years and involved extensive investigations, complete with forensic analysis of lab notebooks. The result is a great deal of pressure on scientists to be honest and to live up to the standards set for them. Nothing will get scientists' attention faster than threatening to take away the research funding that is as essential to the research as air is to the researchers themselves.

The current scientific culture makes us more optimistic when we compare the practice of science at the beginning of the twentieth century and the practice of science at the beginning of the twenty-first century. Students passing through the higher education system in biomedical sciences receive signals from all levels of the people surrounding them that scientific rigor and honesty are not just expected but required. In rare cases in which someone does something wrong, the ripples expand into shock waves throughout the scientific community. A few years ago a graduate student was suspected of falsifying data, and his mentor immediately took steps to address the scientific community at large, as well as to contact NIH and bring them into an investigation of what had happened. The student was found to have committed scientific fraud and much publicity accompanied the withdrawal of some very important publications from some outstanding journals. The flurry of activity that took place, and the attitudes of horror expressed in lab after lab after lab around the country, offered an example of the kind of reinforcement for the norm of honesty in the field. Anyone growing up scientifically in this kind of atmosphere is growing up in a highly moral climate. This does not mean that we think no one will ever commit fraud again. It does not mean that we think that no one will make mistakes. However, it does mean that the scientists of the future are being trained in an atmosphere that fosters integrity, and to us that seems a hopeful thing for the science of the future.

One of the hazards that we face in the near future, as our body of knowledge grows and changes, is the danger of building laws based on things that are not actually true. This could happen as it did before, when the science behind eugenics was simply bad (Box 15.2).

It could also happen not because of fraud, but rather because we have failed to perceive just how complex some problems are. Even with the best efforts of good scientists, if the situation is terribly complex, there will be limits to our ability to really understand the underlying contributions to many complex traits, including behavioral and psychiatric traits. We are not arguing that the genetic components are not there. Rather, we are arguing that we must be very careful not to build policies based on oversimplified views. We must be very careful that any policies that develop are based on very good science and an adequately sophisticated understanding of the complex factors involved. It is so easy to catch a first glimpse of understanding of a situation that desperately needs help, and to then leap to erroneous conclusions that harm rather than help.

So in terms of the science, we would caution that we not build policy on bad science, and

that we not build policies about things until we know enough to build intelligent policies that will be constructive for society and the individuals involved. We would caution that policy makers be certain they really understand the things they target with their laws. We would also caution that we not set policies in stone today that are fixes for temporary problems that are in the process of solving themselves through the natural progression in our scientific knowledge about the problems.

The other factor that contributed to the eugenics movement, along with bad science, was quite a display of questionable ethics. So one of the most heartening things we see coming out of the Human Genome Project has been the development of the program in *Ethical, Legal, and Social Implications* of genome science technologies, affectionately known as *ELSI*. The National Human Genome Research Institute spends millions of dollars each year on ELSI research, in addition to a variety of educational programs aimed at trying to help improve the ability of the public to understand and contribute to the discussion of key ELSI issues.

ELSI projects touch on a variety of issues. We offer here just a sampling of some of these projects. Researchers in Maryland have looked at philosophical and policy issues relative to the integrity of the human genome and efforts to alter it. A group in Texas has studied the ethical, legal, and social aspects of preimplantation diagnosis. A researcher in Kentucky investigated the legal and social impact of identity testing on families. A project at the University of Michigan, called Engaging Minority Communities in Genetic Policy Making, worked with 15 different minority community organizations and a series of focus groups to identify genetic issues of special concern and importance to African-American and Hispanic populations. A study in Massachusetts looked at issues of importance to genetic dialogues between the scientific community and evangelical Christians. A group at the Hastings Center in New York brought together

---

BOX 15.2

IS GENETIC DETERMINISM STILL AROUND?

Among many flaws in the eugenics movement was the concept of *genetic predetermination*. This concept is particularly problematic when we look at complex traits, especially in the area of behavioral genetics. A few years ago, when Scott was a professor at a medical school, he heard a very famous physician give a lecture to a medical school class on human genetics. This fellow started his lecture by saying: "I want you to understand that genetics is not just an important course – it will be the most important course you are ever going to take, because everything you see is going to be genetic. Now, I see you don't believe that. You probably think that, at some point, you are going to be in an emergency room dealing with a gunshot wound, plugged by a police officer in the middle of a bad drug deal in the middle of the South Bronx. You think that this is not genetics, but you are wrong because I will argue to you that it was that person's genes that led him to be dealing drugs in the first place." In his book *On Human Nature*, Edward Wilson says, "The question is no longer whether human social behavior is genetically determined. It is to what extent. The accumulated evidence for a large hereditary component is more detailed and compelling than most persons, including even geneticists, realize. I will go further, it is decisive."

In evaluating whether these statements are oversimplifying something terribly complex, we would refer you back to the discussions of MAOA in Chapter 9. A simple initial view of the data offers the view of a gene defect that correlates with a behavior, but the fact that something contributes to a trait does not mean that it is sufficient to be the sole cause of the trait. Here we encounter one of the pitfalls to be avoided if we are to avoid wandering back into the kinds of scientific errors that plagued the eugenics movement. That pitfall is oversimplification of a complex problem. Errors of interpretation can happen because someone is taking too simplistic an approach to a complex problem, but the worrisome thing is that such errors can also happen because someone has not figured out the right questions to ask. Fortunately, when we look at work going on in behavioral genetics, we see smart, gifted researchers with an appreciation for the complexity of the problems they are studying. For now, we can't ask for more.

---

groups including bioethicists and behavioral geneticists to develop resources with which to educate the public about behavioral genetics and to explore the associated ethical, legal, social, and scientific issues. A researcher in Oregon asked how culture and social class affect communication about genetic information in breast cancer families. Other projects have tackled legal and policy issues relative to insurance and use of genetic information, issues involved in genetic manipulation of the germline, and efforts to enhance ELSI education at universities. As this book is coming out, researchers are investigating a wide variety of other topics, including patient attitudes towards prenatal testing and preimplantation diagnostics of genes involved in later onset traits such as breast cancer, eugenics concerns of the disability rights community, and issues of trust and mistrust of research in minority communities. The recurring theme we see in the NIH descriptions of these projects is the interactive nature of the projects. Clearly, major efforts are being made to engage representatives from a variety of different communities within and outside of science, and to bring feedback from the community back into the scientific process. At the same time, major efforts are being made to improve the level of genetic education and awareness in the population. What we see in all of this is not a remote, paternalistic scientific society clinging to the right to decide things for everyone else; rather, we see a scientific community making major efforts to bridge gaps and bring the rest of the community into the decision-making process.

One of the other areas to look at with concern is the area of public policy. The people who write the laws may be earnest in their efforts to write helpful laws, but do they know enough about some of the complex issues? Sometimes yes. When we see nuclear transfer technology being tossed out along with cloning human beings, for apparently semantic reasons, we end up concerned that some of the dialogue and education processes in which the scientific community has invested may not be succeeding in putting across all of the important information and issues. On the other hand, we see progress. The Genetic Information Nondiscrimination Act prevents insurers and employers from discriminating based on genetic information. This means that an insurance company cannot deny you insurance or charge you more because you have a particular genotype, and that an employer cannot fire you or pay you less because you have a particular genotype.

As with most of the technologies that have come out of a variety of scientific and engineering endeavors, the emerging technologies offer tremendous potential for good or for ill, depending on whether or not they are used wisely and with respect for the rights of individuals to determine their own fates. It will fall to your generation and the next to figure out how to control this technology. The decisions cannot be made by the scientists alone, but they also cannot be made without input from the scientists. The good news is that people, scientists, and community members alike

are talking very seriously about these problems. On many topics the feelings are strong and people hold views at substantial extremes from each other. However, we view the whole tapestry of this discussion, including input from the extremists on both ends of the range of perspectives, as very healthy. The more we as a society discuss the ethical implications of this new genetics, the better off we shall be.

15.3 FANTASIES – A TALE OF OUR GENETIC FUTURE

Any sufficiently advanced technology is indistinguishable from magic. —**Arthur C. Clarke in Profiles of the Future, an Inquiry into the Limits of the Possible**

It's the year 2102, and the activities in the delivery room are routine as the newborn infant Alain puts forth his first lusty cry of protest at the cold, bright environment that has just replaced the dark warmth that is all he has ever known. His hastily read genome sequence is implanted into a chip under his skin, and stem cells from his cord blood are transferred down the hall to the tissue engineering lab. There the cells will be remanufactured, through a combination of gene therapy and developmental induction, to become developmental precursor cells that can differentiate within his ears to replace the damaged cochlear hair cells responsible for a form of hereditary deafness that was detected in his genome sequence. Metabolic testing shows that he is not subject to any of the most common metabolic disorders that would potentially call for altering an infant's diet. His parents are a bit confused by some of the more complex predictions arising from his sequence, probabilities that shift his chances of various problems depending on a variety of environmental factors that may be hard to control, but they have been assured that GENI (gene environment interaction) counseling will be available to the family at critical stages as their son grows to assist in minimizing some of these potential hazards. By the time that Alain's happy parents are ready to take their new treasure home, they have in hand a list of the two common over-the-counter medications that won't work correctly in Alain's body and the 18 allergens that they should eliminate from Alain's home environment to help avoid asthma. The new family heads happily home, knowing that modern genetic medicine has provided Alain with the sequence information that will let his doctor optimize his health care at every step along the way. They are content that the medical system is busy repairing his hearing defect before it can ever have a chance to affect his interactions with the world.

In the 1930s, Allen's deafness met with brutality and disrespect. In 2006, Alan's deafness faced limited options in a supportive environment. At the beginning of the twenty-second century, Alain may not ever even realize that he was born deaf and may never encounter the perspective that some might offer: that curing his deafness has deprived him of the opportunity to participate in an important, alternative culture, that in eliminating his deafness his doctors have joined in a process of genocide against the deaf culture. Is it a fantasy that the twenty-second century will erase traits that were considered tragic a couple of centuries before? Will future medicine offer the ability to whisk away the previously unsolvable problems with little more trouble than we now expend on a headache or strep throat?

Actually, we suspect that the most fantastic things in Alain's future are things we cannot talk about because they have not been dreamed up yet. And we also suspect that the day will come when traits such as epilepsy, "imbecility," and deafness that would have been sterilized and institutionalized 100 years ago or that would have been struggled with by the schools and hospitals of today, will indeed be dealt with so efficiently that they will become a dispensable point of curiosity, commented on much as we now will remark with wonder that a baby was born with a full head of hair or some baby teeth already in place. We have to wonder what loss of insight will accompany this freedom from adversity, and we have to wonder if anyone will elect to decline to have such differences washed away in a wave of engineered stem cells. We also have to wonder at how substantial will be the disparities in access to these wonders.

When Scott and Julia started graduate school, obtaining even a small piece of a human gene was out on the cutting edge of the most advanced genetic science. Now people talk about a thousand-dollar genome, a customized rendition of the order of As, Cs, Gs, and Ts in your genome to be worn in an implanted microchip that costs less than the down payment on most new cars. Some of the things we foresee in our wildest dreams will likely not come to pass. In some cases our current projections will be blocked by some unforeseen technical wall that will stop things from progressing in a direction we think they should go. In other cases some other new capabilities will arise that will take us in a direction that we cannot currently imagine. We live in an era of computers that can talk to us and send information over wireless network systems, and watch the development of the first computers built into the human body using currents that flow through the skin. These would have been an unimaginable form of magic to those who drew the cave paintings or those who used a stylus to press cuneiform letters into tablets made of clay from the Tigris and Euphrates rivers. In fact, it would have been incomprehensible to our great-great-grandparents who had not yet seen an automobile or a telephone or penicillin. We live in an era when people in need of organ transplants die for lack of organ donations. We laugh at the absurdity of a scene in a *Star Trek* movie in which Dr McCoy pronounces surgery barbaric and gives a woman a new kidney by having her take a pill. Somewhere between the cave paintings and the wild fantasy of Dr McCoy's kidney pill lie the realities that will bring about medical miracles we can only vaguely anticipate.

Current advances already give us views of where this is heading. Researchers have built a new heart from stem cells and seen it start beating; as we start the second decade of the twenty-first century, such a heart is not yet ready to keep anyone alive, because this heart grown in a lab does not yet have all of the necessary components (like valves) but even a couple of generations ago the concept that cells dripped into a membranous sac might turn themselves into a beating heart would have seemed as impossible as Dr McCoy's kidney pill.

Some of our dreams for the outcomes of genome science and stem cell technology run far field. Will we reach the day when genetic defects will simply be repaired at birth before they go on to cause cancer or heart disease or Alzheimer disease much later in life? Will we someday be able to give humans copies of the genes that let goldfish regenerate tissues that can't be repaired in humans? Will we someday find parents signing up to modify their children to have the ability to detect and follow the earth's magnetic field the way birds do? Can studies of animals taken into space point towards the genetic modifications we would need to truly adapt a human being to long-term existence in free fall without complications like bone density reduction? Could we, or more importantly, should we turn on and off the right combination of genes to grow gills in addition to lungs? Will we ever have enough wisdom to tinker with "improving" the human form? Currently, some of these ideas are the stuff of science fiction, topics whose social implications are tackled by authors such as Nancy Kress in *Beggars in Spain* or Lois McMaster Bujold in *Falling Free*.

Others of our dreams appear on the near horizon of our scientific view. Some of these ideas are actually being worked on currently by research groups actively trying to find out which genes in fish are responsible for regeneration of tissues that don't regenerate in humans. We already see companies working on a variety of gene-based strategies for enhancing the effectiveness of chemotherapeutic agents or protecting sensitive non-cancerous cells from higher doses of chemotherapy. Similarly, we see possible near-future breakthroughs in gene-based treatments for heart disease, neurodegenerative disorders, cystic fibrosis, and more. And we see gene therapy improving the sight of children who had been watching their visual abilities drain away as they aged.

However, the term "breakthrough" looms as a large unknown. Research proceeds as a series of baby steps punctuated by occasional leaps. It is usually not possible to predict how many baby steps will be needed before the next leap will occur, but it also is usually possible to predict that it will occur. As more leaps move us forward into new technologies and ideas we cannot even guess at yet, being educated about genetics is one of the best ways to ensure that you will be in a position to understand the implications of those breakthroughs.

So here is one of our cautions for the future. If you want a say in preventing the mistakes of the past, you must engage in dialogues that will take place as society struggles to integrate major changes in ways that are beneficial and that do not create new problems to replace the problems just solved. And you must educate yourself so that you can base your contributions to the discussion on facts, and not on whatever sound bite you pulled off of a news headline, and that education needs to encompass both the science and the ethics. Checking out what your legislators think about these issues, what laws they are proposing, which laws they support or reject, can help you to decide whether they are helping to move policies in the directions you want them to go. And you can add your views to the process by contacting your legislators' offices to tell them when you agree or disagree with how they are voting on policy legislation that touches on these topics. Only through understanding the issues will we avoid the pitfalls of the past so that the best and brightest promises of today will carry us to the treasure pot at the end of the genomic rainbow. Surely there are wonders waiting there for our children and grandchildren if we can negotiate all of the ethical, legal, and social landmines, keep everyone engaged in the dialogue, and not succumb to unreasoning fears.

And so ends our book. In fact, we have only brushed the surface of this deep, complex topic. We hope that some of what we have told you helped you to understand some things about yourself and your family. We hope that you have come away with questions that will lead you to further explore some of the topics we touched on. We think of the chapters as letters from us to you. If you get the chance, write to us in care of the publisher or send an e-mail message to Scott at rsh@stowers.org or Julia at richj@umich.edu to let us know what you think.

We hold these truths to be self-evident, that all men are created equal, that they are endowed by their Creator with certain unalienable Rights, that among these are Life, Liberty, and the pursuit of Happiness. —**Thomas Jefferson in The Declaration of Independence**

Short Essays

1. Autonomy in determining our own medical fates is one of the hallmarks of current ethical views. One of the gray zones that poses especially interesting questions is the issue of determining the medical fates of children. What ethical issues must be weighed and balanced when determining the medical fate of a child when the decision of the parents is at odds with society's standards for how such a child should be cared for? As you consider this question please read "Four scenarios" by Raanan Gillon in *Journal of Medical Ethics*, 2003;29:267–8.

2. In the case of Tay–Sachs, couples who are carriers electing not to marry or have children substantially reduced the rate of Tay–Sachs in some communities. If couples in which one of the partners has epilepsy were willing to forgo having children for fear of passing along epilepsy, is there any expectation that such a decision would have any significant impact on the rate of epilepsy in their communities? As you consider this question please read "Ethical, legal and social dimensions of epilepsy genetics" by Sara Shostak and Ruth Ottman in *Epilepsia*, 2006;47:1595–602.

3. Population-based genetic screening has led to substantial reductions in some severe genetic traits in specific communities. Are there any downsides to programs of this kind, and what might be done to minimize such negative aspects to the programs? As you consider these questions, please read: "Can population-based carrier screening be left to the community?" by Aviad E. Raz in *Journal of Genetic Counseling*, 2009;18:114–18.

4. When it was found that some breast cancer mutations were especially linked to one racial/ethnic group, researchers shifted their focus away from individual families and onto that population. When studies focus on one particular ethnic group, what are the consequences for other groups? As you consider this question, please read: "Ashkenazi Jews and breast cancer: The consequences of linking ethnic identity to genetic disease" by Sheri L. Brandt-Rauf and her colleagues in *American Journal of Public Health*, 2006;96:1979–88.

Resource Project

As people developed their attitudes towards eugenics, they were not just taking their cues from the popular press and their neighbors. Go to www.eugenicsarchive.org/eugenics/ and write a paragraph on some of the factors that influenced people to accept eugenics as being ethically and scientifically valid.

Suggested Reading

Videos

The Treasure, sign language poetry by Ella Mae Lentz (2006, Dawn Sign Poetry).
Sound and Fury (2000, New Video Group, Inc.).

Articles

"Synthetic biology: new engineering rules for an emerging discipline" by Ernesto

Andrianantoandro, Subhayu Basu, David K. Karig, and Ron Weiss in *Molecular Systems Biology*, 2006; art. no.0028 (published online: 16 May 2006).

"Misusing the Nazi analogy" by Arthur L. Caplan in *Science*, 2005;309:535.

"The physician–scientist, the state, and the oath: thoughts for our times" by Barry S. Coller in *Journal of Clinical Investigation*, 2006;116:2567–70.

"Getting married in China: Pass the medical first" by Therese Kesketh in *British Medical Journal*, 2003;326:277–9.

"Gene therapy fulfilling its promise" by Donald B. Kohn and Fabio Candotti in *New England Journal of Medicine*, 2009;360:518–21.

"The code of codes: scientific and social issues in the Human Genome Project" by F. D. Ledley in *New England Journal of Medicine*, 1993;329(8);584–5.

"Where science meets society" by Alan I. Leshner in *Science*, 2005;307(5711):815.

"Stem cells: A revolution in therapeutics – recent advances in stem cell biology and their therapeutic applications in regenerative medicine and cancer therapies state of the art" by M. Mimeault, R. Hauke, and S.K. Batra in *Clinical Pharmacology and Therapeutics*, 2007;82:252–64.

"A Hippocratic Oath for life scientists: A Hippocratic-style oath in the life sciences could help to educate researchers about the dangers of dual-use research" by James Revill and Malcolm R. Dando in *EMBO Reports*, 2006;7:S55–S60.

"Canine morphology: hunting for genes and tracking mutations" by Abigail L. Shearin and Elaine A. Ostrander in *PLoS Biology*, March 2, 2010;8(3): e1000310.

Books

Memoir upon the formation of a deaf variety of the human race by Alexander Graham Bell (1884, University of Michigan Library digital collection).

A Fair Chance in the Race of Life: The Role of Gallaudet University in Deaf History, edited by Brian H. Greenwald and John Vickrey Van Cleve (2008, Gallaudet University Press).

Everyone Here Spoke Sign Language by Nora Ellen Groce (1985, Harvard University Press).

The Mask of Benevolence by Harlan Lane (1999, Vintage Books).

A Journey Into the Deaf-World by Harlan Lane, Robert Hoffmeister, and Ben Bahan (1996, Dawn Sign Press).

Eugenic Nation: Faults and Frontiers of Better Breeding in Modern America by Alexandra Minna Stern (2005, University of California Press).

Answers to Study Questions

Chapter 1

Questions on pages 37–38

1. If the vital spark theory were true then in any cross the progeny would exhibit the phenotype of the female parent. Mendel's observations show clearly that both parents contribute to the offspring's phenotype.

2. If the homunculus theory were true then in any cross the progeny would exhibit the phenotype of the male parent. Mendel's observations show clearly that both parents contribute to the offspring's phenotype.

3. A square represents a male and a circle represents a female.

4. A deceased individual is marked with a diagonal line through the symbol.

5. An organism that has two different alleles on the two different copies of the gene.

6. An individual in whom the information content is the same on the two copies of the gene that the individual carries.

7. A genotype is a piece of information, like a recipe, and the phenotype is the set of characteristics that are produced as a result of the cell using that information. The phenotype results from the cell making use of the information encoded by the genotype.

8. One allele comes from each parent and one allele is passed along to each offspring.

9. Alleles are different forms of a gene that contain different information content.

10. A dominant allele is one that will produce a phenotype even if it is heterozygous. Thus only one copy is needed for its phenotype to be observed.

11. A recessive allele is one that can only produce an abnormal phenotype if both copies of the gene are mutated. The abnormal phenotype is not observed when heterozygous with a copy of the normal allele.

12. In some families deafness is inherited as a dominant trait, and in other families it appears to be a recessive trait. Although it was not discussed in the chapter, it can also be inherited in an X-linked recessive manner, a mode of inheritance that we will look at later.

13. Something is considered a syndrome if there are multiple different signs and symptoms involving different organs or systems in that body that come together to form the definition of the trait.

14. Monozygotic twins are formed from a single fertilized egg that produces two separate embryos that develop together in the womb at the same time. Dizygotic twins are siblings formed from separate fertilized eggs that develop together in the same womb at the same time. Thus monozygotic twins share all of the same genetic information, but fraternal twins do not share any more genetic information than any two siblings who were born at different times.

15. Pleiotropy is the property of a mutation that can affect multiple seemingly unrelated characteristics. There are many different ways in which a mutation can have pleiotropic effects. One of the most obvious

is that a mutation may have very different effects when expressed in very different tissue types. So the symptoms resulting from a mutation that results in production of a defective gene product in the brain might have neurological consequences, while the same mutation in the same gene, when expressed in the endocrine system, might affect levels of circulating hormones with effects on growth or metabolism.

16. We can figure out the expected frequency of offspring with different phenotypes by using a Punnett square to assign genotypes to sperm and eggs and then figure out the frequency with which different combinations of genotypes will occur in the progeny.

Kate

Dan		C	c
	C	CC	Cc
	c	cC	cc

On average, we expect that one-half of their children will be carriers and three-quarters of their children will not develop cystic fibrosis.

17. No, the proband is not always someone with the trait. Someone who is at risk may be the first one identified, even though they might be unaffected and remain that way throughout life. This will often happen in the case of someone who consults a doctor or genetic counselor because they are concerned about whether they have or will develop the trait that is present in their family.

18. They share the same intrauterine environment so they have all of the same influences on early developmental events. This is especially important when studying traits that are present at birth or that have early onset. In some cases, twin studies include the spouses of the twins as controls for environmental effects playing a role in later-age onset disease.

19.

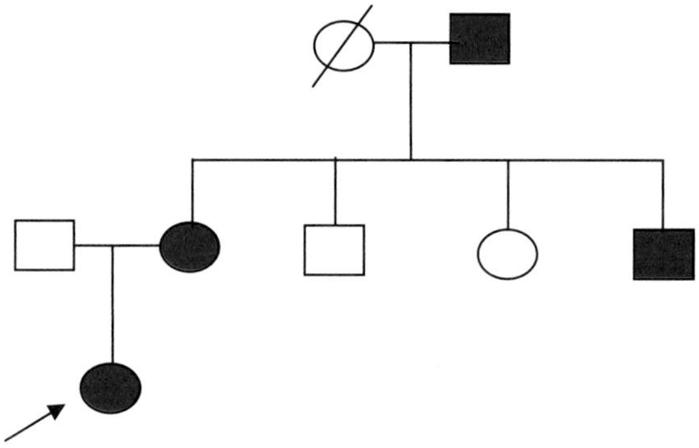

20. A trait may be considered a disease if it has substantial health consequences or is otherwise detrimental to the welfare of the individual. Genetic diseases may be considered a sub-group of the broad category of genetic traits.

Chapter 2

Questions on pages 77–78

1. An autosome is any chromosome other than the X or the Y.
2. A nucleolus is a region within the nucleus where high levels of ribosomal RNAs are transcribed and assembled.
3. Mitochondria provide the energy that runs the cell.
4. T pairs with A.
5. G pairs with C.
6. We mean that the sequence of one strand of a DNA double helix can be predicted based on the sequence of the other strand using the DNA base-pairing rules.
7. They happen less than once every 10,000,000,000 replicated bases.

8. An RNA primer is used to start off the synthesis of Okazaki fragments on the lagging strand during DNA replication.

9.

T	C	C	A	T	G	C	A	G	G	G	T	A	G	A

5′ ... 3′

| A | G | G | T | A | C | G | T | C | C | C | A | T | C | T |

3′ ... 5′

10. Semi-conservative replication takes place when the two strands of the double-helix separate and two new strands are generated that are complementary to one of the old strands. Knowing that this is the mechanism tells us that each cell gets a DNA helix that contains one old strand and one new strand.

11. A nucleosome is the "bead" on the beads-on-a-string version of chromatin, and it is made up of a core of histone proteins with DNA wrapped around it.

12. A histone is a highly conserved protein that forms part of the core of a nucleosome.

13. DNA is very long, thin, and fragile and would break if it were not protected by being packed into a highly ordered structure on a framework of proteins.

14. Maurice Wilkins, Alex Rawson Stokes, Herbert Wilson, Rosalind Franklin, and Raymond Gosling.

15. By putting the chromosome pictures into pairs and arranging them in order, it makes it much easier to tell if a chromosome is missing and to tell whether there are small aberrations in the structure of a chromosome.

16. When we use one primer, our only template for every round of amplification is the original starting template, so on every round we simply make one more copy from the original template. If we use two primers, then each round of replication creates new templates from which the next round can be made. So on every round of replication, there are more templates from which to copy.

17. As pair to Ts through the use of two hydrogen bonds, while Gs pair to Cs through use of three hydrogen bonds. The amount of heat that can break apart the weak bonding in an AT-rich region is not adequate to break apart the stronger bonding in the GC-rich region, but as the temperature increases the more strongly bonded G/C-rich regions start to come apart.

18. Fluorescence in situ hybridization. A specific sequence of DNA labeled with fluorescent dyes hybridizes to a spot on the chromosome that contains sequences homologous to the probe.

19. All children of an affected woman will be affected.

20. No children of an affected man will be affected.

Chapter 3

Questions on page 112

1. A promoter is a region of DNA sequence immediately adjacent to the point at which transcription of a gene begins. Multiple short sequences called regulatory elements are scattered across the promoter region. Binding of transcription factors to the regulatory elements results in regulation of transcription of the gene adjoining the promoter. The orientation of a promoter is important, and it does not direct transcription of the gene if the direction of the promoter is reversed.

2. An enhancer is a region of DNA sequence that is required for initiation of transcription. It is located somewhere in the vicinity of the gene, or even inside of

one of the introns of the gene, and proteins bound to the enhancer bind to proteins in the transcription activation complex to start transcription. An enhancer is not directional, so the sequence works even if it is flipped over so that its orientation on the chromosome is backwards relative to its normal orientation.

3. An inducible gene is one whose expression can be turned on or substantially turned up in response to a molecule such as a hormone from within the body or in response to a stimulus from outside the body such as a drug or other environmental factor.

4. A housekeeping gene is a gene that is expressed in all of the different types of cells in the body.

5. A hormone is a molecule produced by one cell that travels to another cell and in the process signals changes in gene expression and other functions on the part of the target cell.

6. RNA polymerase generates an RNA copy of the DNA sequence in a gene by adding bases to the growing RNA chain based on the use of base-pairing rules.

7. A transcription factor is a protein that binds to DNA for the purpose of regulating gene expression. A regulatory element is a short DNA sequence to which the transcription factor binds. The interaction of the transcription factor with the regulatory element alters the level of transcription of the gene.

8. Ectopic expression is expression of a gene in a location where it is not normally expressed or at a time when it is not normally expressed. This can lead that cell type to produce a gene product and carry out functions it would not normally carry out. In the case of ectopic expression of a developmental regulatory gene, ectopic

expression can result in the presence of cell types or even whole organs in parts of the body that normally do not have such cells or organs.

9. Examples of similarities between DNA and RNA: the genetic letters are bases, the bases are not connected directly to each other but are connected to a backbone that holds the strand together, they use the same base-pairing rules except that RNA pairs A with U instead of A with T, both use the base-pairing rules as the basis for copying information from a strand of DNA, and DNA and RNA have slightly different structures to their backbones.

10. Examples of differences between DNA and RNA: DNA uses T where RNA uses U, DNA is a permanent record of the genetic information and RNA is a temporary repository of the information that is used to carry information to the ribosome but is eventually degraded by the cell, RNA pairs A with U instead of A with T, and DNA is usually double-stranded while RNA is usually single-stranded.

11. Examples of aspects of gene regulation that can be controlled by transcription factors include whether or not the gene is being expressed, the quantity of expression, the cell type in which expression takes place, the timing of expression (developmentally), the timing of expression (in terms of the cell cycle), or responses of the gene to a signal such as a hormone.

12.

5'	U	G	C	A	A	G	C	A	G	U	C	U	A	G	A	3'
3'	A	C	G	T	T	C	G	T	C	A	G	A	T	C	T	5'

13. Things that function in trans can move away from the gene in question to act at

other genes or to act on the other copy of the gene on the other chromosome; an example of something trans acting is a transcription factor that can bind to both promoters of the gene being regulate. Things that act in cis are physically tied to the gene in question and can only act on the copy of the gene that it adjoins; an example of something cis acting is a promoter, where any sequence differences in that promoter only affect the transcription of the copy of the gene that is contiguous with that promoter and do not affect transcription of the other copy.

14. RNA polymerase is a single protein that carries out the assembly of the RNA molecule. The transcription initiation complex is a combination of the polymerase plus multiple accessory proteins plus the DNA to which it binds in preparation for beginning transcription.

15. Lactase persistence is the healthy phenotype not associated with medical symptoms, which many would consider normal. However, lactose intolerance is the most common phenotype, when we look across the world's populations. Lactose intolerance is also the ancestral genotype/phenotype present before the lactase persistence genotype arose, so we might consider that the genetic alteration causing lactase persistence results from the mutant (or more recent) genotype.

16. Methylation of a cytosine, if it is in a regulatory region such as a promoter region, can affect regulation of transcription. It is not considered a mutation because a methyl cytosine within the coding sequence is read by the polymerase as if it were a non-methylated cytosine.

17. Because the individual we are evaluating might not have experienced the eliciting stimulus needed to bring on the characteristics of the trait. For example, we cannot tell if someone has the potential for favism if they have never eaten fava beans.

18. The hormone alters the receptor which then travels to the nucleus to act as a transcription factor.

19. The amount of RNA can also be affected by the rate at which the RNA is being degraded.

20. A "knock-out" animal has had a gene removed or inactivated; a "knock-in" animal has a gene or a mutation within a gene added.

Chapter 4

Questions on pages 140–141

1. An mRNA is a messenger RNA that is produced in the nucleus and travels to the cytoplasm where ribosomes use it to build a protein whose amino acid content is based on the codons in the mRNA. An mRNA includes a 5′ untranslated region, a section of coding sequence and a 3′ untranslated region.

2. *Met Gly Leu Pro Arg Asp Cys Stop*
   — — — — — — — —.

3. The exons remain a part of the final spliced mRNA. The introns are removed from the transcript and discarded.

4. The amino acids are specified by codons and the genetic code is considered redundant because there are more codons than there are amino acids, so some amino acids are specified by more than one codon.

5.

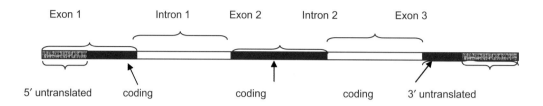

6. Splicing removes the two introns and a polyA tail is added.

7. A codon is a group of three neighboring DNA bases that are read by the ribosomal machinery as a single piece of information that determines which amino acid will be added next to the growing protein chain. The piece of information is read by the ribosome by pairing with the anti-codon of the tRNA. Each codon specifies an amino acid, except for the three stop codons that provide a signal to the ribosome to stop translating the RNA.

8. The normal product of the *situs inversus* gene gives the information needed for the body to place the heart on the left side of the body. In the absence of a functional *situs inversus* gene product, the body does not know which side to put the heart on, so in each individual there is a random chance that the heart will be on the left or on the right.

9. Ectopic expression is expression of a gene in a location where it is not normally expressed or at a time when it is not normally expressed. This can lead that cell type to produce a gene product and carry out functions it would not normally carry out. In the case of ectopic expression of a developmental regulatory gene, ectopic expression can result in the presence of cell types or even whole organs in parts of the body that normally do not have such cells or organs.

10.

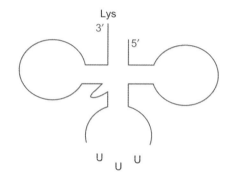

11. It is a sequence in the intron to which the cut end of the intron binds before the other end of the intron is cut to remove the intron. Binding to the lariat site results in forming a structure that looks like a loop with a tail, rather like a rodeo lariat.

12. A modular gene is one that is alternatively spliced so that the cell can use some parts of the gene product in some situations and a different combination of the parts of the gene product in other situations.

13. Examples of ways in which transcription is like translation: Both of them create a long complex molecule based on the use of information contained in a nucleic acid chain (RNA based on DNA information, protein based on RNA information), and the order in which they make additions to the molecule they produce is based on the order in which that information occurs on the nucleic acid chain providing the information. Examples of ways in which transcription differs from translation: They differ in which type of nucleic acid chain they are based on, which aspects of cellular machinery carry out the operation, where in the cell they take place, and which types of molecular building blocks they incorporate into the chain they are creating.

14. Through alternative splicing the cell can create different multiple different mRNAs from one primary transcript, and each of those different mRNAs will result in the production of a different protein. Normally the products of alternative splicing will share some sections of related sequence.

15. A nucleotide change in an intron can sometimes create a new splice site, or destroy an existing splice site, thus changing which exons appear in the final mRNA. It could also alter the lariat site.

16. The cell uses groups of three bases to specify one amino acid. The available combinations of four different bases used in groups of three can produce 64 unique coding units called codons.

17. Many codons that specify a particular amino acid are have related sequences. For instance both arginine codons start with AG, both lysine codons start with AA, and all four glycine codons start with GG. Thus we see that the sequences of codons assigned to particular amino acids is not random.

18. Alternative splicing lets the body make a version of the gene product in most cell types that does not include exon 3 where the mutation is located, but the splicing process in the neuronal cell types keeps exon three in the transcript and includes the corresponding altered amino acid in the final protein product. Thus although the mutation is present on the chromosomes in all cell types, only those cell types that use a splicing pattern that keeps exon 3 in the final transcript will make a protein product that could make the cell ill.

19. Glycine is a very small amino acid, and the much larger phenylalanine might not fit into the space normally occupied by glycine in that protein.

20. Glutamic acid has an extra negative charge, and histidine has an extra positive charge, and changing the charge can affect function.

Chapter 5

Questions on pages 192–193

1. A gain-of-function mutation results in the protein acquiring a function or characteristic that it did not have before. The mutation results in the gene product actively doing something new rather than simply being inactivated and failing to carry out a function. An example of a gain-of-function mutation might be a mutation in a regulatory region that causes the gene to be expressed at a different time during development or in different tissue. A gain-of-function mutation usually shows a dominant mode of inheritance, but not always. Sometimes people will classify a mutation that makes a protein that not only fails to function, but even inactivates the normal versions of that same protein in heterozygotes, as being gain-of-function mutations, but there is another term that is specific for this particular class of mutations: they are called antimorphs.

2. A loss-of-function mutation is a change to the genotype that results in either failure

to make the gene product or results in alteration to the gene product in such a way that it cannot carry out some or all of its normal functions.

3. A missense mutation is a change to the DNA sequence that results in a change in a codon sufficient to alter the amino acid that is inserted into the growing protein chain by the ribosome. Missense mutations are usually, but not always, point mutations.

4. A nonsense mutation is a change to the DNA sequence that results in changing a codon used to insert an amino acid into one of the three stop codons. The result is truncation of the amino acid because the ribosome terminates translation of the transcript prematurely when it reaches the newly created nonsense mutation.

5. Radiation and chemicals (mutagens) can both cause errors in DNA replication. Errors can also occur naturally as a result of the inability of the cellular machinery to achieve perfection when carrying out DNA replication.

6. A silent mutation is a change in the DNA sequence that does not result in a missense or nonsense mutation. This is most often used to refer to a point mutation that fails to alter the protein sequence either because it occurs in a non-coding region or changes the codon in such a fashion that it does not alter the amino acid specified by the original codon. (Remember that all but two amino acids can be specified by more than one codon! So a TTT to TTC mutation that changed the mRNA codon from UUU to UUC would not change the amino acid since UUU and UUC both specify phenylalanine. Two amino acids that are uniquely specified by only one codon are methionine, AUG, and tryptophan, UGG, so any change to one of these two codons would necessarily result in a change in

which amino acid is inserted into the protein chain.)

7. The Ames test detects increases in the rate of mutation in bacteria. The test uses a strain of bacteria that requires histidine to grow. The bacteria are plated out on growth media that lacks histidine and a disc soaked with the test chemical is dropped onto the surface of the plate. If the test chemical is capable of bringing about mutation of the DNA in the bacterial genome, some cells will acquire a mutation that restores the ability of the bacteria to grow on media that lack histidine; if the chemical causes mutations then the cells with the mutated genomes will begin to grow and form colonies on the plate in the vicinity of the chemical-soaked disc. A chemical that causes a mutation that allows bacterial cell growth in the Ames test is called a mutagen, and we now know that chemicals that turn out to be mutagens usually also turn out to be carcinogens, where carcinogens are chemicals that can cause types of mutations that result in cancer. Knowing that mutagens are usually carcinogens (and vice versa) is important since the cheap, simple, and highly predictive Ames test is often the first way a chemical is tested for potential carcinogenicity.

8. Lactose interolerance results from changes in regulatory sequences that control transcription of the lactase gene.

9. A sequence tagged site is a piece of sequence in a computer that describes the actual sequence of DNA at a specific point on a chromosome. Historically, once a group found a gene, other groups could work on that gene only if a physical copy of the gene was transferred from one research group to another. Sequence tagged sites made it possible for other research groups to acquire information about the gene sequence in the form of a sequence tagged

site and then use that information as the basis for synthesizing their own copy of the gene without any shipping of biological materials from one research group to another. Posting of sequence tagged sites in publicly accessible databases allows research groups worldwide to gain access to new information about genes and genomes as soon as the information is posted online.

10. An *in silico* experiment is an experiment, such as testing of theoretical models, carried out on a computer instead of at a lab bench.

11. Different fields or research groups use the term polymorphism differently. Different definitions that are used include: any change to the DNA that occurs at a frequency of 1% or more, any change to the DNA sequence without regard to frequency, or changes to the sequence that are benign and do not affect the phenotype. For most studies the first definition is preferable because it allows researchers to focus on changes in the DNA sequence that are common enough that it is possible to apply statistical tools to their study, and such statistical approaches do not work well for changes that are very rare. In some cases where we are not dealing with data on populations, the term can be used effectively to discuss changes that are rare or of unknown frequency in the population, but in the case of changes that are not altering the phenotype a preferable, and suitably neutral, term would be sequence variant.

12. Different fields or research groups use the term mutation differently. Different definitions that are used include: the process of altering the DNA sequence, a change in the DNA sequence that has a functional impact on the phenotype, or any change in the DNA sequence. Thus, there are cases in which the terms *mutation* and

polymorphism end up being used to describe the same kinds of changes to the DNA. By including descriptors it is possible to distinguish benign polymorphisms from causative mutations.

13. The Duchenne muscular dystrophy gene covers the largest chromosomal region of any known human gene at more than 2,000,000 bases in length. However, the titin gene produces the largest transcript at more than 87,000 bases and the largest gene product with a length of more than 34,000 amino acids.

14. A tandem repeat is the repeated occurrence of a sequence where the copies are placed immediately next to each other, one after the other, along a region of the chromosome. The CA repeat is one of the most common with more than 50,000 clusters of CA repeats scattered around the genome.

15. The mutation rate, which is the rate at which the trinucleotide repeat changes length, increases as a man ages. This mutation rate can be assayed by using PCR to make copies of the region of DNA containing the trinucleotide repeat from individual sperm, followed by use of agarose gels to measurement of the length of the PCR fragment.

16. Anticipation is the earlier occurrence of a disease in succeeding generations. In some cases, age at diagnosis will be earlier, not because the disease actually started at an earlier age, but rather because improved technology makes it possible to detect the disease at earlier stages or heightened awareness of the disease leads people to seek diagnostic testing at an earlier age.

17. A transgenic animal is one that has been altered by putting in or taking out genetic material.

18. DMD and BMD are both caused by mutations in dystrophin. DMD results from

mutations that eliminate the gene product or activity. BMD is caused by mutations that affect protein function without eliminating the protein product.

19. The problem of asking whether you want to know when you will die or what you will die of if there is not a solution or remedy for it, if knowing it does not offer you any way to change that fate.

20. A male biological clock is an increasing rate of germline mutation with age in males.

Chapter 6

Questions on page 245

1. A balanced translocation is a pair of abnormal chromosomes that have switched pieces. For example, imagine that the end of chromosome 13's long arm has been traded with a piece of the end of chromosome 15. We now have a copy of chromosome 13 that is missing genes from the end of its long arm, but carries genes from the end of chromosome 15 and vice versa. An individual who is heterozygous for this balanced translocation also carries a normal copy of chromosomes 13 and 15. So all of the genes are present in two copies, but the arrangement has been altered.

2. Interphase, Prophase, Metaphase, Anaphase, and Telophase.

3. Sister chromatids are identical copies of a given chromosome that are attached at their centromeres. They were created by replication and have identical genetic contents. So prior to replication a human cell has 46 chromosomes each of which has but one chromatid. After replication that same cell still has only 46 chromosomes, but each chromosome has TWO sister chromatids. Each chromatid, or sister chromatid, has a single DNA molecule.

4. A spindle is a structure built of microtubules that provided the tracks on which the chromosomes move during cell division.

5. The chromosomes at metaphase are located in the middle of the spindle, at a region called the metaphase plate.

6. The spindle poles or centrioles.

7. During pre-meiotic S phase the chromosomes replicate.

8. It separates the two daughter cells.

9. The homologous chromosomes need to be "matched" (paired), locked (recombined), and "moved" (homologs are separated to opposite poles at anaphase I).

10. Only two (non-sister) chromatids may participate in a single recombination event.

11. When the female is still a fetus!

12. • In males all four products of meiosis become sperm. In females a given meiosis produces only one egg.
    • Males begin meiosis at puberty and continue to perform meiosis and make sperm throughout their lives. Females begin meiosis before birth and then complete meiosis a relatively small number of times throughout their lifetime.
    • Male meiosis is a continuous process, stopping only if an error is made. Female meiosis has several built in delays, one of which (prophase arrest) can last for five decades!!!
    • Male meiotic spindles have centrioles; female spindles do not.

13. Fertilization.

14. To lock homologs together.

15. QRS, qrs, Qrs, and qRS.

16. We might not be detecting the recombination because there were two crossovers between the two markers

(A and B) being followed. If both those crossovers involved the one chromatid that we recovered, then no exchange event will be observed (it will still be AB or ab). If we tested an additional marker that lies between A and B (such as C), we might expect to be able to detect such "double cross over events" between A and B in ACB/acb parents by the recovery of AcB or aCb recombinants.

17. Uniparental heterodisomy has one copy of each of the two homologs present in one parent plus no copies of that chromosome from the other parent. Uniparental isodisomy has two copies of one of the chromosomes from one parent and no copies of the other homolog from that parent and no copies of that chromosome from the other parent.

18. This individual could pass along different combinations of normal and abnormal chromosomes. The child who receives only normal copies of the chromosomes is expected to most likely have no problems. The child with the balanced translocation will most likely have a problem from it only if the parent also has a problem from it. But the two abnormal combinations of chromosomes that could be passed along are different, with each one show an excess of what they other has a shortage of, and vice versa.

19. The checkpoint tests to see whether the chromosomes are all paired, and if the cell fails the test it does not go on to complete meiosis and produce sperm.

20. If the problem is caused by a deletion of the Prader–Willi critical region, then if a brother and sister who carry the mutation both have children, one would pass along only a maternal copy of the region and one would pass along only a paternal copy of the region.

Chapter 7

Questions on pages 268–269

1. Because only one X is active, no matter how many X chromosomes a cell possesses. In XX females or XXY males the other X is inactivated and becomes a Barr body. In XXX females both additional X chromosomes are inactivated.

2. We already know that males do just fine with only one copy of the X, and we know that females with two copies of the X inactivate one to arrive at effectively having only one copy.

3. Sex linkage is a mode of inheritance associated with genes on the X chromosome. The unusual pattern of inheritance derives from the fact that females have two X chromosomes, but males have only one copy of the X. This difference, and the fact that males always receive their Y from their father, leads to a pattern of inheritance in which males always receive their sole X from their mother and females always receive their father's X. Thus a male carrying a recessive sex-linked trait like colorblindness cannot pass it on to his sons, but all of his daughters will be carriers.

4. A Barr body is the physical manifestation of an inactive X chromosome. It is a dark mass (or brightly staining dot, depending on what staining method is used) of chromatin that is easily observed in nuclei of cells carrying more than a single X chromosome.

5. None.

6. A single Barr body is present in the nuclei of XX women and XXY men.

7. Three Barr bodies are present in XXXX women.

8.

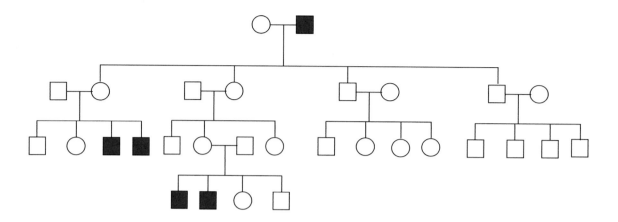

9.

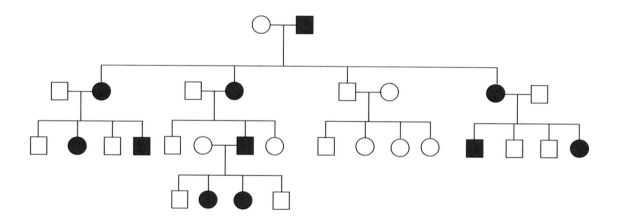

10. The pseudoautosomal region contains genes and sequences common to both the X and Y chromosomes.

11. The pseudoautosomal region is used for pairing and recombination of the X and Y chromosomes so that they undergo correct segregation of the sex chromosomes into daughter cells during meiosis. There are actually several other pseudoautosomal regions on the X, but these do not recombine with the Y chromosome.

12. Pairing and recombination (crossing over) occurs at high frequency in the pseudoautosomal region.

13. Genes on the mammalian X are up-regulated to achieve an expression level equivalent to what would have been produced from two copies of the chromosome.

14. There are genes in the pseudoautosomal regions that are usually required in two doses.

15. There is no sign of XIST at the eight-cell stage but there is a lot of it by the blastocyst stage when there are about ten times that many cells.

16. This will not affect risk to her son, since the X chromosome is not inactivated in a male and inactivation in the mother does not affect which X she passes along.

17. No, gene expression on the X is up-regulated to restore the level of gene expression one would expect from two copies.

18. Because one of the definitive tests for X-linked dominant inheritance is to look at the phenotypes of the daughters of an affected male, but many X-linked dominant traits are lethal in males or show such a severe phenotype in males that they do not have children.

19. If her father is affected and her mother is a carrier, then there is a 50% chance that she will have the defect on both copies of the X chromosome.

20. It is a region that contains sequences that are present on both the X and the Y, and it is involved in pairing between the X and the Y during meiosis.

Chapter 8

Questions on pages 296–297

1. Gonadal sex in a male is determined by the presence of testes. In a female gonadal sex is determined by the presence of ovaries. Both the testes and the ovaries arise from structures called the indifferent gonads. If the SRY gene is present and expressed the indifferent gonads become testes; if SRY is absent, they become ovaries.

2. Primary sex characteristics include penis and scrotum. Secondary characteristics include body hair distribution, greater upper body strength, narrow hip structure, and ability to add lean muscle mass.

3. Primary sex characteristics are vagina, cervix, uterus, fallopian tubes and clitoris. Secondary characteristics include breasts, shortage of body hair, less upper body strength, less ability to add lean muscle mass, increased body fat, and the menstrual cycle.

4. Sex is the physical set of primary and secondary sexual characteristics. Gender is the term for the sex with which the individual identifies; this is an issue of sense of self that is completely separate from the individual's physical characteristics, although in many individuals the gender identity agrees with the gonadal and somatic sex.

5. Researchers find that homosexuality is shared more often by identical twins than by fraternal twins. They also find that the brother of a gay man is more likely to be gay than is a man drawn at random from the general population. They mapped a gene to chromosome Xq28, and the data seem to suggest the existence of a gay gene there, although their findings cannot explain all cases of male homosexuality; however, a second study failed to map a "gay" gene to this location, leaving a lot of questions about whether the discrepancy between the two studies is due to something technical, or whether it is telling us something important about the genetic underpinnings of sexual orientation.

6. The SRY gene is the primary determinant of male gonadal sex, and the gene product of this gene initiates the process of male sexual differentiation.

7. The indifferent gonad is the fetal tissue that is destined to become either male or female gonads; early in development, these cells are indifferent, having no inclination towards differentiating in either male or female directions. During the seventh to

eighth week of development, the indifferent gonad stops being indifferent, proceeding towards male sexual development and becoming testes if the signal from the SRY gene is received during this brief window of time, or developing towards female sexual development and becoming ovaries if no signal is received during this window of time.

8. The Müllerian Inhibitory Factor (MIF) produced by the testes starting during the eighth week of development.

9. XY individuals with mutations in (or deletion of) a single gene on the Y chromosome, the SRY gene, develop as females. Similarly translocating just the SRY gene to the tip of the X chromosome can induce male development in individuals with two X chromosomes. For sex determination, all that matters is the SRY gene.

10. In XXY males the second X appears to become activated in the male germline at puberty causing cell death and testicular atrophy. However, the inactive X chromosome is re-activated in the female germline. So part of the reason Turner syndrome females are sterile may be that human females require two active X chromosomes for fertility. Another possibility is that two copies of genes present in the pseudoautosomal region are required for fertility.

11. They have testicles because they have a Y chromosome and an SRY gene that signals to the indifferent gonad that it should become testes. They lack other male sexual characteristics because the lack of a functional androgen receptor keeps the cells in their bodies from detecting that the testes are sending out testosterone as a signal to carry out male somatic sexual development. They lack cervix, uterus or fallopian tubes because the testes do produce MIF, which inhibits the development of the Müllerian

ducts. (Note that the Müllerian ducts give rise to the cervix, uterus and fallopian tubes.)

12. They do not have testicles because they lack the SRY gene product that provides the signal telling the indifferent gonad to become testes. Because there was no SRY signal during the seventh to eighth week of development, the indifferent gonad proceeded to develop into ovaries which began producing high enough levels of estrogen to bring about development of the cervix, uterus, and fallopian tubes. In the absence of testes, no MIF is produced and the cervix, uterus, and fallopian tubes develop.

13. An increase in levels of SOX9 signals development of male sex, and decrease in levels of SOX9 signals development of female sex.

14. SRY causes SOX9 to increase, and absence of SRY causes SOX9 to decrease.

15. SRY acts to bring about sex determination through a series of downstream events, and intervening in those events downstream of SRY can get the same effect as intervening with SRY itself.

16. If the brothers in the study also shared some other second property besides sexual orientation then it might actually be the trait that has been mapped.

17. She would be considered male by the karyotypic and gonadal definitions of sex, but would be considered female in terms of secondary sexual characteristics. Of greater importance, she is of female gender and considers herself to be female.

18. Twin studies suggest a genetic component to the trait.

19. If a trait will not develop until later in life, then we cannot tell whether an individual who currently lacks the trait is an individual who will lack the trait over the course of their lifetime. If we are trying to use the phenotype as an indicator of the probability

that someone has the causative genotype, the chance of erroneously classifying someone will be proportional to the chance that they will eventually develop the trait even though they do not have it currently.

20. In the absence of the signal from SRY the outcome will be female sex.

Chapter 9

Questions on page 340

1. A child producing very low levels of MAOA might have reduced tolerance for an abusive environment and be more likely to develop serious problems in such an environment. At this time, the body of information on the subject is quite limited, and the study has not been replicated, so such screening should not be started until we are more certain of what particular test results mean and what the test results predict.

2. Ethically, there are major problems with any study of prison populations where prisoners might not be participating of their own free will, and might not be willing to stand up for themselves and say, "I don't want to participate." Scientifically, individuals in a prison population may well have very different reasons from a free population for giving the answers they give, which can potentially influence what answers they give. An imprisoned individual may very well not trust that things they tell researchers would be kept confidential and might limit information they give to protect themselves against having that information get to others in the prison or to authorities in the legal system, so it is hard to tell how accurate the information is.

3. Certain diseases or traits can be called digenic because they are caused by the simultaneous presence of mutations in each of two different genes. This means that more than one gene may need to be targeted/modified to alleviate disease progression. This is often difficult, particularly when it is not clear how each gene plays a role in a particular disease.

4. Different genes of the retinoid cycle can cause a related set of diseases by affecting different steps in the same pathway. This cycle is responsible for processing and delivering the vitamin A derivative used by rhodopsin when it detects light. Knocking out genes can block production of the products of the biochemical pathway, but different byproducts can accumulate depending on which step in the pathway is blocked.

5. Inheritance in cardiovascular disease is multifactorial. Cardiovascular disease can be a plumbing problem (blocked blood vessels or damaged valves), an electrical problem (aberrant timing of heart beats), or a muscular problem (damage to the heart muscle). Many of the loci involved in cardiovascular disease have minor effects, increasing risk or decreasing risk without being a simple outright cause of disease. Genes whose products are involved in transport or processing of lipids can affect cholesterol levels. Reactive oxygen species can play a role in ischemic heart disease so genetic variants in genes that process reactive oxygen species can affect cardiovascular risk. Channel genes affect cardiac rhythms, as can many other kinds of genes. But in most people a combination of many of these factors is going on, with multiple genetic variants that each cause only a small variation in the system combining to increase or decrease any one person's risk.

6. The risk to immediate family members of an affected individual is higher than for the general population. The risk is much lower for second-degree relatives. The risk is

higher when more than one family member is affected. The more severe the expression of the trait, the greater the risk of recurrence of the trait in relatives.

7. Viruses, bacteria, and parasites. Viruses have the general structure of a genome (made of DNA or RNA) packaged in a protein coat; even though they can cause infections that result in production of many new copies of the virus, they need a host cell to provide the machinery that copies the viral genome and produces new copies of the viral protein coat. Bacteria are cells that possess all of the machinery needed for replication of the bacterial genome and cellular structures. Parasites are much more similar to the cells of the human host. Cells like bacteria that lack a nucleus are considered prokaryotes, and the nucleated host cells are considered eukaryotes. Parasites are also eukaryotes and have a nucleus. Parasites can be especially difficult to treat because of their similarity to human cells.

8. Yes. In fact there are many instances. One example is pregnancy, where byproducts of a defective biochemical pathway in the mother can cause developmental problems in the fetus she is carrying. Another common example is organ transplant rejection, in which the tissue transplantation antigens that result from the donor's genotype may be incompatible with the tissue transplantation antigens recognized and accepted by the recipient's immune system.

9. Variable expressivity occurs for most diseases, although the variability can be large for some traits and small for others. Sometimes individuals with big differences in severity of disease because they have different mutations in the same gene. They can also have the same disease caused by mutations in different genes, or have the same primary cause of disease

with different modifier mutations or environmental exposures.

10. Unfortunately although many genes have been identified that play a role in simple Mendelian traits, the underlying causes of many of the common complex traits have not been worked out and huge numbers of sequence variants have been identified for which phenotypic consequences are not yet known. In addition, many sequence variants turn out to be predictive of increased risk of a trait, but only some of the individuals with that variant develop the trait. Also, many factors can influence expressivity so that for some variants we do not know whether the trait will be severe or whether onset will be late or early. Thus huge amounts of information can emerge from a whole genome sequence for which we do not have a clear interpretation.

11. Because studies of children with high and low levels of MAOA expression indicate a probability that children will go on to a good or poor outcome, but sometimes children with low levels of MAOA expression go on to a result typical of a high level of expression.

12. The affected individuals are all males, and the genotype gets passed through the women but they are not affected. In sibships with affected men, half the men are affected.

13. The behaviors of the different men in the family are not all the same. This is a frequent problem with efforts to classify individuals in families where we are evaluating psychological and behavioral traits. We end up having to make a judgment regarding whether rape and assault with a pitchfork both constitute the same behavioral phenotype.

14. Some modifiers have a very specific interaction with a single gene product, while others affect generalized processes in the cell such as aging or oxidative damage.

15. If we were dealing with a modifier gene we expect that the two genes would likely segregate independently of each other so that some affected family members would show earlier onset and some would show later onset, but in each family the age at onset is consistent.

16. Each is the product of the additive effects of multiple factors/genes.

17. For a quantitative trait the effect is simply additive, but for the threshold trait the phenotype is not affected by the first several genes/factors but only starts to manifest after the accumulation of enough of an additive effect to cross the threshold.

18. The normal PRPH2 protein can form a tetramer, and effectively replace the missing ROM1, but L185P can only make dimers and thus cannot replace the missing ROM1.

19. With more than 120 different loci, many of which are recessive, the chance is small that two randomly selected individuals with retinitis pigmentosa will have this trait because of mutations in the same gene. Clearly, sometimes they will turn out to have defects in the same gene but it is more likely that they will each have a defect in a different gene.

20. Because one kind of mutation in the rhodopsin gene can cause retinitis pigmentosa and a different kind of mutation in the rhodopsin gene can cause congenital stationary night blindness.

Chapter 10

Questions on pages 364–365

1. Cancer is a state in which cells that are not supposed to be dividing are dividing. Mechanistically, cancer happens when cells keep advancing into the cell cycle when they should be sitting in G0.

2. G0 is a resting state that takes the cell outside of the normal cell cycle and leaves it sitting quietly without dividing.

3. Bone marrow, lining of the gut, skin.

4. A benign tumor is a dense, well-defined, and tightly bordered mass that stays confined to the region it is in and does not invade surrounding tissues.

5. The ability of cells to invade into the surrounding tissue or metastasize to other locations in the body.

6. Metastatic cells have gained the ability to leave the site of the original tumor and its surrounding tissue and move out into other tissues and organs elsewhere in the body.

7. A tumor suppressor gene encodes a protein that serves to keep cell division in check, controlling regulation of cell growth by keeping the cell from dividing when it is not supposed to.

8. A proto-oncogene is a gene whose product normally functions to actively promote cell division at a specific time (for example, early embryogenesis) or in a specific cell type (for example, normally proliferating cells of the immune system). Mutations in, or chromosomal rearrangements of, a proto-oncogene create what are called oncogenic mutants that express these genes at an inappropriate place or developmental stage, causing abnormal cell proliferation and thus tumors. Proto-oncogenes can be converted to tumor-causing oncogenes by mutation, re-arrangement or gene amplification.

9. According to the two-hit model, individuals who are born with a defect in one copy of a tumor-suppressor gene develop without cancer until mutation of the second copy of that tumor-suppressor gene results in a cell that is no longer making any functional gene product in the cell that has had two hits on the same cancer gene.

10. Sandy had one tumor develop, but Gary had multiple tumors develop, and we think that the number of tumors is a very approximate indicator of the number

of times a mutation happened in the retinoblastoma gene. We expect that Sandy might be the only one in her family who would have retinoblastoma, and we expect that this would be the case even if she came from a large family. We expect Gary to have a family history of retinoblastoma with a two-hit autosomal dominant mode of inheritance. We expect this result because Sandy was born with only normal copies of the retinoblastoma gene, and she had to end up with both copies of the gene knocked out in the same cell before she would develop a tumor of this kind; it is highly unlikely that both copies of the same gene would be hit in the same cell, but over the extent of a whole population, rare cases like Sandy will be seen where two hits happen in one cell in an individual who did not inherit a mutation. It is substantially less likely that the same thing would happen to that same person a second time. In the case of Gary, we predict that he already has one copy of the gene knocked out, so it only takes one hit on the second copy in any one cell to produce a tumor; this suggests that he has a hereditary form of retinoblastoma and will have a family history. We predict the autosomal dominant mode of inheritance for Gary because that is the known mode of inheritance for individuals with a hereditary form of the trait.

11. Because we expect defects in a DNA repair gene to lead to instability of genotypes and resulting tumors in many different tissues. Also, many different cancer genes could be impacted by a DNA repair defect, and different cancer genes have their own phenotype, including tissue specificity.

12. The only thing protecting the individual carrying the defect in the colon cancer gene is the good copy of the gene present on the other chromosome. Exposure to the chemical would increase the risk of losing the good copy of the colon cancer gene in one or more cells, leading to uncontrolled growth and tumor formation. If a similar loss of one copy of the colon cancer gene happened in one cell in the individual who did not inherit the colon cancer defect, there would still be a second normal copy of the gene available to keep regulation of cell growth under control. While the individual with the colon cancer gene defect might be fated to eventually incur such a loss of the good copy, exposure to a chemical that induces deletions (or other types of mutations for that matter) might be expected to hasten the day on which the good copy is lost and a tumor starts to develop, or to increase the number of new tumors that arise.

13. Most cells in the body of the CML individual do not have the cancer-causing mutation. All that is needed to bring about CML is a single translocation event in one cell. A translocation event of this kind can take place in a single cell in a child or adult and result in leukemia in an individual who was not born with a leukemia mutation.

14. The translocation events break the genes at different places within each gene, so that patients with CML have a larger fusion protein than patients with ALL, and the larger fusion protein associated with CML has many exons of the BCR gene that are not present in the smaller ALL fusion protein.

15. Mutation, deletion, resolution of trisomy in individual cells, or mitotic recombination.

16. Smoking, UV exposure, and radon gas.

17. They predict how fast or slow a prostate tumor will grow.

18. Use PCR to detect the presence of the cancer cell's translocation in shed cells that show up in the bloodstream.

19. Deletions, insertions, translocations, and aneuploidy.

20. It inhibits the BCR-ABL tyrosine kinase, and is used to treat CML caused by the BCR-ABL fusion.

Chapter 11

Questions on pages 400–401

1. An ordered set of genetic markers or loci, where the order of those markers in the genome has been determined by looking at the rates at which recombination events fall between the different pair-wise combinations of markers.

2. Any visible or measurable difference between two individuals, or between the product or activity of the two copies of a gene that exist within an individual.

3. Blood groups, proteins, microsatellite repeats, or single nucleotide polymorphisms. You could also have listed structural abnormalities in a chromosome visible under the microscope, a genetic disease, a cosmetic trait such as eye color, or a difference in a level of enzyme function.

4. It may mean that we have found the location of the gene or it may mean that we have found the gene itself.

5. There are much greater numbers of SNPs and they can be assayed more rapidly because they can be assayed by PCR. You might also have mentioned that they are cheaper to assay and that they are less likely to show new mutations.

6. They were some of the first genetic markers available, so lots of disease genes (and other genes) were tested for whether they co-segregated with the blood group markers.

7. If the parent in any family is homozygous then we cannot tell which of the two alleles in that parent was the one that was passed along to each of the children.

8. r2.

9. The parents are homozygous so we cannot tell which of the two alleles was passed along to the next generation.

10.

11.

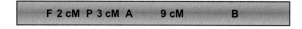

12. The recombination fraction tells us how far apart two items are while the LOD score tells us what level of confidence we can have that the recombination fraction is right.

13. They may be located on different chromosomes, or they may be located far apart on the same chromosome.

14. That they share sequences in common with each other and that they likely represent large regions of shared synteny. We expect that they are not identical in sequence, and that although there may be many genes arranged in approximately the same order, there may well be genes out of order between the two or even chunks of DNA within these chromosomes that are inverted relative to each other or shifted in position relative to the other genes on this syntenic group.

15. The human genome had been sequenced and a large fraction of the human genes had been discovered. So in 1980 once they found the location they then had to start doing laboratory experiments to begin identifying the genes located in the vicinity of the mapped locus. In the twenty-first century the researchers on the progeria project were able to look up the list of genes in the region by connecting to a database that lists the genes and their locations.

16. Helpful information would include what the gene function is, what pathways the gene acts in, and where or when the gene is expressed.

17. Find an animal model that has the same phenotype as the one you are studying in human. Do positional candidate cloning in the animal model. Once the gene has been found, use bioinformatics and sequence homology to help identify the human counterpart of the animal gene. Screen for mutations in human samples to see whether the gene that causes the phenotype in the animal is involved in the human phenotype.

18. Because once you get very, very close to the gene there are potentially multiple genes that are all so close that recombination is unlikely to fall between the gene and the disease locus. So an allele in the disease gene will co-segregate with the phenotype but alleles in other genes that are in linkage disequilibrium with the disease gene will also co-segregate. Thus we normally want to find some additional piece of evidence to confirm that the gene we sequenced is the right one. This can take the form of finding a mutation that co-segregates in another family or mutations in this gene that occur at a higher rate in individuals with the disease phenotype as compared to controls who do not have the phenotype.

19. There are many differences between the mouse and human that affect what the gene product interacts with and the environment in which it functions.

20. The rate of recombination relative to the physical distance is different near the centromeres, near the telomeres, and where there are recombination hot spots. So if we average values across large regions the correlation comes out to be about 1 cM to 1

million bases, but if we look at any one local region the correspondence can be off by a lot.

Chapter 12

Questions on page 450

1. There are approximately 20,000–30,000 genes. The number 20,000 refers to messenger RNAs that encode proteins and that have been validated as real genes. If we consider all the categories of genes, and not just the messenger RNAs, the number of human transcribed genes is over 30,000. It is likely that more non-coding genes will be identified. It was originally thought that the human genome must have thousands of genes to be a complex multi-cellular organism but it turns out the genome size is very similar in higher eukaryotes.

2. Traditionally, naturally occurring versions of things are not patentable. Thus your genome sequence is unique and continuously evolving and is not expected to be patentable. However, some specific uses of the information such as some types of tests that detect mutant sequences are likely to be patentable. The debate regarding what aspects of human genetic information can be patented is ongoing.

3. The reference sequence serves as a standardized piece of information so that when different researchers want to exchange information about the sequence they have a shared point of reference. This is taken from multiple sequences that have been pooled so that the reference sequence does not come from just one person's genome.

4. Deserts are regions lacking transcribed sequences or anything that looks like a known gene. However, if a phenotype maps to a gene desert that region is considered to contain a gene even if we have not yet figured out the nature of the gene.

5. Because mitochondria are replicated with each cell division during the growth and healing processes of the organism, mitochondria can accumulate mutations that were not inherited. A mutation that happens in one copy of the mitochondrial chromosome on one day is likely to be different from the mutation that happens in a different copy of the mitochondrial chromosome on another day, so the cells gradually accumulate novel mutations as the individual ages. The result is heteroplasmy – presence of more than one mitochondrial sequence simultaneously in the same cell. Mitochondrial heteroplasmy has been found to be associated with a variety of late onset diseases such as Alzheimer disease and Parkinson disease. Mitochondrial heteroplasmy is distinct from predominant mitochondrial mutations that show a mitochondrial mode of inheritance.

6. The transcription factors that turn on expression of a gene are made in one tissue but not in another, so the gene will be expressed in those cells that make the needed transcription factors and will not be expressed in cells that do not make those transcription factors.

7. HapMap is a project aimed at identifying variability in the human population. The HapMap project has identified large numbers of SNPS, and evaluated frequencies of SNPs and haplotypes in each of four different populations with ancestry from different parts of the globe.

8. Many high throughput technologies have enabled researchers and doctors to begin to explore many unknowns with regard to sequence information and downstream function. A dozen or more markers can be analyzed in one multiplex PCR reaction. Chip-based technologies can be used to assay expression of a whole genome worth of genes. Finally automated sequencing systems and Next Generation Sequencing as well as bioinformatics and improved database technologies has allowed for the ability to compare and examine multiple genomes in an extremely efficient process.

9. Whole genome analysis can be used to identify new disease-causing genes or correlation between genetic mutations and disease.

10. Pharmacogenomics is the use of genomic approaches to study pharmacologic responses. Many early pharmacogenomics projects were hypothesis-driven projects. Now with the advent of genome-wide technologies, a much more global view of the problem becomes possible. Pharmacogenomics can identify individuals who respond well to a drug and individuals who do not. It can also predict drug efficacy among different patient populations. These approaches could increase life expectancy in many devastating diseases.

11. Exome sequencing determines the sequence of all of the exons in the genome. Whole genome sequencing determines the complete genomic sequence.

12. Copy number variation can cause an altered phenotype even if there is no mutation. In the case of CMT1A, addition of one copy of the gene, and an accompanying increase in the amount of gene product made, leads to the neuropathy.

13. In the case of HNPP, the trait results from losing a copy of PMP22, while CMT1A results from adding a copy.

14. A genome-wide association study.

15. Allele A might be located right next to the allele that is responsible.

16. It is a test for whether two things occur together more frequently than we would expect by chance alone.

17. Because we often screen hundreds of thousands of markers and the number of statistical tests done is huge.

18. Programmable robotics, chip-based technologies, and use of multi-well plates are three of them.
19. Migration, selection, and random drift.
20. 3.7 million places show SNPs that are heterozygous.

Chapter 13

Questions on pages 484–485

1. A karyotype is an individual's collection of chromosomes. The term also refers to a laboratory technique that produces an image of an individual's chromosomes. The karyotype is used to look for abnormal numbers or structures of chromosomes. The traditional process for karyotyping involves adding a dye to metaphasic chromosomes.
2. Preimplantation genetic testing is a technique used to identify genetic defects in embryos created through *in vitro* fertilization (IVF) before pregnancy. Preimplantation genetic diagnosis (PGD) refers specifically to when one or both genetic parents has a known genetic abnormality and testing is performed on an embryo to see if it also carries the genetic abnormality present in the affected or carrier parent(s).
3. *In vitro* fertilization is commonly referred to as IVF. IVF is the process of fertilization by manually combining an egg and a sperm sample in a laboratory dish. When the IVF procedure is successful, the process is combined with a procedure known as embryo transfer, which is used to physically place the embryo in the uterus.
4. Alpha fetoprotein (AFP) is a protein normally produced by the liver and yolk sac of a fetus. AFP levels decrease soon after birth. Elevated levels of AFP are associated with neural tube defects in the fetus, including spina bifida (defective closure of the spine) and anencephaly (absence of

brain tissue). Information on AFP levels can be combined with information on other maternal serum proteins to estimate the probability that the fetus has one of the viable forms of aneuploidy such as Down syndrome. When AFP levels are elevated, a more specific diagnosis is attempted, using ultrasound and amniocentesis.

5. A SNP is a single nucleotide polymorphism, which is also sometimes called a point mutation. By studying the frequency of a SNP in case and control populations we can ask whether a SNP is associated with a particular trait. By screening a large number of SNPs from throughout the genome we can evaluate association of any of hundreds of thousands of SNPs in one experiment.
6. The initial prenatal screening of pregnant women involves a combination of ultrasound and maternal blood screening. Prenatal testing done in response to risk factors, symptoms, or screening includes *amniocentesis* or *chorionic villus sampling* (CVS).
7. One way is to perform fluorescence in situ hybridization or FISH. This method uses a specific probe from one specific chromosomal region to determine whether a particular sequence has been deleted or duplicated. An alternative approach is deep sequencing but it is still too expensive for routine diagnostic testing in a clinical environment.
8. Most common reasons include females of advanced maternal age, a previous child with a genetic disorder, chromosomal anomaly in a parent or parents that are possible carriers of a genetic trait.
9. Risk estimation can make use of information on maternal serum levels of alpha-fetoprotein (AFP), estriol, beta-HCG, and dimeric inhibin (DIA).
10. No. A variety of environmental factors and genotypes at additional loci in the genome can all influence the outcome.

11. They can test for alleles that affect disease risk, for alleles that indicate things about ancestry or region of origin, and for alleles that can predict how well someone will respond to a particular medication.

12. A screen is done on a population of people that is not done in response to a symptom or test result. A test is done on an individual in response to specific symptoms and characteristics of that person.

13. We can treat phenylketonuria, and treatment needs to start immediately after birth. Huntington disease is not one of the traits included in prenatal screening because we do not normally do screening or testing on children for traits that will not develop until late in life unless knowing the result will affect medical management of their case before they turn 18.

14. Sometimes the research group that identified the gene and its involvement in the trait can do the test on a "research basis."

15. Genetic testing can evaluate forensic evidence or archaeological samples. It can also be used for paternity testing or to establish the identity of an individual.

16. AFP levels indicate risk of a trait such as Down syndrome, but AFP levels vary over the course of a pregnancy so we can only tell that an AFP level is normal if we are comparing it to the normal levels for that stage of the pregnancy.

17. The baby forms from the cells on the inside of the embryo but the cells on the outside of the embryo do not become part of the baby so there is no damage if a couple of those cells are missing.

18. Because many pieces of information on the Internet are not true, and if you do not have a validated source of information you do not know whether the information you are finding is valid.

19. Some psychiatric symptoms may appear first in some individuals, although other individuals may be quite free of them.

20. Genetic matchmaking is the process of including genetic risk factors in matching people up with potential mates. For some traits, such as the very severe, early-onset Tay–Sachs disease, the use of genetic matchmaking in whole communities has succeeded in reducing the frequency with which individuals are born with the trait.

Chapter 14

Questions on pages 509–510

1. Gene therapy is the insertion of genes into an individual's cells and tissues to treat diseases, such as hereditary diseases where deleterious mutant alleles are replaced with functional ones.

2. ADA deficiency is the cause of Severe Combined Immune Deficiency.

3. The choice of gene therapy vector depends on the type of therapy but some considerations are: adeno-associated virus (AAV) integrates into the genome to give stable, long-term expression, and it can infect both dividing and non-dividing cell types. AAV, however, cannot accommodate really large genes, and the processes for producing it are less efficient than for some of the other vectors. Adenoviruses (AV) can take larger inserts but tend to produce a much stronger immune reaction than the adeno-associated viruses. Other viruses offer differences in sizes of genes that can be accommodated, how immunogenic they are, and whether they can infect particular cell types.

4. There are problems associated with viral delivery systems for gene therapy. Nanotechnology offers delivery systems that

do not attract the kind of immune response associated with viral vector delivery systems and can give more effective introduction of gene therapy than delivery of naked DNA. Therapeutic genes delivered to cells in this manner are able to drive cellular production of a gene-encoded protein through normal processes.

5. Gene therapy techniques to reduce the amount of a specific RNA can lead to reducing the amount of gene product being made. This is most useful if a product produced is toxic or detrimental to the cell. Small interfering RNA technology can reduce the amount of transcript coming from the offending gene. The sequence of this small RNA is complementary to the sequence of the mRNA produced by the disease gene allele. Because of the sequence complementarity, the small interfering RNA can bind to the mutant transcript and get the cell to destroy the RNA coming from the disease allele.

6. The problems with immune responses to the viral vector system that happen with systemic treatment are not seen in the eye. These problems are not seen in the eye because the human immune system does not carry out surveillance of the tissues inside of the eye.

7. During gene therapy, if the gene integrates into a region of junk DNA between genes, it might have little effect other than curing the cell's metabolic defect. However, if the transgene integrates into a gene in the chromosome and disrupts it, the consequences will depend on which gene gets disrupted. If it disrupts one copy of a gene encoding an enzyme involved in metabolism, it may have little effect or perhaps at the worst it will kill that one single cell. However, if the gene therapy agent is delivered *in vivo* into a large number of existing cells in a human organ, each cell becomes a separate integration event. So even if a large number of cells

receive transgenes that integrate safely between genes, it would take only one transgene integration into certain kinds of cancer genes to cause a problem.

8. In all cases, the patients enrolled should have a greater benefit than risk. This would need to be carefully weighed on a disease and person-by-person basis. In general factors to consider include type of disease, progression, benefit or life expectancy of the patient.

9. Gene therapy to kill tumor cells or gene therapy to improve effectiveness of traditional cancer therapies.

10. Gene therapy to target a downstream pathology instead of correcting the primary defect or addition of more of something that is already present to restore a normal level of gene product.

11. Because this is a simple form of gene replacement therapy. RPE65 is an autosomal recessive trait in which neither copy of the gene is making functional RPE65, and getting RPE65 into the one limited tissue type called the RPE was all that was needed to start getting a therapeutic response. Also, because this treatment is taking place in the eye we do not have to worry about some of the immune responses that complicate systemic forms of gene therapy.

12. Because Lancelot had a naturally occurring version of a severe human trait, and the demonstration that Lancelot was cured by gene therapy became the basis for starting gene therapy trials in humans.

13. ADA gene therapy was selected because of the severity of the trait. The gene was known and had been cloned. The gene was small enough to be put into a gene therapy vector. The target cells, white blood cells, are accessible, and therapeutic success did not require that 100% of the cells be treated.

14. The particular gene therapy construct used led to leukemia in some participants, but knowing the features of the construct that led to leukemia has allowed the development

of newer versions of ADA gene therapy that are not expected to cause leukemia so gene therapy efforts have continued.

15. Accumulation of ammonia in the blood because nitrogen cannot be processed all the way through the urea cycle.

16. The administration of a construct administering a growth factor that assists the heart muscle to heal.

17. In the case of someone who is hemizygous for a particular gene, they are making the necessary protein but they are not making enough. Gene therapy can be used to boost level of gene product back to normal levels.

18. By putting a gel into the break that contains a gene product that induces bone growth such as a bone morphogenic protein.

19. In treating cancer we are trying to kill cells rather than heal them. For some metabolic traits it is enough to successfully treat a lot of cells without treating all of them, but in the case of cancer if even a few cells escape the gene therapy effort then the cancer can grow back.

20. It can be used to produce a protein outside of the body that is then used in a traditional manner as a drug.

Glossary

acetylation: attachment of an acetyl chemical group ($CH_3COO–$) onto a molecule.

ADA: *see* adenosine deaminase

adenine: one of the bases that make up DNA and RNA, usually abbreviated by the symbol A.

adenomatous polyposis coli (APC): the APC gene is a tumor suppressor gene that, when defective, causes familial adenomatous polyposis (FAP).

adenosine deaminase: an enzyme responsible for one of the steps in the pathway that breaks down adenosine and deoxyadenosine; also the name of the gene that encodes this enzyme. The ADA gene was the target for the first human gene therapy experiments.

agarose gel electrophoresis: a process that separates DNA molecules according to size as they move through a porous matrix in the presence of an electrical current.

age-dependent penetrance: the extent to which a late-onset trait manifests itself over the course of aging with an increasing chance of being manifested as the individual grows older.

AIS: *see* androgen insensitivity syndrome

allele: different forms of a gene are called alleles; alleles of a given gene differ in terms of their DNA sequence. Homo- or heterozygosity for some alleles results in differences in the phenotype, but other alleles have a different sequence without causing any change in the phenotype.

allelic: being alternative genetic variants of the same gene. Things that are allelic to each other cannot complement each other in a complementation test.

allelic association: *see* linkage disequilibrium

allelic heterogeneity: *see* genetic heterogeneity

alpha-fetoprotein (AFP): a protein secreted by a developing fetus and found in the mother's serum. This is one of several proteins whose levels can be used to predict the chance that the developing fetus has a trait such as Down syndrome, trisomy 18, or neural tube defects.

alternatively spliced genes: genes in which one subset of all of the exons in a gene is used in one splice variant and a different subset of the exons is used in a different splice variant. Alternative splicing allows the cell to make multiple different proteins from one gene and thus to make more different proteins than there are genes.

American eugenics movement: a disastrous twentieth-century political movement in the United States that claimed the objective of "improving" the American population by enforced sterilization, stringent laws against immigration, and laws against interracial marriage. This movement was founded on both erroneous science and problematic ethics.

Ames test: a biological assay for mutagenicity – the potential of a chemical compound to bring about a change in the sequence of an organism's DNA. Although many items that test positive in the Ames test turn out to be carcinogens, the test itself tells us only that the item is a mutagen.

amino acids: the chemical building blocks of which proteins are made. Proteins are standardly made up of twenty different amino acids. Amino acids all have two chemical groups in common – an amino group and a carboxyl group – but each different amino acid has a unique third component that is sometimes referred to as an R group and that has the different chemical properties that give the different amino acids their distinct functional properties. Amino acids are encoded by the codons of the genetic code. A few rare proteins use an alternative twenty-first amino acid called selenocysteine but there is no separate codon specific to selenocsyteine.

amniocentesis: a technique used for prenatal diagnosis of a large number of genetic abnormalities. A needle is used to remove a small amount of amniotic fluid that surrounds the developing fetus. This fluid contains fetal cells that are used for tests such as karyotyping, PCR amplification of DNA to be used in sequencing, and tests for a variety of biochemical processes.

amnion: a membrane that surrounds the developing fetus. The amnion is composed of cells derived from the developing embryo and is thus genetically identical to the embryo.

amniotic fluid: fluid that fills the amnion and provides a cushion for the developing fetus.

anaphase: the stage in mitosis when sister chromatids separate and move to opposite poles. The term anaphase I is used to refer to the stage in the first meiotic division when homologous chromosomes separate and begin to move towards opposite poles. The term anaphase II is used to refer to the stage in the second meiotic division when sister chromatids separate and move to opposite poles.

anaphase I: anaphase of the first meiotic division.

anaphase II: anaphase of the second meiotic division.

androgen receptor: protein that detects presence of testosterone and signals its presence to the cells that carry the receptor on their surface.

androgen receptor (AR) gene: gene that encodes the androgen receptor. Mutations that eliminate production of the receptor or render it non-functional result in complete androgen insensitivity syndrome (CAIS). Mutations that reduce the effectiveness of the androgen receptor but do not eliminate its activity can result in partial androgen insensitivity (PAIS). Mutations that cause repeat expansion result in spinal bulbar muscular atrophy. Some missense mutations have also been observed in prostate cancer.

aneuploid: possessing an incorrect number of chromosomes. The chromosome constitution of a normal human being is 46 chromosomes, composed of 22 pairs of homologous chromosomes and the sex chromosome pair (XX or XY.) Any deviation from this chromosome composition, such as possessing 45 or 47 (or more) chromosomes, is considered to be aneuploid. Normal human gametes possess 23 chromosomes, one from each pair. Gametes with more or fewer chromosomes than 23 are aneuploid.

aneuploidy: the state of being aneuploid.

***Antennapedia* gene:** gene responsible for key steps in the development of the fruit fly *Drosophila malanogaster*. A mutant form of the Drosophila antennapedia gene can result in legs growing out of the fly's head where antennae would normally be. The antennapedia gene encodes a transcript factor.

anticipation: a phenomenon in which a trait manifests at an earlier age in succeeding generations. Some cases of anticipation are real, in traits such as Huntington disease, but apparent anticipation can also result from increased surveillance of succeeding generations or improved technical abilities allowing diagnosis of a trait at an earlier stage.

anticodon: the three base sequence on a tRNA molecule that undergoes Watson–Crick base pairing with the three base codon on the mRNA. A tRNA with a particular anticodon always carries the same amino acid as other tRNAs with that same anticodon.

antimiscegenation laws: laws prohibiting interracial marriage.

APC: *see* adenomatous polyposis coli

artificial selection: in plant or animal breeding, artificial selection occurs when breeders select organisms with a specified phenotype to be the parents of the next generation, resulting in an increased representation of the selected phenotype in subsequent generations.

autosomal dominant: a pattern of inheritance displayed by dominant mutations in genes contained on the autosomes. The most notable features of this type of inheritance are that affected individuals always have an affected parent, children of both sexes are equally likely to display the trait, the trait can be passed on by parents of either sex, and about half of the progeny of an affected individual are expected to express the trait.

autosomal recessive: a pattern of inheritance displayed by recessive mutations in genes contained on the autosomes. The most notable features of this type of inheritance are that affected individuals often do not have an affected parent, children of both sexes are equally likely to display the trait, and about one-quarter of the children produced by the mating of two people that are heterozygous for an autosomal recessive mutation will be affected.

autosome: those chromosomes that are present in identical pairs in both sexes: chromosomes 1-22 (or all the chromosomes except the X and the Y chromosome).

bacteria: living, microscopic cells. They are prokaryotes so they have no membrane bound organelles or nuclear membrane. Thus the bacterial chromosome lives in the cytoplasm along with the other organelles.

bacteriophage: a virus that infects bacteria.

balanced translocation: a chromosomal aberration in which parts of two nonhomologous chromosomes have been interchanged, but no material has been lost or gained. This may also be referred to as a balanced rearrangement. Individuals with a balanced translocation still have the correct number of copies of each gene. In some cases the individual will be free of any phenotypic effects from the translocation. In other cases there may be an altered phenotype if the translocation has interrupted the sequence of a gene or has joined pieces of two different genes together as happens with the Philadelphia chromosome.

band patterns: a pattern of dark and light bands present along the length of a chromosome in a characteristic pattern that is unique to that chromosome and can be used to identify it. A chromosomal banding pattern has been likened to a bar code.

banding: a process by which metaphase chromosomes on a glass slide are treated with dyes that leave a characteristic and reproducible pattern of dark and light bands.

Barr body: a cytologically visible mass in the nucleus corresponding to the inactive X chromosome. Nuclei from XY males do not display a Barr body, but nuclei from XX females and XXY males do possess one.

base pair (*abbrev.* bp): Watson–Crick-paired nucleotides in a double-stranded DNA molecule. A forms bonds with T, and G forms bonds with C.

BCR–ABL: a fused gene created when a chromosomal translocation brings together a piece of the BCR gene on chromosome 22 and a piece of the ABL gene on chromosome 9 to result in a chromosome of recognizable structure called the Philadelphia chromosome which is seen in both chronic myelogenous leukemia (CML) and acute lymphoblastic leukemia (ALL.) The translocation creates a gene that has the promoter and 5′ end of the BCR gene with the rest of the coding sequence coming from the 3′ end of the ABL gene. The ABL gene is involved in cellular division and differentiation, and the translocation removes a portion of the ABL gene that normally carries out negative regulation of ABL functions. With this negative regulatory domain removed, ABL becomes an oncogene and leads to cell division in cells that should be quiescent.

B-DNA: the structure of a double-helix that is twisted in the direction most often found in human DNA *in vivo* (called a right-handed helix), with about 10.5 bases per turn.

benign polymorphism: sequence variant that does not alter the phenotype. This can include not only silent changes that leave the protein sequence unaltered but can also include changes such as missense mutations that change the amino acid sequence without changing the protein function.

benign sequence variant: another term for benign polymorphism.

benign tumor: a tumor that remains locally confined and does not grow and invade surrounding tissues or other parts of the body. Benign tumors are often, but not always, harmless.

binary traits: traits that have only two states.

bioethics: the study and analysis of ethical issues in the realm of the life sciences including biology, medical sciences, population sciences such as epidemiology, and social sciences. Bioethics may deal with issues such as moral quandaries encountered in research, policy development and application, or medical ethics.

bioinformatics: the field of study that uses approaches from biology, mathematics, computer science, and information technology to develop new methodologies, expand theoretical understanding, and apply bioinformatic methodologies to the investigation of biological data sets.

biomarker: any biological variable that is associated with a biological characteristic or variation in that characteristic.

bivalent: two paired homologous chromosomes, usually physically connected by chiasmate exchange at the sites where recombination has taken place, during the first division of meiosis.

blending: a theory that the phenotype of a child will be intermediate between the phenotypes of the two parents as a result of blending of features from the parents.

bp: *see* base pair

CAIS: complete androgen insensitivity syndrome.

cancer: a genetic disease resulting in uncontrolled cell division. Beginning with a single cell, this rapid division results in the formation of one or, in the cases of metastatic cancers, many tumors. Cancer results from mutations in the cellular genes that control cell division and/or in the genes whose products carry out DNA repair.

carbohydrate: hydrated carbon, or any of a variety of organic molecules made up of carbon, hydrogen, and oxygen in which there are two hydrogens per oxygen. Examples include sugars, starches, and cellulose. Carbohydrates serve as an energy source in the diet.

carrier: individual who carries one defective copy of a recessive gene where both copies must be defective for the trait to be manifested; *see also* obligate carrier.

causative mutation: a change in the DNA sequence that results in a change in the phenotype.

cDNA: a DNA copy of an RNA molecule. A cDNA can be cloned to allow study of the RNA transcript and the genomic DNA sequence implied by it.

cell cycle: a process by which the cell moves through a regular, programmed series of changes that result in duplication of the contents of the cell followed by separation of the two resulting daughter cells. The stages of the cell cycles include a growth period (G1), a period in which its DNA is replicated (S), and a second growth period (G2) before reentering mitosis (M). Many cells in the body are sitting in a permanent "parking lot" called G0, where they do not advance through the cell cycle.

cell cycle regulator: a protein that regulates the cell cycle. Mutations in these genes are important causes of inherited and sporadic cancers.

cell division: the process by which a cell divides into two new "daughter" cells after having duplicated all of its contents and apportioned them into those daughter cells.

centimorgan: a measure of distance defined by a rate of recombination of 1 per 100 meiotic events. In humans, 1 centimorgan (or 1% recombination) is loosely approximate to a million bases although this is not strictly true since the rate of recombination is not the same at all points along a chromosome and, for any one region, the recombination rate in males may differ from the recombination rate in females.

central dogma of molecular biology: a model that describes the direction of information flow in the cell going from DNA to RNA to protein. Although most cellular information flow goes in this direction, some RNA viruses can use reverse transcriptase to carry out a "backwards" step in the information pathway, going from RNA to DNA.

centriole: a microtubular structure within the centrosome.

centromere: the site on a chromosome by which the chromosome attaches to the cellular machinery that pulls it toward the poles of the spindle during mitosis and meiosis. A set of proteins bound to the centromeric DNA sequences creates a structure called the *kinetochore.*

centrosome: a structure found at the poles of mitotic spindles in both sexes and at the poles of meiotic spindles in males only. Centrosomes serve to organize the microtubule fibers into spindles.

chaperone proteins: proteins that interact with other proteins to assist with protein folding and assembly.

checkpoint: a point in the cell cycle where the cell assesses its ability to continue the cell cycle. There are, for example, DNA damage-sensitive checkpoints in G1 and G2 that are activated by damage to DNA. The activation of one of these checkpoints will halt the cell cycle until that damage is resolved. Checkpoints of this kind play a key role in eliminating aneuploid sperm, but a similar checkpoint for aneuploidy does not exist in the production of eggs, resulting in higher rates of aneuploidy in eggs as compared to sperm.

chiasma: a recombination event that is visible by microscopy during late prophase of meiosis I.

chiasmata: more than one chiasma.

chimera: an organism that is a mixture of two genetically different types of cells. For example, if a new mutation happens in a very early embryo, so that the resulting baby has the mutation in some cells but not in others, the baby would be considered chimeric.

chorion: an extra-embryonic membrane, also derived from cells of the early embryo, that will go on to become the placenta. Because the chorion and the fetus are both derived from cells of the early embryo, they are genetically identical. Thus genetic analysis of cells from the chorionic villi (structures in the chorion) provides information about the genetic composition of the embryo.

chorionic villus sampling: a technique used for prenatal diagnosis of a large number of genetic abnormalities. A cannula inserted through the center of the cervix is used to sample a small amount of tissue from the chorionic villi, a tissue that will go on to form the placenta. This tissue divides mitotically very actively, and thus metaphase cells for karyotyping can be obtained quickly by letting the small number of cells from the tissue sample grow in a culture dish until there are enough cells present to carry out the test. The tissue can also be subjected to other tests such as PCR amplification of DNA from the cells followed by re-sequencing of a specific gene.

chromatid: either of the two full-length DNA molecules produced by replicating a eukaryotic chromosome. Thus, prior to replication, each chromosome is composed of only one chromatid. Following replication, each chromosome is composed of two identical chromatids (denoted *sister chromatids*). At mitosis, or in meiosis II, the two sister chromatids separate from each other and move to opposite poles.

chromatin: the mixture of DNA and proteins that comprises a chromosome.

chromosome: a chromosome is a DNA molecule, plus associated proteins, that is capable of stable meiotic and mitotic segregation. Chromosomes carry a single centromere. Most chromosomes consist of linear DNA molecules and thus also possess two telomeres at their ends. There are rare exceptions to this structure, such as ring chromosomes, which do not normally occur in humans. Prior to replication, each chromosome is composed of only one chromatid. Following replication, each chromosome is composed of two identical chromatids (denoted *sister chromatids*). Thus before replication a chromosome has one copy of each gene, and after replication a chromosomes has two copies of each gene.

chromosome painting: a process by which each pair of homologous chromosomes is stained with a single color. Each chromosomal pair is painted a different color than the other chromosomes.

chromosome segregation: the process of transmitting the two copies of a chromosome into the daughter cells (mitosis) or the two homologs of a chromosome into daughter cells (meiosis.)

cilia: hair-like extensions from the surface of a cell. Cilia play a role in moving fluids and small particles in tissues such as the lungs.

cis: on the same side. In genetics and molecular biology this may be used to refer to processes that are physically linked on the same molecule, such as a promoter mutation that can influence the regulation of the copy of the gene located on the same chromosome with it but not the copy located on the other chromosome.

clinical trial: a research study designed to evaluate use of a new therapy or device, or a new application of an existing therapy or device, in a human population; *see also* Phase I clinical trials, Phase II clinical trials, Phase III clinical trials, and Phase IV clinical trials.

clone: a genetically identical copy of an organism or piece of DNA, where getting the identical copies of a piece of DNA requires that the DNA be spliced into a cloning vector.

cloning: unfortunately for every biologist trying to communicate about cloning, this term has multiple meanings. In the parlance of movie script writers, journalists, and politicians, the term means creating genetically identical copies (clones) of a living organism. In the jargon of cell biologists, cloning means isolating populations of genetically identical cells derived from a single progenitor cell. Geneticists and molecular biologists use the term to mean splicing a piece of DNA into a vector, and then introducing that vector into a host organism (usually the bacterium *Escherichia coli*) and letting the replication machinery of the host make copies of the cloned DNA molecule. Cloning an organism gets you additional copies of the organism. Cloning a piece of DNA gets you copies of the piece of DNA.

cloning vector: a DNA molecule such as a plasmid or a viral chromosome that can be used to "clone" other DNA molecules in a given host cell. The cloning vector provides a set of essential functions that are not present on the targeted piece of DNA that is being cloned. Cloning vectors must be able to replicate in the host cell. Plasmids used as cloning vectors also carry a selectable gene, usually one that confers resistance to an antibiotic, that allows the scientist to select only host cells carrying the plasmid. They must also possess restriction enzyme cut sites that allow other DNA molecules to be inserted into the vector.

CNV: *see* copy number variant

code: *see* genetic code

coding sequence: the part of sequence of a gene or transcript that is used by ribosomes to direct the synthesis of the protein product of a gene.

coding strand: the strand of DNA that has the "same" sequence as the RNA transcript made from that section of DNA (the difference being that the DNA copy at T and the RNA copy has U).

co-dominant: this term (or co-dominance) refers to cases where the phenotype of the heterozygote is distinguished easily from either homozygote. Excellent examples can be found for genetic markers such as RFLPs or microsatellite repeat polymorphisms, which are considered to be co-dominant. Another example is the gene for the ABO blood group. In this case individuals heterozygous for both the A and the B alleles express both the A and the B surface antigens on their blood cells rather than just A or just B. (Note, however, that not all alleles of a given gene are co-dominant. In the case of the ABO blood group, the O allele is fully recessive. Thus, AO individuals are phenotypically identical to AA individuals.) The term co-dominant has also been extended to describe dominant mutations that exert a stronger phenotype in homozygotes than in heterozygotes.

codon: three contiguous bases in an mRNA molecule that specify the addition of a specific amino acid to the growing amino acid chain as the ribosome moves along the mRNA.

complementary: describes a sequence on one strand of nucleic acid that can pair with a sequence on another strand of nucleic acid by using the base-pairing rules: A pairs with T or U, and G pairs with C.

complementation test: one of the most elegant tools available to a geneticist. This test allows us to determine whether or not two recessive mutations that exhibit the same or similar phenotypes when homozygous are in the same or different genes. Suppose that there are two genes (gene A and gene B) required for normal odor detection (smell). Fully recessive mutants that result in a diminished ability to distinguish certain odors when homozygous are known for both genes (i.e. both aaBB and AAbb individuals show similar defects in odor recognition). So if two people with this disorder marry, what do we expect for their children? If both people are aa homozygotes then all of their children will have an impaired sense of smell. But if Dad is AAbb and Mom is aaBB then all their children will be AaBb and thus have a normal sense of smell. This very simple test allows us to determine whether or not Mom's mutation and Dad's mutation were in the same or different genes. The critical components of his test in human population is that both mutations must be fully recessive and relatively rare. Imagine the confused result if the dominant alleles A and B are common, so Dad was AaBB and Mom was aaBB. In this case some children would be affected while others were not. Such misapplications of the complementation test can lead to erroneous conclusions.

compound heterozygote: an individual heterozygous for two different loss-of-function mutations in the same gene (e.g., m1m2, where both m1 and m2 are different mutations in the same gene).

conditional traits: traits that are manifested only if some eliciting environmental event is present in addition to the necessary genotype.

cone cells: photoreceptor cells in the eye that detect colored light and that are responsible for our vision in the bright light of daytime or a well-lit room. Each cone detects either red, green or blue light.

congenital: in-born; present at birth. Congenital traits may be genetic and/or environmental in origins.

consanguinity: marriages between genetically related individuals. Also known as "inbreeding."

contiguous gene syndromes: cases where a single deficiency or deletion of a region of DNA removes several essential genes and thus produces a complex phenotype.

copy number variant (CNV): sequence variant characterized by presence of extra numbers of copies or reduced copies of a sequence in one genome as compared to another. Often refers to differences found upon comparison to the reference sequence for the organism.

co-segregation: transmission of alleles of two different genes into the same gamete so that they are passed along together to the same progeny. If alleles segregate away from each other they will go into different gametes and be transmitted to different progeny.

coupling relationship: consider the case of an individual who is heterozygous at both of two closely linked genes A and B. Thus their genotype is AaBb. The term coupling relationship refers to arrangement of alleles of the A and B genes on the same homolog. The genotype AB/ab is not in the same coupling relationship as the genotype Ab/aB.

crossing over: the process of genetic recombination that exchanges DNA between two copies of a chromosome.

crossover: a recombination event that exchanges a region of DNA on one homolog of a chromosome for the same region of DNA on the other homolog of that chromosome, resulting in trading alleles of the genes located in the region that has been exchanged.

cytokine: proteins released by cells of the immune system that carry signals to other cells in the course of regulating functions of the immune system. Some classes of these molecules may be produced by multiple different cell types in the body. Cytokines include interleukins and chemokines.

cytokinesis: the final step in mitosis or meiosis when the cell completes the process of dividing into two separate cells.

cytology: the study of cells, especially their structure, function, and pathology. Cytology generally makes use of microscopes to view cellular structures and processes, but may also make use of other imaging techniques.

cytoplasm: material between the outside of the nuclear membrane and the inside of the membrane surrounding the cell in which the organelles such as mitochondria and ribosomes are suspended. Protein production and energy production both occur in the organelles of the cytoplasm.

cytosine: one of the bases that make up DNA and RNA, usually abbreviated by the symbol C.

deletion: a type of mutation that removes one or more contiguous base pairs within a region of DNA. The term can be used to refer to the removal of one or more bases within a gene, but it can also refer to much larger aberrations that remove hundreds, thousands, or millions of bases, thus removing an entire gene or group of linked genes.

denature: to separate a double-stranded DNA molecule into two separate single strands.

dendrimer: complex molecule characterized by repeated branching from a central core or spine.

deoxyribonucleic acid: *see* DNA

development: the processes by which an organism grows and changes through different stages in the life span, including the dynamic process leading to or brought about by growth, migration, or morphogenesis of cells or tissues.

dicentric: a chromosomal rearrangement in which a given chromosome now possesses two centromeres. Although such chromosomes should be very unstable, often times one of the two centromeres appears to be inactivated.

digenic: a trait resulting from the combined effects of defects in two different genes.

diplo-: used in reference to gametes carrying two copies of the same chromosome (e.g., "diplo-X ova" would mean an egg carrying two copies of the X chromosome).

diploid: carrying two copies of each chromosome except the sex chromosomes for which either two Xs or an X and a Y will be present in a diploid cell.

diplotene/diakinesis: the stage in male meiosis when paired homologs begin to repel each other. At this stage they are held together only by the recombination events that occurred earlier during meiosis.

DM protein kinase (DMPK): the protein that leads to myotonic dystrophy when a triplet repeat expansion occurs in the gene that encodes it.

DNA: deoxyribonucleic acid. The biochemical material within the cell that encodes the genetic blueprint. DNA, the hereditary material, is made up of the four types of molecules called bases, named adenine, cytosine, guanine, and thymine (abbreviated A, C, G, and T) that are linked together by a sugar (deoxyribose)-phosphate backbone. In our cells, DNA is normally found as a *double helix* composed of two DNA strands.

DNA chips: silicone chips to which DNA has been bound. The most common types of chips are microarrays, used to assay changes in gene expression of large numbers of genes all represented on one chip, or sequencing chips that are used to re-sequence genes of known sequence to evaluate whether there are any sequence variants in the sample being tested.

DNA ligation: the act of sealing two double-stranded DNA molecules together at their ends, carried out by an enzyme called DNA ligase.

DNA methyltransferase: an enzyme that puts methyl groups (CH3–) onto DNA molecules.

DNA polymerase: an enzyme capable of replicating a DNA molecule.

DNA repair: the process of repairing damage done to a DNA molecule. A number of different types of damage can occur within a DNA molecule, including double strand breaks, single strand breaks, incorrectly incorporated bases, and chemically modified bases. Different classes of DNA repair systems exist to repair these various types of damage.

DNA replication: the process of replicating a double-stranded DNA molecule.

dominant: a form of inheritance in which only one copy of the causative allele needs to be present for the phenotype to be manifested. That causative allele would be referred to as a dominant allele.

dominant negative mutation: a mutation that is inherited in a dominant manner and whose protein product performs a negative action such as poisoning the cell or interfering with normal copies of the protein.

double helix: the double-stranded, helical structure of DNA.

double-stranded: a term used to describe DNA or RNA occurring in a form that has paired two strands of nucleic acid according to the base-pairing rules. DNA is normally double-stranded, and RNA is normally single-stranded but some instances of RNA–RNA pairing and DNA–RNA pairing are observed. Some RNA molecules such as tRNAs have some double-stranded regions and some single-stranded regions.

DMD gene: encodes the dystrophin protein. Defects that eliminate the gene product or produce a non-functional gene product result in Duchenne muscular dystrophy. Mutations that result in a protein product that is present but working sub-optimally result in Becker muscular dystrophy.

Drosophila melanogaster: the common fruit fly. Genetic studies on this organism laid the ground work for our understanding of genetics and continues providing genetic breakthroughs of importance to basic science, human biology, and medicine.

duplication: a chromosomal rearrangement that "duplicates" a region of DNA.

dystrophin: the protein product of the DMD gene. The complete absence of this protein leads to Duchenne muscular dystrophy. However, some types of DMD gene mutations that apparently allow the production of some amount of partially functional dystrophin can lead to a less severe type of dystrophy known as Becker muscular dystrophy (BMD.)

dystrophy: wastage or atrophy of the muscles, as in Duchenne muscular dystrophy (DMD) and myotonic dystrophy (DM).

E. coli: see *Escherichia coli*

ectopic expression: expression of a gene in a cell type in which that gene is not normally expressed or at a time when it is not normally expressed. In especially dramatic cases this can result in the existence of organs or structures, such as eyes or legs, in locations on a body at which they would not normally occur, but ectopic expression can also refer to situations in which the location of expression is aberrant even if there is no such dramatic phenotypic manifestation of the expression.

electron transport: a chemical process that couples an electron donor to an electron receptor, resulting in the production of ATP which is the key source of energy inside of living cells.

electrophoresis: the movement of molecules in response to an electrical current. In molecular biology, this term is used to refer to the separation of RNA, DNA, or protein molecules by using an electric current to propel these molecules through a porous gel. Under these conditions smaller molecules move faster than larger molecules.

ELSI: acronym for ethical, legal and social issues, a term that arose in the course of the Human Genome Project that is used in the fields of genetics and genomics.

encephalopathy: brain dysfunction. It may result from genetic factors or from environmental effects such as medications or infections.

endemic: illness present in a region on a continuous basis that is maintained without need for input from other regions.

enhancer: a region of DNA in the vicinity of a gene that enhances expression of that gene when bound by

proteins. An enhancer may influence level of expression or where or when the gene is expressed. The action of an enhancer normally affects only the copy of the gene that is on the same chromosome with it.

enzymes: macromolecules, usually proteins, that act to catalyze biochemical reactions. Enzymes act as a catalyst that participates in the reaction without being used up by the reaction so that after the reaction is done the enzyme is still available to participate in the reaction again through interaction with new substrates.

epidemic: illnesses in a region occurring at a rate that is higher than is usual for that region or under the conditions that prevail.

epigenetic: affecting gene function or expression without a change in the sequence of the DNA.

epigenetic changes: phenotypic effect that can be transmitted from a cell to its descendant or from an organism to its descendant but that is not caused by a change in the DNA sequence. An epigenetic phenotype may result from chemical alterations to the DNA that do not alter the DNA sequence, notably methylation or from alterations to the histones that are part of the chromatin structure.

epistasis: suppression of the phenotype of one gene through the actions of another gene.

epistatic: having the property of masking the ability of another gene to manifest its associated phenotype.

Escherichia coli: bacteria that are common inhabitants of the human gut. Some types of *E. coli* are important causes of some human diseases but also of great importance as a model organism used widely in research laboratories for purposes such as cloning.

EST: *see* expressed sequence tag

euchromatin: the region of a chromosome that is rich in active genes, as opposed to heterochromatin, from which few genes are actively expressed.

eugenics: the political philosophy of "improving" the human race by selective breeding and enforced sterilization that was based on false scientific claims. It caused a large amount of injustice and raised important ethical issues that lead to ethical and legal advances in genetics.

Eugenics Records Office: a US Government-sponsored office in New York that existed during much of the first half of this century. This office directed research into human breeding and provided recommendations to the government.

eukaryote: an organism whose cells possess a nucleus, as opposed to prokaryotes, such as bacteria, in which the chromosome is found in the cytoplasm. Humans are eukaryotes.

evolution: change in phenotype over the course of successive generations through a combination of processes that introduce genetic variation and other processes that select for organisms that pass their genotypes along to successive generations. Factors affecting introduction of variation may include mutation in response to environmental factors, but some level of variation goes on even if there are no mutagens because of naturally occurring errors in replication of genetic information. Factors affecting which organisms pass along their genotypes to the next generation include natural selection, sexual selection, artificial selection, migration, and random drift. Many independent lines of evidence support the existence of evolution.

exon: a stretch of a eukaryotic gene that is included in the final processed mRNA molecule. The *ex*on is included in the final transcript and *ex*ported from the nucleus.

expanded repeats: simple sequence repeats that have more repeat units present in the offspring than in the parent, or that have more repeat units present in a daughter cell than in the progenitor cell.

expressed sequence tag: also known as EST; a piece of information that indicates the sequence of a short region within a transcript. ESTs are derived by reading short runs of sequence from cDNAs created by reverse transcribing RNA. An EST is considered a tag because an EST that contains only part of the sequence of a gene can serve as a unique identifier of gene.

expression profile: the description of the level of transcript present for each of the tens of thousands of genes being expressed in an organism. It measures the result of two main influences on the amount of transcript present – the rate at which the gene is transcribed and the rate at which the transcript is degraded and eliminated from the cell.

facultative heterochromatin: chromosomes or regions of heterochromatin that can exist as heterochromatin (i.e., highly condensed and transcriptionally inactive) but are not always in heterochromatic form. An inactivated X chromosome, or Barr body, is an excellent example of facultative heterochromatin.

fetal hemoglobin: also called HbF, a form of hemoglobin that transports oxygen in the fetus. Fetal hemoglobin is expressed before birth and binds oxygen with higher affinity than adult hemoglobin. A treatment for sickle cell anemia involves turning back on expression of fetal hemoglobin.

FISH: *see* fluorescence in situ hybridization

fluorescence in situ hybridization (FISH): a technique used to visually map DNA sequences onto metaphase chromosomes using DNA–DNA hybridization of fluorescently labeled "probe" DNA.

FMRI: the gene defined by the fragile X mutation. This gene is required for proper development of the face, brain, and testes. Triplet repeat expansion in this gene leads to both fragile X syndrome and a cytologically detectable fragile site.

founder effect: the presence of a given allele at a higher frequency within a genetically isolated population where many of the current members of the population are descended from one or more of the small number of individuals that "founded" that population and who carried that allele.

fragile site: a site on a metaphase chromosome where breaks or stretching are often observed under the microscope. Many, if not most, fragile sites are thought to be the result of triplet expansion mutations.

frameshift mutation: a mutation that alters the reading frame of a message. As an example, consider mutations that insert (or delete) 1 or 2 bp. As a result of such mutations, the normal sequence is read in the correct frame up until the codon bearing the insertion or deletion. However, the insertion or deletion of one or two bases shifts the reading frame of the rest of the message so that a completely different string of amino acids is produced. Because the frame shift will result in the use of a new stop codon at a different position, it is common for a frame shift to alter the length of the protein produced in addition to causing major changes in the amino acid sequence beyond the point of the mutation. A frame shift can result from insertion or deletion of any number of bases that is not a multiple of three.

FTI: farnesyl transferase inhibitor. This is one of the key drugs being used in the progeria clinical trial.

G0 (G zero): the stage in the cell cycle occupied by cells that are no longer dividing. Cells in this stage contain unreplicated and greatly decondensed chromosomes that are carrying out active transcription of RNA. G0 cells comprise most of the cells in our body.

G1 (G one): the stage in the cell cycle that follows mitosis but precedes replication. Each chromosome is composed of a single chromatid. During this stage the decondensed state of the chromosome that occurs at telophase provides an opportunity for gene expression. Cells in G1 must pass through the start point in the cell cycle in order to continue cell division. Passage through the cell cycles start point is under very tight genetic control.

G2 (G two): the stage in the cell cycle that follows replication (S phase) but precedes mitosis. Cells in G2 have replicated their DNA and thus each chromosome is composed of two sister chromatids. Cells in G1 must pass through one or more "check points" in the cell cycle in order to continue cell division.

gamete: a sperm or an egg.

gametogenesis: the process by which male or female gametes are produced.

GenBank: a database located at the National Center for Biotechnology Information into which researchers deposit sequence information for the sequences of genes, chromosomal regions, cDNA representations of transcript sequences, and annotation including references to published information about the sequence and notation of where functional items are within the sequence such as splice sites. Often in publications researchers use the GenBank identification numbers to assist readers in correctly identifying which genes or sequences the paper is presenting.

gender: there are multiple different definitions of gender. Of importance to this book, it can be defined as a set of societal roles, self-perceptions, tastes, or views that are traditionally associated with being male or female. A dictionary would also tell us that gender can refer to one's sex or that it refers to "a formal classification by nouns that are grouped or inflected."

gene: one could broadly define a gene as a region of DNA that is transcribed and for which that transcript contributes to some aspect of the phenotype of that organism. This definition approximates Mendel's definition of a gene as a unit of heredity. A common usage of the term gene refers to a region of DNA that that can be transcribed into mRNA that can be used to direct the synthesis of a protein, but this definition clearly leaves out the non-coding genes. There are other definitions, but these two usually suffice.

gene desert: a large region of chromosomal sequence that apparently contains no genes.

gene families: a group of genes with similar sequences. Members of gene families may have related functions.

gene regulation: control of the rate at which a gene is transcribed.

GeneClinics: a database containing articles about human genetic disorders including information on genetic testing for those traits.

genetic: hereditary, describes something passed from one generation to the next on the basis of transmission of genetic information from one generation to the next.

genetic background: this term refers to the full array of genotypes present in an individual. Putting the same mutation onto different genetic backgrounds can result in differences in the phenotype. Genetic background is probably as critically important as it is difficult to assess.

genetic code: the cellular "dictionary" that allows 64 possible three-base codons to specify a group of 20 amino acids plus three stop sequences.

genetic markers: any difference between the two homologous regions of DNA that can be assayed. Many genetic markers are differences in the DNA sequence between the two copies of the chromosome but a genetic marker can also be a difference in a protein, such as a blood group protein.

genetic segregation: the process of transmitting different alleles and their accompanying phenotypes to different progeny as a direct result of the processes of chromosome segregation. If two items are near each other on the same piece of DNA then they will pass together into the progeny unless a recombination event falls between them.

genome: most often, the complete DNA content of an organism. In the case of RNA viruses, the genome would refer to the RNA molecules that are replicated.

genome scan: a survey of genetic markers spread across the whole genome. A genome scan will usually involve screening of hundreds of markers.

genome-wide association study (GWAS): a large-scale test for association using large numbers of different markers closely spaced all across the genome.

genotype: the allelic composition of a given individual for one or more genes (e.g., the genotype of one person might be Aa rather than the AA or aa genotypes found in other individuals or the genotype under consideration might be something complex such as $AaBBccTtY^+Y_{fn}Z3Z5$).

genotypic heterogeneity: the situation in which one phenotype can result from any of multiple different genotypes. This can include allelic heterogeneity (multiple different alleles at the same locus that can cause the trait) or locus heterogeneity (multiple different loci that can cause the same trait).

germ line: the lineage of cells that produce the gametes, sperm or eggs.

Gleevec: the drug called imatinib mesylate or STI571, used to treat a variety of cancers including chronic myelogenous leukemia.

gonadal sex: the sex of an individual as indicated by whether the individual possesses ovaries or testes.

gonads: ovaries or testes.

guanine: one of the bases that make up DNA and RNA, usually abbreviated by the symbol G.

GWAS: *see* genome-wide association study

haploid: possessing only one complete copy of the genome. For example, a human sperm bearing 23 chromosomes, one from each chromosomal pair, is haploid.

haplotype: the arrangement of sequence variants along a chromosome. Because each individual has two allelic variants at each position it is not obvious for a pair of adjoining heterozygous markers which alleles are on the same chromosome together, but evaluation of which alleles are present in parents, siblings or children of that individual can assist in telling which alleles are present on the same chromosome and which are on the opposite chromosomes.

helicase: a protein that acts at the DNA replication fork to unwind the helix.

hemizygote: an individual carrying only a single copy of a given gene or chromosomal region. For example, males are hemizygotes for any gene that is only present on the X chromosome (i.e., for the vast majority of their X chromosomal genes, excepting only those genes that are also present on the Y chromosome). Individuals heterozygous for an autosomal deletion are sometimes referred to as hemizygotes in terms of the genes removed by that deficiency.

hemizygous: the state of being a hemizygote.

hemoglobin: the protein in red blood cells that carries oxygen.

hereditary: something that is passed from one generation to the next as a result of the transmission of genetic information from one generation to the next.

heredity: transmission of a characteristic from parent to child through transmission of alleles that cause or contribute to the characteristic.

heritability: an estimate of the extent to which the variation in phenotype for some trait within a population is determined by genetic variation. Expressed as a number from 0 to 1.0, the higher the heritability the greater the role that genotype plays in determining that trait in that population under the conditions prevailing at the time heritability is evaluated. Comparisons of heritability values between two populations are essentially meaningless and provide no information about genetic differences between those populations because the same trait can have stronger environmental components in one population than in another.

heterochromatin: the region of a chromosome that contains few active genes, is rich in highly repeated simple sequence DNA (so-called satellite DNAs), and is usually very highly condensed during interphase.

heterogeneous nuclear RNA: the primary transcript from genes prior to splicing and polyadenylation.

heteroplasmy: presence of multiple different mitochondrial genotypes in the same cell as the result of mutation. Heteroplasmy accumulates over the course of aging and has been implicated in some late-onset traits such as Parkinson disease.

heterozygote: an individual or organism that possesses two different alleles for a given gene.

heterozygote advantage: a condition where a heterozygote survives better than either homozygote. Sickle cell anemia may exemplify such a case in that the same mutant allele of the hemoglobin gene that causes sickle cell disease in homozygotes confers resistance to malaria in heterozygotes. Thus in areas in which malaria is epidemic, the heterozygote may do better than either homozygote.

heterozygous: the state of being a heterozygote.

histones: any of five different related proteins that complex with DNA to form chromatin. Histones from the protein component (histone core) of the nucleosomes.

HNPCC: hereditary nonpolyposis colon cancer.

hnRNA: *see* heterogeneous nuclear RNA

holandric: the pattern of inheritance exhibited by mutations in genes located only on the Y chromosome.

homologous: in evolutionary terms, homologous refers to things that share an evolutionary origin, and the practical consequence of the shared ancestry of DNA sequences is that the sequences are similar. How similar they are will be related to the amount of time since the ancestral divergence. In molecular biology the term homologous may be found referring to sequences that resemble each other or can pair with each other without regard to consideration of the ancestral origins.

homologs: the two different copies of the same chromosome present in the same cell. Two homologs of a chromosome have the same genes in the same order, but have allelic differences at many points along the chromosome.

homozygote: an individual or organism that possesses two identical alleles for a given gene.

homozygous: the state of being a homozygote.

homunculus (*pl.* homunculi): a tiny human hypothesized by early microscopists to be present in sperm, based on erroneous interpretation of images seen under very early microscopes.

hormone: a chemical messenger secreted by one cell type that induces a change in the state or behavior of one or more other groups of cells or tissues (e.g., testosterone and estrogen).

host cell: a bacterial, yeast, or animal cell used to propagate clones, infectious agents such as viruses, or plasmids carrying elements such as drug resistance genes. Host cells used to make copies of clones, plasmids, or bacterial viruses are bacterial cells that have special properties that allow growth under laboratory conditions but limit the ability of the cell to grow outside of the laboratory. Some clones, called yeast artificial chromosomes (YACs), are produced in yeast cells. In some cases, human or animal cells in culture are used to host a clone that is being used to express the protein product.

housekeeping genes: genes that are expressed in all cells under all conditions. Many of these genes carry out the basic functions that are needed for all cell types no matter what their specialized functions.

human genome project: 13-year-long project carried out by six countries with objectives to: determine sequence of the human genome, create maps of human genetic markers, identify all of the genes, develop and disseminate new genome-related technologies and investigate the ethical, legal and social issues associated with use of genomic technologies and information.

human immunodeficiency virus (HIV): the virus that causes AIDS.

Huntington disease: an autosomal dominant form of neurodegeneration characterized by typical choreiform movements. Huntington disease results from expansions of the simple trinucleotide repeat CAG within the coding sequence of the HD gene which is located on chromosome 4. Onset is most often in middle age but some of the largest repeat expansions can result in a juvenile-onset form of the disease.

hybridization: the process of bringing together single strands of DNA and/or RNA that have sequence compositions that can come together with each other to form a double-stranded nucleic acid by pairing of single strands using the base-pairing rules.

hydrophilic: water-loving. Refers to molecules or regions of molecules that can interact with water through hydrogen bonding. Hydrophilic molecules are soluble in water and contain charged or polar regions that assist with the interactions with water molecules. An example of a hydrophilic molecule is sodium chloride, or table salt.

hydrophobic: water-fearing. Refers to uncharged, nonpolar molecules that cannot interact with water molecules through hydrogen bonding. An example of a hydrophobic molecule is fat, which forms droplets or separates from the water entirely. In proteins, the hydrophilic regions face outwards into the water while hiding the hydrophobic portions of the molecule inside away from the water.

hypolactasia: deficiency of lactase, especially in the intestines.

iatrogenic: a condition resulting from medical treatment. Examples of iatrogenic conditions include infection following surgery or anaphylactic shock in response to penicillin.

identical by descent (IBD): two sequences are identical by descent if they are the same sequence in both individuals as a result of those two individuals having inherited that same sequence from a shared ancestor.

identical by state (IBS): two sequences are identical by state if they happen to have the same sequence but not as a result of inheriting the sequence from a shared ancestor.

imprinting: a germline process that "presets" or predetermines the potential of a transmitted gene or chromosome to be expressed in a pattern that is different depending on whether that copy of the gene was inherited from the mother or from the father. These parent-of-origin specific expression patterns occur without any change in the actual sequence of As, Gs, Cs, and Ts in the DNA. For example, in kangaroo females, and in the extraembryonic membranes of XX human fetuses, it is always the paternally derived X chromosome that is chosen to be inactivated. Imprinting presumably reflects a modification of the DNA, such as methylation, or proteins, such as histones, in such a way as to preset activity in the embryo.

in silico: in the environment of a silicone chip. Refers to experiments carried out on computers and in virtual environments instead of in a test tube or in an organism.

in situ **hybridization**: literally means "on site" or "in place" hybridization. When used to map cloned DNA sequences to chromosomes, this term refers to the hybridization of a labeled DNA probe to chromosome spreads affixed to a glass microscope slide (the DNA in the chromosomes has been denatured into separate strands). The resulting hybridization of the probe to the homologous region on the chromosomes results in the labeling of one site on a specific chromosome or two sites on a pair of homologs. The position of that signal can be identified by examining the banding pattern of the chromosome. When *in situ* hybridization is used to assess transcription activity, a labeled DNA probe containing only the coding strand of the DNA is hybridized to sections of cells from various tissues, again affixed to a glass microscope slide. The formation of DNA:RNA duplex molecules results in the labeling or staining of these cells, indicating that that gene is transcribed there.

in utero: in the uterus.

in vitro: literally "in glass". Experiments performed in a test tube, flask, petri dish or other such lab container.

in vitro **fertilization**: the process of creating zygotes in a test tube by combining sperm with eggs outside of the human body. Eggs are obtained by hormonally stimulating a woman's ovaries to produce ova. The resulting eggs are removed from the woman and then mixed in the laboratory with sperm obtained from the prospective father. The resulting zygotes are then allowed to go through several cell divisions to produce early embryos. One or two cells from these embryos can be removed safely, allowing the performance of a variety of PCR-based genetic tests. Once "healthy" embryos are identified, they can be introduced into the uterus of the prospective mother.

in vivo: in the living organism.

independent assortment: the observation that for the heterozygote AaBb, where the A and B genes are unlinked, all four classes of gametes (AB, Ab, aB, ab) are produced at equal frequency. Thus whether a given gamete carries the A allele or the a allele does *not* influence which allele of the B gene (B or b) that same gamete will carry.

indifferent gonads: structures in the early embryo that can develop as either testes or ovaries.

inducible gene: a gene whose expression can be turned on or turned up by an inducing agent.

induction: turning on the transcription of a gene that was not being transcribed or increasing the transcription of a gene.

informative: useful information about inheritance coming from a genetic marker.

informed consent: consent by a patient to receive treatment or participate in research that meets legal and regulatory standards for the types of information the individual has received that let us consider their consent to be fully informed. This process requires that the individual be notified of what will be done, why it will be done, and the risks of doing it so that they can make a fully informed decision about whether to participate.

inheritance: the passing of a characteristic from one generation to the next.

insertion mutation: a mutation that results when one or more bases is inserted into the DNA sequence.

interphase: a stage of the cell cycle between cell divisions when the nuclear envelope is intact and the contents of the nucleus appear as an indistinct fuzz rather than as individual chromosomes. Interphase encompasses the resting state, G0, as well as the stages of metabolism and synthesis that lead up to cell division, including G1, S, and G2.

intron: a stretch of a eukaryotic gene removed from the final processed RNA molecule through splicing, as opposed to exon sequences that remain during RNA processing. The *in*tron is removed from the transcript and thus stays *in* the nucleus.

invasiveness: a property of tumors that denotes the ability of tumor cells to spread into the surrounding normal tissues.

inversion: a type of chromosomal aberration in which some or all of a chromosome arm has been inverted with respect to the centromere.

junk DNA: DNA that carries out no useful functions for the host organism. Some of the noncoding sequences are stretches of sequence that are repeated. Some of them are repeated a few times, whereas others are repeated 1000, 50,000, or even a million times. As we learn more, we discover that more and more of the supposed junk DNA is actually doing things, but for many of these sequences the function has not yet been found.

karyotyping: the analysis of human chromosomes by viewing under a microscope. Metaphase cells are broken onto glass slides in such a way that all 46 chromosomes are well separated within a small area. The chromosomes are then stained in a way that leaves a characteristic pattern of bands that can be used to identify and distinguish the different chromosomes from each other. Individual chromosomes and chromosome pairs are then identified and scrutinized for errors in number (e.g., trisomy 13, 18, or 21, or monosomy for the X) or structure (chromosome aberrations such as large deletions or translocations).

kb: kilobase.

kilobase (*abbrev.* kb): one thousand bases.

kinetochore: the DNA-protein complex assembled at the *centromere* that allows chromosomes to attach to, and move along, the *microtubules* in the mitotic or meiotic spindle. The kinetochore is known to include proteins called *motor proteins* (kinesins and dyneins) that function to pull or push the chromosomes along the spindle.

lactase: an enzyme that cleaves lactose into galactose and glucose.

lactase persistence: continued production of lactase after weaning or on into adult life.

lactose: a sugar made up of one glucose and one galactose.

lactose intolerance: inability to metabolize lactose, resulting in abdominal gas and pain.

lagging strand: the strand of DNA that is being synthesized in the direction that points away from the direction in which the replication fork is moving.

leading strand: the strand of DNA that is being synthesized in the same direction in which the replication fork is moving.

lethal mutation: a mutation that creates a sufficiently severe defect as to be incompatible with life.

library: in genetic terms, a collection of clones representing part or all of a genome or, in the case of cDNA libraries, a population of DNA clones corresponding to a set of mRNA molecules.

ligase: an enzyme that can link two DNA molecules together at their ends.

linkage disequilibrium: also known as allelic association; the association in a given population of individuals of a particular allele at one gene, or genetic marker, with a specific allele of some other gene or genetic marker (or in some cases a disease gene allele) at an unexpectedly high frequency. Consider the case of a disease-causing mutation and a separate genetic marker (A), both of which are segregating in a given population. If 90% of individuals in the population that carry the disease-causing mutation also carry allele A1 of genetic marker A, but only 10% of the general population has allele A1, then allele A1 and the disease-causing mutation are considered to be in linkage disequilibrium or allelic association with each other.

lipid: water-insoluble, hydrophobic compounds such as fats and waxes.

liposomes: a microscopic spherical particle in which a single layer of lipids surrounds an internal aqueous environment. Liposomes are used to deliver biochemical payloads such as gene therapy agents into a cell without the use of a viral vector.

locus: a specific site or position on a chromosome as determined by recombinational mapping. Often used to refer to a gene at that position.

locus heterogeneity: *see* genetic heterogeneity

LOD score: a statistical tool used in the determination of recombination frequencies in human populations. In brief, a LOD score is a measure of the probability of observing a given pedigree, or set of pedigrees, assuming that the map distance that has been identified as separating two markers (or genes) is the actual distance between them. The acronym LOD is an abbreviation of "log of the odds." A LOD score is the logarithm base 10 of the odds ratio obtained by taking the probability of obtaining observed data for any recombination frequency (called Q or theta) and dividing that probability by the probability of obtaining the observed data if the recombination fraction is actually 0.5 (that is to say that the two markers are unlinked). For the human genome, a LOD score of 3.0 or greater is required to consider that the evidence for linkage between a marker and a gene is highly significant (likely to be true.)

loss of heterozygosity (LOH): a genetic event often observed in cells heterozygous for some recessive mutation that causes the loss of the normal or wild-type allele. As such the cell is now either homozygous or hemizygous for the recessive allele and will likely express the mutant phenotype. LOH can occur by mutation or deletion of the normal allele as a result of mitotic recombination or by loss and reduplication of the entire chromosome that carries the mutant allele. LOH is thought to play a pivotal role in the development of many, if not most, types of cancers.

luciferase: an enzyme present in bioluminescent organisms that oxidizes luciferin to create light.

malignant: a characteristic of a tumor, indicating the ability of cells of that tumor to spread to other sites of the body and colonize new tumors and/or the ability of that tumor to invade surrounding normal tissues (i.e., a tumor that is metastatic and/or invasive).

MAOA: *see* monoamine oxidase A

map unit: a measurement of recombination frequency. One map unit equals one percent recombination.

masking: covering of one trait by another trait; *see* epistasis.

meiosis: the process by which haploid gametes are produced. Basically, homologous chromosomes pair and recombine during meiotic prophase. At anaphase I of the first meiotic division, entire homologs move to opposite poles of the spindle. The first meiotic division, *meiosis I*, in humans thus creates two *haploid* daughter cells, each with 23 chromosomes. (Realize that whether a cell is haploid or diploid depends on the number of chromosomes it has, not the number of chromatids.) Each product of meiosis I has 23 chromosomes (as determined by counting the number of centromeres), as compared to the diploid number of 46, so it is haploid. The second meiotic division, *meiosis II*, is then essentially a haploid mitosis, creating four haploid daughter cells, each with 23 chromosomes.

meiosis I: the first meiotic division.

meiosis II: the second meiotic division.

melanin: pigment responsible for coloration of skin, hair and parts of the eye.

melanoma: a type of skin cancer.

membrane: a lipid bilayer that surrounds cells and eukaryotic organelles.

messenger RNA: also known as mRNA, is the final, processed RNA copy of the coding strand of a gene that has had introns removed and a polyA tail added. The mRNA molecule is transported to the *ribosomes* in the cytoplasm where it is used to direct the synthesis of the protein product of the gene.

metaphase: in *mitosis*, the stage in human cell division where all 46 chromosomes are individually balanced between the two poles of the spindle, with one sister chromatid attached to each pole of the spindle. In *meiosis I*, the stage at which each pair of homologous chromosomes (bivalents) are balanced between the two poles of the spindle, such that the centromeres of the two homologs are each attached to the opposite pole. This stage is referred to as metaphase I. In *meiosis II*, the stage at which all 23 chromosomes are individually balanced between the two poles of the spindle, with one sister chromatid attached to each pole of the spindle.

metaphase I: the metaphase of first meiotic division.

metaphase II: the metaphase of the second meiotic division.

metaphase plate: an imaginary line across the equator of the spindle where chromosomes are normally aligned.

metastasis: the process of a tumor spreading to new sites in the body. A characteristic of malignant tumors.

methylation: the process of adding a chemical group called a methyl group (CH3-) onto a molecule.

microarray: an array of a large number of different genes assembled so that DNA sequences from each gene are placed in a spot on the surface of a small slide or chip that ends up containing between hundreds and tens of thousands of different genes. Microarray experiments can use one test to simultaneously evaluate the level of expression of every gene represented in the microarray. Use of microarrays that are duplicates of each other allows comparisons between experiments so that it is possible to evaluate differences in gene expression between different cell types, during different stages of development, or under different environmental conditions.

microRNA (*abbrev.* miRNA): single-stranded RNA molecules of about 21 to 22 bases in length that play a role in gene regulation by binding to the 3′ untranslated region of a specific mRNA to block translation.

microsatellite: a region of genomic DNA composed of dinucleotide or trinucleotide repeats (e.g., CACACACACACACACACACACA). These sequences are dispersed widely throughout the genome and are extremely polymorphic in terms of the number of copies of the repeating unit. As such, they have proven incredibly valuable as genetic markers for mapping human genes.

microtubules: fibers that make up the meiotic and mitotic spindles. Chromosomes attach to these fibers and move along them as the chromosomes migrate to the poles.

MIF: *see* Müllerian inhibiting factor

miRNA: *see* microRNA

mismatch repair: a process of repair of damage to DNA caused by incorporation of the wrong base during replication or by chemical damage to one or more bases. Defects in this process play a role in the origin of some types of cancers.

missense mutation: a mutation that changes a codon in such a way as to direct the incorporation of a different amino acid at that site in the protein.

mitochondria: plural of mitochondrion.

mitochondrial chromosomes: circular piece of DNA found in each mitochondrion, consisting of more than 16,000 bases of DNA encoding 37 genes.

mitochondrial inheritance: the pattern of inheritance exhibited by mutation in the genome of the mitochondria in individuals in whom the mitochondrial mutation predominates in the population of mitochondrial DNA. In general, all or most children of affected mothers will be affected. Because sperm donate few, if any, mitochondria to the zygotes, paternal transmission is not observed. This mode of inheritance is not seen for heteroplasmy that develops late in life in somatic cells as a result of the gradual accumulation of many separate mutations over the course of a lifetime.

mitochondrion: cellular organelle responsible for energy production. Mitochondria possess their own small circular DNA genomes. Mitochondria are thought to have arisen from bacterial parasites during the early stages of eukaryotic evolution. Some of the mitochondrial proteins are encoded by the genes of the mitochondrial chromosome, and some of the mitochondrial proteins are encoded by the genes contained on the chromosomes in the cell nucleus.

mitosis: the process by which a cell divides to produce two identical daughter cells. Unlike meiosis, mitosis does not change the ploidy of a cell. The human cell starts out diploid, with 46 chromosomes (identifiable by the presence of 46 centromeres), and the two daughter cells both carry 46 chromosomes. In addition, chromosomes do not usually pair or recombine prior to mitosis.

modulatory factor: a factor that does not cause a trait but does cause variation in some aspect of the trait such as severity or how early the trait starts.

molecule: a chemical structure consisting of at least two atoms held together by covalent chemical bonds.

monoamine oxidase A (MAOA): an enzyme involved in the breakdown of neurotransmitters. The effects of a mutation in this gene on violent aggression have been described.

monosomy: possessing only one copy of a given chromosome pair. Turner syndrome (45, XO) is an example of monosomy for the X chromosome.

mosaicism: being composed of two different genotypes. Consider a case in which an error in the first mitotic division of an embryo resulted in an XX and XO daughter cell. The resulting baby would be composed of normal 46 XX cells and 45 XO cells. This girl would be a mosaic for Turner syndrome. 46XX females may also be considered mosaics with respect to X chromosomal genes because they carry a population of cells in which only their father's X is active and a population of cells in which only their mother's X is active.

mRNA: *see* messenger RNA

MSY region: male specific region of the Y chromosome. This is the portion of the Y chromosome outside of the pseudo-autosomal region that does not share any sequences with the X chromosome.

Müllerian ducts: structures present in the early embryo of both sexes. Unless their development is inhibited by the presence of Müllerian inhibiting factor, the Müllerian ducts will go on to produce the cervix, uterus, and fallopian tubes.

Müllerian inhibiting factor (MIF): a hormone produced by the developing testes that inhibits the development of the Müllerian ducts.

multifactorial inheritance: a pattern of inheritance in which the final phenotype is determined both by multiple genes and by the interaction of alleles of those genes with the environment.

multi-hit hypothesis: the hypothesis that additional mutations contribute to the progression of cancer beyond the initial mutation(s) that cause the cancer to start.

multipoint analysis: genetic analysis of alleles of multiple adjoining genetic markers. This form of analysis takes into account data from multiple markers at once.

mutagen: a chemical or physical agent that can alter the DNA sequence.

mutation: a stable and heritable change in the base sequence of a DNA molecule.

natural selection: natural selection refers to any environmental factors that make it more likely that some organisms will survive and produce progeny as compared to others. Natural selection affects the frequency of a trait within a population according to whether that trait makes it more or less likely that the individual with the trait will survive and produce progeny.

neural tube: a structure in the early embryo that eventually forms the brain and the spinal cord.

neural tube defects: errors in the formation or closure of the neural tube that result in disorders ranging from incomplete closure of the spine (spina bifida) to an absence of the brain (anencephaly).

neuropathy: a disease or abnormality of peripheral nerves, often, but not always, first noticed in the form of tingling or numbness in the extremities.

neurotransmitter: biochemical messenger that carries a signal from one nerve cell to another.

non-coding strand: the DNA strand from which the RNA transcript is made through use of the base-pairing rules. This strand is also referred to as the *template strand*.

nondisjunction: in meiosis I, the segregation of two homologous chromosomes to the same pole. In meiosis II and mitosis, the separation of two sister chromatids followed by their migration to the same pole of the spindle. The end result of nondisjunction is the production of aneuploid daughter cells.

nonsense mutations: mutations that convert a codon that specifies an amino acid into a stop codon. Such mutations cause a halt in translation and produce a truncated protein product.

Northern blotting: a tool for determining the presence or absence of a particular mRNA molecule among a population of RNA molecules. The process of transferring a population of size-fractionate RNA molecules that have been transferred onto a filter or membrane. The resulting Northern blot can then be hybridized to a labeled DNA probe to result in signal bound to the member at the point where sequences from the fractionated RNA are homologous to the sequences from the probe.

nuclear pore: large protein complex that extends through the nuclear membrane to facilitate the transport of other molecules into and out of the nucleus.

nuclease: an enzyme that can cut a nucleic acid molecule.

nucleic acids: a molecule consisting of multiple nucleotides joined together into a chain, held together by a sugar phosphate backbone to which bases are attached. In DNA the sugar is deoxyribose and the bases are A, C, G, and T. In RNA the sugar is ribose and the bases are A, C, G, and U.

nucleolus: a structure within the nucleus where very high levels of transcription of ribosomal RNA takes place. It is visible as a large irregularly shaped dense region within the nucleus.

nucleosome: genomic DNA wrapped around a histone core to form a bead-like structure that is the smallest unit of the complex chromatin structure the packs DNA down into a very small volume inside a eukaryotic cell. The nucleosome core consists of two molecules each of histones H2A, H2B, H3 and H4.

nucleotides: the basic building blocks of nucleic acids, composed of a base (A, T, G, C, or U) a sugar, and a phosphate.

nucleus: the structure in the cell that contains the genomic DNA. This is the site of transcription and replication.

nullo: used in reference to gametes carrying no copies of a given chromosome (e.g., the term nullo-21 ova would mean an egg lacking chromosome 21).

obligate carrier: an individual who is necessarily a carrier of one defective copy of a gene for which both copies must be defective for the trait to be manifested. An individual of either sex can be a carrier for a recessive autosomal trait. An XX woman or an XXY male can be a carrier for an X-linked recessive trait; *see also* Carrier.

Okazaki fragment: a short RNA molecule synthesized by the cell to serve as a temporary primer that can pair with DNA to form double-stranded regions needed as points of initiation of DNA polymerization during the synthesis of DNA.

oligonucleotides: a chain of multiple nucleotides.

oncogene: a gene that can cause uncontrolled growth of a cell, as a result of a mutation in a proto-oncogene involved in regulation of cell growth.

oocyte: a female meiotic cell.

oogenesis: development of a female gamete called an oocyte or egg.

open reading frame (ORF): a stretch of DNA that could encode a continuously translatable stretch of mRNA (i.e., a stretch of DNA whose transcript does not include a stop codon).

organelle: an organelle is a microscopic structure inside of a cell that carries out a specialized function. In a eukaryotic cell, many organelles such as the nucleus and mitochondria are bounded by membranes, but ribsosomes are not membrane-bound.

origin of replication: a site on a chromosome or plasmid that is sufficient to initiate DNA replication at that site.

ornithine transcarbamylase (OTC): one of the proteins in the urea cycle that carries out a series of biochemical reactions aimed at converting non-secretable ammonia into urea, which can be secreted.

OTC: *see* ornithine transcarbamylase

oxidative stress: a potentially harmful state that results when oxidative agents such as free radicals exceed the antioxidant levels available to the cell, resulting in oxidative damage to molecules in the cell.

p53: a gene whose product is normally involved in suppression of cell growth, and that is mutated in multiple different kinds of cancer.

PAIS: partial androgen insensitivity syndrome.

palindrome: a DNA palindrome is a sequence that reads the same in the 5′ to 3′ direction of one strand and on the 5′ to 3′ direction of its complementary strand. Thus a DNA palindrome is not like a word palindrome with the forward and reverse letters all in one line (Madam I'm Adam); instead the DNA palindrome does the first half of the palindrome on one strand and the second half of the palindrome on the complementary strand.

PAR: *see* pseudoautosomal region

paralogous regions: regions of the genome that have a set of related genes that occur in the same order in both regions. The model to explain this is that both paralogous regions are derived from an ancestral piece of DNA that duplicated, with the duplicates ending up in different regions of the genome and gradually diverging through mutation.

partial aneuploidy: aneuploidy involving only part of a chromosome, which may occur in cases such as unbalanced translocations.

PCR: *see* polymerase chain reaction

pedigree: a diagram of the individuals in a family and their relationships to each other. Standardized symbols are used in drawing pedigrees, including circles for females, squares for males, filled symbols for affected individuals, a diagonal line through symbols for deceased individuals, and a small diagonal arrow to mark the proband.

penetrance: a measure of the degree to which a given genotype is correlated with the phenotype.

phage: a virus that infects bacteria. The term phage is an abbreviation for bacteriophage.

phage library: a collection of phages bearing different DNA insertions derived from the genome of interest. For example, a human phage DNA library consists of a bacterial virus, each carrying a piece of human DNA inserted into the viral genome.

pharmacogenetics: the study of the genetic components of the interactions of a drug with a cell or organism.

pharmacogenomics: use of information from throughout the whole genome to study the interactions of a drug with a cell or organism.

phase I clinical trials: experiments primarily aimed at evaluating safety and dose responses of experimental treatments in human populations. Phase I clinical trials may also gain limited information on effectiveness or side effects. These trials are usually carried out in small groups of less than a hundred subjects, who may be patients or normal controls.

phase II clinical trials: experiments to study efficacy of a therapy when used for a particular purpose, using a few hundred subjects who are patients with the trait being treated. Phase II trials continue additional safety evaluation.

phase III clinical trials: once earlier phases of clinical trials have provided preliminary evidence of safety and effectiveness, phase III clinical trials use expanded numbers in the range of hundreds to even thousands of patients to allow statistical evaluation of safety, efficacy, and risk–benefit ratio, as well as detection of rare side effects, prior to the authorization of the therapy for use in clinical, non-experimental settings.

phase IV clinical trials: experiments involving continued monitoring or testing of a therapy in human populations after the treatment has come to market. Phase IV clinical trials evaluate long term effects and are especially important for therapies that will be used over extended periods of time, that have come to market through rapid approval mechanisms that involved limited evaluation prior to licensing.

phenocopy: a phenotype present in a member of a family when the underlying cause of that phenotype is different from the cause of that trait in the other family members.

phenotype: the manifestation of a trait in terms of physical or mental characteristics that are determined by genes, e.g., height, hair color, or disease.

phenotypic heterogeneity: variations in phenotypic features that result from a particular genotype. In some cases heterogeneity can take the form of a difference in age at diagnosis of the same trait, but in other cases heterogeneity may manifest as completely different traits, such as testicular feminization and prostate cancer that can both result from mutations in the androgen receptor gene.

Philadelphia chromosome: a translocation between chromosomes 9 and 22, which can be notated t(9;22) (q34;q11), that has fused together the BCR and ABL genes. It leads to either acute myeloid leukemia (AML) or chronic myelogenous leukemia (CML) depending on exactly where the fusion occurs within the two genes.

plasmid: a circular DNA molecule, often used as a cloning vector, that can replicate in a bacterial or yeast cell.

pleiotropy: having effects on multiple different aspects of the organism's function, development, or appearance, with the example of albinism affecting not only hair, skin, and eye color but also affecting eye development and function.

ploidy: the number of haploid genomes possessed by a cell. To determine the number of chromosomes, count the number of centromeres, not the number of sister chromatids, since ploidy is not determined by differences in numbers of chromatids present at different stages of chromosome replication. Cells with two copies of each chromosome are called *diploid*. Gametes with one copy of each chromosome are called *haploid*. In humans the diploid number is 46 and the haploid number of chromosomes is 23. Cells possessing three copies of each chromosome are *triploid*. *Triploidy* occurs reasonably frequently among human conceptions, but is incompatible with fetal normal development and results in spontaneous abortion. Cells carrying too few or too many copies of just one or more chromosomes are called *aneuploid*. Types of aneuploidy involving only single chromosomes, rather than the whole genome, include trisomies and monosomies.

point mutation: a mutation that changes one DNA base to a different base, affecting only one base and not adding or removing anything.

polar body: one of the nonfunctional products of female meiosis.

polyA tail: a long stretch of As added to the 3′ end of an mRNA molecule.

polygenic inheritance: the type of inheritance observed when a phenotype is determined by defects in two or more genes. Formally this includes digenic and triallelic inheritance but frequently refers to more complex situations involving defects in even larger numbers of genes.

polyglutamine repeats: a repeat of CAG in frame with the reading frame of the coding sequence of a gene that encodes a string of glutamines in the protein. A polyglutamine repeat acts as if it is sticky, binding to a specific set of other proteins in the cell.

polymerase chain reaction (PCR): the process of rapidly amplifying a defined region of DNA by sequential steps of denaturation and replication.

polymorphism: a variant form of a gene. A gene or genetic marker with many variant forms is considered to be highly polymorphic. Some fields use the term polymorphism only when it is present in at least 1% of a population; *see also* benign polymorphism.

polypeptide: a chain of multiple amino acids.

positional candidate cloning: cloning of a gene based on a combination of information regarding the location of the gene and functional information on the genes within that region being evaluated as candidates.

positional cloning: cloning of a gene based simply on its location within the genome, something that happens especially in cases where the gene being sought turns out to be a gene about which there was little or no prior functional information available.

primary sexual characteristics: gonads (testes, ovaries) and external genitalia (penis, scrotum, clitoris, vagina).

primer: short single-stranded, synthetic piece of DNA or RNA such as those used to prime the synthesis of DNA in PCR reactions or DNA sequencing.

prions: an infectious agent that consists only of an abnormally folded version of the prion protein, with no nucleic acid content. The abnormal prion protein is thought to induce endogenous copies of the protein to fold abnormally and adopt the prion form. Prions are implicated in causing spongiform encephalopathies in many different organisms, including kuru in humans, scrapie in sheep, and mad cow disease in cattle.

proband: the individual that brought a family or kindred to the geneticist's attention. In some non-genetic studies proband may simply refer to a study subject without reference to a family context.

probe: a DNA or RNA molecule labeled by a radioactive or fluorescent tag that is used in DNA:DNA or DNA:RNA hybridization reactions.

prokaryote: an organism that lacks a nucleus and that does not make membrane-bound organelles such as mitochondria. The two domains of prokaryotes are bacteria and archaea.

promoter: a region of DNA next to the transcription start site that controls expression of the gene; the site at which RNA polymerase binds to initiate transcription. A promoter is directional in its activity and does not direct transcription of the gene if it is put in backwards.

pronucleus: one of the two gametic nuclei in a fertilized egg. The female pronucleus came from the egg and the male pronucleus came from the sperm.

prophase: the first step in mitosis and meiosis I, during which chromosome condensation occurs. In meiosis I, pairing and recombination also occur during prophase.

protease: an enzyme that cuts up other proteins. Most proteases are specific for a site on a protein or set of proteins.

protease inhibitor: a chemical that can inhibit the activity of a specific protease.

protein: a complex molecule composed of a linear array of amino acid building blocks.

proteomics: study of the global array of proteins produced by an entire organism.

proto-oncogene: a gene normally required only in cells undergoing rapid division. Inappropriate expression of these genes can play an important role in tumor formation.

PSA: prostate specific antigen.

pseudoautosomal region (PAR): a region of homology between the X and the Y chromosomes. Meiotic recombination in the PAR located at the tips of the short arms of the X and Y chromosomes is required for proper meiotic segregation of the X and Y chromosomes.

pseudogene: a non-functional relative of a functional gene, identifiable as a region of sequence that is highly similar to that of a known gene but that has been disabled by mutations such as nonsense or frameshift mutations.

Punnett square: a grid that displays the possible genotypes of the gametes produced by both parents and is used to estimate the possible genotypes of their offspring.

pure-breeding: in reference to Mendel's pea experiments, a group of plants that produce only offspring like themselves even if bred repeatedly over many generations.

Quad test: measures levels of α-fetoprotein (AFP), estriol, β-HCG, and dimeric inhibin in maternal serum and compares those values to known normal levels to determine the probability that the fetus has Down syndrome, trisomy 18, or other major problems.

rearrangement: one of a class of mutations that alters the sequence of genes along a chromosome or interchanges genetic material between nonhomologous chromosomes. Such mutations include deficiencies (deletions), dicentrics, duplications, inversions, rings, translocations, and transpositions.

receptor: a protein that receives and reports signals from the environment or from another cell. Receptors may be bound to the cell membrane or may be present in the cell cytoplasm.

recessive: a mutation is said to be recessive if its expression is "masked" by the normal or wild-type allele (i.e., it does not produce a phenotype when heterozygous with the normal allele). A recessive mutation manifests the accompanying phenotype only when the mutation is homo- or hemizygous.

recombination (also called crossing over and exchange): the process by which two homologous chromosomes exchange genetic material. This process requires the breakage and rejoining of two homologous DNA molecules followed by "healing" of the break.

recombination fraction: the fraction of gametes carrying a recombination event that falls between two loci. The recombination fraction measures real recombination events when the two items are located on the same chromosome, but can evidence independent assortment of chromosomes when the two markers are located on separate chromosomes. Due to multiple exchange events along the same pair of homologs, when two loci are on the same chromosome but far apart, the recombination fraction will underestimate the actual number of recombination events that fell between the two loci.

reference sequence: a known sequence against which re-sequenced sequences can be compared to evaluate whether there are sequence variants in the new sequence that differ from the reference sequence.

regulatory elements (REs): components of a eukaryotic gene that determine when and where that gene will be expressed. The function of these elements is mediated by the transcription factor proteins that bind to them.

repair-deficient mutations: mutations in genes whose protein products are required for one or more DNA repair processes. Such mutations impair the ability of the cell to repair damage to its DNA. Some types of repair-deficient mutations in humans strongly predispose individuals who carry such mutations to develop cancer.

repeat expansion: the process by which simple sequence repeats become longer.

replication: the process of making a copy. In genetics this word usually refers to the process by which DNA is copied to make two identical copies of each chromosome, where each new chromosome has one old strand of DNA and one new strand of DNA.

replicative senescence: the point at which a cell permanently stops dividing or has lost the ability to keep dividing.

re-sequence: for genes for which the sequence is already known, it is possible to re-sequence that same gene from a new DNA sample to evaluate whether the sequence in the new DNA sample is identical to the reference sequence already known for that gene.

restriction endonuclease: also called restriction enzyme. An enzyme that recognizes a specific short DNA sequence and cuts the DNA at that point.

restriction enzyme: a protein that cleaves DNA molecules at specific sequences (e.g., GGATCC). Restriction enzymes are critical tools in the analysis of DNA molecules.

restriction fragment length polymorphism (RFLP): a difference in the pattern of cleavage of a given DNA sequence by a restriction enzyme. An RFLP results from a difference in sequence between individuals where the sequence difference falls within a restriction enzyme recognition sequence. Some individuals have a sequence at that point that can be cut by the enzyme and others have a sequence that the enzyme does not recognize or cut.

retrovirus: a virus whose genome is composed of RNA.

reverse transcriptase: an enzyme that creates a complementary DNA copy of an RNA molecule. Reverse transcriptase is an essential component of retrovirus replication.

reversion: a second mutation that reverses the phenotypic effects of a previously induced or isolated mutation in the same gene. Some precise reversions actually reverse the original mutation, whereas others simply compensate for the effect of the first mutation. Reversion events are usually quite rare.

RFLP: *see* restriction fragment length polymorphism

rhodopsin: a protein present in rod photoreceptors responsible for receiving faint light signals and initiating a biochemical signal that is transmitted to nerves that inform the brain that that photoreceptor cell has detected a photon of light.

ribonucleic acid (RNA): RNA is made up of the four chemical bases, adenine, cytosine, guanine, and uracil (abbreviated A, C, G, and U) connected by a ribose-phosphate backbone. RNA serves as a transient copy of the permanent information stored in the DNA of the cell.

ribosomal proteins: proteins that serve as components of a ribosome.

ribosomal RNA: a class of RNA molecules that serve as components of a ribosome. Several classes of rRNA molecules play different roles in determining the structure and the function of the ribosome.

ribosome: a structure located in the cytoplasm that carries out the process of *translation*. Ribosomes are composed of ribosomal proteins and ribosomal RNAs. Ribosomes attach to an mRNA molecule at the 5' end and translate it sequentially by reading each *codon*.

ribozyme: RNA molecule that can catalyse a chemical reaction such as splicing of another RNA molecule.

ring chromosome: a type of chromosomal aberration in which the two ends of the DNA molecule have been fused so that the chromosome forms a circle.

RNA: ribonucleic acid.

RNAi: RNA interference, a naturally occurring mechanism by which the cell uses very short pieces of RNA complementary to a transcript to target that transcript to be degraded. Experimental systems have been developed that use the RNAi system to reduce expression of genes in a sequence specific manner.

RNA polymerase: an enzyme, composed of many protein subunits, that carries out transcription through synthesis of an RNA copy of the coding strand that is complementary to the template strand.

RNA processing: the chemical modifications of an RNA transcript, including splicing out introns and addition of a polyA tail.

rod cells: photoreceptor cells in the eye that detect faint light, such as starlight. Rod cells use rhodopsin to detect light and cannot distinguish colors.

RPE65: a protein encoded by the RPE65 gene that is expressed in the retinal pigment epithelium of the eye. Mutations in RPE65 cause an early childhood form of blindness, and gene therapy for RPE65 has resulted in visual improvement in affected individuals.

rRNA: *see* ribosomal RNA

S phase: the period in the cell cycle when DNA replication takes place.

satellite DNA: repeated sequences that occur a large number of times over long regions such as the centromeres.

screening: analyzing samples or information from a population without specifically selecting individuals because of specific symptoms or diagnosis.

secondary sex characteristics: physical differences between the two sexes other than gonads and external genitalia. Secondary sexual characteristics include breast development, facial and body hair, and skeletal and muscular differences.

segregation: the separation of two homologs of a chromosome, or two alleles of a gene, into separate gametes at the end of meiosis.

selection: *see* artificial selection, natural selection, or sexual selection

semiconservative replication: the process by which a double-stranded DNA molecule is copied by making complementary copies of the two strands of the original DNA molecule. After replication is completed, each of the two new double-stranded DNA molecules consists of an old strand and a new strand.

sequence: the order of bases in a nucleic acid molecule or the order of amino acids in a protein.

sequence tagged site (STS): a piece of information consisting of a small piece of sequence that serves as a unique identifier of a gene or chromosomal region.

sequence variant: a change in the sequence of a genome as compared to the reference sequence for that genome.

sex: physical structures of the body associated with being male or female. Sex may be defined based on categories including genotype, karyotype, primary sexual characteristics or secondary sexual characteristics, and in some individuals will be different depending on which category is used. Some individuals may simultaneously possess both male and female secondary sexual characteristics.

sex chromosome: the X and Y chromosomes are considered sex chromosomes.

sex-linked dominant: a pattern of inheritance exhibited by dominant mutations of the X chromosome. All daughters of affected males are affected.

sex-linked recessive: also called X-linked recessive. A pattern of inheritance exhibited by recessive mutations in genes located on the X chromosome. Unless the mother is also a carrier, all children of affected males are normal. Half the sons, but none of the daughters, are affected in matings of normal males to carrier females.

sex-limited trait: a trait that can be expressed in only one sex. Some sex-limited traits can manifest in individuals of the other sex if the hormone environment is shifted towards that found in the sex that usually manifests the trait, as may occur in some therapeutic situations.

sexual identity: a key component of gender. The sex or set of sex roles with which an individual identifies. This is usually, but not always, congruent with biological sex.

sexual orientation: the pattern of sexual and emotional attraction to the same sex, the opposite sex, or both sexes. Heterosexual orientation is attracted to the opposite sex. Homosexual orientation is attracted to the same sex. Bi-sexual orientation is attracted to both sexes.

sexual selection: competition among members of one sex for the right to mate with specific individuals of the other sex, with selection of sexual partners on the basis of their phenotype influencing which genotypes get passed along to the next generation and leading to a gradual shift in the characteristics of the population.

shared synteny: a situation found when genes that are found in a syntenic group (located together on the same chromosome) in one species are found in a syntenic group in another species. When we see a group of genes that are near each other in human that are also found near each other in mouse we talk in terms of shared synteny between the two species for those genes and regions of DNA.

silent mutation: a mutation within a coding sequence that does not alter the protein sequence that will result from the use of that coding sequence to direct protein synthesis. Silent mutations are the result of the redundancy of the genetic code and occur because in some cases the new codon encodes the same amino acid that was encoded by the previous codon.

single nucleotide polymorphism (SNP): point mutation.

single-stranded: a nucleic acid strand that is not paired with another strand.

siRNA: *see* small interfering RNA

sister chromatids: the two identical copies of a DNA molecule produced by semiconservative replication. Prior to mitotic anaphase, sister chromatids usually remain tightly connected along their lengths, especially at their centromeres. The separation of sister chromatids heralds the onset of anaphase in mitosis. During the first meiotic anaphase, sister chromatids separate along their arms, but the sister chromatids remain connected at the centromeres. Separation of sister centromeres occurs only at anaphase II.

situs inversus: a condition in which the heart and other organs in the chest and abdomen are reversed in their positions, so that the whole arrangement of organs is the mirror image of the normal arrangement.

small interfering RNA (*abbrev.* siRNA): molecules used to cause the cell's RNA interference system to degrade the RNA with which the siRNA hybridizes, resulting in reduced expression of the targeted gene.

small nuclear RNA (*abbrev.* siRNA): small RNA molecules found in the nucleus that participate in molecular complexes that carry out functions such as RNA splicing and polyadenylation.

SNP: *see* single nucleotide polymorphism

snRNA: *see* small nuclear RNA

somatic cell: any cell of the body except those of the germ-line that lead to production of eggs or sperm.

somatic sex: the sex assignment that would be made to an individual based on the secondary sexual characteristics of the body, as opposed to genetic sex (based on genotype) or gonadal sex (based on having ovaries or testes.)

SOX9: a transcription factor (made by the SOX9 gene) that, among other things, regulates expression of the gene encoding the anti-Müllerian factor that suppresses development of the Müllerian ducts in males. Defects in SOX9 lead to a phenotype that includes sex reversal.

spermatocyte: a meiotic cell in a male.

spermatogenesis: developmental of the male gamete called the sperm.

spindle apparatus: a structure made of microtubules that forms during mitosis or meiosis. After assembly at the metaphase plate, chromosomes move along the spindle fibers to reach the opposite poles of the cell.

splice site: the point in the sequence of the RNA molecule at which splicing takes place. Splice site sequences are found at the exon–intron boundaries and are essential for correct splicing of the transcript. The minimum feature of a canonical donor splice site (at the 5' end of an intron) is that the first two bases inside of the intron must be the bases GU and that the last two bases of the canonical receptor splice site (at the 3' end of the intron) must be the bases AG. Additional sequences surrounding the splice site are also important to the splicing. The canonical splice site sequences are those used by the major spliceosome to carry out most splicing reactions, but a minor spliceosome can use other splice site sequences to remove some rare introns in a very small number of genes.

splice variants: different final mRNA molecules derived from alternative splicing of the same primary transcript. Such splice variants will often share some exons but not others, although in theory, it would be possible to end up with a completely different set of exons in one splice variant than in the other.

splicing: the process by which introns are removed from a primary transcript (the RNA copy of the coding strand of a gene) so that the exons are joined together in the mRNA molecule in the order that they are found on the chromosome.

SRY: a protein encoded by the SRY gene on the Y chromosome in males. SRY stands for sex determining factor Y, and it has also been called TDF, for testis determining factor. The SRY protein is a transcription factor that upregulates expression of SOX9 and leads to a series of cellular and biochemical events that both promote development of male-specific tissues and suppress development of some female specific tissues. Defects in this gene result in sex reversal.

start codon: the first codon to be used by the ribosome when translating a message. The first codon used is an AUG, so the first amino acid is a methionine, but the ribosome does not always select the first AUG and may sometimes use a later AUG on the RNA as the first codon to be translated.

stem cell: a cell that has the potential for self-renewal and the capability to differentiate to produce many different cell types. Embryonic stem cells have the capability to differentiate into any cell type in the body, but other cell types with more limited differentiation capability can serve as precursors to many different cell types within a particular lineage.

stop codon: the codons UAA, UGA, or UAG, which cause the termination of translation. Because there are no tRNA molecules with a complementary *anticodon*, these codons cannot be translated. When any of these three codons occur "in frame" on an mRNA molecule, synthesis of the protein stops and the mRNA and the protein are released from the ribosome.

structural protein: proteins that form the structures that make up the organelles, cells, tissues, and organs. Structural proteins can usually be thought of as distinct from enzymes that catalyze biochemical reactions. However, some proteins that may be thought of as part of a cell structure are also capable of carrying out an enzymatic or signaling function under some circumstances.

STS: *see* sequence tagged sites

syndrome: a complex set of traits or symptoms that occur together. Syndromes are complex traits that often involve multiple different types of cells or organs in the body.

synteny: *see also* shared synteny. Synteny describes the physical localization of genes on the same chromosome. A syntenic group of genes in a particular organisms will be characterized by a particular placement, order and spacing on the chromosome that we expect to see replicated in other members of the same species, but in some cases such as the presence of a large inversion, the order of genes in a syntenic group may be altered.

tandem repeats: repeated occurrence of the same sequence more than once in a row. Some tandem repeats involve repetitions of simple short sequences, so that a repetition of CA, as in CACACACA, or the repetition of GGGAGA, as in GGGAGAGGGAGAGGGAGA, are tandem repeats. It is possible to find one or more copies of a large complex sequence, one after the other, but it is common in those cases if the core repeated sequence is quite long, to find that mutation has introduced small changes so that the recurring sequence is only approximately duplicated.

telomerase: the enzyme that extends telomeric DNA by adding new telomere DNA onto the ends of the chromosome.

telomere: a region of repeated DNA sequences, and associated proteins, that is found at the end of eukaryotic chromosomes.

telophase: the last stage during mitosis and meiosis II. The chromosomes have reached the poles and the nuclear envelope reforms around them.

template strand: in *replication*, the DNA strand being copied by DNA polymerase. In *transcription*, the DNA strand being copied by RNA polymerase. Sometimes also called the non-coding strand.

testing: analyzing samples or information from individuals in response to specific characteristics of those individuals. For instance a medical test on an individual may be ordered by a doctor after the doctor has evaluated the individual's symptoms, medical history, and other characteristics. For contrast, *see also* screening.

testosterone: the primary male sexual hormone responsible for important aspects of sexual development in males.

TFM: testicular feminization, also known as complete androgen insensitivity syndrome.

threshold trait: a form of quantitative trait where many genes contribute to the trait, and rather than showing gradually increasing severity in proportion to the number of defective genes, the individual shows no signs of the trait until a critical number of genes, or threshold, has been passed at which point the individual has the full trait. The trait may show increased severity with the presence of additional genes beyond the number needed to reach the threshold.

thymine: one of the bases found in DNA and usually abbreviated by the symbol "T." Thymine is not found in RNA.

Tiresias complex: a term coined by Dr Nancy Wexler to describe the dilemma that occurs when we have to choose whether or not to receive potentially dire information if there is nothing we can do if the result turns out badly.

tissue specificity: the property of being present in some cells, tissues or organs but not others.

titin gene: the gene that produces the largest transcript and protein product. It is sometimes considered the largest gene.

topoisomerase: an enzyme that controls the topology of DNA as it unwinds the super-coiled DNA ahead of the advancing polymerase during DNA replication.

trans: on the opposite side. Something that acts in trans can act not only on the molecule of DNA from which it originates but also on the other copy.

transcription: the process by which the enzyme RNA *polymerase* makes an RNA molecule complementary to the sequence of the *template strand* of the DNA. Transcription begins with binding of the RNA polymerase to the promoter region 5' to the region to be transcribed.

transcription factor: other proteins, in addition to RNA polymerase, that bind to regulatory sequences in the DNA to affect transcription of a gene.

transcriptome: the complete array of transcripts that can be produced by a genome.

transfer RNA: also called tRNA. The RNA molecule that provides specificity to the translation process. Each type of tRNA molecule carries a specific anticodon complementary to the codon on the mRNA and an appropriate amino acid. By matching the anticodon and the codon, the appropriate amino acid is added to the lengthening peptide chain.

transformation: the process of introducing exogenous DNA into a cell with a resulting affect on phenotype.

transgene: a gene put into a transgenic animal.

transgenic animals: animals, usually mice, that have been genetically altered.

translation: the process by which a ribosome makes a protein by "reading" the sequence of an mRNA codon by codon. To carry out translation, three steps must happen: initiation, elongation, and termination.

translocation: a chromosomal aberration in which material from one chromosome is moved onto another chromosome. In the case of *reciprocal translocation*, material is exchanged between two non-homologous chromosomes in a reciprocal fashion.

triallelic: describes the situation that exists when a genetic trait results from defects in three different alleles. The examples that have been seen involve two defective alleles in one gene plus a defective allele in another gene in the case of Bardet–Biedl syndrome. This can also be referred to as digenic triallelic.

trinucleotide repeat expansion: simple sequence repeat, with a repeat unit of three, where there are more repeat units present in the offspring than in the parent, or that have more repeat units present in a daughter cell than in the progenitor cell.

triple test: a test used in screening of blood samples from pregnant women to detect abnormal levels of three fetal proteins, alpha fetoprotein, human chorionic gonadotropin, and estriol.

triploid: a cell carrying three copies of each chromosome. Triploid conceptions do occur in humans, but are always non-viable.

trisomy: possessing three copies of a given chromosome.

trisomy 13: a form of aneuploidy that leads to severe malformations of the nervous system, as well as a host of other problems, including heart defects and facial anomalies. A large fraction of cases of trisomy 13 result in spontaneous miscarriage, and of live-born children it is generally lethal by 6 months of age.

trisomy 18: a form of aneuploidy that leads to a characteristic set of malformations along with significant neurological deficiencies. There are usually severe cardiac problems and survival beyond 6 months is very rare. A large fraction of cases of trisomy 18 result in spontaneous miscarriage, and of live-born children it is often, but not always, lethal before 1 year of age.

trisomy 21: form of aneuploidy responsible for Down syndrome, which is characterized by a broad spectrum of features resulting from the presence of an extra copy of chromosome 21, or in some cases a subset of the whole syndrome present in individuals with an extra copy of part of chromosome 21. Features include short stature, mental retardation, flattened facial appearance and epicanthic folds of the eyes, skeletal defects, joint laxity, heart defects, and early onset of Alzheimer disease. While individuals with Down syndrome had previously been considered severely retarded with a short life span and limited prospects, advances in medical care for features such as heart defects have extended life spans and computer-based learning systems have shown that these children can achieve far more than was possible when they were simply institutionalized. Risk of producing a child with Down syndrome increases dramatically with maternal age (*see* maternal age effect).

tRNA: *see* transfer RNA

tumor: a mass of rapidly and inappropriately dividing cells arising from a single progenitor cell.

tumor promoter gene: a gene whose protein product acts to cause continued cell growth on the part of a cell that should not be growing continuously, which can lead to cancer.

tumor suppressor gene: a gene whose protein product acts to prevent inappropriate cell division. Mutation "knockouts" of tumor suppressor genes are a key element in the etiology of many types of cancers.

two-hit hypothesis: the hypothesis that it takes two separate events to lead to cancer in a cell, with one of those events sometimes being the inheritance of a mutation in one copy of a cancer gene, and the second event being a mutation in a second copy of that same gene, leading to uncontrolled growth on the part of the cell that has mutations in both copies of the gene.

uniparental disomy: having both copies of a chromosome pair come from one parent, with no copy of that chromosome originating with the other parent. This can include uniparental isodisomy, in which the copies are identical rather than homologous, and uniparental heterodisomy, in which the individual has both of the homologous copies from one parent.

uracil: a base found in RNA but not DNA, and usually abbreviated by the symbol "U".

vaccine: a weakened or killed version of a pathogen which when administered elicits an immune response against the live pathogen.

variable expressivity: variation in the phenotype that occurs in the case of someone who has the phenotype. This may include variation in features such as severity, quantity, duration, or how early the trait starts.

variable number of tandem repeats (VNTR): when a sequence is repeated many times, errors in replication of that sequence can lead to the presence of a different number of copies in different individuals. Repeated sequences that have this property of variation in number of repeat copies in a population are called VNTRs.

vector: A specially engineered plasmid, virus, or artificial chromosome into which a piece of DNA can be cloned so that it can be propagated and studied.

virus: an infectious organism composed of protein and nucleic acid that can replicate only when its genome is inserted into an appropriate host cell. The virus itself does not possess all of the machinery needed to synthesize its components and must use the host cell machinery to complete replication.

VNTR: *see* variable number of tandem repeats

wild type: the typical naturally occurring variant of a gene or phenotype.

X0: also called 45 X0. This is a form of aneuploidy responsible for Turner syndrome in females, who are sterile and may display a number of other phenotypes, including short stature.

X-autosome translocation: a translocation involving the X chromosome and an autosome.

X-linked recessive inheritance: also celled sex-linked recessive inheritance, a pattern of inheritance exhibited by recessive mutations on the X chromosome. Unless the mother is also a carrier, all children of affected males are normal. Half the sons, but none of the daughters, are affected in matings of normal males to carrier females.

XX(SRY) male: a 46 XX individual in which the Y chromosomal SRY gene has been translocated onto the tip of one of the two X chromosomes. These individuals develop into sterile males.

XXY male: a form of aneuploidy that results in a male bearing an extra X chromosome. The trait that results, called Klinefelter syndrome, results in small testicles, reduced fertility, learning disability, and a variety of additional features that can result from decreased testicular hormone production. The expressivity of the syndrome is highly variable.

XY(del SRY) female: a 46 XY individual bearing a Y chromosome lacking the SRY gene. Such individuals develop into sterile females.

XYY male: a form of aneuploidy that results in a male bearing an extra Y chromosome.

Z-DNA: the structure of a double-helix sometimes found in nature that is twisted in the opposite direction most often found in human DNA *in vivo*. Z-DNA is considered a left-handed helix and it has a zig-zag structure to it that sharply contrasts with the smooth turns of the right-handed B-DNA helix.

zygote: the product of the fusion of the two pronuclei following fertilization.

zygote rescue: one of several processes by which a zygote is saved from death by the gain or loss of a chromosome to restore proper copy number.

Index